The Water, Climate, and Food Nexus

Mohamed Behnassi · Abdulmalek A. Al-Shaikh ·
Ameenah Gurib-Fakim · Mirza Barjees Baig ·
Mohammed Bahir
Editors

The Water, Climate, and Food Nexus

Linkages, Challenges and Emerging Solutions

Springer

Editors
Mohamed Behnassi
Economic, Social and Environmental Council (ESEC) and Center for Environment, Human Security and Governance (CERES)
Ministry of Higher Education and Innovation
Rabat, Morocco

Ameenah Gurib-Fakim
John Wesley School of Leadership
Carolina University
Winston-Salem, NC, USA

Mohammed Bahir
High Energy and Astrophysics Laboratory
Faculty of Sciences Semlalia
Cadi Ayyad University
Marrakech, Morocco

Abdulmalek A. Al-Shaikh
Prince Sultan Institute for Environmental, Water and Desert Research
King Saud University
Riyadh, Saudi Arabia

Mirza Barjees Baig
Prince Sultan Institute for Environmental, Water and Desert Research
King Saud University
Riyadh, Saudi Arabia

ISBN 978-3-031-50961-2 ISBN 978-3-031-50962-9 (eBook)
https://doi.org/10.1007/978-3-031-50962-9

This Springer imprint is published by the registered company Springer Nature Switzerland AG
The registered company address is: Gewerbestrasse 11, 6330 Cham, Switzerland

Paper in this product is recyclable.

Foreword by Dr. Rachael McDonnell

It has always seemed incongruous that water security and food security challenges have not been tackled together with integrated policies, investments, and technologies. This result of different approaches that have had a more singular focus and this has been captured in the important *FAO 2021 State of Land and Water Report: Systems at Breaking Point* (FAO 2021) with clear evidence given of the alarming trends in both the use and deterioration in quality and quantity of water and land systems around the world. The resulting losses in non-renewable water resources and of productive land from poor food and water systems management highlight the outcomes of opposing policies, governance, and practices in many countries. Limited investments and management in the development and maintenance of water systems have ensured both wastage and under-utilization of different water resource options that could have supported both direct consumption and agricultural practices. In both water and food security developments, the increasing impacts of climate change on precipitation variability and intensity, as well as changes in demography, socio-economic, and political systems are ensuring pressure on land and water ecosystems is more intense than ever, and many are stressed to critical breaking points.

Thus, the need for new thinking, evidence, policies, and science across the water/food/climate nexus has never been more urgent. The chapters in this book provide important insight into the nexus linkages, challenges, and emerging solutions in different countries, with many of case studies coming from countries facing some of the greatest challenges to delivering security. The approaches considered fall across a broad spectrum from on-farm technologies through to political ecology. They highlight just how complex and multidimensional integrated approaches need to be adopted to face the upcoming crisis threatening many countries over the next decade.

Several chapters have highlighted that climate change is changing the base operating conditions for water management, whether for a water utility operator or a farmer. Large- and small-scale storage systems, groundwater reserves, desalinization, recycled water, and smart-data systems are all part of a plethora of technology-based approaches for managing the increasing impacts of droughts and floods, and changes in timings of the growing season. The provision of new data such as from soil moisture sensors, AI-enhanced climate information systems, and precipitation forecasting can help ensure water is managed and used with an understanding of near-term conditions. Many of the book chapters highlight the advances in data science and in operations that are already supporting this. Climate-smart agriculture (CSA) can also bring genetic innovations and updates in on-farm practices that can offset some of the emerging challenges from limited water availability. Conservation agriculture is just one of many evapotranspiration/soil moisture-saving approaches in CSA for adaptation and resilience building, and various chapters highlight developments that reflect the local context.

Technologies in both water and agriculture management need to be supported by policy and investment drivers. The increasing role of blended finance and impact investment to small and medium enterprises is bringing important scaling of developments in bundled services and technologies. This increasing engagement of the private sector, alongside the traditional national agricultural research and extension services, can be one of several conduits for delivering nexus solutions at the scale needed to address the emerging crises.

These efforts need to be supported by policy drivers and coherence that really balance water, food, and climate security. Tradeoff analyses across policies of different ministries are needed, based on both present and future conditions, so integrated planning is possible, and adjustments made to ensure incentives and drivers for the delivery of one are not at the expense of the other. In the interim report *Turning the Tide* from the Global Commission on the Economics of Water, the major recommendations highlight some of the priorities needed in water and food policy realignment '...we must phase out some USD 700 billion of subsidies in agriculture and water each year, which tend to generate excessive water consumption and other environmentally damaging practices. We must drastically reduce leakages in water systems ('non-revenue water') that cost billions annually, by prioritizing sustained maintenance efforts' (GCEW 2023). Some chapters of the book touch on the importance of policy instruments and the political economy to help realize nexus securities.

I hope that this publication will be able to offer answers to students and specialists familiar with water, food and climate nexus issues with new insights and solutions. These need to be scaled at an unprecedented pace and coverage if there is any hope of achieving the Sustainable Development Goals 2 and 6, as well as all the others that are dependent on healthy water and food systems.

Dr. Rachael McDonnell
Rome, Italy
R.Mcdonnell@cgiar.org

References

FAO (2021) The state of the world's land and water resources for food and agriculture—systems at breaking point. Synthesis report 2021. Rome. https://doi.org/10.4060/cb7654en

Global Commission on the Economics of Water (GCEW) (2023) Turning the tide: a call to collective action. https://watercommission.org/wp-content/uploads/2023/03/Turning-the-Tide-Report-Web.pdf

Dr. Rachael McDonnell is the Deputy Director General of the International Water Management Institute (IWMI), where she leads 170 researchers across 12 offices to address global development challenges related to food, land, and water systems under a climate crisis.

Foreword by Prof. Dr. Eddy J. Moors

The world is facing a water crisis, and the challenges it poses are largely created by the global population growth and the related increase in food and energy production and use, with climate change exacerbating the problems more and more severely. It is evident that, because these drivers are interconnected, solutions will have to be sought in a holistic manner. The water, energy, food, and environment (WEFE) nexus approach will help in our search for sustainable development pathways.

The present book describes a number of options such as the use of alternative water resources, e.g., re-use of (un)treated wastewater, water harvesting, demand management, etc. The book also describes experiences with these options in certain regions. An important question will be: 'How can we implement these innovations?.'

In 2020, UN Water launched the accelerator framework for the Sustainable Development Goal 6 'Ensure access to water and sanitation for all' with five pillars: data, finance, governance, innovations, and capacity development, showing the need for the water sector to include aspects such as finance and governance in their thinking. In 2023, the UN created the Water Action Agenda to speed up implementation of urgently needed improvements in the water sector. From the related UN 2023 Water conference, it became clear that engaging with other sectors is of paramount importance to implement this Water Action Agenda. The WEFE nexus will be an important tool to ensure the involvement of different sectors.

As Albert Einstein once said, 'We can't solve problems by using the same kind of thinking we used when we created them.' We need something else than business as usual. We need capacity; we need another way of working for water and sanitation access; and we need innovative approaches. Yes, we all agree that efforts must be accelerated, but despite the development and the political endorsement of the global acceleration framework, it remains unclear how in practice we will fulfill the justified and legitimate expectations of those without water and sanitation access.

The two billion people who lack access to safely managed drinking water aren't helped by declarations, promises, or goals—they need access to water and sanitation, and they need it soon, if not now. The many people around the world who suffer in damaged environments need healthy ecosystems for their survival and their livelihoods. And the many people whose security is threatened by water-related conflict at local, national, and international levels need peace. To deliver for them, we must change our approach.

The fastest way to implement a new approach is through the development of capacity. Investments in hardware are not sufficient. There is also a need to invest in the less concrete parts, to ensure for example operation and maintenance, but also to create knowledge to take the right decisions, to develop investment plans, to propose adjustment of or new policies; in short: create local capacity for a long-term sustainable development of our water resources.

Capacity development is about transferring knowledge to individuals, creating leadership, i.e., taking the responsibility to use this knowledge. Capacity development is also about institutions, creating enabling conditions, facilities, and organizations that allow for the implementation of innovations, measures, and policies needed to achieve Sustainable Development Goals in which water is key.

Taking groundwater as an example. It is clear that groundwater can play a crucial role in solving some of the pressing issues in our water supply. The development of a sustainable use of the available groundwater resources requires investments next to an appropriate governance system. In numerous reports and conferences, this has been highlighted. It seems a no-brainer that also capacity will be required to implement and, even more important, maintain the infrastructure required, next to an increase in the awareness of the value of water. However, rarely the needs and possibilities for financing capacity are mentioned.

It should be noted that capacity building is often required at all levels, for example, at primary schools for awareness raising, at vocational training for operation and maintenance, and at academic level for local leadership. The latter is important as, in some areas, there are still a lot of unknowns surrounding safe yield, recharge requirements, water quality, and other matters that may threaten a sustainable use of these groundwater resources. Sharing such knowledge between water experts is not enough. For the implementation of technical and social innovations, decision makers across different sectors and politicians should also be involved.

Additionally, as conditions keep on changing, with climate change being just one of them, capacity to adapt to these conditions is required. Therefore, long-term sustainability can only be achieved by creating creative thinkers in the region that are capable of taking leadership and are able to handle new developments and conditions as well as the present needs. Attracting and keeping these local (potential) water leaders in the water sector also require good human resource management with career perspectives for man and women.

Is it not surprising that although it is evident from literature that investing in education gives the highest return on investment for society, it still seems difficult to find financing for the required capacity development in the water sector!

I hope this book will help to show the way forward and that together we are able to make the urgently needed actions to improve the present and future management of our precious water.

Prof. Dr. Eddy J. Moors
Rector IHE Delft Institute for Water Education
Delft, Netherlands
e.moors@un-ihe.org

Foreword by Prof. Shahbaz Khan

Climate change will have major impacts on the availability of water for growing food and on crop productivity in the decades to come, warns a new FAO survey report, which summed up current scientific understanding of impacts, highlights knowledge gaps and areas for attention. The Climate Change, Water, and Food Security book is a comprehensive survey of existing scientific knowledge on the anticipated consequences of climate change for water use in agriculture.

The major impacts of climate change on water for farming include reductions in river runoff and aquifer recharges in the Mediterranean and the semi-arid areas of the Americas, Australia, and southern Africa—regions that are already water-stressed. In Asia, large areas of irrigated land that rely on snowmelt and mountain glaciers for water will also be affected, while heavily populated river deltas are at risk from a combination of reduced water flows, increased salinity, and rising sea levels.

Meanwhile, an acceleration of the world's hydrological cycle is anticipated as rising temperatures increase the rate of evaporation from land and sea. Rainfall will increase in the tropics and higher latitudes but decrease in already dry semi-arid to mid-arid latitudes and in the interior of large continents. A greater frequency in droughts and floods will need to be planned for, but already water-scarce areas of the world are expected to become drier and hotter.

Even though estimates of groundwater recharge under climate change cannot be made with any certainty, the increasing frequency of drought can be expected to encourage further development of available groundwater to buffer the production risk for farmers.

And the loss of glaciers—which support around 40% of the world's irrigation—will eventually impact the amount of surface water available for agriculture in key producing basins.

Increased temperatures will lengthen the growing season in northern temperate zones but will reduce the length almost everywhere else. Coupled with increased rates of evapotranspiration, this will cause the yield potential and water productivity of crops to decline.

So, to respond to these new challenges, actions can be taken by national policy-makers, regional and local watershed authorities, and individual farmers. One key area requiring attention is improving the ability of countries to implement effective systems for 'water accounting'—the thorough measurement of water supplies, transfers, and transactions in order to inform decisions about how water resources can be managed and used under increasing variability. Water accounting in most developing countries is very limited, and allocation procedures are nonexistent, ad hoc, or poorly developed. So, helping developing countries acquire good water accounting practices and developing robust and flexible water allocations systems will be a first priority.

At the farm level, growers can change their cropping patterns to allow earlier or later planting, reducing their water use and optimizing irrigation. Yields and productivity can be improved by shifting to soil moisture conservation practices, including zero- and minimum tillage. Planting deep-rooted crops would allow farmers to better exploit available soil moisture.

Mixed agroforestry systems also hold promise. These systems both sequester carbon and also offer additional benefits such as shade that reduces ground temperatures and evaporation, added wind protection, and improved soil conservation and water retention.

However, small-scale producers in developing countries will face an uphill struggle in adopting such strategies. Farm size and access to capital set the limits for the scope and extent of adaptation and change at farm level. Today, many developing world farms produce yields far below their agro-climatic potential. Therefore, greater precision and focus are needed to understand the nature, scope, and location of climate change impacts on developing country water resources for agriculture, the report says, adding: 'Mapping vulnerability is a key task at national and regional levels.'

Prof. Shahbaz Khan
Director of UNESCO Multisectoral
Regional Office for East Asia
and UNESCO Representative to China
DPRK, Japan, Mongolia, and ROK
Jakarta, Indonesia

Acknowledgements

I have been honored to share the editorship of this volume with my colleagues: Dr. Abdulmalek A. Al-Shaikh (Director of the Prince Sultan Institute for Environmental, Water, and Desert Research, and Chair of the UN Prince Sultan International Water Prize; King Saud University, Saudi Arabia); Prof. Dr. Ameenah Gurib-Fakim (an eminent Scholar and the Former President of the Republic of Mauritius); Prof. Dr. Mirza Barjees Baig (Prince Sultan Institute for Environmental, Water, and Desert Research, King Saud University, Saudi Arabia); and Dr. Prof. Mohammed Bahir (High Energy and Astrophysics Laboratory, Faculty of Sciences Semlalia, Cadi Ayyad University, Marrakech, Morocco). I seize this opportunity to thank all of them for their valuable collaboration during the publishing process. The real value of this volume should be, however, credited to the chapters' authors, whose works had been accepted for publication after a rigorous peer review and proofreading. Their collaboration, reactivity, and engagement during the process were indeed very remarkable and impressive. Therefore, any shortcomings are undoubtedly the editors' responsibility.

The chapters published in this volume are also the result of the invaluable contribution made by reviewers, who generously engaged their time and energy to provide insight and expertise regarding the volume's chapters, thus enhancing their quality. On behalf of my co-editors, I would specifically like to acknowledge, with sincere and deepest thanks, the following reviewers:

- Prof. Dr. Mohammed Abdelbaset Hessane. Laboratory of Geosciences, Environment and Associated Resources, Faculty of Sciences, Sidi Mohamed Ben Abdellah University, Morocco.
- Prof. Dr. Kamel.zouari. Laboratory of Radio-Analysis and Environment (LRAE), National School of Engineers, Sfax, Tunisia.
- Prof. Dr. Najiba Chkir. Laboratory of Radio-Analysis and Environment (LRAE), National School of Engineers, Sfax, Tunisia.
- Dr. Muhammad Irfan. School of Economics and Management, Xiamen University, Malaysia.

- Prof. Dr. Ihtsham Ul Haq Padda. Head, Department of Economics, Federal Urdu University for Arts Science and Technology (FUUAST), Islamabad, Pakistan.
- Dr. Irfan Ahmad Baig. Dean, Faculty of Social Science and Humanities, Department of Agribusiness and Applied Economics, MNSAU—Multan. Pakistan.
- Dr. Manzoor Ahmad Malik. Director (R), Pakistan Council of Research in Water Resources, Islamabad, Pakistan.
- Dr. Christian Siderius. Grantham Research Institute on Climate Change and the Environment, London School of Economics, Houghton Street, WC2A 2AE, London, UK.
- Eng. Dr. Isaiaka Toyin Busari. Ministry of Agriculture and Rural Development, Kwara State, Nigeria.
- Eng. Dr. Clement Kirotich Kiptum. School of Engineering, University of Eldoret, Eldoret, Kenya.
- Dr. Asif Sardar. National Center of Industrial Biotechnology, PMAS-Arid Agriculture University, Rawalpindi, Pakistan.
- Prof. Dr. Rafiq Islam. Soil, Water, and Bioenergy Resources, The Ohio State University South Centers, USA.
- Prof. Dr. Özlem Korkut. Atatürk Üniversity, Faculty of Engineering, Erzurum, Turkey.
- Dr. Adiqa Kausar Kiani. Federal Urdu University of Arts, Science and Technology, Islamabad, Pakistan.
- Dr. Muhammad Abid. State Key Joint Laboratory of Environment Simulation and Pollution Control, School of Environment, Tsinghua University, Beijing, China.
- Dr. Asim Jilani. Center of Nanotechnology, King Abdulaziz University, Jeddah, Saudi Arabia.
- Dr. Muhammad Rizwan. University of Chinese Academy of Sciences, Beijing, China.
- Dr. Muhammad Aamir Shakoor. Department of Agricultural Engineering, Bahauddin Zakariya University (BZU), Multan. Pakistan.
- Dr. Ijaz Ahmad. Director, Remote Sensing Agricultural Wing, SUPARCO, Islamabad, Pakistan.
- Dr. Amer Masood. Department of Space Science, University of the Punjab, Lahore, Pakistan.
- Dr. Syed Muhammad Hasan Raza. Department of Space Science, University of the Punjab, Lahore, Pakistan.
- Dr. Muhammad Amin. Institute of Geo-Information and Earth Observation, Pir Mehar Ali Shah University of Arid Agriculture, Rawalpindi, Pakistan.
- Dr. Fiaz Ahmad. Physiology and Chemistry Section, Central Cotton Research Institute, Multan, Pakistan.
- Dr. Omar Jouzdan. Expert—Soil and Water Sciences—Visiting Researcher, Lancaster University—UK.
- Prof. Dr. M. M. Abd El-Azeim. Vice-Dean, Faculty of Agriculture, Department of Soil and Water Sciences. Minia University, Egypt.
- Prof. Dr. Muhammad Arshad. King Abdulaziz University, Jeddah, Saudi Arabia.

- Dr. Ammari Abdelhadi. The National Higher School of Hydraulics, Blida, Algeria.
- Aftab Ahmad Sheikh. Institute. of Soil Chemistry and Environmental Sciences, Kala Shah Kaku, Pakistan.
- Prof. Dr. Rafiq Islam. Program Director—Soil, Water, and Bioenergy Resources, Ohio Sustainability Institute, College of Food, Agriculture and Environmental Sciences (CFAES), The Ohio State University, USA.
- Dr. Asif Sardar. National Center of Industrial Biotechnology, PMAS-Arid Agriculture University, Rawalpindi, Pakistan.
- Prof. Dr. Özlem Korkut. Faculty of Engineering, Atatürk Üniversity, Erzurum, Turkey.
- Dr. Adiqa Kausar Kiani. Federal Urdu University of Arts, Science and Technology, Pakistan.
- Dr. Khondker Iftekhar Iqbal. Faculty of Arts and Social Sciences, Universiti Brunei Darussalam, Brunei Darussalam.
- Prof. Dr. Syed Serajul Islam. Department of Political Science, Lakehead University, Orillia, ON, Canada.
- Prof. Dr. Shaikh M. Kais. Department of Sociology, University of Rajshahi, Rajshahi, Bangladesh.
- Dr. Farhat Konain Shujahi. National University of Modern Languages, Islamabad. Pakistan.
- Dr. Muhammad Rizwan. Researcher, University of Chinese Academy of Sciences, Beijing, China.
- Dr. Muhammad Mubeen. Department of Environmental Science, COMSATS University Islamabad, Vehari Campus, Pakistan.
- Prof. Dr. Hamza Farooq Gabrial. NUST Institute of Civil Engineering (NICE), School of Civil and Environmental Engineering (SCEE), National University of Sciences and Technology (NUST), Islamabad, Pakistan.
- Prof. Dr. Muhammad Zaman. Faculty of Agricultural Engineering and Technology, Department of Irrigation and Drainage, University of Agriculture, Faisalabad, Pakistan.
- Dr. Abed A. Alataway. Prince Sultan Institute for Environmental, Water, and Desert Research, King Saud University, Riyadh, Saudi Arabia.
- Dr. Turki K Alasiri. Prince Sultan Institute for Environmental, Water, and Desert Research, King Saud University, Riyadh, Saudi Arabia.
- Engr. Ali Wafa Abu-Risheh. Prince Sultan Institute for Environmental, Water, and Desert Research, King Saud University, Riyadh, Saudi Arabia.

Mohamed Behnassi

About the Publishing Institutions

The Center for Environment, Human Security and Governance (CERES), Morocco

CERES, previously the North-South Center for Social Sciences (NRCS), 2008–2015, is an independent and not-for-profit research institute founded by a group of Moroccan researchers and experts in 2015 and joined by many partners worldwide. It aspires to play the role of a leading think tank in the Global South and to serve as a reference point for relevant change processes. Since its creation, CERES managed to build a robust network involving various stakeholders such as researchers, experts, Ph.D. students, decision makers, practitioners, journalists, etc. from different spheres and scientific areas. These achievements are being rewarded by the invitation of CERES members to contribute to global and regional assessments and studies (especially Ipbes, Medecc, EuroMeSco, etc.) and the invitation of the Center to become a member of the MedThink 5+5, which aims at shaping relevant research and decision agendas in the Mediterranean Basin. The Center has organized so far five international conferences and several training/building capacity workshops, provides expertise for many institutions and publishes numerous books, scientific papers, and studies which are globally distributed and recognized. These events and publications cover many emerging research areas mainly related to the human-environment nexus from multi-dimensional, multiscale, interdisciplinary, and policy-making perspectives. Through its initiatives, the CERES attempts to provide expertise, to advance science and its applications, and to contribute to effective science and policy interactions.

The Prince Sultan Institute for Environmental, Water and Desert Research (PSIEWDR), King Saud University, Riyadh, Kingdom of Saudi Arabia

PSIEWDR was established in 1986 to conduct scientific research related to environmental issues and water resources. It also engages with vital issues related to the problem of aridity and the desert environment. It conducts development initiatives for the country's desert areas, particularly programs for combating desertification in the Arabian Peninsula. PSIEWDR designed and carried out two major water harvesting and storage programs, including the construction of purpose-built infrastructure, throughout the Kingdom of Saudi Arabia using novel techniques and equipment. The institute actively applies remote sensing technologies using advanced satellite image processing systems and GIS to study the country's environment and natural resources. In 2007, the institute published *The Space Image Atlas of the Kingdom of Saudi Arabia*, and it is currently developing *The Environmental Atlas of the Kingdom of Saudi Arabia*. The institute has been the primary sponsor of the biennial International Conference on Water Resources and Arid Environments (ICWRAE) held in Riyadh, Saudi Arabia, since 2004. The institute hosts the General Secretariat of the Prince Sultan Bin Abdulaziz International Prize for Water (PSIPW) which honors scientists all over the world for their innovative water-related research. PSIPW, in turn, has many agreements with various international water associations as well as a close partnership with the United Nations. PSIPW and the United Nations Office of Outer Space Affairs (UNOOSA) jointly produce and maintain the International Space4Water Portal, an online hub for all stakeholders involved in utilizing space technologies for water resources applications.

Contents

Editors and Contributors

About the Editors

Dr. Mohamed Behnassi has been nominated in July 2022 as Senior Environmentalist Expert by the Economic Social and Environmental Council (ESEC), Rabat, Morocco. Prior to this, he served as Full Professor (since 2004) and Head of Public Law in French Department at the College of Law, Economics, and Social Sciences of Agadir, Ibn Zohr University. He holds a Ph.D. in International Environmental Law and Governance (2003), a M.Sc. in Political Sciences (1997), and a B.A. in Administration (1995) from Hassan II University of Casablanca. He obtained a Diploma in International Environmental Law and Diplomacy from the University of Eastern Finland and UNEP, 2015. He is also Alumnus of the International Visitors Leadership Program of the Department of State, USA. Dr. Behnassi is currently Founding Director of the Center for Environment, Human Security and Governance (CERES). From 2015 to 2018, he was Director of the Research Laboratory for Territorial Governance, Human Security and Sustainability (LAGOS). Recently, he was appointed as Expert Evaluator for the National Center for Scientific and Technical Research (CNRST/Morocco) and selected twice (2019–2024) as Assessment Scoping Expert and Review Editor by the Intergovernmental Science-Policy Platform on Biodiversity and Ecosystem Services (IPBES) and Member of the Mediterranean Experts on Climate and Environmental Change (MEDECC). He was also selected by the Intergovernmental Panel on Climate Change

(IPCC) as Expert Reviewer of the 1st Order Draft of the Synthesis Report (SYR) of the IPCC VI Assessment Report (AR6). Accordingly, he was among the Lead Authors of the 1st Assessment Report (MAR1): *Climate and Environmental Change in the Mediterranean Basin—Current Situation and Risks for the Future* (MEDECC, 2021). He is Senior Researcher in the areas of international law and politics of environment and human security where he published considerable number of scientific papers and book chapters in addition to 20 books, including recent ones on: *Food Security and Climate-Climate Smart Food Systems—Building Resilience for the Global South* (Springer, 2022); *The Climate-Conflict-Migration Nexus from a Human Security Perspective* (Springer, 2022); and *Social-Ecological Systems in the Era of Risks and Insecurity—Pathways to Viability and Resilience* (Springer, 2021). Dr. Behnassi serves as a reviewer for many global publishers (such as Routledge and Springer) and scientific journals with high-impact factor. He has organized many international conferences covering the above research areas, managed many research and expertise projects, and is regularly requested to provide scientific expertise nationally and internationally. Other professional activities include social compliance auditing and consultancy by monitoring human rights at work and the sustainability of the global supply chain.

Dr. Abdulmalek A. Al-Shaikh is Director of the Prince Sultan Institute for Environmental, Water, and Desert Research at the King Saud University (KSU) in Riyadh, Saudi Arabia. He earned his Ph.D. in Arid Land Studies from the University of Arizona, USA, in 1983. He also serves as General Secretary of the Prince Sultan bin Abdulaziz International Prize for Water (PSIPW). The prize is awarded bi-annually at the United Nations Headquarters. Being Director and Research Chair, he oversees the Research Programs at the Institute. He is the team leader of many research projects concerning the environment, desertification, climate change, and water harvesting in Saudi Arabia including the King Fahd Project for Water Harvesting and Storage in Saudi Arabia, Prince Sultan Project for Villages and Hamlets Rehabilitation in Saudi Arabia, the Space Images Atlas of Saudi Arabia and the Environmental Atlas of Saudi

Arabia. Being Chairman of the international conferences on water resources and arid environments, he successfully organized them at the King Saud University. He has brought world fame water scientists to Saudi Arabia and published nine proceedings of organized international conferences. He has published extensively in journals of international repute.

Dr. Ameenah Gurib-Fakim is Professor, a Biodiversity Scientist, and Entrepreneur. She has served as the 6th and First Female President of the Republic of Mauritius (2015–2018). Prior to that, she has been Managing Director of the Centre International de Développement Pharmaceutique (CIDP) Research and Innovation as well as Professor of Organic Chemistry with an endowed chair at the University of Mauritius. Since 2001, she has served successively as Dean of the Faculty of Science and Pro-Vice-Chancellor (2004–2010). She has also worked at the Mauritius Research Council as Manager for Research (1995–1997). Dr. Gurib-Fakim earned a B.Sc. in Chemistry from the University of Surrey (1983) and a Ph.D. from the University of Exeter, UK (1987). During her academic journey, she has participated in several consultation meetings on environmental issues organized by international organizations. Between 2011–2013, she was elected and served as Chairperson of the International Council for Scientific Union—Regional Office for Africa and served as Independent Director on the Board of Barclays Bank of Mauritius Ltd. between (2012–2015). As Founding Member of the Pan African Association of African Medicinal Plants, she co-authored the first-ever African Herbal Pharmacopoeia. She has authored and co-edited 30 books, several book chapters, and scientific articles in the field of biodiversity conservation and sustainable development. She has lectured extensively across the world, is a member of the Editorial Boards of major journals, and has served on technical and national committees in various capacities. Elevated to the Order of the Commander of the Star and Key by the Government of Mauritius in 2008, she has been admitted to the Order of the Chevalier dans L'Ordre des Palmes Academiques by the Government of France in 2010 and is the recipient of 5 D.Sc. (s). As Elected Fellow of several academies and societies, she received several international prizes, including the 2007 l'Oreal-UNESCO Prize for Women in Science and the African Union Commission Award for Women in Science, 2009. She was elevated to the Order of GCSK by the Government of Mauritius and received the Legion d'Honneur from the Government of France in 2016. In 2017, she received both the lifelong achievement award of the United States Pharmacopoeia-CePat Award and the American Botanical Council Norman Farnsworth Excellence in Botanical Research Award. In 2018, she received the Order of St. George at the Semperopernball, Dresden, Germany. In 2019, she received the 'Trailblazing Award for Political Leadership' by the World Women Leaders Council in Iceland. In 2020, she was elected Honorary President of the International and Engineering Institute and received their 2020 5th IETI Annual Scientific Award. She also received the IAS-COMSTECH Ibrahim Memorial Award from the WIAS in Jordan. In 2021, she received the Benazir Bhutto Lifetime Achievement Award, the Obada Prize and the RUFORUM Recognition Prize 2021. In 2021, she has been appointed as Distinguished Professor at the John Wesley School of Leadership, Carolina University, USA. In June 2016, she was in the Forbes List for the 100 'Most Powerful Women in the World' and first among the Top 100 Women in Africa Forbes List 2017, 2019. She is honored as one of the Foreign Policy's 2015 Global Thinkers.

Dr. Mirza Barjees Baig is a Professor at the Prince Sultan Institute for Environmental, Water, and Desert Research, King Saud University, Saudi Arabia. He earned his MS degree in International Agricultural Extension in 1992 from the Utah State University, Logan, Utah, USA, and was placed on the 'Roll of Honor.' He completed his Ph.D. in Extension for Natural Resource Management from the University of Idaho, USA, and was honored with the '1995 Outstanding Graduate Student Award.' Dr. Baig has published extensively on the issues associated with natural resources in national and international journals. He has also made oral presentations about agriculture and natural resources and the role of extension education at various international conferences. Food waste, water management, degradation of natural resources, deteriorating environment, and their relationship with society/

community are his areas of interest. He has attempted to develop strategies to conserve natural resources, promote the environment, and develop sustainable communities. Dr. Baig started his scientific career in 1983 as a researcher at the Pakistan Agricultural Research Council, Islamabad, Pakistan. He served at the University of Guelph, Ontario, Canada, as the Special Graduate Faculty from 2000 to 2005. He served as Foreign Professor at the Allama Iqbal Open University (AIOU), Pakistan, through the Higher Education Commission from 2005–2009. He served as Professor of Agricultural Extension and Rural Society at the King Saud University, Saudi Arabia, from 2009–2020. He serves as well on the Editorial Boards of many international journals and is a member of many international professional organizations.

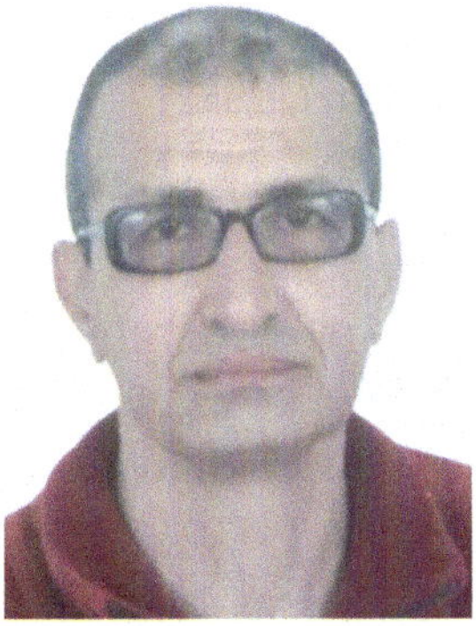

Prof. Dr. Eng. Mohammed Bahir, Ph.D. Hydrogeological Engineer, has published over 200 research/professional papers and reports. He co-authored more than a dozen books and book chapters. He supervised and/or co-supervised more than 20 Ph.D. students. He is a reviewer of many international scientific journals with high-impact factors, including *Environment, Development, and Sustainability Journal, Environmental Earth Sciences Journal, Hydrological Sciences Journal, Sécheresse, Gaia, Communicaçoés International, Science of the Total Environment, Marine and Freshwater Research Journal, Proceedings of the Indian National Science Academy, and Groundwater for Sustainable Development*. He also coordinates more than a dozen research projects.

The Asia Research Awards, in association with Times of Research and the World Research Council, announce that Professor Mohammed Bahir has been selected as the recipient of the prestigious ASTRA 2023 (Asia's Science, Technology, and Research Awards).

Contributors

Chérifa Abdelbaki Pan-African University Institute for Water and Energy Sciences (PAUWES), Tlemcen, Algeria;
Laboratoire EOLE, Université de Tlemcen, Tlemcen, Algeria

Sultan Adeel Department of Business Administration, Sukkur IBA University, Sukkur, Pakistan

Ijaz Ahmad Centre of Excellence in Water Resources Engineering, University of Engineering and Technology, Lahore, Pakistan

Mushtaque Ahmed Department of Soils, Water and Agricultural Engineering, College of Agricultural & Marine Sciences, Sultan Qaboos University, Muscat, Oman

Waqar Akram Department of Accounting, Finance, and Economics, University of Waikato, Hamilton, New Zealand

Waad Al-Aghbari Department of Soils, Water and Agricultural Engineering, College of Agricultural & Marine Sciences, Sultan Qaboos University, Muscat, Oman

Ahmed Al-Busaidi Department of Soils, Water and Agricultural Engineering, College of Agricultural & Marine Sciences, Sultan Qaboos University, Muscat, Oman

Abdulmalek A. Al-Shaikh Prince Sultan Institute for Environmental, Water & Desert Research, King Saud University, Riyadh, Saudi Arabia

Yahya Al-Yahyaei Department of Marine Science and Fisheries, College of Agricultural and Marine Sciences, Sultan Qaboos University, Muscat, Oman

Muhammad Arshad Department of Irrigation and Drainage, University of Agriculture, Faisalabad, Pakistan

Muhammad Ashfaq Institute of Agricultural and Resource Economics, Faculty of Social Sciences, University of Agriculture, Faisalabad, Pakistan

Muhammad Awais College of Earth and Environmental Sciences, University of the Punjab, Lahore, Pakistan

Willis Awandu Department of Civil and Structural Engineering, Masinde Muliro University of Science and Technology, Kakamega, Kenya

Nadia Badr Environmental Sciences Department, Faculty of Science, Alexandria University, Alexandria, Egypt

Mohammed Bahir High Energy and Astrophysics Laboratory, Faculty of Sciences Semlalia, Cadi Ayyad University, Marrakesh, Morocco

Mirza Barjees Baig Prince Sultan Institute for Environmental, Water and Desert Research, King Saud University, Riyadh, Kingdom of Saudi Arabia

Larbi Barhazi Department of Geography, LADES, FLSH-M, Hassan II University of Casablanca, Mohammedia, Morocco

Kenan Barik Department of Soil Science and Plant Nutrition, University of Atatürk, Erzurum, Turkey

Saira Batool Center for Integrated Mountain Research, University of the Punjab, Lahore, Pakistan

Mohamed Behnassi Economic, Social and Environmental Council (ESEC), Rabat, Morocco;
Center for Environment, Human Security and Governance (CERES), Rabat, Morocco;
Ministry of Higher Education and Innovation, Rabat, Morocco

Halima Belarbi Laboratoire EOLE, Université de Tlemcen, Tlemcen, Algeria

Soufiane Bouhafa Department of Geography, LADES, FLSH-M, Hassan II University of Casablanca, Mohammedia, Morocco

QingLin Chen Environment and Sustainability Research Cluster, School of Social Sciences, Nanyang Technological University, Singapore, Singapore

Rachida El Morabet Department of Geography, LADES, FLSH-M, Hassan II University of Casablanca, Mohammedia, Morocco

Otman El Mountassir High Energy and Astrophysics Laboratory, Faculty of Sciences Semlalia, Cadi Ayyad University, Marrakesh, Morocco

Hasan Er Department of Biosystem Engineering, University of Bingöl, Bingöl, Turkey

Faraz Ul Haq Centre of Excellence in Water Resources Engineering, University of Engineering and Technology, Lahore, Pakistan;
Center for Applied Earth Science and Engineering Research, The University of Memphis, Memphis, United States;
Department of Civil Engineering, The University of Memphis, Memphis, United States

Wenresti Gallardo Department of Marine Science and Fisheries, College of Agricultural and Marine Sciences, Sultan Qaboos University, Muscat, Oman

Ameenah Gurib-Fakim Women Leadership, John Wesley School of Leadership, Carolina University, Winston-Salem, USA;
Port Louis, Republic of Mauritius

Rokiatou Haidara Pan-African University Institute for Water and Energy Sciences (PAUWES), Tlemcen, Algeria

Jan W. Hopmans Soil Science and Irrigation Water Management, University of California Davis, Davis, USA

Zakir Hussain Metropolitan College of New York, New York, NY, USA;
Grand Asian University, Sialkot, Pakistan

Md Saidul Islam Associate Professor of Sociology, Nanyang Technological University, Singapore, Singapore

Edwin Kimutai Kanda Department of Civil and Structural Engineering, Masinde Muliro University of Science and Technology, Kakamega, Kenya

Ashvini Kannan Environment and Sustainability Research Cluster, School of Social Sciences, Nanyang Technological University, Singapore, Singapore

Noor Muhammad Khan Centre of Excellence in Water Resources Engineering, University of Engineering and Technology, Lahore, Pakistan;
Civil Engineering Department, University of Engineering and Technology, Lahore, Pakistan

Roohul Abad Khan Department of Civil Engineering, King Khalid University, Abha, Saudi Arabia

Susan Namaemba Kimokoti Department of Peace and Conflict Studies, Masinde Muliro University of Science and Technology, Kakamega, Kenya

Pavneet Kaur Kingra Department of Climate Change and Agricultural Meteorology, Punjab Agricultural University, Ludhiana, Punjab, India

Rakhshanda Kousar Institute of Agricultural and Resource Economics, Faculty of Social Sciences, University of Agriculture, Faisalabad, Pakistan

Surinder Singh Kukal Punjab Water Regulation and Development Authority, Government of Punjab, Chandigarh, India

Yasemin Kuslu Department of Agricultural Structures and Irrigation, University of Atatürk, Erzurum, Turkey

Lynette Loh Environment and Sustainability Research Cluster, School of Social Sciences, Nanyang Technological University, Singapore, Singapore

Syed Amer Mahmood Department of Space Science, University of the Punjab, Lahore, Pakistan

Muhammad Sohail Amjad Makhdum Department of Economics, Government College University, Faisalabad, Pakistan

Hidayat Mohammedi Pan-African University Institute for Water and Energy Sciences (PAUWES), Tlemcen, Algeria

Adeel Mukhtar School of Public Policy, Oregon State University (OSU), Corvallis, USA

M. M. M. Najim Southeastern University of Sri Lanka, Oluvil, Sri Lanka

Muhammad Nawaz Khan Islamabad Policy Research Institute (IPRI), Islamabad, Pakistan

Safdar Ali Shirazi Department of Geography, University of the Punjab, Lahore, Pakistan

Josephine Toh Environment and Sustainability Research Cluster, School of Social Sciences, Nanyang Technological University, Singapore, Singapore

Abbreviations and Acronyms

ABHT	Tensift Basin Hydraulic Agency
ASAL	Arid or Semi-Arid Land
AVA	Agri-Food and Veterinary Authority of Singapore
AWD	Alternate Wetting and Drying
CFA	Continuous Flow Analyzer
CNRST	Moroccan Center for Scientific and Technical Research
COA	Court of Arbitration
COD	Chemical Oxygen Demand
CSA	Climate-Smart Agriculture
CWANA	Central and West Asia and North Africa
CWP	Crop Water Productivity
CWU	Consumptive Water Use
DCR	Diversified Crop Rotation
DIs	Drought Indices
DOAJ	Directory of Open Access Journals
ET	Evapotranspiration
FACE	Free Air Carbon Enrichment
FAO	Food and Agriculture Organization
GCM	Global Climate Models
GHG	Greenhouse Gas
GMWL	Global Meteoric Water Line
GPR	Ground Penetrating Radar
GWA	Genome-Wide Association
GWAVA	Global Water AVailability Assessment
GWP	Global Warming Potential
ICARDA	International Center for Agricultural Research in the Dry Areas
ICIMOD	Integrated Mountain Development
ICJ	International Court of Justice
IIASA	International Institute for Applied System Analysis
IMD	India Meteorological Department
IPCC	Intergovernmental Panel on Climate Change

IWQI	Irrigation Water Quality Index
IWRM	Integrated Water Resources Management
IWT	Indus Water Treaty
JGCRI	Joint Global Change Research Institute
KNAP	Kenya National Adaptation Plan
LMWL	Local Meteoric Water Line
MENA	Middle East and North Africa
NCAR	National Center for Atmospheric Research
NCCRS	National Climate Change Response Strategy
NCEP	National Center for Environmental Prediction
NDMA	National Disaster Management Authority
NDMC	National Drought Monitoring Center
NHPC	National Hydroelectric Power Corporation
NWHSA	National Water Harvesting Authority
ONEE	National Office of Electricity and Water
PA	Precision Agriculture
PI	Precision Irrigation
PLF	Precision Livestock Farming
PMD	Pakistan Meteorological Department
QTL	Quantitative Trait Locus
RAS	Recirculating Aquaculture System
RCM	Regional Climate Models
RDI	Reconnaissance Drought Index
RWHT	Rainwater Harvesting Techniques
SDGs	Sustainable Development Goals
SDI	Subsurface Drip Irrigation Systems
SI	Supplemental Irrigation
SWAT	Soil Water Assessment Tool
TSM	Total Suspended Matters
UIB	Upper Indus Basin
UM6P	Mohamed VI Polytechnic University of Benguerir
UNFCCC	United Nations Framework Convention on Climate Change
VIC	Variable Infiltration Capacity
VRT	Variable Rate Application
WAPDA	Water Allocation and Distribution Authority
WEFE	Water, Energy, Food, and Ecosystems
WHO	World Health Organization
WMO	World Meteorological Organization
WQI	Water Quality Index
WU	Water Use
WUE	Water Use Efficiency
WWTP	Wastewater Treatment Plant

List of Figures

Chapter 2

Chapter 3

Chapter 4

Chapter 5

Chapter 8

Chapter 9

Chapter 11

Chapter 12

Chapter 13

Chapter 14

Chapter 15

Chapter 16

Chapter 17

Chapter 18

Chapter 20

List of Tables

Chapter 8

Chapter 9

Chapter 11

Chapter 12

Chapter 13

Chapter 14

Chapter 15

Chapter 16

Chapter 17

Chapter 19

Chapter 20

Chapter 1
The Water, Climate, and Food Nexus: Linkages, Challenges and Emerging Solutions—An Introduction

Mohamed Behnassi, Abdulmalek A. Al-Shaikh, Ameenah Gurib-Fakim, Mirza Barjees Baig, and Mohammed Bahir

Abstract Water, as an increasingly scarce natural resource, is still a vital resource for both development and security. This resource is coming under an enormous strain because of a myriad of factors, including the increased demands of a growing global population and economy, environmental and climatic changes, and the irresponsible use. The decline of water resources is currently putting food systems under pressure. Indeed, the twin challenge is currently providing an ever-growing human population with sufficient and nutritious food, while facing environmental/resource limits and climate change. Thus, with the escalating water scarcity, a sustainable transformation of the global food system is unreachable without a revolution, especially in the area agricultural water use. In this perspective, the agricultural sector is being transformed to greater heights by new technological developments such as

M. Behnassi (✉)
Economic, Social and Environmental Council (ESEC), Rabat, Morocco
e-mail: behnassi@gmail.com

Center for Environment, Human Security and Governance (CERES), Agadir, Morocco

Ministry of Higher Education and Innovation, Rabat, Morocco

A. A. Al-Shaikh · M. B. Baig
Prince Sultan Institute for Environmental, Water & Desert Research, King Saud University, Riyadh, Saudi Arabia
e-mail: aasheikh@ksu.edu.sa

M. B. Baig
e-mail: mbbaig@ksu.edu

A. Gurib-Fakim
Women Leadership, John Wesley School of Leadership, Carolina University, Winston-Salem, USA
e-mail: aguribfakim@gmail.com

Port Louis, Republic of Mauritius

M. Bahir
High Energy and Astrophysics Laboratory, Faculty of Sciences Semlalia, Cadi Ayyad University, Marrakesh, Morocco
e-mail: bahir@uca.ac.ma

M. Behnassi et al. (eds.), *The Water, Climate, and Food Nexus*,
https://doi.org/10.1007/978-3-031-50962-9_1

smart farming, precision agriculture, and climate-smart agriculture, which will enable this primary sector to move to a higher level of farm productivity, profitability, and resource-use efficiency while maintaining sustainability and fostering food security and resilience. This process remains dependent, however, on the extent to which different sectors involved in water resources management work in integrated and coordinated ways. This involves a shift in governance systems and the definition of roles and responsibilities for everyone engaged in the process of water resource generation, management, and usage. Against this background, the present introductory chapter presents the framework and content of this book, which addresses timely and future-oriented topics. In the first set of chapters, the water, food, and environmental/climate security nexus is explored theoretically and by reference to empirical research covering many regions and sectors. In another set of chapters, the impacts of climate change on water resources and water-stressed regions are identified along with their implications for food systems and security. Other chapters of the volume identify the emerging solutions to the nexus challenges, mainly adaptation and mitigation options, governance and management approaches, technological and economic solutions, innovative farming and water management practices, etc. Most chapters scheduled for publication are based on empirical research particularly done in water-constrained and climate vulnerable countries from Asia, Africa, and the MENA region and provide policy-oriented inputs and recommendations to guide change processes at multiple scales.

Keywords Nexus · Water scarcity · Climate risks · Food production · Smart techniques

1 The Volume's Background, Scope, and Objectives

Water is a basic necessity, without which no individual or nation can survive. Every nation therefore needs water security, which encompasses all dimensions of human health, livelihood and well-being while being essential for both food and energy production. However, even if over 70% of the Earth's surface is covered by water, 97.5% of this potential is salt water, and less than 1% of the world's freshwater is accessible for human use; thus, the real issue is the limited amount of freshwater available for all. Based on this reality, it is currently perceived that water will be the oil of the future since this increasingly scarce natural resource is still a vital resource for both development and security. Like oil, water supplies, especially clean and easily accessible ones, are coming under an enormous strain because of a myriad of factors, including the increased demands of a growing global population and economy (especially in the areas of agriculture, industry, and energy production), environmental and climatic changes, and the irresponsible use of water resources (i.e., overuse, loss, and pollution). Furthermore, water, unlike oil, has no substitute, which makes it even more precious and subject to competition at all levels and among a growing list of actors. Due to an increase in water scarcity, coupled to a growing demand, intra- and interstate tensions began to rise in many regions over the control of the limited available freshwater resources.

The United Nation's 2030 Agenda for Sustainable Development has been shaped in line with the above challenges. It now legally acknowledges commitment to a sustainable future for all countries, the promotion of which puts food systems under pressure. Indeed, the twin challenge is currently providing an ever-growing human population with sufficient and nutritious food, while facing environmental/resource limits and climate change. With the escalating degradation of freshwater resources, a sustainable transformation of the global food system is unreachable without a revolution, especially in the area of agricultural water use.

The Intergovernmental Panel on Climate Change (IPCC) has already stated in its assessment reports that the vulnerability of freshwater resources is being increased by climate change, with drastic social, economic, and ecological consequences. Climate change will have major impacts on the water cycle, resulting in greater climatic and hydrological instability, with serious consequences for societies and their water and food security. The IPCC anticipates an increased frequency and severity of droughts due to declining precipitation trends in certain areas, while others will have higher prevalence of floods and typhoons due to increasing precipitation intensity. Moreover, the rise in sea levels due to global warming will lead to melting of the polar ice caps and increased saltwater intrusion, which will in turn affect water quality. Global warming also increases the amount of water that the atmosphere can carry, therefore leading to more frequent and intense rainfall when the air cools. Although rainfall can contribute to freshwater resources, heavy rainfall can also lead to accelerated movement of water from the atmosphere back to the sea, thereby decreasing humans' ability to store and use it. According to the World Bank, fluctuation in rainfall alone could push more than 12 million people into poverty, while climate change could increase global malnutrition by up to 25% by 2080. In this broad context, marginalized and poor communities remain the ones most affected by water-related disasters, but they do not have sufficient resources to cope due to factors such as poor environmental practices, rapid and unplanned urbanization in dangerous zones, and government incapacity.

In all cases, changes in the hydrological cycle due to natural and anthropogenic dynamics are expected to threaten existing water infrastructure, making societies more vulnerable to water-related disasters and resulting in decreased water security. The concept of 'water security' is not simply about dealing with dwindling water supplies. It has been defined by the UN-Water as *"the capacity of a population to safeguard sustainable access to adequate quantities of acceptable quality water for sustaining livelihoods, human well-being, and socio-economic development, for ensuring protection against water-borne pollution and water-related disasters, and for preserving ecosystems in a climate of peace and political stability"*. A comprehensive definition of water security goes beyond water availability to issues of access, which involve fundamental individual rights, national sovereignty over water, equity and affordability, and the role of states and markets in water's allocation, pricing, regulation, and distribution.

In addition to the above, more countries facing serious water scarcity will be unable to ensure a sustainable domestic food security because agriculture, which is still the primary source of food production, will deeply suffer from water scarcity. Moreover, the agriculture sector will be affected by environmental and climatic changes in various ways—including biodiversity loss, land degradation, and changes in typical temperatures, precipitation, and atmospheric conditions—and these effects are unevenly distributed over the globe. On average, global agriculture uses around 70% of the available freshwater resources. In arid climate zones such as the Middle East and North Africa (MENA), the agricultural water consumption can even be up to 90%. Global demand for water is expected to grow by 50% in the year 2030 and most of this demand will be in cities. This will require new approaches to non-conventional water management systems (such as wastewater collection and management, desalinization, and rainwater harvesting, storage and use).

To face water scarcity, the agricultural sector has been transformed to greater heights by new technological developments such as smart farming, precision agriculture, and climate-smart agriculture, which will enable this primary sector to move to a higher level of farm productivity, profitability, and resource-use efficiency while maintaining sustainability and fostering food security and resilience. The importance of smart-farming applications in irrigation water management to increase farm productivity by overcoming climate change, using less human energy, increasing efficiency, and improving management techniques through precision farming technologies is now widely acknowledged. Many researches acknowledge that inadequate water management practices, limited storage capacity, climate-induced water stress, among others, have all contributed to making water management the most pressing problem of the twenty-first century. Therefore, to offer sustainable, cost-effective, and equitable water services based on advanced water resources, effective water management may have a significant impact on water insecurity when considering water-related issues such as climate change, pollution, misuse, and inappropriate governance settings. In order to effectively manage water resources, many different sectors must work in integrated and coordinated ways. This involves a shift in governance systems and the definition of roles and responsibilities for everyone engaged in the process of water resource generation, management, and usage.

Responsible consumption and production is the stated goal of the Sustainable Development Goal 12. Striving for such a stated goal will ultimately lead to sustainable consumption and production in all sectors of the economy and will lead to conservation of water and energy and reduce concerns regarding the achievement of water and food security. For instance, many opportunities currently exist in the area of food production using limited but highly treated wastewater. Treated wastewater is a good source of different plant nutrients for agricultural lands. It can improve soil fertility and crop productivity and minimize the inputs of fresh water and synthetic fertilizers. Indeed, in regions suffering from water scarcity, wastewater is no longer considered as waste to be disposed but as an integral part of potential water resources. The use of wastewater in agriculture started in Australia, France, Germany, India, the United Kingdom, and the United States of America in the late of the nineteenth century. It is used for irrigation in treated and untreated forms, varying by geographic and economic contexts, although with the majority in untreated form in developing countries.

Against this background, this contributed volume is addressing timely and future-oriented topics. In the first set of chapters, the water, food, and environmental/climate security nexus is explored theoretically and by reference to empirical research covering many regions and sectors. In another set of chapters, the impacts of climate change on water resources and water-stressed regions are identified along with their implications for food systems and security. Other chapters of the volume identify the emerging solutions to the nexus challenges, mainly adaptation and mitigation options, governance and management approaches, technological and economic solutions, innovative farming and water management practices, etc. Most chapters scheduled for publication are based on empirical research particularly done in water-constrained and climate vulnerable countries from Asia, Africa, and the MENA region and provide policy-oriented inputs and recommendations to guide change processes at multiple scales.

2 The Volume's Content

This volume is a collation of the water, climate, and food nexus-specific case studies from many regions in the Global South presenting various multi-scale perspectives on the most pertinent nexus related linkages, challenges, and solutions. In the following part, chapters' key content, findings, lessons learned, and recommendations are presented.

In Chap. 2, *The Water, Food, and Environmental Security Nexus*, Muhammad Sohail Amjad Makhdum, Rakhshanda Kousar, Muhammad Ashfaq, and Mohamed Behnassi consider the decreasing quality of water due to anthropogenic activities and over-pumping of underground water to meet increasing demands as growing factors of global water scarcity and insecurity. In the area of food, despite the increase of production due to biological and technological advancements and the progress made in terms of economic development and growth, food security is still unachieved in many parts of the world since the sufficient and quality food is not available to all for many reasons, including access and governance challenges. Regarding the natural environment, it is also on the verge of deterioration and degradation due to anthropogenic activities, including the over and inefficient use of natural resources and unsustainable development. Based on this, the authors analyze the different dimensions of water, food, and environment from a nexus perspective, given their interdependence. The authors intend to provide a better understanding of the linkages between these areas, and thus guiding public policies aiming at fighting poverty, food insecurity, and vulnerability while promoting a healthy environment for present and next generations.

In Chap. 3, *Water and Food Security in the Middle Eastern and Northern African Countries*, Waqar Akram, Zakir Hussain, and Sultan Adeel consider water as a growing concern in the MENA region, entailing both risks and opportunities. This situation of water insecurity, especially when combined with food insecurity, is further compounded by fast-changing politico-socio-economic and environmental conditions, thus making it more challenging for policymakers. According to the authors, water shortage, food security, and environmental problems are intertwined in MENA countries and are more than ever-changing targets. The climate of this region is arid and near arid and the drought cycle is shortened from three years to annual and often brings floods. In this context, the authors believe that there are many solutions to the MENA region's water resource management and associated problems. Accordingly, it is imperative to use water resources efficiently to ensure environmental sustainability and obtain distributive justice, thus contributing to social contract or integration. This argument also includes delivering water input reliably and affordably to ensure constructive relationships between service providers and water users and help promote the renewed social contract. Implementing these solutions, however, needs clear incentives to bring about water management changes, including conservation, allocation, and management of riparian issues and water conflicts among MENA countries. Any failure to implement policies addressing water challenges may severely affect the region's well-being and fragile political stability. For the authors, MENA countries should act with urgent attention in strengthening water security, food security, and a sustainable environment instead of waiting for doomsday for impending water crises leading to water conflicts. To do this, the authors provide a way forward in addressing these issues.

In Chap. 4, *The Water-Energy-Food (WEF) Nexus in Kenya: Climate Change Impacts and Adaptation Strategies—A Review*, Willis Awandu, Edwin Kimutai Kanda, and Susan Namaemba Kimokoti consider such a nexus as a novel concept aiming at integrating three key drivers of development and human security. The authors believe, by reference to the case of Kenya, that the effective management of these three resources requires careful assessment of synergies, conflicts, and trade-offs which are inherent in the nexus, especially in the uncertain context of climate variability and change. In Kenya, over 75% of agricultural activities are rain-fed, and thus risky in the face of the erratic temporal and spatial distribution of rainfall. Similarly, energy reliability in the country is low due to the dependence on hydropower sources, which are vulnerable to climate risks. Therefore, the authors consider climate adaptation mechanisms for building resilience in cropping systems, water service, and energy provision as key elements for the improvement of livelihoods. To this end, they highlighted the importance of the WEF nexus in the face of climate change impacts as an approach which can help address sustainably and holistically the three key sectors through policy, legal and institutional frameworks and initiatives. The authors has also developed assessment models and tools to monitor the achievement of targets under the WEF nexus.

In Chap. 5, *The Impact of Climate Change on Groundwater Resources in Northwestern Morocco*, Mohammed Bahir used the results of 9 campaigns—1990, 1995, 1997, 2009, 2015, 2016, 2017, 2018 and 2019—to assess the quality of groundwater in the Essaouira region, Morocco in the context of a changing climate. His hydrogeochemical study shows that the groundwater of the Cenomanian–Turonian aquifer presents the Cl–Ca–Mg, Cl–Ca, Cl–Na, and HCO_3–Ca mix facies with the dominance of the Cl–Ca–Mg mix facies, and Cl–Ca. Moreover, the analysis of the correlations established between the concentrations of major elements has shown that the mineralization of groundwater is controlled by the phenomenon of the dissolution of the evaporitic minerals (halite, gypsum and/or anhydrites) and carbonates (dolomite), by the reverse ion exchange phenomenon and by the marine intrusion, especially at the Plio-Quaternary aquifer. This study, focusing on the spatio-temporal evolution of the groundwater quality in the study area, shows a gradual deterioration in time and space. However, the author believes the Essaouira basin is more vulnerable to climate change because its recharge is entirely dependent on meteoric waters.

In the same vein, Mohammed Bahir, in Chap. 6, *Assessment of the Climate Change Impact on the Past and Future Evapotranspiration and Flows from a Semi-Arid Environment*, recalls that access to drinking water for the greatest number of populations, securing this often over-exploited and poorly managed resource, controlling agricultural and industrial use of water, and protecting the environment represent major challenges, especially for developing countries. His investigation focused on the assessment of the climate change impact on the hydrological regime within the Essaouira basin, Morocco for the period 2020–2050. For this purpose, the author used Rural Genius GR2M model to simulate flows based on rainfall and evapotranspiration. The Mann–Kendall and Pettitt tests were used to study the homogeneity and trend sense of the time series studied. According to the author, the investigation of the links between rainfall and flows for the period 1978 to 2005 in the Essaouira basin shows the existence of a cause-and-effect relationship between these two parameters, and this rule remains valid for all zones under arid and semi-arid climate. Therefore, these findings can serve as a basis for water resource protection and management in the Essaouira watershed by constructing hill reservoirs along the Igrounzar, Zelten, and Ksob Wadi.

In Chap. 7, *Evolution of Historical and Future Precipitations and Temperatures Within Essaouira Basin Under Climate Change Effect*, Mohammed Bahir, Otman El Mountassir, and Mohamed Behnassi consider Morocco as one of the countries most vulnerable to the impacts of climate change, particularly in regions with arid and semi-arid climates. In this context, the authors used a climatological analysis (rainfall and temperature) and groundwater level, salinity, and isotopic methods to better understand the relationship between climate change and water availability. The annual precipitation analysis results using the Nicholson rainfall index graphical method and statistical tests, namely Pettitt test and Mann–Kendall tests, revealed an overall negative trend for the basin of 12 to 16% from 1978 to 2015. This decrease in average annual precipitation is accompanied by an increase in temperatures with a very significant extent of warming of 1.2 °C in the downstream part of the study area and 2.3 °C in its upstream region. This reflects the effect of the continentality of temperatures in the study area. It is recommended for local and global policies to integrate targets such as the reduction of greenhouse gas emissions, the use of clean and renewable energy, and the education of the population on the swift adoption of appropriate behaviors towards a warmer and drier climate, especially for people living in areas with semi-arid, arid, and Saharan climate.

In the same perspective, Yasemin Kuslu and Kenan Barik examined in Chap. 8, *The Climate Change Phenomenon Using Temperature and Precipitation Observations by Reference to The Case of Erzurum.* The authors recall that the most important indicators of the climate change are the long-term changes in the spatial and temporal precipitation and temperature patterns, and this is being increasingly felt by extreme weather events and changes in temperature and precipitation trends. Data collected about these aspects provide important clues about climate change. For this purpose, the authors used different statistical analysis methods. And since hydro-meteorological observations are data covering a series of time, hence they may be intermittent, short-term, irregular, and skewed, the authors applied homogeneity tests before the statistical analysis. Applications were made on the temperature and precipitation values obtained from observation stations that show slight differences from each other and represent three sub-climates in the city of Erzurum, Turkey, which has a semi-arid main climate. The authors found that there is an increase in the temperature trend and a decrease in the precipitation trend for all three stations. Therefore, the statistical significance of the change in temperature and precipitation trends requires urgent action to be taken against climate change, especially in terms of adaptation.

In Chap. 9, *Effect of Climate Change on Sea Water Intrusion in the Essaouira Basin Coastal Aquifer*, Mohammed Bahir, Otman El Mountassir, and Mohamed Behnassi claim that sea-level rise, during which salt water penetrated coastal underground aquifer systems and mixed with fresh water, has gradually contributed to the decline in water quality. This is happening in a context where groundwater is becoming an important source of both potable and irrigation water uses in many parts of the world. The authors show in their hydrogeochemical study that the seawater begins to invade the freshwater of the Plio-quaternary aquifer of Essaouira basin, Morocco. This intrusion, demonstrated by ionic ratios and corroborated by stable isotopes and the combined use of oxygen-18 contents and chlorides, has a mixing rate that vary between 12.8 and 15.9%. Stable isotopes approach highlights that the recharge of Plio-Quaternary and Turonian is ensured by precipitation of Atlantic origin without significant evaporation and that these waters have been subjected to contamination by seawater.

Accordingly, in Chap. 10, *Analysis of Groundwater Regimes Utilizing Hydrogeological Modeling Under Climate Change Scenarios*, Muhammad Awais, Muhammad Arshad, Jan W. Hopmans, Mirza Barjees Baig, and M. M. M. Najim consider the groundwater, given its multiple advantages, as one of the most imperative natural resources increasingly used worldwide for a variety of purposes (agricultural, domestic, and industrial usage), especially in a context marked by the decline of surface water resources in terms of availability and quality. However, climate change is currently affecting global natural resources, particularly the water cycle, with many consequences for groundwater resources as well. In this scenario, the authors think groundwater flow models could be very useful tools for the efficient management of groundwater resources if properly calibrated and validated. The models can help manage the aquifer system and its related features in real time; however, the authenticity and reliability of downscaled values of the same models are still questionable.

Therefore, for the assessment of climate change impacts on groundwater resources, the authors believe that it is imperative to understand the extent and magnitude of groundwater vulnerability to droughts, over-exploitation as well as deterioration in its quality. And to understand the spatial and temporal availability of groundwater, a better quantification of the regional water budget is also required. This information can be a good source of evidence for the groundwater managers dealing with present and future climate change regimes. In sum, the future sustainable availability of groundwater is highly dependent on long-term climate trends and their possible implications on groundwater recharge and water level fluctuations.

In the same perspective, Mohammed Bahir, Otman El Mountassir, and Mohamed Behnassi in Chap. 11, *Hydrogeochemical Processes Regulating the Groundwater Quality and its Suitability for Drinking and Irrigation Purpose in a Changing Climate in Essaouira, Southwestern Morocco*, consider groundwater as vital for the water supply and environmental protection, especially in semi-arid and desert regions. Based on this, the authors undertake an integrated assessment in the Cenomanian–Turonian aquifer during the campaigns 2017, 2018, 2019, and 2020 using, in a combined way, the Water Quality Index (WQI), Irrigation Water Quality Index (IWQI), and geochemical and isotopic ($\delta^{18}O$, $\delta^{2}H$ and $\delta^{3}H$) tools. The combination of major element and stable isotope geochemistry allowed the authors to understand the hydrodynamic functioning of the Cenomanian–Turonian aquifer and to clarify the geologic factors that control its water chemistry and mineralization process. Taking all the findings into consideration, and in order to decrease the soil degradation process in the Essaouira region, the authors call for a continuous monitoring of groundwater, control of salinity through proper management, implementation of more efficient irrigation methods, a good leaching regime, maintenance of low groundwater levels, as well as awareness raising of farmers.

In Chap. 12, *Drought Assessment in Potwar Region, Punjab Pakistan during 1981–2019*, Saira Batool, Syed Amer Mahmood, and Safdar Ali Shirazi focus on drought, given its character as a key disaster that affects numerous segments of the natural environment and economy throughout the world. To this end, they computed Drought indices (DIs), Standard Precipitation Index (SPI), and Reconnaissance Drought Index (RDI) for Potwar region (PR) in Punjab-Pakistan, using DrinC software. According to the findings, the drought situation of 12, 9, 6, and 3 months was estimated on temporal basis. DIs obtained by deciles technique showed that, for the last 39 years, 8 years are with drought severity in a cycle and are occurring every 2 to 7-years just the once repetitively. The RDI and SPI index showed the analogous trends as of deciles. Though, for RDI and SPI, the extremely dry and severely dry class was merely two years and the rest of the drought-affected years with respect to deciles were normally and intermediately dry. SPI is better than to deciles for a better understanding of drought severity. Regression analysis revealed that the RDI and SPI indices are mutually interrelated, and if first 3-month precipitation is obtainable, one can forecast yearly RDI. The authors believe that their investigation and its findings are valuable to devise future development plans to tackle vulnerable drought incidents and mitigate its socio-economic impacts.

In Chap. 13, *Managing Agricultural Water Productivity in a Changing Climate Scenario in Indo-Gangetic plains*, Pavneet Kaur Kingra and Surinder Singh Kukal recall that climate change has a significant impact on the hydrological cycle and warming scenarios along with increased uncertainty in rainfall behavior may lead to increase in crop water requirements and decrease in water availability for irrigation. Consequently, the groundwater resource is being depleted at alarming rates worldwide, especially in south-east Asian region. Due to increased frequency and intensity of extreme weather events, the agriculture has become highly vulnerable to climatic risks, thus endangering food security and stimulating the over-exploitation of natural resources. For the authors, such dynamics should orient research towards improving crop water productivity rather than yields and climate-smart agriculture seems the viable option in this perspective. On-farm water management, rainwater harvesting, groundwater development, advanced techniques of irrigation, breeding for resistance to droughts and floods as well as construction of dams for water storage are some of the practices the authors consider important to manage water scarcity. Similarly, improving soil moisture retention, changing cropping calendars, encouraging crop diversification, irrigation management such as deficit irrigation, supplemental irrigation, alternate wetting and drying in rice, etc. are very important on-farm practices for enhancing crop water productivity. In addition to this, some policy measures such as climate proofing structures, reallocation of water among different sectors, and crop insurances should be implemented by governments. Moreover, improved weather forecasting can play a very crucial role to minimize climatic risks in agriculture and crop simulation and hydrological modeling are techniques, which can assist in tactical decision making for improving crop water productivity and better management of water resources. Remote sensing and geospatial techniques can also be used successfully for improved hydrological monitoring at regional level. Overall, the authors believe that there is a dire need of taking quick actions to enhance water-use efficiency and save this precious resource for sustaining agriculture and achieving food security in future.

From a different perspective, Ahmed Al-Busaidi, Mushtaque Ahmed, Wenresti Gallardo, Waad Al-Aghbari, and Yahya Al-Yahyaei focus, in Chap. 14, on *A Sustainable Method of Production Towards Food Security Using Aquaponics* by reference to the *Case Study from Oman*. The authors claim that Gulf countries need to look for innovative and sustainable food production for local and expatriate populations in line with the Sustainable Development Goal (SDG)-12. In this regard, an opportunity exists for food (mainly fish and vegetable) production using limited but highly treated wastewater, thus contributing to sustainable consumption and production (SCP) in the region. Therefore, the objective of the study, conducted by the authors at the Sultan Qaboos University in Oman, was to evaluate the effect of tertiary treated wastewater on fish growth, and subsequently the effect of the produced effluent coming from fish tank on grown crops. According to the findings, the authors noticed that tanks with treated wastewater got higher concentrations of dissolved oxygen due to algae growth and more salts content due to minerals added from treated wastewater compared to fresh water alone. Therefore, lettuce and bean growth was much better and got higher values of chlorophyll content compared to plants in control tanks.

For heavy metal analysis, all waters got similar values but, in some samples, the concentrations of B, Cu, Mn and Zn were higher in treated wastewater compared to fresh water and that was reflected in lettuce roots. For the edible part, lettuce grown in treated wastewater got higher value of Fe and B compared to control. Similar concentrations were found with bean plants with higher values in treated wastewater compared to freshwater. However, low concentrations of heavy metals were found in the edible parts of all plants in all treatments. Fish analyses showed that all tested heavy metals were within the safe limit. For the authors, the positive aspect of this system is that it will help protect the environment by reusing treated wastewater and reducing fertilizer applications. Moreover, farmer income will increase since both fish and crops will be produced with minimum resources.

In the same perspective, Chérifa Abdelbaki, Nadia Badr, Hidayat Mohammedi, Rokiatou Haidara, and Halima Belarbi investigate, in Chap. 15, the *Wastewater Reuse for Agriculture as an Adaptation Measure to Water Scarcity and Climate Change* by reference to the *Case Study of Ain Temouchent, Algeria*. The authors consider Algeria among the African countries most affected by water stress. Therefore, considering irrigation using reclaimed water maybe an appropriate solution to secure and enhance agricultural production. In their chapter, the authors determined the water quality of the treated wastewater produced from the Wastewater Treatment Plant (WWTP) of Ain Temouchent—a semi-arid region located in the northwestern part of Algeria—including physico-chemical parameters, heavy metals, and microbiological contaminants. According to the findings, and except of phosphate ions ($PO_4{}^{-3}$), the other physico-chemical parameters and heavy metals were below the recommended norm of WHO and FAO and Algerian standards for irrigation purposes. However, the results of the microbiological analyses indicated that the number of fecal coliforms and intestinal nematodes are above the WHO norms for treated wastewater intended for irrigation. On the other hand, according to Algerian norms, the authors believe that irrigation of the fruit trees through drip irrigation technique is likely to reduce the risk of contamination while preserving the health of the consumers.

In Chap. 16, *Water Quality and its Health Impact in the Prefecture of Mohammedia, Morocco: A Review*, Rachida El Morabet, Larbi Barhazi, Soufiane Bouhafa, Mohamed Behnassi, and Roohul Abad Khan investigate the links between water quality and health. Focusing on the case of the prefecture of Mohammedia, Morocco, which is a booming region with the extension of intensive irrigated agriculture and industrial zone, in addition to the accelerated urbanization and the creation of the New Zenata city within the prefecture, the authors believe that this development is accompanied by an increased degradation of water resources, which affects, in turn, the local socio-ecological system. In this chapter, the authors review and analyze recent literature about the qualitative and quantitative evolution of water associated with global change, with a focus on data from both institutional reports and field surveys. The groundwater was found to have higher concentrations in indicating permeation of wastewater (presence of E-coli); whereas surface water sources were contaminated and crossed the permissible limits for safe drinking water quality at point of meeting wastewater discharge. Also, bacteriological concentration in groundwater validated groundwater contamination, especially in the vicinity of landfills. The findings indicate a significant spatial variability in the quantitative (piezometry) and qualitative (physio-chemical, biological, and hydro-geomorphological quality) distribution in the prefecture of Mohammedia. Also, the number of water-borne diseases validated the impact of water quality on consumer's health. The authors think that further studies about the overall quality analysis of water resources based on the same parameters are needed so they can serve as a reference for policy making in the field of sustainable water resources management in the study area.

In Chap. 17, *Basic Planning Principles of Roof Precipitation Harvesting Systems*, Hasan Er and Yasemin Kuslu expose the precipitation harvesting techniques and the various methods to use these techniques on buildings' roofs in order to manage water scarcity, which is becoming one of the biggest problems for mankind. The authors define precipitation water harvesting as the collection and accumulation of rainwater and runoff water, which help supply the water required for both plant/animal production and domestic consumption. This water supply method has provided drinking water to many historical cities since ancient times—archaeological findings show that rainwater harvesting dates back to 6000 BC—and the practice has been used since humans began to live and grow crops in arid areas. Currently, while in many arid countries a large part of rainwater is lost through evaporation or turning into wastewater, worldwide awareness and importance of rainwater harvesting are increasing. In addition, since the water obtained is free of charge, the authors believe that protecting natural water resources can be easily integrated into existing water network systems. Also, low operating and maintenance costs make precipitation harvesting extremely attractive for the management of water in times of scarcity. With the collection of precipitation water, purposes such as preventing soil erosion and floods, providing quality irrigation water, feeding groundwater, and saving network water can be achieved. Based on these advantages, the authors describe the elements of roof precipitation water collection systems, including the basic principles that should be considered when planning.

From a governance perspective, in Chap. 18, *A Political Economy of Water Security: The Case of Singapore*, Md. Saidul Islam, Ashvini Kannan, Chen QingLin, Josephine Toh, and Lynette Loh recall the need for every country to ensure its water security, which encompasses all dimensions of human health, livelihood and well-being, and food and energy production. In the past, all industrial nations strived to ensure their water security through early substantial investments in infrastructure, institutions, and capacity to manage water and wastewater. In this regard, the case of Singapore is important to understand the political economy of water security. With its high dependence on neighboring nation for water supply, and with limited land to collect and store rainwater, the country encountered drought, flood, and pollution since its independence in 1965. Over the years, Singapore adopted integrated, effective, robust, and cost-efficient approaches with strategic investments in research and technology to treat, recycle, and supply water. Today, the country is recognized internationally not only as a model city for integrated water management but also as an emerging global hydro-hub. Touching on the causes of water insecurity in the world today, the authors delineate the Singapore story of how 'water scarcity' in the nation has been transformed into a 'water opportunity'. To bring critical insights, the authors also compare the Singapore case with two other similar nations: Saudi Arabia and Israel.

In the same vein, Muhammad Nawaz Khan and Adeel Mukhtar focus, in Chap. 19, on *Water Management in Pakistan: Challenges and Way Forward*. For the authors, water management is the biggest challenge being confronted by Pakistan due to inadequate water management practices, insufficient storage capacity, irrigation inefficiency, population explosion, over-exploitation of groundwater, climate-induced water stress, and India's water hegemony as upper riparian. However, this critical situation does not favor a good management of water resources through, for instance, water conservation and storage strategies, thus increasing water insecurity in the country. Therefore, the country would not be water secure until it implements stringent water management measures, in addition to building dams. In this context, the authors attempt to investigate the importance of water management practices for the future water security of Pakistan as well as its socio-economic development. While doing so, the authors also highlight challenges faced by the water sector in the country and conclude with some doable policy recommendations.

In Chap. 20, *Climate Change, Water Variability, and Cooperation Along Transboundary River Basins in Perspective of Indus Water Treaty*, Faraz ul Haq, Ijaz Ahmad, and Noor Muhammad Khan recall that transboundary rivers can be the cause of water conflict between nations and that's why treaties and agreements have been made to settle these conflicts. In the Indian sub-continent, western rivers are becoming a matter of concern for Pakistan in terms of water quantity, especially that many dams had been constructed by India on these rivers and many more are planned or under construction. In this work, the authors undertake a comparison between the viewpoints of both countries regarding transboundary interventions. For India, variations in river flows and water shortages in Pakistan are due to its negligence and India is doing nothing to violate the Indus water treaty and is just utilizing its water allocation on western rivers. For Pakistan, India is stealing its water supplies and constructing the water-controlling structures only as a political exercise to improve political power. Therefore, the authors believe that the maintenance of transboundary aquifers and groundwater management, transboundary cooperation in watershed management, and integrated water resources management can lead to sustainable development between both countries. The authors provide in their work information on transboundary conflicts and treaties in addition to future concerns to both researchers and policymakers.

Building on all chapters presented above, this volume attempts to explore the linkages, synergies and trade-offs among the water, climate, and food nexus as a step ahead in understanding the complexity of such a nexus and its challenges. Indeed, many related challenges were tackled by the authors, especially the impacts of climate change on water security and the contribution of water scarcity in the decline of food production and security. The volume equally focuses on the solutions side, covering the relevance of sustainable water management, the multiple uses of nonconventional water resources as an adaptation strategy, the use of advanced technologies and models to assess the quality and quantity of water resources in the context of climate change, innovative governance settings in the area of water management, etc. This may help fill the huge science-policy-practice gap in the nexus areas. Additionally, since this volume has been compiled at a time when the nexus is being recognized at a global level, we hope it will contribute to ongoing work, including, the IPBES nexus assessment, which is currently underway.

Dr. Mohamed Behnassi has been nominated in July 2022 as a Senior Environmentalist Expert by the Economic Social and Environmental Council (ESEC), Rabat, Morocco. Prior to this, he served as a Full Professor (since 2004), and Head of Public Law in French Department at the College of Law, Economics, and Social Sciences of Agadir, Ibn Zohr University. He holds a Ph.D. in International Environmental Law and Governance (2003), a MSc. in Political Sciences (1997), and a B.A. of Administration (1995) from Hassan II University of Casablanca. He obtained a Diploma in International Environmental Law and Diplomacy from the University of Eastern Finland and UNEP, 2015. He is also an Alumnus of the International Visitors Leadership Program of the Department of State, United States of America. Dr. Behnassi is currently the Founding Director of the Center for Environment, Human Security and Governance (CERES). From 2015 to 2018, he was the Director of the Research Laboratory for Territorial Governance, Human Security and Sustainability (LAGOS). Recently, he was appointed as Expert Evaluator for the National Center for Scientific and Technical Research (CNRST/Morocco) and selected twice (2019–2024) as an Assessment Scoping Expert and a Review Editor by the Intergovernmental Science-Policy Platform on Biodiversity and Ecosystem Services (IPBES) and Member of the Mediterranean Experts on Climate and Environmental Change (MEDECC). He was also selected by The Intergovernmental Panel on Climate Change (IPCC) as Expert Reviewer of the 1st Order Draft of the Synthesis Report (SYR) of the IPCC VI Assessment Report (AR6). Accordingly, he was among the Lead Authors of the 1st Assessment Report (MAR1): *Climate and Environmental Change in the Mediterranean Basin—Current Situation and Risks for the Future* (MEDECC, 2021). He is a Senior Researcher in the areas of international law and politics of environment and human security where he published considerable number of scientific papers and book chapters in addition to 20 books, including recent ones on: *Food Security and Climate-Climate Smart Food Systems—Building Resilience for the Global South* (Springer, 2022); *The Climate-Conflict-Migration Nexus from a Human Security Perspective* (Springer, 2022); and *Social-Ecological Systems in the Era of Risks and Insecurity—Pathways to Viability and Resilience* (Springer, 2021). Dr. Behnassi serves as a reviewer for many global publishers (such as Routledge and Springer) and scientific journals with high impact factor. He has organized many international conferences covering the above research areas, managed many research and expertise projects, and is regularly requested to provide scientific expertise nationally and internationally. Other professional activities include social compliance auditing and consultancy by monitoring human rights at work and the sustainability of the global supply chain.

Dr. Abdulmalek A. Al-Shaikh is the Director of the Prince Sultan Institute for Environmental, Water, and Desert Research at the King Saud University (KSU) in Riyadh, Saudi Arabia. He earned his Ph.D. in Arid Land Studies from the University of Arizona, USA in 1983. He also serves as the General Secretary of the Prince Sultan bin Abdulaziz International Prize for Water (PSIPW). The Prize is awarded bi-annually at the United Nations Headquarters. Being the Director and Research Chair, he oversees the Research Programs at the Institute. He is the team leader of many research projects concerning the environment, desertification, climate change, and water harvesting in Saudi Arabia including the King Fahd Project for Water Harvesting and Storage in Saudi Arabia, Prince Sultan Project for Villages and Hamlets Rehabilitation in Saudi Arabia, the Space Images Atlas of Saudi Arabia and the Environmental Atlas of Saudi Arabia. Being the Chairman of the international conferences on water resources and arid environments, he successfully organized them at the King Saud University. He has brought world fame water scientists to Saudi Arabia and published 9 proceedings of organized international conferences. He has published extensively in journals of international repute.

Dr. Ameenah Gurib-Fakim is Professor, a Biodiversity scientist, and an entrepreneur. She has served as the 6th and First Female President of the Republic of Mauritius (2015–2018). Prior to that, she has been the Managing Director of the Centre International de Développement Pharmaceutique (CIDP) Research and Innovation as well as Professor of Organic Chemistry with an endowed chair at the University of Mauritius. Since 2001, she has served successively as Dean of the Faculty of Science and Pro Vice Chancellor (2004–2010). She has also worked at the Mauritius Research Council as Manager for Research (1995–1997). Dr. Gurib-Fakim earned a BSc in Chemistry from the University of Surrey (1983) and a Ph.D. from the University of Exeter, UK (1987). During her academic journey, she has participated in several consultation meetings on environmental issues organized by international organizations. Between 2011–2013, she was elected and served as Chairperson of the International Council for Scientific Union—Regional Office for Africa, and served as an Independent Director on the Board of Barclays Bank of Mauritius Ltd. between (2012–2015). As a Founding Member of the Pan African Association of African Medicinal Plants, she co-authored the first ever African Herbal Pharmacopoeia. She has authored and co-edited 30 books, several book chapters and scientific articles in the field of biodiversity conservation and sustainable development. She has lectured extensively across the world; is a Member of the Editorial Boards of major journals, has served on Technical and national committees in various capacities. Elevated to the Order of the Commander of the Star and Key by the Government of Mauritius in 2008, she has been admitted to the Order of the Chevalier dans L'Ordre des Palmes Academiques by the Government of France in 2010 and is the recipient of 5 DSc (s). As Elected Fellow of several academies and societies, she received several international prizes, including the 2007 l'Oreal-UNESCO Prize for Women in Science and the African Union Commission Award for Women in Science, 2009. She was elevated to the Order of GCSK by the Government of Mauritius, and received the Legion d'Honneur from the Government of France in 2016. In 2017, she received both the lifelong achievement award of the United States Pharmacopoeia-CePat Award and the American Botanical Council Norman Farnsworth Excellence in Botanical Research Award. In 2018, she received the Order of St George at the Semperopernball, Dresden, Germany. In 2019, she received the 'Trailblazing award for political leadership' by the World Women Leaders Council in Iceland. In 2020, she was elected Honorary President of the International and Engineering Institute and received their 2020 5th IETI Annual Scientific Award. She also received the IAS-COMSTECH Ibrahim Memorial Award from the WIAS in Jordan. In 2021, she received the Benazir Bhutto Lifetime Achievement Award, the Obada Prize and the RUFORUM Recognition Prize 2021. In 2021, she has been appointed as Distinguished Professor at the John Wesley School of Leadership, Carolina University, USA. In June 2016, she was in the Forbes List for the 100 'Most Powerful women in the world' and 1st among the Top 100 Women in Africa Forbes List 2017, 2019. She is honoured as one of the Foreign Policy's 2015 Global Thinkers.

Dr. Mirza Barjees Baig is a Professor at the Prince Sultan Institute for Environmental, Water, and Desert Research, King Saud University, Saudi Arabia. He earned his MS degree in International Agricultural Extension in 1992 from the Utah State University, Logan, Utah, USA, and was placed on the 'Roll of Honor'. He completed his Ph.D. in Extension for Natural Resource Management from the University of Idaho, USA, and was honored with the '1995 outstanding graduate student award'. Dr. Baig has published extensively on the issues associated with natural resources in national and international journals. He has also made oral presentations about agriculture and natural resources and the role of extension education at various international conferences. Food waste, water management, degradation of natural resources, deteriorating environment, and their relationship with society/community are his areas of interest. He has attempted to develop strategies to conserve natural resources, promote the environment and develop sustainable communities. Dr. Baig started his scientific career in 1983 as a researcher at the Pakistan Agricultural Research Council, Islamabad, Pakistan. He served at the University of Guelph, Ontario, Canada as the Special Graduate Faculty from 2000 to 2005. He served as a Foreign Professor at the Allama Iqbal Open University (AIOU), Pakistan through the Higher Education Commission from 2005 to 2009. He served as a Professor of Agricultural Extension and Rural Society at the King Saud University, Saudi Arabia from 2009 to 2020. He serves as well on the Editorial Boards of many international journals and is a member of many international professional organizations.

Prof. Dr. Eng. Mohammed Bahir, Ph.D., Hydrogeological Engineer, has published over 200 research/professional papers and reports. He co-authored more than a dozen books and book chapters. He supervised and/or co-supervised more than 20 Ph.D. students. He is a reviewer of many international scientific journals with high impact factors, including Environment, Development, and Sustainability Journal, Environmental Earth Sciences Journal, Hydrological Sciences Journal, Sécheresse, Gaia, Communicaçoés International, Science of the Total Environment, Marine and Freshwater Research Journal, Proceedings of the Indian National Science Academy, and Groundwater for Sustainable Development. He also coordinates more than a dozen research projects.

The Asia Research Awards, in association with Times of Research and the World Research Council, announce that Professor Mohammed Bahir has been selected as the recipient of the prestigious ASTRA 2023 (Asia's Science, Technology, and Research Awards).

Chapter 2
The Water, Food, and Environmental Security Nexus

Muhammad Sohail Amjad Makhdum, Rakhshanda Kousar, Muhammad Ashfaq, and Mohamed Behnassi

Abstract Water, food, and environment are not only essential components of human life and existence but are also important for the economy. Given their interdependence, it is increasingly imperative to study them from a nexus perspective. Water is essential for daily life, industry, and agriculture and originates from varied underground and surface sources. The decreasing quality of water due to anthropogenic activities—such as pollution induced by industrialization, unsustainable agriculture and the wide use of toxic materials—contributes to water insecurity worldwide. Moreover, the underground water is being over-pumped in large quantities across in many regions, which results in water scarcity. Biological and technological advancements including hybrid seeds, improved irrigation facilities, efficient fertilizers use, effective disease and pest control, and efficient farm machinery in agricultural sector globally have contributed to increase the food production. Also, progress is being made in terms of economic development and growth but the sufficient and quality food is not available to the entire population for many reasons, including access and governance issues. Increasing food prices, unemployment, lack of opportunities and consumption smoothing make it difficult for the poor to access sufficient food. The natural environment is also on the verge of deterioration and degradation due to anthropogenic activities. Over and inefficient use of natural resources and unsustainable development have harmed the environment, sometimes irreversibly.

M. S. A. Makhdum
Department of Economics, Government College University, Faisalabad, Pakistan
e-mail: sohailmakhdum@gcuf.edu.pk

R. Kousar (✉) · M. Ashfaq
Institute of Agricultural and Resource Economics, Faculty of Social Sciences, University of Agriculture, Faisalabad, Pakistan
e-mail: rakhshanda.kousar@uaf.edu.pk

M. Ashfaq
e-mail: mashfaq@uaf.edu.pk

M. Behnassi
Economic, Social and Environmental Council (ESEC), Rabat, Morocco

Center for Environment, Human Security & Governance (CERES), Agadir, Morocco

Ministry of Higher Education and Innovation, Rabat, Morocco

M. Behnassi et al. (eds.), *The Water, Climate, and Food Nexus*,
https://doi.org/10.1007/978-3-031-50962-9_2

Hence, this chapter aims at analyzing the different dimensions of water, food, and environment from a nexus perspective for a better understanding of the linkages between them and thus guide public policies aiming at fighting poverty, food insecurity and vulnerability while promoting a healthy environment for present and next generations.

Keywords Water · Food · Environment · Security · Poverty · SDGs

1 Introduction

The world has undergone substantial technological, industrial, agricultural, and human development (Hoff 2011). However, many societies are still facing the problems of uneven distribution of water resources, food insecurity, and environmental degradation. A large part of the world population, especially in the Global South, is food insecure with the environmental degradation and water scarcity further worsening the situation. Moreover, the severity of extreme events like floods, droughts, weather extremities, and irregular and untimely rains, and the prolongation of cropping seasons etc. are exacerbating existing vulnerabilities (Kundzewicz et al. 2007). The world has faced natural disasters like tsunamis, floods, earthquakes with human and material casualties similar to those of the World Wars I and II while the rapid and unsustainable industrialization continue to increase the environmental degradation. Human societies have altered the natural resources, deviated the rivers, harmed the flora and fauna, and spread the urbanization which overuses land and natural resources. Clean water is gradually becoming insufficient for drinking and other human activities including agriculture. Chemical pollution and industrial and household wastes have deteriorated water quality. Overall, the environment is at risk; emissions of greenhouse gases, pollution, melting of glaciers, sea-level rise, and increase of global temperature are negative externalities of unsustainable development. Keeping in view the prevailing conditions, the upcoming decades will be mostly marked by food and water deficit, an overpopulated and warm planet, and ecological imbalance (Meadows et al. 1972; WEF 2011).

Water, food, and environment are three important pillars of human security and development. Many sustainable development goals confirm their importance such as: the SDG 2 about zero hunger, which depicts the scarcity of good quality and sufficient quantity of food; the SDG 6 about clean water and sanitation, which intends to combat the unavailability and lack of access to clean water; the SDG 13 about climate action to combat global warming; and the SDGs 14 and 15 which can also be considered under the context of environmental issues (Fig. 1).

Given the interlinkages between water, food, and the environment, a nexus approach is necessary to capture interrelated dynamics and implications and develop response mechanisms. The nexus approach is recently gaining importance as it takes into account the collective intervention of different aspects under consideration (Zidjaly 2012). This, in turn, helps assess the multiplier effects of the different

Fig. 1 Sustainable development goals (SDGs). *Source* United Nations[1]

phenomena in the changing nature (Kuure et al. 2018). Many studies have considered different nexuses, such as the food-energy-water nexus (Cai et al. 2018; Nie et al. 2019; Scanlon et al. 2017) and the water-food-energy-environment nexus (Momblanch et al. 2019; Sood et al. 2019). Within the same perspective, this chapter aims to discuss the water-food-environment (WFE) security nexus with the objective to highlight relevant policy implications and provide tools to manage the nexus related challenges. First, it gives a brief overview about the core issues under study, and then focuses on the WFE security nexus by formulating tools and inputs for policy-making processes.

2 Water Security

Water is a prerequisite for the existence and evolution of human species on the planet. It is a vital component of life and a significant resource, which supports different types of human activities. However, its availability and accessibility are a great challenge (Chen and Trias 2020). Grey and Sadoff (2007) defined water security as, "the availability of an acceptable quantity and quality of water for health, livelihoods, ecosystems and production, coupled with an acceptable level of water related risks to people, environments, and economies". Water is found in abundance in lakes, rivers, oceans, glacier, and aquifer covering around 71% of the surface of Earth (Gleick 1993), making it a plentiful resource. However, the quality and purity of the water are deteriorating over time making it increasingly unsuitable for human consumption. Moreover, the global distribution of water resources is not uniform:

[1] https://www.un.org/development/desa/disabilities/envision2030.html.

97.5% of the total water on earth is saline, out of which 96.6% is in the form of oceans and 1% resides as underground water; the remaining 2.5% is present in the form of fresh water as surface water and soil moisture (Stephens et al. 2020).

Water is the source of good health and hygiene (Merchant et al. 2003) and positively associated with happiness (Nadeem et al. 2020). The lack of sanitation and hygiene leads to diseases like malaria, diarrhea, skin problems, and cough in Amassoma, Nigeria as reported by Raimi et al. (2018). Globally, 1.1 billion people lack access to clean water and 2.5 billion lack sanitation facilities (UNICEF 2012), resulting in 7% addition to diseases and 19% child mortality (Ustun and WHO 2008). The United Nations (2010) has declared the access to clean water and sanitation as a basic human right.

Water is an essential element for running industries, environment and living organisms. It helps in metabolism and maintains body temperature (Hanslmeier 2011). Water gets polluted when impurities in the form of chemicals, salts, heavy metals, solid household wastes and other soluble and insoluble materials, making water unfit for industrial, agricultural and human use. Water stress areas around the globe are likely to increase on daily basis and more than 60% of the population is expected to face this problem by 2025 (Arnell 1999). The point of concern is that these are irreversible changes, being caused by human activities including industry, economy and development but at a very high cost. It is not possible to undo these changes and restore our resources and environment.

Developed and industrialized countries have already invested in water security by all appropriate means and have generally managed to ensure the continuous availability of quality water. For many developing countries, which are water resource poor, they are increasingly suffering the adverse consequences in the form of increased poverty, economic insecurity, and decline in health conditions and overall living standards. Figure 2 shows that most of these countries in Africa and Asia will be on the verge of water scarcity in 2040. Moreover, the water demand is increasing, and it is expected that around 40% of the world population will suffer from water scarcity by 2050 (OECD 2013); such a population will reach 9.7 billion (UN 2015) and require 60% more food by the same year (Alexandratos and Bruinsma 2012). Hence, all concerned countries, especially in the Global South, should develop appropriate governance frameworks, including the implementation of adequate water policies, keeping in view the future needs of the water use and the water scarcity imperative. This will help maintain their human security while achieving development and sustainability.

3 Food Security

Food security, unlike water security, is linked to the access and availability of sufficient and quality food (Barrette 2010). "Food security exists when all people, at all times, have physical and economic access to sufficient, safe and nutritious food to meet their dietary needs and food preferences for an active and healthy life" (World

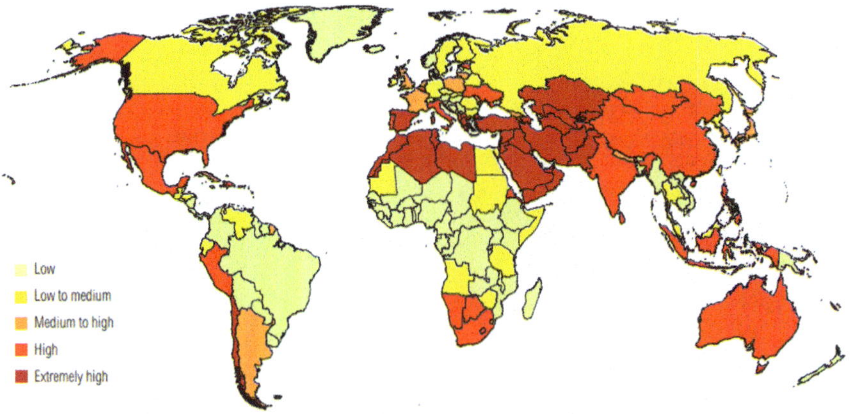

Fig. 2 Water stress in 2040. *Source* Luo et al. (2015)

Food Summit 1996). Food insecurity is the condition when, "people do not have social, physical and economic access to food" (FAO 2009). Here, there are two basic aspects: one is the ability of a person to purchase enough food to meet the dietary requirements; and the other is the capability of the economy to meet food demands through the production of nutritious and quality food. The current situation, however, does not seem to meet the SDG 2 (zero-hunger) as hunger is still a growing challenge for the international community with 746 million people still suffering from acute food insecurity and 1.25 billion from moderate food insecurity (FAO et al. 2020). The comparison of the most food-insecure continents and the world is shown in Fig. 3, which confirms that food insecurity mostly prevails in the Global South.

The progress made by humanity in many areas—such as science, technology, modes of organization, etc.—have provided numerous endowments including better seeds, fertilizers, pest and disease control, plant protection measures, modern irrigation systems, farm machinery, redistribution and transportation systems, etc. However, despite these achievements, food insecurity still prevails, giving rise to doubts about the relevance of current development efforts (Staupe-Delgado 2019). Therefore, the need to adjust existing response mechanisms with the scope and amplitude of the food insecurity challenge is pressing.

4 Environmental Security

The natural environment plays an important role in supporting life, provides essential resources for economic development, and serves as a reservoir for wastes. However, the big challenge being faced by both developed and developing countries during recent decades is how to ensure development while reducing at maximum negative

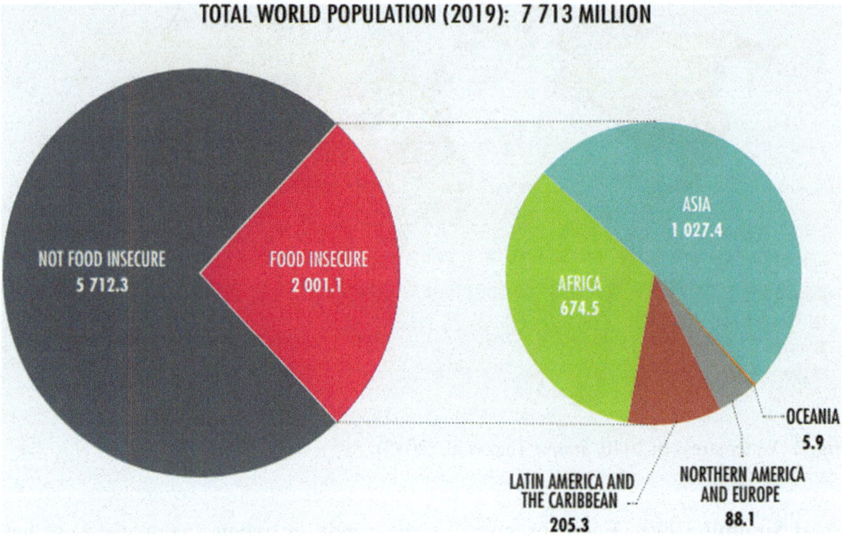

Fig. 3 Food insecurity status of the world and most affected continents. *Source* FAO et al. (2020)

environmental externalities (Usman et al. 2020a). This challenge has been insufficiently managed since the world is still facing many environmental problems such as biodiversity loss, deforestation, desertification, air, land and water pollution, and most importantly climate change. Moreover, industrialization and overuse of natural resources to achieve higher levels of economic growth and consumption have resulted in increased ecological and carbon footprint (Destek and Sarkodie 2019; Saidi and Omri 2020; Salahuddin et al. 2020) and environmental damage (Shahbaz et al. 2016). These environmental costs are beyond the self-regulating capacity of ecosystems and jeopardize the ecological balance.

In this perspective, the concept of 'environmental security' has been developed and linked with the protection of the environment and its components—such as the atmosphere, land, water, and biodiversity—in addition to the preservation of the ecological balance. In order to do so, there is a need to develop new value systems and practices such as recycling, shifting to renewable energies and energy sobriety, reforestation, optimal use of resources, changing consumption preferences and habits, eco- and climate-friendly technologies, etc. Accompanying these shifts with appropriate governance systems is also imperative. The main goal of next decades should be the preservation of environmental quality while enhancing the sustainability of economic development (Usman et al. 2020b).

5 The Water, Food, and Environmental Security Nexus (WFE)

According to Merriam-Webster dictionary, the word ‘nexus’ was used in 1663 for the first time, meaning: ‘connection’ or ‘link’. Hence, the nexus consists of connecting or linking two or more things together. In the scientific literature, the nexus approach studies the interdependence and linkage of different concepts (De Laurentiis et al. 2016; Smajgl et al. 2016). In this perspective, water, food, and environment are increasingly perceived as interconnected where actions taken in each area may have multidimensional effects on other areas and the security of one area depends on the security of others (Lele et al. 2013) (Fig. 4).

According to Rostow’s Growth Model (Rostow 1960), *The drive to maturity* puts extra burden on the resource use and allocation. Savings ultimately lead to investments, increasing the volume of economic activity, industrialization, and promoting urbanization. There is a dire need for the inputs to manufacture products for domestic consumption, replace imports, and increase export to earn foreign deposits. An economy reaching this stage often overlooks environmental degradation and over depletion of natural resources, and ultimately reaches a critical point where interference is needed to safeguard the environment, its resources and the ecological balance.

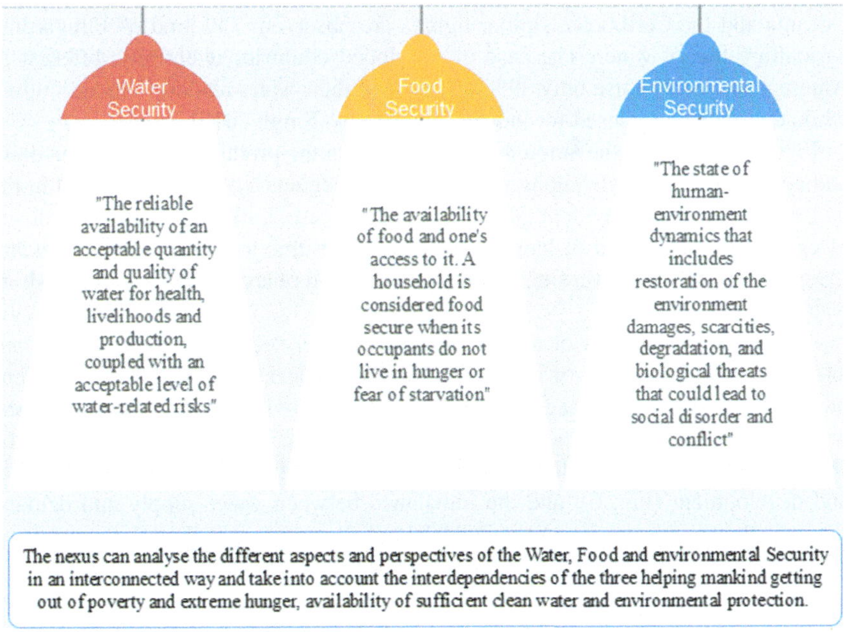

Fig. 4 Components of the WFE Nexus. *Source* Developed by the authors

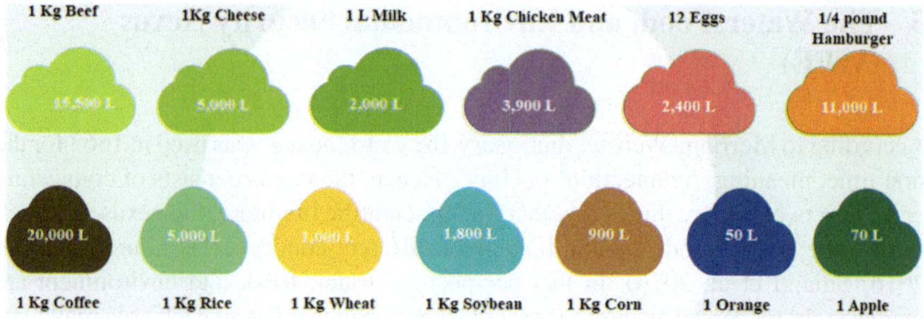

Fig. 5 Water use for food production. *Source* Developed by the authors based on data from Hoekstra and Chapagain (2008) and Singh (2017)

Water, food, and environment are the interconnected components of all kinds of life on earth. In the WFE nexus, water holds a central position since it is an essential ingredient for the production of food, maintaining life, and keeping environment secure. In this context, agriculture as a human activity is the major user of water, especially in developing countries. Out of total available water, Asian agriculture consumes 81%, the remaining 11 and 7% are consumed by industry and households, respectively (Singh 2017). The statistics regarding Latin America are almost similar for agriculture whereas for industry and domestic uses the percentages are 10% and 19%, respectively. In Africa, the percentages are 8%, 4% and 10%, respectively. For Oceania and the Caribbean, similar figures are observed: 73% and 69% regarding agriculture use of water. The case of developed countries is entirely different as America and Europe use only 39% and 32% of their water resources in agriculture while the large part is used for industrial purposes (Singh 2017).

Figure 5 highlights the amount of water used for the production of different dairy and agricultural products, along with the interdependence of water and food in the underlying nexus. It is evident that 1 kg of beef, cheese, milk, chicken meat, a dozen of eggs, and 0.25 pound of hamburger jointly consume as much amount of water (39,800 L) as required by a single human for a period of around 34 years as drinking water.[2]

The different concerns or aspects related to water security include: continuous and uninterrupted supply of clean and sufficient water for drinking and hygiene purposes; ecosystems' management which, in turn, provides services and resources; water-related hazards, including floods and droughts, contamination of water by chemicals and wastes, and climate change; water requirements of economic activities and development (Fig. 6); and the imbalance between water supply and demand for agriculture, industry, and household. These concerns are behind the interrelated challenges of ensuring water and food security and environmental sustainability. Water and food insecurity are alarming since a large number of individuals (1.5–3

[2] Per day water intake by adult male is 3.7 L and 2.7 L for a woman (Sawka et al. 2005), and the average calculated is 3.2 L/day. The rest of calculation is: $(39{,}800/3.2)/365 = 34.07$.

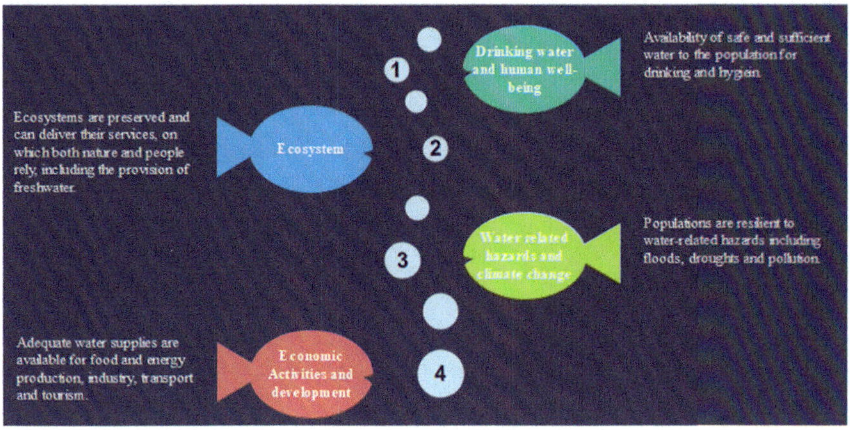

Fig. 6 Aspects of water security. *Source* Developed by the authors based on data from: www.watercooperation2013.org

billion) are currently relying on groundwater resources, which are at increased risk (Kundzewicz et al. 2007).

Excessive use of water resources causes serious threats to the environment like waterlogging, salinity, and damage to infrastructures and ecosystems (Han 2003; Jiang 2009; Liu and Xia 2004). Weather conditions affect the production of crops, and hence food security. Environmental degradation and climate change negatively affect water supply and crop and livestock production (Wollenberg et al. 2016). The decline in the quality and quantity of freshwater adversely affects biodiversity and ecosystems' functioning, which in turn creates undesirable changes in phenotype as well as genotype of living organisms. On the other hand, deterioration of freshwater resources spurs land degradation and loss of agroecosystem productivity. Fertile lands become less fertile and barren, thus increasing the risk of food insecurity. Consequently, all these aspects when combined affect human health and environment (WHO 2003). Such situations can be differentiated into two types: short-term effects when the environment can quickly recover from the harm; and severe long-term effects where they exceed the self-recovery capacity of the environment. Water pollution, in addition to climate change impacts, belong to the second category given their cumulative and long-lasting effects on water and environment (Inyinbor et al. 2018).

FAO (2017) reported that food security is a chronical issue and the number of undernourished individuals around the globe has increased at an alarming rate, especially in Asia and Africa which are considered the most affected regions. Insufficient availability of freshwater makes agricultural land unable to meet increasing food demands. Figure 7 displays water withdrawals for agriculture, industry, and domestic consumption, which are highest in Asia and Africa; two regions already at the verge of water insecurity. Top water-consuming countries in these two regions are India, China, USA, Russia, Indonesia, Nigeria, and Brazil. Russia and Indonesia consume

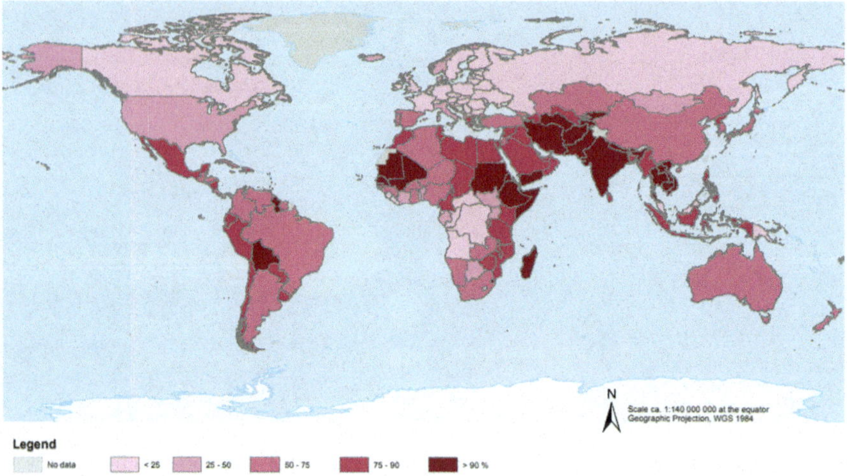

Fig. 7 Ratio of water withdrawal for agriculture, industry, and domestic consumption. *Source* FAO AQUASTAT (2015)

4%, and Nigeria and Brazil consume 3% individually leaving behind 52% water for the rest of countries (Singh 2017).

As mentioned earlier, water holds a central position in the WFE nexus as it is the key resource for food production, energy generation, daily life and ecosystems' functioning (Fig. 9). In addition, not only water scarcity and degradation affect the environment and the climate, rather these two can also affect water availability and quality (Bates et al. 2008; IPCC 2019). Indeed, there is a strong relationship between environment, climate, and water cycle. In case of low rainfall for instance, there will be less soil moisture required for the growth of crops and natural vegetation, underground water level will fall, and less water will be available for industry and domestic uses. Climate change is enhancing environmental degradation and increasing global temperatures, the latter has tendency to evaporate most of surface waters, resulting in an increased gap between water supply and water demand and growing vulnerability of agricultural sector (FAO 2011). Similarly, the gap between food production and demand and provision of agricultural raw material to the industry is widening (OECD and FAO 2018). The projections indicate increased food demand for 2050 (FAO 2017), whereas the production may decline by one-third (Ray et al. 2015). A study by Chai et al. (2020) in China has proved that the economic growth process exacerbates water and food insecurity and environmental damage, a fact which makes economic growth appears a non-desirable goal (Fig. 8).

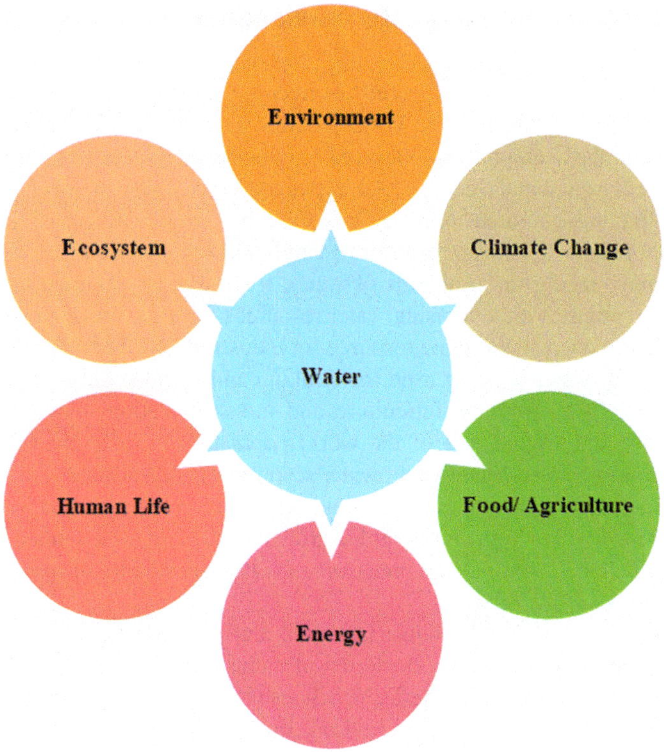

Fig. 8 Position of water in the WFE nexus. *Source* Developed by the authors

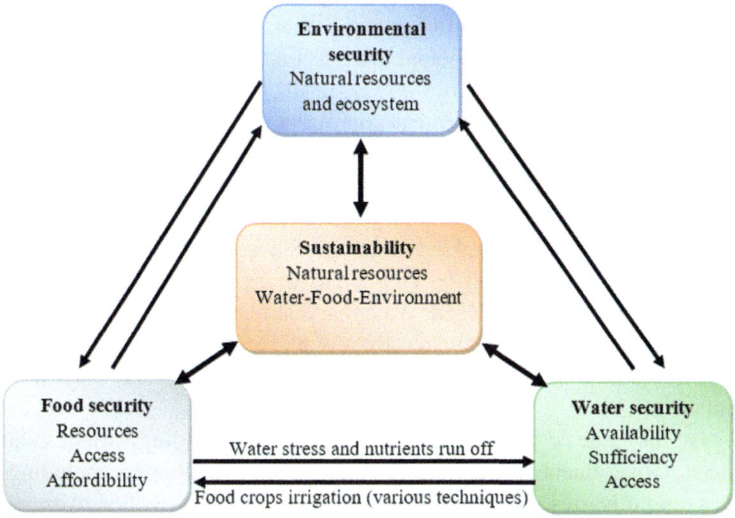

Fig. 9 WFE nexus and sustainability. *Source* Developed by the authors

6 Conclusions and Policy Recommendations

Adopting a nexus approach to the study of interdependent issues is relatively recent and, in most cases, fruitful. The WFE security nexus, which is the focus of this chapter, particularly emphasizes the need to consider the interconnection of water, food, and environmental security. As explained above, any positive or negative changes affecting one area imply effects for other areas with possible tradeoffs, and this should be considered both in research and policy making. Today's world is increasingly facing the problems of water, food, and environmental insecurity. Developing countries are overusing water resources for agricultural purposes while developed ones overuse the same resource for industrial development purposes. The least amount of water is being used for drinking and hygiene purposes. Therefore, economic activities—mainly agriculture and industrialization—have deteriorated water quality and around half of the world population is suffering from scarcity of clean drinking water. Meanwhile, water scarcity and degradation and arable land shortage due, among other factors, to rapid urbanization have limited food production. Combined with a high demographic growth and rising food demands, the food insecurity challenge continues to undermine the human security of many countries, especially in Asia and Africa.

In addition, unsustainable industrialization and intensive agriculture, urbanization, greenhouse gas emissions, depletion of natural resources, and pollution have put unbearable burden on ecosystems, which results in increased environmental insecurity. The air we breathe, the water we drink, the food we take include carbon, heavy metals, chemicals, pesticides and other wastes and are not fit for a healthy life and sustainability. The damage being caused to the environment is increasingly irreversible and many thresholds are being transgressed. The Planet Earth is the only home to human beings. Safeguarding such a planet and its balance means, inter alia, ensuring human existence, well-being, and sustainability.

From the above discussion, policy makers in all countries, especially in the Global South, should develop and implement appropriate water policies and governance frameworks keeping in view the future needs of the water use and the water scarcity imperative. Existing water infrastructure should be improved, and new technologically advanced infrastructure should be developed, such as modern irrigation techniques, including drip and sprinkler irrigation. Surface water reservoirs like dams, rivers, canals and lakes should be employed for irrigation purpose. Pumping of groundwater should be discouraged. Water harvesting techniques should be adopted in order to save extra sources of water. Both public and private investment in irrigation infrastructure should be encouraged. In this way, underground water level could be maintained.

To solve the issue of food insecurity, seeds of high yielding varieties should be introduced. Government should ensure the availability of inputs including seeds, fertilizers, and other resources. Food prices should be fixed by the government in a way that poor and vulnerable individuals and households could purchase healthy and accessible food. Individuals, households, and industries should be informed through awareness programs about the severity of WFE security issues and encouraged to adopt eco-and climate-friendly practices and technologies, including the reduction of resource use and negative externalities such as wastes and carbon emissions.

References

Alexandratos N, Bruinsma J (2012) World agriculture towards 2030/2050: the 2012 revision

Arnell NW (1999) Climate change and global water resources. Glob Environ Change 9:S31–S49

Barrett CB (2010) Measuring food insecurity. Science 327(5967):825–828

Bates BC, Kundzewicz ZW, Wu S, Palutikof JP (eds) (2008) Climate change and water. Technical paper of the intergovernmental panel on climate change (IPCC). IPCC Secretariat, Geneva. www.ipcc.ch/publication/climate-change-and-water-2/

Cai X, Wallington K, Shafiee-Jood M, Marston L (2018) Understanding and managing the food-energy-water nexus–opportunities for water resources research. Adv Water Resour 111:259–273

Chai J, Shi H, Lu Q, Hu Y (2020) Quantifying and predicting the water-energy-food-economy-society-environment nexus based on Bayesian networks-a case study of China. J Clean Prod 256:120266

Chen C, Trias APL (2020) Water security in southeast Asia: regional, national, and sub-national challenges

De Laurentiis V, Hunt DV, Rogers CD (2016) Overcoming food security challenges within an energy/water/food nexus (EWFN) approach. Sustainability 8(1):95

Destek MA, Sarkodie SA (2019) Investigation of environmental Kuznets curve for ecological footprint: the role of energy and financial development. Sci Tot Environ 650:2483–2489. https://doi.org/10.1016/j.scitotenv.2018.10.017

FAO (2009) The state of food insecurity in the world, economic crises—impacts and lessons learned, United Nations, Rome, Italy. ISBN 978-92-5-106288-3

FAO (2011) The state of the world's land and water resources for food and agriculture: managing systems of risk. Earthscan/FAO, London/Rome. Retrieved from: www.fao.org/nr/solaw/solaw-home/en/

FAO. (2017). What is climate-smart agriculture? Infographic. FAO. Retrieved from: www.fao.org/3/a-i7926e.pdf

FAO AQUASTAT (2015) Proportion of total water withdrawal withdrawn for agriculture. Retrieved from: http://www.fao.org/nr/water/aquastat/maps/WithA.WithT_eng.pdf

FAO, IFAD, UNICEF, WFP, WHO (2020) The state of food security and nutrition in the world 2020. Transforming food systems for affordable healthy diets. FAO, Rome

Gleick PH (1993) Water in crisis. Pacific Institute for Studies in Dev., Environment & Security. Stockholm Env. Institute, Oxford University Press, 473

Grey D, Sadoff CW (2007) Sink or swim? Water security for growth and development. Water Policy 9(6):545–571

Han Z (2003) Groundwater resources protection and aquifer recovery in China. Environ Geol 44(1):106–111

Hanslmeier A (2011) Water in the universe. Astrophysics and Space Science Library. Springer Science+Business Media B.V. https://doi.org/10.1007/978-90-481-9984-6_2

Hoekstra AY, Chapagain AK (2008) Globalization of water: sharing the planet's freshwater resources. Blackwell Publishing, Oxford

Hoff H (2011) Understanding the nexus. Background paper for the Bonn 2011 conference: the water, energy and food security nexus. Environment Stockholm Institute

Inyinbor AA, Adebesin BO, Oluyori AP, Adelani ATA, Dada AO, Oreofe TA (2018) Water pollution: effects, prevention, and climatic impact. In: Water challenges of an urbanizing world, 33

IPCC (2019) IPCC special report on the ocean and cryosphere in a changing climate. Geneva, IPCC. www.ipcc.ch/srocc/
Jiang Y (2009) China's water scarcity. J Environ Manage 90(11):3185–3196
Kundzewicz ZW, Mata LJ, Arnell NW, Doll P, Kabat P, Jiménez B (2007). Freshwater resources and their management. In: Parry ML, Canziani OF, Paluticof JP, van der Linden PJ, Hanson CE (eds) Climate change 2007: impacts, adaptation, and vulnerability. Contributions of the working group II to the fourth assessment report of theintergovernmental panel on climate change
Kuure L, Riekki M, Tumelius R (2018) Nexus analysis in the study of the changing field of language learning, language pedagogy and language teacher education. AFinLA-e: Soveltavan kielitieteen tutkimuksia (11):71–92
Lele U, Klousia-Marquis M, Goswami S (2013) Good governance for food, water and energy security. Aquatic Procedia 1:44–63
Liu C, Xia J (2004) Water problems and hydrological research in the Yellow River and the Huai and Hai River basins of China. Hydrol Process 18(12):2197–2210
Luo T, Young R, Reig P (2015) Aqueduct projected water stress country rankings. Technical Note. World Resources Institute
Meadows DH, Meadows DL, Randers J, Behrens WW (1972) The limits to growth. N Y 102(1972):27
Merchant AT, Jones C, Kiure A, Kupka R, Fitzmaurice G, Herrera MG, Fawzi WW (2003) Water and sanitation associated with improved child growth. Eur J Clin Nutr 57(12):1562–1568
Momblanch A, Papadimitriou L, Jain SK, Kulkarni A, Ojha CS, Adeloye AJ, Holman IP (2019) Untangling the water-food-energy-environment nexus for global change adaptation in a complex Himalayan water resource system. Sci Tot Environ 655:35–47
Nadeem AM, Rafique MZ, Bakhsh K, Makhdum MSA, Huang S (2020) Impact of socio-economic and water access conditions on life satisfaction of rural farmers in Faisalabad district of Pakistan. Water Policy 22(4):686–701
Nie Y, Avraamidou S, Xiao X, Pistikopoulos EN, Li J, Zeng Y, Zhu M et al (2019) A food-energy-water nexus approach for land use optimization. Sci Tot Environ 659:7–19
OECD (2013) Water security for better lives. OECD Publishing. https://doi.org/10.1787/9789264202405-en
OECD and FAO (2018) OECD-FAO agricultural outlook 2018–2027. OECD Publishing/FAO, Paris/Rome. https://doi.org/10.1787/agr_outlook-2018-en
Raimi M, Vivien OT, Odipe OE, Owobi OE (2018) The sources of water supply, sanitation facilities and hygiene practices in oil producing communities in central senatorial district of Bayelsa state, Nigeria. MOJ Public Health 7(6):337–345
Ray DK, Gerber JS, MacDonald GK, West PC (2015) Climate variation explains a third of global crop yield variability. Nat Commun 6(5989):1–9. https://doi.org/10.1038/ncomms6989
Rostow W (1960) The five stages of economic growth–a summary. Sociol Dev:1
Saidi K, Omri A (2020) The impact of renewable energy on carbon emissions and economic growth in 15 major renewable energy-consuming countries. Environ Res:109567. https://doi.org/10.1016/j.envres.2020.109567
Salahuddin M, Habib MA, Al-Mulali U, Ozturk I, Marshall M, Ali MI (2020) Renewable energy and environmental quality: a second-generation panel evidence from the sub-Saharan Africa (SSA) Countries. Environ Res:110094. https://doi.org/10.1016/j.envres.2020.110094
Sawka MN, Cheuvront SN, Carter R (2005) Human water needs. Nutr Rev 63(suppl_1):S30–S39
Scanlon BR, Ruddell BL, Reed PM, Hook RI, Zheng C, Tidwell VC, Siebert S (2017) The food-energy-water nexus: transforming science for society. Water Resour Res 53(5):3550–3556
Shahbaz M, Mahalik MK, Shah SH, Sato JR (2016) Time-varying analysis of CO_2 emissions, energy consumption, and economic growth nexus: statistical experience in next 11 countries. Energy Policy 98:33–48. https://doi.org/10.1016/j.enpol.2016.08.011
Singh VP (2017) Challenges in meeting water security and resilience. Water Int 42(4):349–359
Smajgl A, Ward J, Pluschke L (2016) The water–food–energy nexus–realising a new paradigm. J Hydrol 533(1):533–540

Sood A, Nicol A, Arulingam I (2019) Unpacking the water-energy-environment-food nexus: working across systems, vol 186. IWMI

Staupe-Delgado R (2019) The water–energy–food–environmental security nexus: moving the debate forward. Environ Dev Sustain:1–17

Stephens GL, Slingo JM, Rignot E, Reager JT, Hakuba MZ, Durack PJ, Rocca R (2020) Earth's water reservoirs in a changing climate. Proc R Soc A 476(2236):20190458

Usman M, Kousar R, Makhdum MSA (2020a) The role of financial development, tourism, and energy utilization in environmental deficit: evidence from 20 highest emitting economies. Environ Sci Poll Res 27(34):42980–42995

Usman M, Kousar R, Yaseen MR, Makhdum MSA (2020b) An empirical nexus between economic growth, energy utilization, trade policy, and ecological footprint: a continent-wise comparison in upper-middle-income countries. Environ Sci Poll Res 27(31):38995–39018

Ustun PA, World Health Organization (2008) Safer water, better health: costs, benefits and sustainability of interventions to protect and promote health. World Health Organization

UNICEF, Indonesia (2012) Ringkasan Kajian: Air Bersih, Sanitasi, dan Kebersihan. UNICEF Indonesia, Jakarta

United Nations (2010) Resolution A/RES/64/292. United Nations General Assembly. July 2010

United Nations (2015) World population prospects: the 2015 revision, key findings and advance tables. Department of Economic and Social Affairs, Population Division. Working Paper No. ESA/P/WP.241. http://esa.un.org/unpd/wpp/

WEF (2011) Water security: the water-energy-food-climate nexus. World Economic Forum initiative

World Food Summit (1996) Rome declaration on world food security

World Health Organization (2003) McMichael AJ, Campbell-Lendrum DH, Corvalán CF, Ebi KL, Githeko AK, Scheraga JD, Woodward A (eds) Climate change and human health: risks and responses. ISBN 92 4 156248 X

Wollenberg E, Vermeulen SJ, Girvetz E, Loboguerrero AM, Ramirez-Villegas J (2016) Reducing risks to food security from climate change. Glob Food Sec 11:34–43. https://doi.org/10.1016/j.gfs.2016.06.002

Zidjaly NA (2012) Scollon, Ron. The encyclopedia of applied linguistics. https://doi.org/10.1002/9781405198431.wbeal1043

Muhammad Sohail Amjad Makhdum earned a Ph.D. in Agricultural Economics from Christian-Albrechts-Universität zu Kiel (CAU), Germany in 2016. Currently he works as an Assistant Professor of Agricultural Economics at the Department of Economics, Government College University Faisalabad, Pakistan. He has authored several articles in renowned journals. Areas of his research include Agriculture, Environment, Water, Food, Land, and Health Economics. His specific areas of research include Impact Evaluation, Household Welfare, Gender Issues, Land Use and Rights, Poverty, Carbon Footprint, Energy Consumption, and Microfinance.

Dr. Rakhshanda Kousar is an Assistant Professor at the Institute of Agricultural and Resource Economics, University of Agriculture Faisalabad, Pakistan. She received her Ph.D. from Christian-Albrechts-Universität zu Kiel (CAU), Germany in 2014. Her research interests include Gender Issues, Rural Welfare, Land, Environment, Labor, and Livestock.

Dr. Muhammad Ashfaq served as a Professor in Agricultural Economics at the Institute of Agricultural and Resource Economics, University of Agriculture Faisalabad, Pakistan. He has a wide range of research experience in different areas including Agriculture, Gender Empowerment, Crop Productivity, Rural Welfare, Land, Environment, and Climate Change. He has authored several articles and books, and successfully completed numerous national and international research projects.

Dr. Mohamed Behnassi has been nominated in July 2022 as a Senior Environmentalist Expert by the Economic Social and Environmental Council (ESEC), Rabat, Morocco. Prior to this, he served as a Full Professor (since 2004), and Head of Public Law in French Department at the College of Law, Economics, and Social Sciences of Agadir, Ibn Zohr University. He holds a Ph.D. in International Environmental Law and Governance (2003), a MSc. in Political Sciences (1997), and a B.A. of Administration (1995) from Hassan II University of Casablanca. He obtained a Diploma in International Environmental Law and Diplomacy from the University of Eastern Finland and UNEP, 2015. He is also an Alumnus of the International Visitors Leadership Program of the Department of State, United States of America. Dr. Behnassi is currently the Founding Director of the Center for Environment, Human Security and Governance (CERES). From 2015 to 2018, he was the Director of the Research Laboratory for Territorial Governance, Human Security and Sustainability (LAGOS). Recently, he was appointed as Expert Evaluator for the National Center for Scientific and Technical Research (CNRST/Morocco) and selected twice (2019–2024) as an Assessment Scoping Expert and a Review Editor by the Intergovernmental Science-Policy Platform on Biodiversity and Ecosystem Services (IPBES) and Member of the Mediterranean Experts on Climate and Environmental Change (MEDECC). He was also selected by The Intergovernmental Panel on Climate Change (IPCC) as Expert Reviewer of the 1st Order Draft of the Synthesis Report (SYR) of the IPCC VI Assessment Report (AR6). Accordingly, he was among the Lead Authors of the 1st Assessment Report (MAR1): *Climate and Environmental Change in the Mediterranean Basin—Current Situation and Risks for the Future* (MEDECC, 2021). He is a Senior Researcher in the areas of international law and politics of environment and human security where he published considerable number of scientific papers and book chapters in addition to 20 books, including recent ones on: *Food Security and Climate-Climate Smart Food Systems—Building Resilience for the Global South* (Springer, 2022); *The Climate-Conflict-Migration Nexus from a Human Security Perspective* (Springer, 2022); and *Social-Ecological Systems in the Era of Risks and Insecurity—Pathways to Viability and Resilience* (Springer, 2021). Dr. Behnassi serves as a reviewer for many global publishers (such as Routledge and Springer) and scientific journals with high impact factor. He has organized many international conferences covering the above research areas, managed many research and expertise projects, and is regularly requested to provide scientific expertise nationally and internationally. Other professional activities include social compliance auditing and consultancy by monitoring human rights at work and the sustainability of the global supply chain.

Chapter 3
Water and Food Security in the Middle Eastern and Northern African Countries

Waqar Akram, Zakir Hussain, and Sultan Adeel

Abstract Water, both for human use and agriculture, is a concern in the Middle Eastern and North African economies (MENA), entailing risks and opportunities in these countries. Besides, fast-changing politico-socio-economic and environmental conditions make water security challenging for policymakers. Water security, coupled with food security, has become a distinct and more unnerving challenge than ever before. This chapter provides a detailed analysis of water scarcity, food security, and climatic challenges in MENA countries and provides a way forward in addressing these issues. In MENA countries, water shortage, food security, and environmental problems are intertwined. It is imperative to use water resources efficiently to ensure a sustainable environment, efficient (allocative) use of water, and obtain distributive justice, contributing to social contract or integration. This argument also includes delivering water input reliably and affordably to ensure cordial relationships between service providers and water users and help promote the renewed social contract. The climate of MENA is arid and near arid. The drought cycle is shortened from three years to annual and often brings floods. Thus, water security, food security, and environmental issues are more than ever-changing targets; however, some are within reach of humankind. A series of suggested solutions to the MENA region's water resource management and associated problems exist. Implementing these solutions needs clear incentives to bring about water management changes, including conservation, allocation, and address some of the riparian issues and water conflicts among the MENA countries, such as the Nile and other basins. The MENA countries warrant

W. Akram
Department of Accounting, Finance, and Economics, University of Waikato, Hamilton Campus, Hamilton, New Zealand
e-mail: wakram@waikato.ac.nz

Z. Hussain (✉)
Metropolitan College of New York, New York, NY, USA
e-mail: zakir_rana@yahoo.com

Grand Asian University, Sialkot, Pakistan

S. Adeel
Department of Business Administration, Sukkur IBA University, Sukkur, Pakistan
e-mail: sultan.adeel@iba-suk.edu.pk

M. Behnassi et al. (eds.), *The Water, Climate, and Food Nexus*,
https://doi.org/10.1007/978-3-031-50962-9_3

better engaging the civil society (e.g., High Aswan Dam in Egypt) and water users, teaming millions of youths to make the solution work. The debacle of doable policies addressing water challenges can severely affect nations' well-being and fragile political stability. Thus, the strategic question is 'now or never.' The MENA countries should act with urgent attention in strengthening water security, food security, and a sustainable environment instead of waiting for doomsday for impending water crises leading to 'Water Conflict,' quoting the saying of late United Nations Secretary-General Boutros Ghali that the "*3rd world war will be on waters*".

Keywords Food security · Water security · Middle East and North Africa · Water scarcity and strategies

1 Introduction

Water and other natural resources, mainly used for agricultural power generation, are under stress globally (Behnassi et al. 2019) due to the exponential rise in income and population, which is increasing the consumption of natural resources rapidly and stretching the entire world to its natural limits (WEF 2016). Besides, climate change further exacerbates the issues of water and food insecurity globally (Scott 2017). The concern is that the three main sectors providing food and income security are also the top greenhouse gas producers and are at risk due to climate change (Scott 2017).

At a global level, 17 countries that provide homes to nearly one-fourth of the entire population face extreme water crises (WRI 2019); out of these countries, 12 are in the MENA region (Hofste et al. 2019). The World Resources Institute (WRI) further concludes that out of 21 countries in MENA, ten are at extreme baseline water stress levels. In contrast, eight are in the high baseline water stress level, which signifies the importance of this region in mitigating water and food insecurity. The MENA region is dry and hot and faces the severe challenge of water shortage with growing water demand. Such a situation further stresses the present water resources, including the capacity of the agricultural sector to meet the region's food security. A report by the World Bank points out that the region can face a significant economic loss (approximately a 6–14% decline in the current GDP) due to water scarcity (Hofste et al. 2019).

Therefore, the concern for energy production, water, food, and climate has become the global agenda due to the exacerbating challenges. Even the United Nations has included the concern for climate as its 13th Sustainable Development Goal (Climate Action). Further, as the data provides that the majority of extremely and highly water and food stress areas are in the MENA region, and the presence of one-fourth population in it, researchers and policymakers find it essential to answer the strategic question of 'now or never.' Moreover, it is imperative for the MENA countries to strengthen water and food security and strive for a sustainable environment instead of waiting for doomsday for impending water crises leading to '*Water Conflict.*'

To this end, the present chapter explains the intertwined problem of water and food security as well as the environmental problems in the MENA region. The following sections of the chapter include the background of the MENA region, its geographical area, and utilization, followed by the population and education level. The following section reviews the previous work on the MENA region, followed by the economic state and the political economy. Then, the subsequent section provides evidence on the food and water balance in the region, followed by water conflicts, water management, and strategies to address the water issues. The last section concludes the entire work and provides recommendations for future work.

1.1 Background

MENA is widely used in the literature to refer to countries in the Middle East and North Africa region (World Bank 2014). The word is often used in broadcasting, military planning, academics, and business writing. The MENA region is diversified, facing economic and political transformations, and striving for better growth. It is in one of the most economic regions of the world with significant access to economics, an increasingly young and educated population, and some competitive advantages in economic sectors such as renewable energies, tourism, and manufacturing (OECD 2021).

1.2 Geographical Area

As per the definition of Britannica (2021), MENA comprises the countries located on the shores of the Persian Gulf, the Gulf of Aden, and the Red Sea, ranging from the Arabian Sea to the east, the Mediterranean Sea in the south (north in case of North Africa), and the Atlantic shores of Morocco in the west, and encompasses the Arabian, Qatari, and Sinai peninsulas. However, there is no considerate definition of the MENA countries; various global organizations define the MENA region with different countries and territories, and some organizations do not even recognize the MENA region. For instance, the World Bank's 2003 definition of the MENA region includes 21 territories and countries in the Middle Eastern and North African regions (World Bank 2003). These countries include Bahrain, Iran, Iraq, Israel, Jordan, Kuwait, Lebanon, Malta, Oman, Palestinian territories (West Bank and Gaza), Qatar, Saudi Arabia, Syria, the United Arab Emirates, and Yemen from the Middle Eastern region and Algeria, Djibouti, Egypt, Libya, Morocco, and Tunisia in the North African region. The 2021's definition of MENA by the World Bank includes the same countries and territories. Figure 1 provides the marked countries in the MENA region by World Bank. Further, the initiative of OECD with the MENA economies includes 19 countries (OECD 2021): Bahrain, Iraq, Jordan, Kuwait, Lebanon, Oman, Palestinian Authority, Qatar, Saudi Arabia, Syria, United Arab Emirates, and Yemen in the

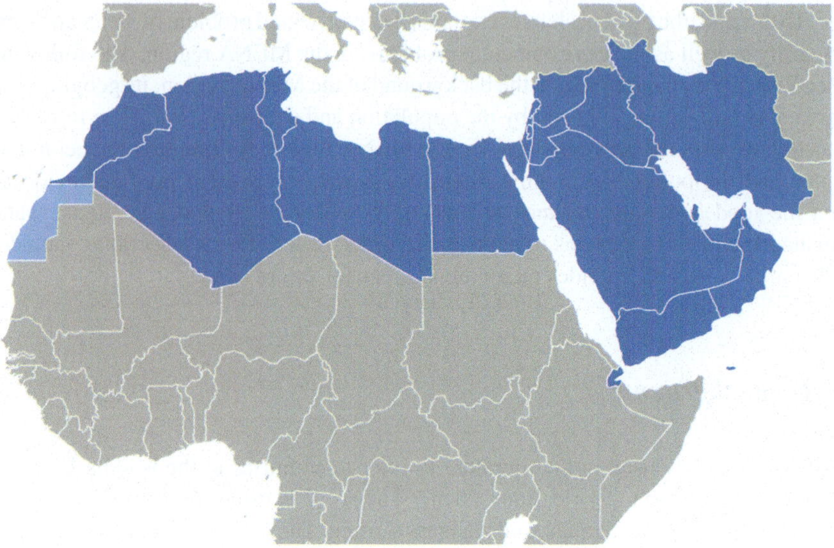

Fig. 1 MENA region by the World Bank. *Source* World Bank (2003, 2021)

Middle Eastern region and Algeria, Djibouti, Egypt, Libya, Mauritania, Morocco, Tunisia in North Africa.

In contrast to the World Bank's definition, the FAO's 2015 definition excludes Iraq and Malta from the MENA region. In contrast, the OECD-FAO 2018–27 outlook report[1] includes Iraq, Mauritania, and Sudan, excluding Malta, Israel, and Djibouti from the MENA region. Other than this, the International Monetary Fund (IMF) and other organizations of the United Nations, such as the UNAIDS, UNICEF, UNHCR, and UNSD, define the difference between some countries. For instance, UNAIDS, UNICEF, and INHCR omit Israel in MENA, UNAIDS omits Palestinian territories, while UNHCR and IMF include Mauritania and Pakistan (IMF only) in the MENA. Figures 2 and 3 provide the marked countries in the MENA region by UNAIDS and IMF.

1.2.1 Middle Eastern Countries

As per the World Bank's definition of the MENA, 15 countries are in the Middle Eastern region (World Bank 2003). The region is located on the shores of the Mediterranean Sea and encompasses the Arabian Peninsula. A brief description of each country is provided below (Britannica 2021).

[1] OECD-FAO Agricultural Outlook 2018–2027: Chapter 2: Middle East and North Africa – Prospects and Challenges. (For more: please see http://www.fao.org/publications/oecd-fao-agricultural-outlook/2018-2027/en/?__cf_chl_captcha_tk__=pmd_o0zL5dijz70GPQO2.r6nGwNbXS2opUszJFcvUfGrAx8-1630928220-0-gqNtZGzNAvujcnBszQR9.

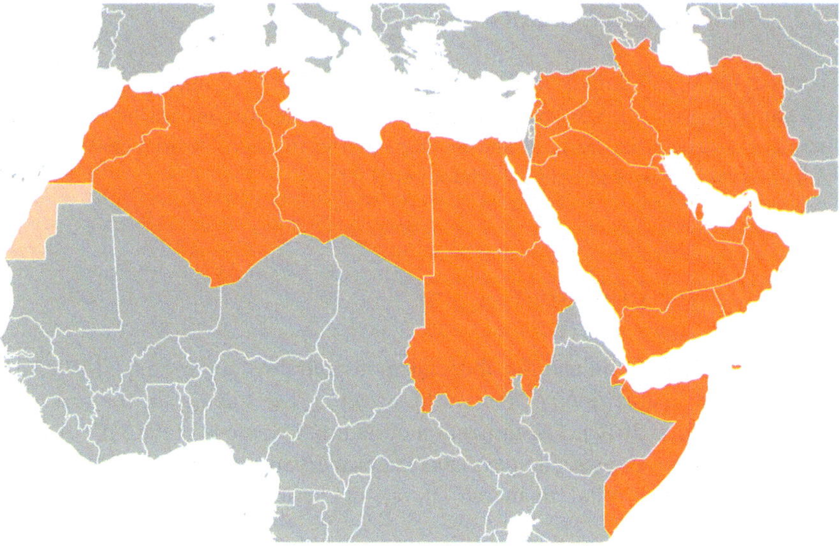

Fig. 2 MENA region by UNAIDS. *Source* UNICEF (2019)

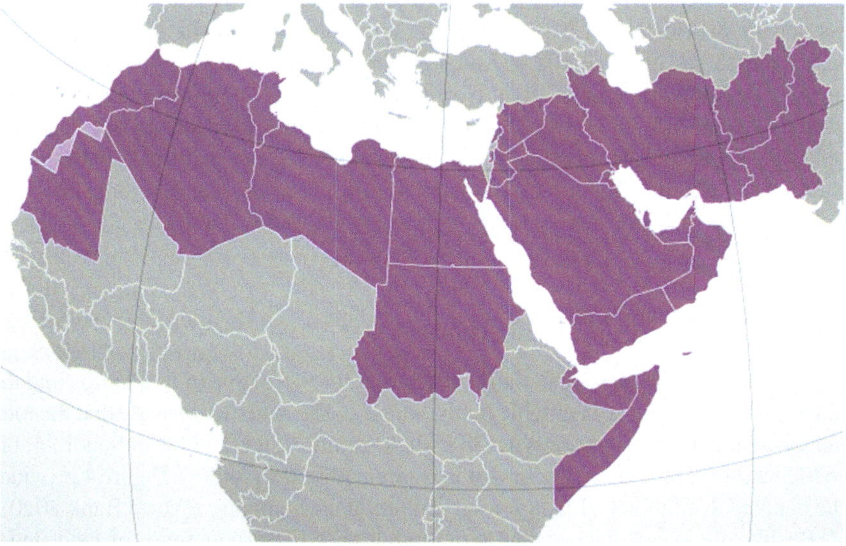

Fig. 3 MENA region by IMF. *Source* IMF (2003)

1.2.2 North African Countries

Likewise, the World Bank's definition includes six countries from Africa in the North African region of MENA (World Bank 2003). The region locates on the Atlantic shores of Morocco in the west and the Red Sea in the east. A brief description of each country is provided below (Britannica 2021).

1.3 Resource Utilization

The geographic location and the abundance of natural resources make MENA one of the most critical regions of the world, affecting the stability of the rest of the economies (OPEC 2012). The utilization of natural resources in the MENA region is diversified. The below section explains the primary natural resources distribution, such as land, labor, capital, finances/FDI, oil, and agriculture in the MENA region.

1.3.1 Land

As per the World Bank (2018a) data, the total land area of the MENA region is 4,333,359.28 sq. miles: 4,108,957.64 sq. miles of rural and 100,890,64 sq. miles of urban land (World Bank 2010). However, FAO (2021) points out that out of the available land, only 1/3 (33.22%) supports agricultural production, out of which only 5% is arable. The remaining land makes up deserts or is utilized under urban settlements. Besides, the total land under forest has increased from 80,512.73 sq. miles in 2000 to 88,517.36 sq. miles in 2018 (World Bank 2018a).

1.3.2 Land for Agriculture

MENA is one of the most challenging regions of the world due to its location and the environment, making it unsuitable for agriculture. The statistics provide that the total land available for agriculture is 1,438,533.59 sq. miles, the highest in Saudi Arabia (670,382.24 sq. miles) and the lowest in Bahrain (33.20 sq. miles). Figure 4 provides the detailed availability of land for agriculture in each country (World Bank 2020).

Besides the dearth of suitable agricultural land, the productivity of land under cultivation has reduced to only 35–30% (FAO 2021). Box 1.1 provides the recent initiatives to address the issue of declining land quality in the MENA region. The primary sources of land degradation are water and wind erosion (removal of the uppermost layer) in rain-fed areas and soil salinity/sodicity and intensive farming in irrigated land areas (FAO 2021).

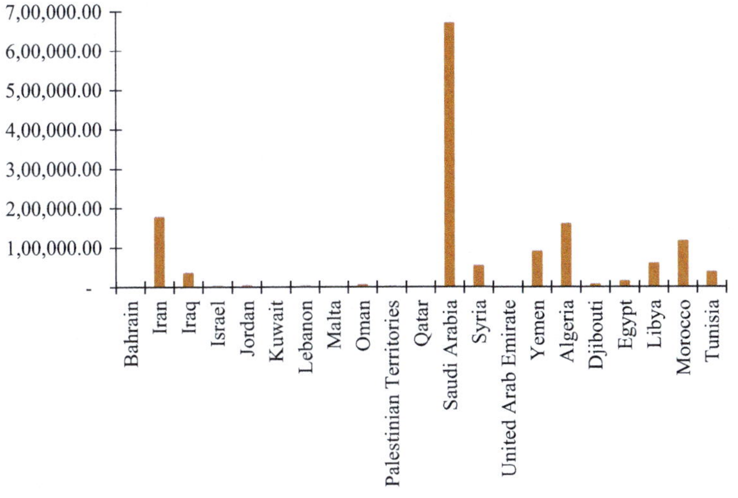

Fig. 4 Land available for agriculture (square miles). *Source* World Bank (2020)

Box 1.1 Initiatives combating the declining land quality in the MENA[2]
Considering the issue of declining productivity in the MENA region, the following two solutions can provide benefits to meet the productivity level.

(a) *Soil Maps*: These documents provide information on soil characteristics, such as soil nutrients, water storage electrical conductivity, soil pH, and organic material. The data is collected in various ways, such as manual field sampling, infrared spectroscopy, and remote sensing. Thus, the information can provide crucial information to farmers and policymakers to decide which crop to plant/sow and how much inputs such as seeds, water, and organic and synthetic fertilizers are required. However, soil maps may bring problems in decision-making if outdated or low-resolution soil maps are utilized.
(b) *Zero Tillage*: This technique helps mitigate the issue arising from the plowing, such as loss of upper layer (organic matter) and moisture which can further pace the water and wind erosion. This technique minimally disturbs the upper layer of the soil. The roots left from the last crop serve the role of soil stabilizer, increasing the fertility and water-holding capacity. Thus, the technique can help maintain land productivity.

[2] Source: http://www.fao.org/3/i9166e/i9166e_Chapter2.pdf, cited in OECD-FAO Agricultural Outlook 2018–2027.

1.3.3 Aquaculture and Fisheries

Due to the presence of the majority of the MENA countries on the shores of the seas, fisheries and aquaculture have greater importance. The region provides one of the most diversified freshwater and marine ecosystems, and it provides livelihood to the locals and services as the source of a nutritious diet. As per the stats provided by FAO (2021), total production from freshwater, marine, and aquaculture included 59 million tons in 2016, which increased from 2.2 million tons in 1996, mainly due to the increase in the captured fisheries. In contrast, during the same time, aquaculture also increased (32% share in 2016). However, regardless of the region's production, it still relies on imports of fish and fish products to meet its national consumption (FAO 2021).

1.3.4 Labor

As per the World Bank 2020 data, the total available labor force is 147.277 million out of the 464.554 million population in the region (World Bank 2020); the highest in Egypt (27.870 million labor, 102.334 million total population) and the lowest in Malta (0.280 million labor, 0.525 million total population). The data shows that, on average, 19.93% are females (highest in Israel, 47.6 percent; lowest in Yemen, 7.9%), and the remaining are males (80.07%). Besides, 14.597% are employed in agriculture (highest in Morocco, 33.25 percent; lowest in Israel, 0.92%), 26.96% in industry (highest in Qatar, 54.00 percent; lowest in Yemen, 10.20%), and 58.29 percent in the services sector (highest in Israel, 81.86%; lowest in Morocco, 43.66%). The data provides that labor participation in the agricultural sector has decreased over the period, sustained in the industry, and increased in the services sector (World Bank 2021). Figures 5 and 6 provide the detailed availability of labor for each sector in each country.

1.3.5 Capital, Finances, and Foreign Direct Investment

The MENA region has remained vital in attracting media attention as one of the most critical hotspots for economic, security, and political vulnerabilities (Miller et al. 2018). In terms of finances and wealth, the countries located in the region are also diversified, ranging from wealthy countries with one of the top natural reserves, such as crude oil, to the poorest economies extravagated by poverty and chronic battles. These issues in developing countries hinder investment from multinational companies (MNCs) operating in the region (Dimitrova et al. 2019). As per the World Bank (2020), the present FDI[3] in the entire region is 66.074 billion USD (highest in Israel—24.283 million USD, lowest in Libya—0.001 million USD, and negative in Yemen, Kuwait, Qatar, and Iraq), which once reached an all-time high of 126.452

[3] Foreign direct investment, net inflows (BoP, current US$) – Middle East & North Africa.

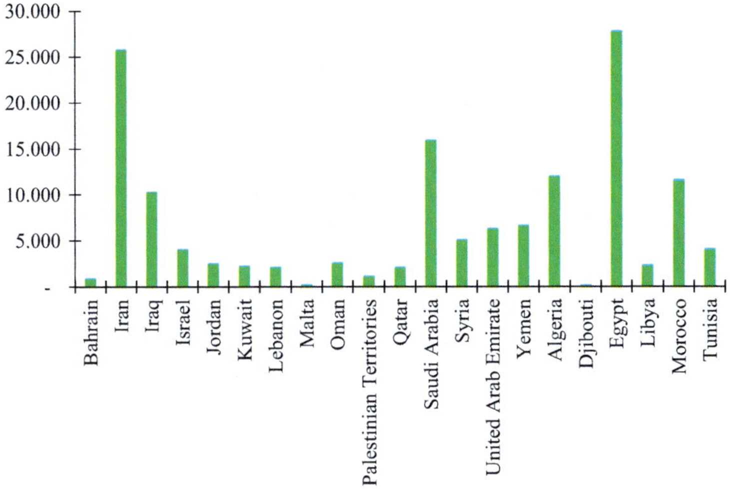

Fig. 5 Total labor force (Millions). *Source* World Bank (2020)

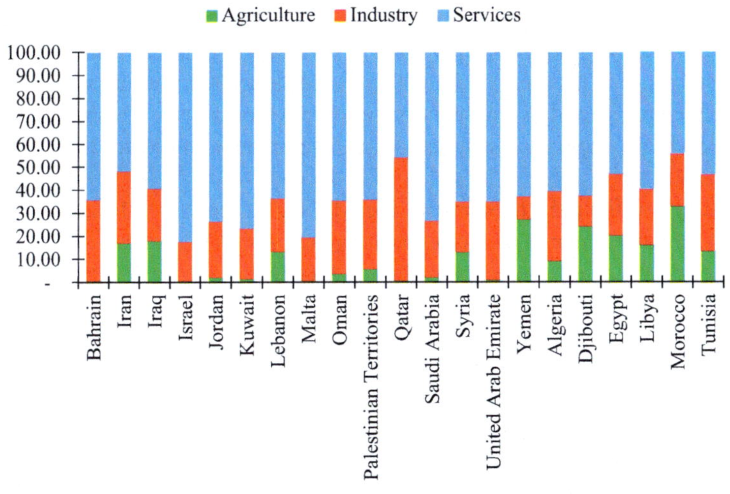

Fig. 6 Total labor force in each sector (percentage). *Source* World Bank (2020)

billion USD in 2007. Figure 7 provides the detailed availability of the FDI in the MENA region.

The present investment (growth and decline) in the MENA region has declined due to chronic fiscal and structural problems, primarily improper infrastructure, prevalent corruption, feeble governance, frail private sector vitality, and inadequate economic divergence outside the oil sector. It has further been exacerbated by the drastic decline

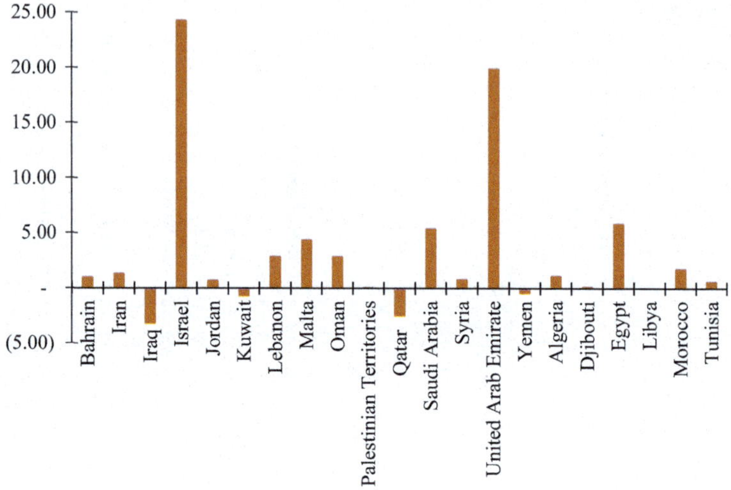

Fig. 7 Foreign direct investment (millions). *Source* World Bank (2020)

in oil prices[4] and the fallout from the Arab Springs uprisings (World Bank 2018a; Saefong 2021). Besides this, the region is also facing political instability coupled with humanitarian disasters, making them least integrated into the world economy despite the presence of two primary trade routes (World Bank 2018b). These issues lead the MENA region to poor economic development and lower FDI (Dimitrova et al. 2019).

Contrary to the present scenario, the MENA region can still attract local and foreign investment due to the presence of natural reserves (gas and oil), the growing young population, and the geopolitical presence (Demirbag et al. 2011).

1.3.6 Agriculture

Similar to the diversified nature of labor force employment, the agricultural sector's contribution to the national economy is diversified (Ahmed et al. 2023). For instance, the agriculture, forestry, and fishing value added to the region's GDP are 5.165%; the lowest in Bahrain (0.31%) and highest in Syria (39.77%). It points to the importance of agriculture in minor oil-producing economies such as Yemen and Syria and lesser importance in most oil-producing economies such as Qatar, Bahrain, and Saudi Arabia. Figure 8 provides the detailed contribution of agriculture, forestry, and fishing in each country (World Bank 2020).

Regardless of the contribution of agriculture to MENA's economy, it plays a vital role in providing raw materials to the manufacturing and servicing industries (Lechtenberg 2018). The irrigation system, mainly in the basins of the Tigris-Euphrates

[4] For more, please see: https://www.bbc.com/news/business-52350082.

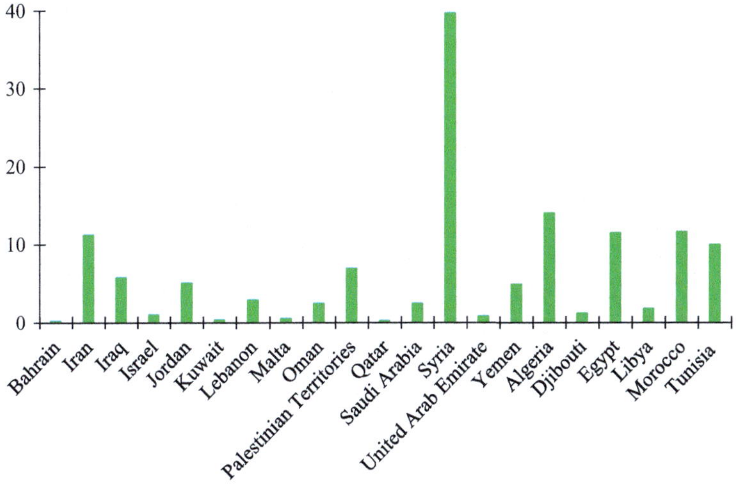

Fig. 8 Agriculture, forestry, and fishing value added (% of GDP). *Source* World Bank (2020)

River Basin, Jordan River Basin, and Nile River Basin, enables the intensive production of crops, supporting the exports of valuable fruits, cereals, and vegetables. However, the food gap in the entire region is broadening, questioning the food security of the MENA economy. Some countries have now focused on increasing agricultural production by introducing policies. For instance, Saudi Arabia's Vision 2030 focuses on food for all populations living in the kingdom by increasing aquaculture, cooperatives, and vegetable and fodder productivity (SPA 2021).[5]

1.3.7 Crude Oil Production

The MENA countries are known for their vast crude oil reserves globally, producing 30.80 million barrels per day. The leading exporter of crude oil and products are Saudi Arabia (also the second-largest oil exporter globally), exporting 12 million barrels per day, followed by Iraq (4.8 million barrels per day), UAE (4.0 million barrels per day), Iran (3.2 million barrels per day), and Kuwait (3.0 million barrels per day). In contrast, Syria is the lowest, with 0.024 million barrels per day (Carpenter 2021). However, oil production is decreasing gradually, and the economy is shifting toward other sources to generate income. The above-given Fig. 4b also identifies the shift in labor participation toward the services sector in the MENA region.

[5] For more: Please refer to: https://www.spa.gov.sa/viewstory.php?lang=en&newsid=2290104.

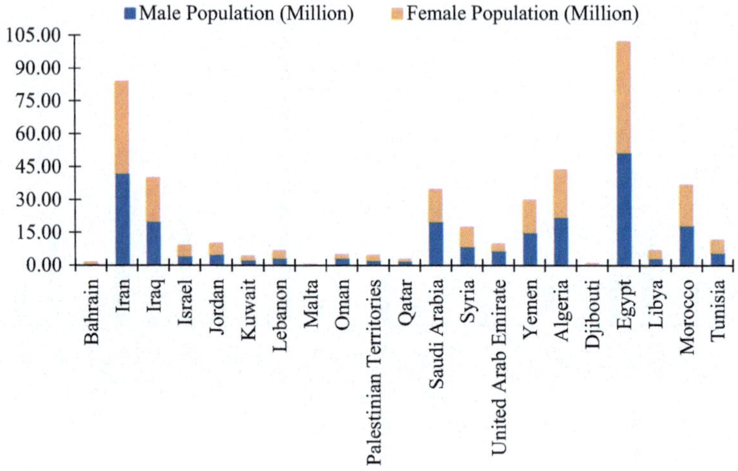

Fig. 9 Total and gender-specific population (millions). *Source* World Bank (2020)

2 Population and Education Level

The population is considered an essential factor in determining the country's growth due to its high role in the labor force. This section explains the population in each MENA economy and their education level.

2.1 *Population in the MENA Countries*

The population level in the MENA region provides that the countries are highly populated, housing 6.3% of the global population. As per the data provided by World Bank (2020), the population in the MENA region is 464.554 million (240.679 million Males, 223.874 million Females). Out of this, 396 million (199 million Males, 196 million Females) live in lower and middle-income[6] countries, which intensifies the population and income inequality in the region. Nevertheless, the population level is highly diverse in MENA, as Egypt (22.03%) and Iran (18.08%) alone encompass more than 40 percent of the region's total population. Figure 9 depicts the gender-specific population in each country.

[6] Data extracted from Population, total – Middle East and North Africa (excluding high income countries). Source: https://data.worldbank.org/indicator/SP.POP.TOTL?locations=ZQ&most_recent_value_desc=true.

trade"; (ii) economic and physical access to food, which ensures "*an adequate supply of food at the national or international level; however, it does not in itself guarantee household level food security*"; (iii) food utilization, which addresses the "*sufficient energy and nutrient intake by individuals as a result of good care and feeding practices, food preparation, diversity of the diet and intra-household distribution of food*"; and (iv) stability of the other three components over time, which states that "*even if one's food intake is adequate today, he or she is still food insecure if he or she has inadequate access to food periodically, risking a deterioration of his or her nutritional status*" (Hameed et al. 2021; World Bank 2023).

Concerning food security in the MENA region, there is a need to look at cereal production to understand its current level of food security.

5.2 Cereal Production in MENA Region

The World Bank (2018) data provides that the cereal production in the region is 70.090 million metric tons, which makes 2.36% of the entire global production (2.965 billion metric tons) for the 6% of the global population. Besides, the per-country total cereal production is also diverse, where only three countries, i.e., Egypt, Iran, and Morocco, collectively produce 75.69% (31.44, 29.44, and 14.81%, respectively) of the region's total cereals production. Djibouti is at the bottom, producing only 18 metric tons of cereals, whereas Bahrain does not produce any. Figure 13 provides the details of cereal production in the MENA region.

Along with cereals, oilseed crops are another type of crop providing energy and protein in the MENA region.

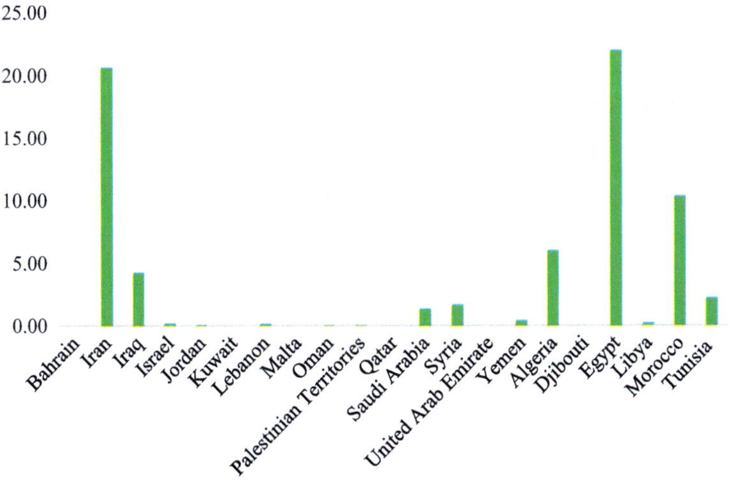

Fig. 13 Cereals production (million metric tons). *Source* World Bank (2018c)

6 Water Balance and MENA

The region is the most water-scare region globally, with only 1.4% of the available freshwater sources housing 6.3% of the entire population (World Bank 2017a). Over 60% of the population in the region has little or no access to drinkable water, while over 70% of the region's GDP is exposed to high or very high-water stress (Al-Zu'bi 2019). The region is also severely affected by climate change and has been subject to an almost continuous drought since 1998. Therefore, water has remained a challenge for the MENA countries (Drake 1997). Nonetheless, this issue is further worsened by ongoing regional instability, climate change, and economic development (World Bank 2017a). These activities will further increase the water demand in the future.

The region is comprised of an area of 4.6[10] million sq. miles, making around 10 percent of the world's total land area (UN-FAO 2014), whereas, in 2020, the region's population reached 464 million with an annual increase of 1.7 percent (World Bank 2018), and expected to reach 586 million by 2030 and 732 million by 2050. It is pertinent to mention that a larger population will require more water. Based on the available water in the region, MENA can be divided into five primary subdivisions (Kandeel 2019):

- **Arabian Peninsula**: Bahrain, Kuwait, Oman, Qatar, Saudi Arabia, the United Arab Emirates, and Yemen
- **Nile Basin**: Egypt
- **Al-Maghreb**: Algeria, Libya, Morocco, and Tunisia
- **Al-Mashreq**: Iraq, Jordan, Lebanon, the Palestinian Territories, and Syria
- **Sahel**: Djibouti.

Besides, the al-Maghreb, al-Mashreq, and Nile Basin countries share common climatic attributes, determining their natural water resource capacity (Kandeel 2019). Further, the available renewable[11] internal freshwater resource in the MENA region is 229.67 billion cubic meters (highest in Iran, 129.50 billion cubic meters; lowest in Malta, 0.05 billion cubic meters; and zero in Kuwait, Bahrain). The MENA region's per capita renewable freshwater resource is 525.747 cubic meters (highest in Iran, 1592.83 cubic meters; lowest in Bahrain, 2.68 cubic meters, and zero in Kuwait). In contrast, it is 604.24 cubic meters in the MENA region, excluding the high-income countries, postulating that the lower middle and middle-income countries in the MENA region have more per capita freshwater available for their people (World Bank 2021). Figure 14 provides the available inland and per-capita water in the region.

Besides the inland water availability, the rainfall is meager in the region as it spans over three deserts[12] worldwide. The average annual rainfall in the region is

---

[10] 1.2 billion Hectares (UN-FAO, 2014).

[11] Refers to internal river flows and groundwater from rainfall in the country.

[12] These three deserts are: (a) Rub'al Khali desert in the Southern Arabia. (b) Sahara Desert in the Africa that includes major part of the Egypt and Libya, and (c) Baidat El-Sham in the Northern Arabia.

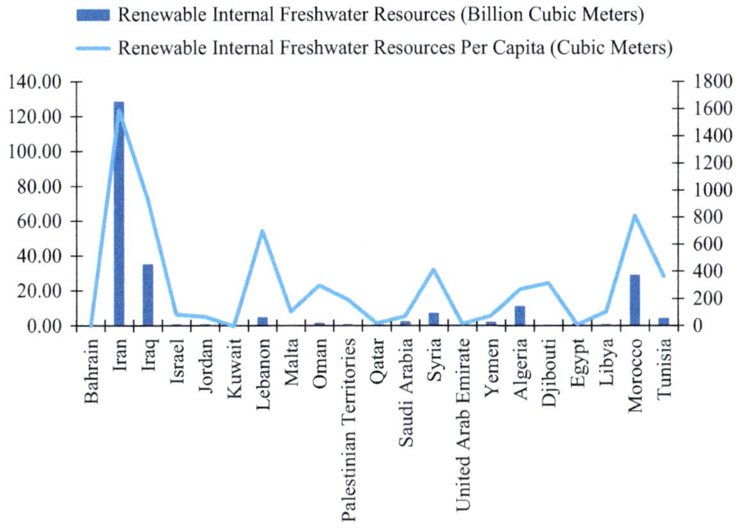

Fig. 14 Renewable internal freshwater resources and per capita (cubic meters). *Source* World Bank (2017b, c)

below 100 mm in 65% of the region, while it is between 100 and 300 mm in 15% of the region; the rest of the area gets more than 300-mm rain annually (Abu Zeid and Abdel-Meguid 2006). Studies on the water conflict and climate change in the region MENA reckon that the water conflict among these countries is escalating due to the worsening climatic conditions. Figure 15 points to the water scarcity in the region.

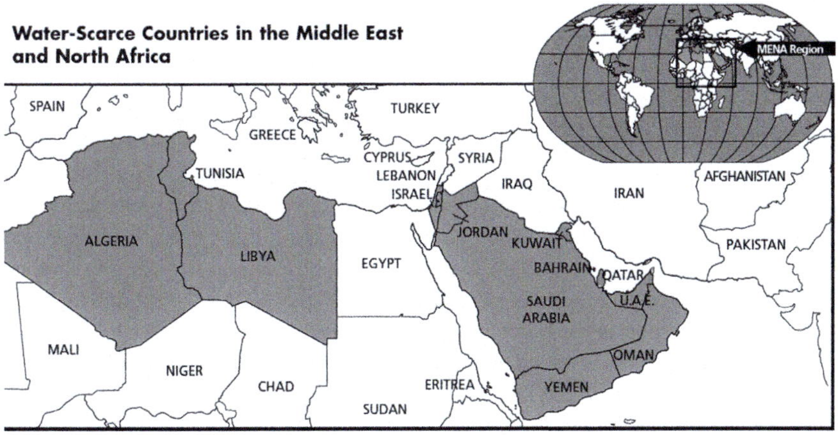

Fig. 15 Water-scarce countries in the Middle East and North Africa. *Note* Shaded countries have less than 1000 cubic meters of renewable freshwater per person yearly. *Source* World Bank (2017a)

The scarcity of anything generates conflicts among the stakeholders. The population living in the countries faces a severe water crisis, leading to potentially disastrous consequences (Selim et al. 2020).

6.1 Water Conflicts

Despite its most significant importance in human living, water has remained the point of conflict among nations. Several countries in the region are currently in a state of geopolitical instability (Zekri and Al-Maamari 2020). The water issue in the MENA region covers three primary basins, including Tigris-Euphrates River Basin, Jordan River Basin, and Nile River Basin.

7 Water Management and Issues

The freshwater withdrawal by the industries[13] in the MENA region is 4.32 percent, which is slightly higher (4.34%) in the lower and middle-income countries. Compared to the industry, the freshwater withdrawal by agriculture[14] in the MENA region is 85.43%, slightly higher (86.04%) in the lower and middle-income countries (World Bank, 2021). Thus, owing to the copious reliance on agriculture in the region, Woertz (2017) argues that agriculture is the primary factor in the water crisis. It withdraws nearly 80% of the total available water.

8 Strategies to Address Water Issues

Before addressing the water issues, it is paramount to understand the need for water management. The climate of the MENA region is predominantly hot, and the region widely relies on rain to irrigate lands and produce food for humans and animals. The total and per capita in-land water is also low in coastal or island countries like Kuwait, Bahrain, and Malta. Water scarcity, coupled with water conflicts among countries such as Turkey, Israel, Jordan, and Egypt, put the MENA region at stake. These issues have made water a precious resource from a widely available commodity. To this end, the present governing bodies in the region need to develop water management strategies to address water issues in the region. The below-given paragraphs elaborate on strategies falling into two subsections, i.e., water consumption strategies and water conservation strategies.

[13] Annual freshwater withdrawal, industry (% of total freshwater withdrawal). Source: World Bank, 2021.

[14] Ibid.

Under the water consumption strategies, the MENA governments must develop the following:

- The first and topmost strategy under water consumption is 'increasing awareness to change consumption patterns.' World Bank data reported the decline of per capita freshwater available for the people in the region, mainly in sub-Saharan Africa and the Arabian Peninsula. It implies that water consumption must be conscious of meeting the present and future demands of the population. Therefore, educating people about sustainable water use is paramount to managing the crisis.
- The second strategy, 'reducing the corporate water footprint,' implies that industrial water consumption nearly accounts for 22% of global consumption. In the MENA region, it is 4.33% (highest in Lebanon, 48.91%, and lowest in Iran, 1.18%). Therefore, the governments in the MENA region must push industries to adopt sustainable manufacturing practices[15] (focused on less water consumption) to produce goods and services. The examples may include: (i) switching to sustainable and renewable energy resources, such as solar, wind, or biogas, to power the production processes; (ii) reducing industrial waste or efficient recycling of the waste to reduce the carbon footprint; and (iii) improving energy efficient materials and machine parts, such as smart sensors and automation, in production processes.
- The third strategy includes 'planning for and reducing the population growth rate.' The population growth rate in the MENA region is 1.718%, implying that the population will increase by 7.98 million yearly, which can post the supply–demand side gap to nearly 65% by 2030. Therefore, reducing the population growth rate will reduce water consumption.
- The last strategy under the consumption strategies includes finding new ways and scaling up innovative water technologies ranging from reusing wastewater, desalination, combined agriculture or aquaculture, and smart irrigation systems. In some countries, more than half of current water withdrawals exceed what is naturally available; 82% of wastewater is not recycled, presenting a massive opportunity to meet water demands (World Bank 2017a). Therefore, finding new ways to recycle would be paramount.

Under conservation strategies, the MENA governments must develop and improve 'ways to conserve the presently available water' for its population:

- The first strategy is to invest in 'inventing new and innovative ways to conserve water in the areas where the aquifers are drying up, and rainwater is increasingly unpredictable.' Governments can focus on providing grants to universities and research institutions to develop improved water catchment and harvesting methods.
- The second strategy is 'developing energy-efficient desalination plants for water conservation.' Presently, desalination is another significant problem in the African continent after food and water security, and the process is costly and energy extensive. Although the Middle Eastern states have developed large energy reserves

[15] Sustainable manufacturing practices aim to reduce the environmental impact of producing goods and services, while also ensuring social and economic benefits.

to build desalination plants due to their strong economies, North African nations cannot do so. Therefore, governments must fund the development of better water desalination plants to conserve water.

- Other than these two broad strategies, the governments in the MENA region can build institutional cooperation, improve distribution infrastructure, develop better policies, and price water appropriately to address water issues. All will collectively affect water consumption and conservation.

9 Conclusions and Recommendations

This chapter highlights the widespread issues of water and food security in the MENA region. The rapid and dynamic changes in political, social, economic, and environmental factors have worsened the water scarcity and food security challenges for policymakers. Both issues are interlinked and pose a more serious and urgent threat than ever before for the future. In particular, water scarcity, climate change, population growth, urbanization, conflict, and displacement are some of the factors that contribute to the challenge of ensuring water and food security for all. The findings from the study suggest that the region has the greatest expected economic losses from climate-related water scarcity, estimated at 6–14% of GDP. Total water productivity in MENA is only about half the world's average, which poses a big question on the current and future economic status of the region. Besides, the region is exceptionally dependent on food imports, especially on wheat and other staple grains. The estimates suggest that the number of Yemenis afflicted by food insecurity reached 24 million— ~83% of the population—in 2021, with 16.2 million needing emergency food. With regard to education, one in every five children in MENA is not in school, and of these out-of-school children, an estimated number of over 3 million would have been in school if the crises had never happened. Equally, by the end of 2017, the armed conflicts in Syria, Iraq, and Yemen brought back the number of regional out-of-school children to its 2007 level of over 14.3 million.

Therefore, policymakers need to adopt a holistic and integrated approach that considers the interlinkages between water, food, energy, and the environment to address these challenges. They also need to foster regional cooperation, invest in innovative technologies, promote water conservation and efficiency, and engage with all stakeholders, especially the youth and women who are most affected by water insecurity. By doing so, the MENA region can turn its water and crisis into an opportunity for peace and prosperity, which may become a case of a "blessing in disguise."

References

Abu-Zeid K, Abdel-Meguid A (2006) Pioneering action in managing the transboundary Nubian Sandstone groundwater aquifer. In: CEDARE 4th World water forum, Mexico City

Ahmed M, Asim M, Ahmad S, Aslam M (2023) Climate change, agricultural productivity, and food security. Global agricultural production: resilience to climate change. Springer International Publishing, Cham, pp 31–72

Al-Shaikh AJ, Jamal MT (2020) Bioaugmentation of halophilic consortia for the degradation of petroleum hydrocarbons and petroleum wastewater treatment. Int J Adv Res Biol Sci 7(9):97–112

Al-Zu'bi I (2019) Middle East and North Africa: Water scarcity is a growing problem across the Middle East. Is this how we solve it? World Economic Forum. Retrieved from https://www.weforum.org/agenda/2019/03/water-scarcity-is-a-growing-problem-across-the-middle-east-is-this-how-we-solve-it/. Accessed on 19 June 2023

Antonelli M, Tamea S (2015) Food-water security and virtual water trade in the Middle East and North Africa. Int J Water Resources Dev 31(3):326–342. https://doi.org/10.1080/07900627.2015.1030496

Antonelli M, Laio F, Tamea S (2017) Water resources, food security, and the role of virtual water trade in the MENA region. In: Environmental change and human security in Africa and the Middle East. Springer, Cham, pp 199–217

Behnassi M, Himangana G, Pollmann O (eds) Human and environmental security in the era of global risks. Springer International Publishing. https://doi.org/10.1007/978-3-319-92828-9

Bindraban P, Koning N, Essers S (1999a) Global Food Security. Initial proposal for a Wageningen vision on food security. Wageningen University, Netherlands

Britannica (2021) Middle East and North Africa. Britannica. Retrieved from https://www.britannica.com/place/Middle-East

Carpenter C (2021) Nanoparticles aid oil recovery during alternating injection. J Petrol Technol 73(09):62–63

Demirbag M, Apaydin M, Tatoglu E (2011) Survival of Japanese subsidiaries in the Middle East and North Africa. J World Bus 46(4):411–425

Dimitrova A, Rogmans T, Triki D (2019) Country-specific determinants of FDI inflows to the MENA region: a systematic review and future research directions. Multinational Bus Rev

Drake C (1997) Water resource conflicts in the Middle East. J Geogr 96(1):4–12

FAO (2021) The state of the world's land and water resources for food and agriculture: systems at breaking point: synthesis report 2021. Food and Agriculture Organization of the United Nations. Retrieved https://www.fao.org/3/cb7654en/cb7654en.pdf

Getachew N, Mesbah M (2015) Middle East and North Africa Region: An Important Driver of World Agricultural Trade, A Report from the Economic Research Service, United States Department of Agriculture (USDA). https://www.ers.usda.gov/webdocs/outlooks/35796/53335_aes88.pdf?v=6740

Gürsoy Sİ, Jacques PJ (2014) Water security in the Middle East and North African region. J Environ Stud Sci 4(4):310–314

Hameed A, Padda IUH, Salam A (2021) Analysis of food and nutrition security in Pakistan: a contribution to zero hunger policies. Sarhad J Agric 37

Hofste RW, Reig P, Schleifer L (2019) 17 countries, home to one-quarter of the world's population, face extremely high water stress

IHS Markit (2021) IHS Markit. Retrieved from https://news.ihsmarkit.com/INFO/press_releases_iframe/view/0/?template=ihsmarkit&years_pagination=1&year=2021

IMF (2003) Challenges of growth and globalization in the Middle East and North Africa. Retrieved from https://www.google.com/books?id=kQTk34bsYecC&pg=PP8. Accessed on 21 Oct 2021

Kandeel A (2019) Freshwater resources in the MENA region: risks and opportunities

Karadirek IE (2017, June) A case study on evaluation of household water security in MENA countries. In: Solutions to water challenges in MENA region: proceedings of the regional workshop, April 25–30, 2017-Cairo, Egypt. Cuvillier Verlag

Lechtenberg D (2018) Use of sewage sludge in cement industry

Mbow C, Rosenzweig C, Barioni LG, Benton TG, Herrero M, Krishnapillai M, Liwenga E, Pradhan P, Rivera-Ferre M-G, Sapkota T, Tubiello FN, Xu Y (2019) Food security. Germany. Retrieved from https://policycommons.net/artifacts/458644/food-security/1431487/. Accessed on 14 June 2023. CID: 20.500.12592/1c7jw6

Miller T, Kim AB, Roberts JM (2018) Index of economic freedom. The Heritage Foundation

OECD (2021). The global recovery is strong but imbalanced. *OECD*. Retrieved from https://www.oecd.org/economic-outlook/december-2021/

OECD-FAO (2018). OECD-FAO Agricultural Outlook 2018–2027. OECD-FAO. Retrieved from https://www.fao.org/3/i9166e/i9166e_Chapter2.pdf

Omidvar N, Ahmadi D, Sinclair K, Melgar-Quiñonez H (2019) Food security in selected Middle East and North Africa (MENA) countries: an inter-country comparison. Food Security 11(3):531–540

OPEC (2012) Market indicators 2012. OPEC. Retrieved from https://www.opec.org/opec_web/en/data_graphs/2234.htm

Saefong MP (2021) Why the blockage of the Suez Canal matters for oil prices? Recuperado el

Scott P (2017) Global panel on agriculture and food systems for nutrition: food systems and diets: facing the challenges of the 21st century: London, UK, 2016 (ISBN 978-0-9956228-0-7), 132pp. http://glopan.org/sites/default/files/ForesightReport.pdf

Selim NA, Saeed AM, Ibrahim MK (2020) Monitoring and controlling bacteria in pharmaceutical industries water system. J Appl Microbiol 129(4):1079–1090

UN-FAO (2014) The state of food insecurity in the world: strengthening the enabling environment for food security and nutrition. UN-FAO. Retrieved from https://www.fao.org/3/i4030e/i4030e.pdf

UNICEF (2019) Seizing the opportunity: ending AIDS in the Middle East and North Africa. Retrieved from https://www.unicef.org/mena/media/5531/file/HIV%20MENA%2029%20Sep%202019%20FINAL%5b2%5d.pdf%20.pdf. Accessed on 21 Oct 2021

Williams R (2015a) Politics and letters: interviews with new left review. Verso Books

Williams TO (2015b) Reconciling food and water security objectives of MENA and sub-Saharan Africa: is there a role for large-scale agricultural investments? Food Security 7(6):1199–1209

Woertz E (2017) Agriculture and development in the wake of the Arab spring. In: Combining economic and political development. Brill Nijhoff, pp 144–169

World Bank (2003). Trade, investment, and development in the Middle East and North Africa. Retrieved from http://documents1.worldbank.org/curated/en/675121468756587109/pdf/multi0page.pdf. Accessed on 21 Oct 2021

World Bank (2010). Land area (sq. km). The World Bank. Retrieved from https://data.worldbank.org/indicator/AG.LND.TOTL.K2

World Bank (2014) Middle East and North Africa. The World Bank. Retrieved from https://www.worldbank.org/en/region/mena

World Bank (2017a) Beyond scarcity: water security in the Middle East and North Africa. The World Bank. https://doi.org/10.1596/978-1-4648-1144-9

World Bank (2017b) Renewable internal freshwater resources per capita (cubic meters)—the Middle East & North Africa. Retrieved from https://data.worldbank.org/indicator/ER.H2O.INTR.PC?locations=ZQ&most_recent_value_desc=true. Accessed on 21 Oct 2021

World Bank (2017c) Renewable internal freshwater resources, total (billion cubic meters)—the Middle East & North Africa. Retrieved from https://data.worldbank.org/indicator/ER.H2O.INTR.PC?locations=ZQ&most_recent_value_desc=true. Accessed on 21 Oct 2021

World Bank (2018a) Global economic prospects. The turning of the tide? World Bank Group, Washington, DC, June 2018

World Bank (2018b) MENA economic monitor. Economic Transformation. World Bank Group, Washington, DC, April 2018

World Bank (2018c) Cereal production (metric tons)—the Middle East & North Africa. Retrieved from https://data.worldbank.org/indicator/AG.PRD.CREL.MT?locations=ZQ. Accessed on 21 Oct 2021
World Bank (2020). Data: The Middle East and North Africa. Retrieved from https://data.worldbank.org/country/ZQ. Accessed on 21 Oct 2021
World Bank (2021) Data: The Middle East and North Africa. Retrieved from https://data.worldbank.org/country/ZQ. Accessed on 21 Oct 2021
World Bank (2023). What is food security? The World Bank. Retrieved from https://www.worldbank.org/en/topic/agriculture/brief/food-security-update/what-is-food-security#:~:text=Based%20on%20the%201996%20World,an%20active%20and%20healthy%20life
World Economic Forum (WEF) (2016) The human capital report 2016: insight report. World Economic Forum. Retrieved from https://www.google.com/url?sa=t&rct=j&q=&esrc=s&source=web&cd=&cad=rja&uact=8&ved=2ahUKEwjazs_uk8X_AhUKVqQEHeqFAeYQFnoECBAQAQ&url=https%3A%2F%2Fwww3.weforum.org%2Fdocs%2FHCR2016_Main_Report.pdf&usg=AOvVaw0gEZWv7uCXXNMParoCG1qH
World Resources Institute (WRI) (2019) 17 countries, home to one-quarter of the world's population, face extremely high water stress. World Resources Institute. Retrieved from https://www.wri.org/insights/17-countries-home-one-quarter-worlds-population-face-extremely-high-water-stress
Wright B, Cafiero C (2010) Grain reserves and food security in MENA countries. In: UC Berkeley conference on agriculture for development-revisited
Zekri S, Al-Maamari A (2020) An overview of the water sector in MENA region. In: Water policies in MENA countries, pp 1–17

Further Reading

Bindraban PS, van Keulen H, Kuyvenhoven A, Rabbinge R, Uithol PWJ (1999) Food security at different scales: demographic, biophysical, and socio-economic considerations (No. 21). AB-DLO

Dr. Waqar Akram is working as a Visiting Research Scholar (Agribusiness Economics) at the School of Accounting, Finance, and Economics (SAFE), University of Waikato, Hamilton Campus. Before this, he worked as a Professor of Economics at the Department of Business Administration at Sukkur IBA University. Dr. Akram has contributed to various academic (and administrative) positions for the last ten years. Overall, in his 15-year career, he was the head of two bachelor's-level joint-degree programs and a coordinator for the MS-PhD program. Most prominently, he introduced and led the innovative education program Agribusiness (2 + 2), a real example of institutional collaboration, industry-academia knowledge sharing, and experiential learning for students. He also remained vigilant in institutional-level entrepreneurship activities. The main activities included aiding multiple companies in Sukkur, Pakistan, to establish and expand their agribusinesses and organizing sessions on social entrepreneurship in collaboration with Oklahoma State University, USA. Related to his research activities, he published his scholarly work in well-reputed national and international journals, mainly in financial management, economic productivity, and resource sustainability. Currently, he is working on food security, value chains, and food away from home.

Dr. Zakir Hussain is an experienced academician and administrator, working as an Adjunct Professor at the Metropolitan College of New York, USA. He has a distinguished academic career in Pakistan, where he served as Dean of the Faculty of Agricultural Economics and Rural Sociology at the University of Agriculture Faisalabad, as Head of the Economics Department and

Dean of the Faculty of Agriculture at the University of Sargodha, and then as Vice Chancellor of Government College (GC) University, Faisalabad. He was also the founding member of Government College (GC) Women University Faisalabad, where he played a key role in setting up the first public sector women's university in Punjab. In a short span of four years, he established a new campus and introduced new faculties that provided the foundation for academic excellence. He emphasized the role of curriculum development and assessment in raising the quality of higher education. Recently, he also served as the Vice-Chancellor of the Grand Asian University, Sialkot. Dr. Hussain obtained his Ph.D. in Economics from Colorado State University, USA, and has published several research papers on various topics related to economic development, poverty alleviation, human capital formation, and social welfare.

Mr. Sultan Adeel is currently working as a Research Assistant at Sukkur IBA University under the Sindh-HEC-funded project covering the value chain analysis of date palms in Sindh. He received his BBA-Agribusiness degree from Sukkur IBA University in 2018 and is currently enrolled in the MS-Management Sciences at the same institute. His master's thesis pivots on the date palm consumption behavior of consumers in Pakistan, with a special focus on marketing agricultural commodities. Along with this, he has given special attention to value chain analysis, food security, climatic changes, and waste management and co-authored scholarly work on various topics related to agribusiness, food security, sustainable agriculture, rural development, and e-waste management.

Chapter 4
The Water-Energy-Food Nexus in Kenya: Climate Change Impacts and Adaptation Strategies—A Review

Willis Awandu, Edwin Kimutai Kanda, and Susan Namaemba Kimokoti

Abstract The water-energy-food (WEF) nexus is a novel concept, which aims at integrating three key drivers of development and human security. Effective management of these three resources requires careful assessment of synergies, conflicts and trade-offs which are inherent in the nexus. In Kenya, the achievement of sustainable development goals relies on the management of these key resources. Climate variability and climate change bring uncertainty to water, energy and food situations in the country. Over 75% of agricultural activities in the country are rainfed and thus risky in the face of the poor temporal and spatial distribution of rainfall. Energy reliability is low in Kenya due to the dependence on hydropower sources which are prone to climate risks. This review highlights the importance of the WEF nexus in the face of climate change impacts. Climate adaptation mechanisms for building resilience in cropping systems, water service and energy provision are key elements for the improvement of livelihoods. It is imperative to address sustainably and holistically the three key sectors through policy, legal and institutional frameworks and initiatives. Assessment models and tools are developed to monitor the attainment of targets under WEF.

Keywords Climate resilience · Energy security · Food security · Livelihoods · Water security

W. Awandu (✉) · E. K. Kanda
Department of Civil and Structural Engineering, Masinde Muliro University of Science and Technology, Kakamega, Kenya
e-mail: willisawandu@mmust.ac.ke

E. K. Kanda
e-mail: ekanda@mmust.ac.ke

S. N. Kimokoti
Department of Peace and Conflict Studies, Masinde Muliro University of Science and Technology, Kakamega, Kenya
e-mail: skimokoti@mmust.ac.ke

M. Behnassi et al. (eds.), *The Water, Climate, and Food Nexus*,
https://doi.org/10.1007/978-3-031-50962-9_4

1 Introduction

The concept of the water-energy-food (WEF) nexus has been of great interest and has attracted the attention of many professional players both in the academic research and policy formulation and development sectors. The concept is aiming at enhancing the integration of these three key economic drivers in the perspective of sustainability, development and human security (Finley and Seiber 2014; Mahlknecht et al. 2020; Rasul 2014; Ringler et al. 2013, 2016). The WEF nexus concept holds the idea that the production and consumption trend of water, energy and food resources are all interlinked and, therefore, affected at the same time. Although the WEF nexus relationship has existed in the past, its initial conceptualization was realized during the Bonn 2011 Nexus conference (Hoff 2011). It is evident that in the nexus approach, water, energy and food are interrelated and connected: impacts in one sector may affect the overall performance in other related sectors. Thus, there is an urgent need to integrate water, energy and food in governance and management in the prescribed framework (Bellfield 2015; Wichelns 2017).

The WEF nexus entails the identification of the interactive nature and of inter-linkages, synergies and trade-offs between water, energy and food components that were in the past considered to have existed separately. For example, water is vital for livelihood such as drinking, irrigation and aquatic life sustenance. Similarly, water is used for energy production such as in hydropower generation, biofuel production, cooling of nuclear and geothermal plants, extraction of fossil fuels, shale gas development, and mining processes. The energy consumption component includes pumping water for food production and irrigated agriculture, desalinization, water purification, water distribution and wastewater treatment (Bizikova et al. 2013). The global statistics show that: food production accounts for over 70% of water use (FAO 2017) and primary energy production; power generation accounts for merely 10% of the water extraction (IEA 2008); the food production and the supply chain account for 30% energy use (FAO 2011); and withdrawal, transportation and sewerage treatment account for barely 8% of energy use (UNU 2014).

A good understanding of the interactive nature between WEF resources in their temporal and spatial scale can enhance resource security and facilitate the inter-sectoral and holistic approaches in decision making that will eventually lead towards the sustainability requirements of the nations. A deeper look into Sustainable Development Goals (SDGs), which are made up of climate action, economic equality, innovation, sustainable consumption, peace and justice etc., cannot be sustainably achieved without ensuring proper management of these natural resources (Bizikova et al. 2013). The concerns addressed in SDGs vastly affect the majority of developing countries since they have a myriad of problems such as national priorities, including a lack of finances, poor data quality, inadequate statistical capacities, unclear indicators and targets (Khalid et al. 2021). Besides, the infrastructural resilience of developing countries is not at par with the Global North and thus partnerships among these nations are desirable.

Kenya, like any other country, aims to address some of the most pressing SDG priorities as indicated in her Vision 2030 flagship document (GOK 2007), including the eradication of extreme poverty, reduction of inequality, promotion of economic growth and development of sustainable cities, while considering climate change, good governance and population growth pressures. The interest has grown globally in the WEF nexus arising from different sectors. Recent literature has identified major challenges and future objectives of the nexus (AbdelHady et al. 2017; Bijl et al. 2018; Garcia and You 2016; Heard et al. 2017; White et al. 2018). Watersheds and water conservation methods have been strongly emphasized by some researchers in modern agriculture with some scholars advocating for a focus on the applicability and operationalization of the WEF nexus based on the mathematical models and optimization.

There are very few publications and projects related to the WEF nexus in Kenya in comparison with other regions like Europe, USA, Asia, and Australia. The existing studies have addressed partial interlinkages in the WEF nexus. There is a progress achieved based on social and environmental issues, with a broad focus on the role played by agriculture in the economy of Kenya. Therefore, this review sought to analyse the WEF nexus in Kenya with relation to climate change and adaptation.

The information was sourced from online databases such as Scopus, Web of Science, Google Scholar, Directory of Open Access Journals (DOAJ), and databases of organizations such as Food and Agriculture Organization (FAO) and the World Bank for peer-reviewed articles, books, and grey literature. The search strategy was executed by means of Boolean operators 'AND and OR' using a string of words and phrases such as water, food and energy, food security, water scarcity, water-food-energy nexus, climate impact, climate adaptation strategies, and Kenya. The literature was screened using these stages: title, abstract and keywords and the full text.

The rest of this document is structured as follows: firstly, the relationship between water and food, water and energy, and energy and food are highlighted; secondly, synergies and trade-offs existing in the WEF nexus are described; thirdly, the impacts of climate change in the water, energy and food sectors are analysed; fourthly, the climate change adaptation strategies are highlighted; and finally, conclusions and recommendations are provided.

2 Water and Food, Water and Energy, and Food and Energy Nexuses

Water availability of the right quality and quantity is critical in agriculture, and consequently impacting on food security. Water scarcity is directly proportional to food insecurity, especially in sub-Saharan Africa where agriculture is the backbone of the economy. Globally, irrigated agriculture accounts for about 70% of water withdrawals (UNESCO 2014).

Power generation is a water-intensive activity in terms of both withdrawals (water removed from a source) and consumption (the volume withdrawn and not returned to the source due to evaporation or transport) (Falchetta et al. 2019). Globally, about 90% of energy generation is water-intensive (UNESCO 2014). Cooling of power plants requires a substantial amount of water. About 43% and 50% of water in Europe and America respectively are used for cooling in power plants (UNESCO 2014). Hydropower generation is directly linked to water availability and water contributes indirectly to the biomass energy. Kenya's electricity generation in 2017 consisted of 44% geothermal, 33% hydropower, and 21% diesel-powered thermal sources (Boulle 2019). Water production requires energy, and thus energy accessibility and reliability are important. Water pumping from low-lying rivers and streams and groundwater from wells and boreholes requires reliable sources of energy.

Food production accounts for 30% of energy use in the world (FAO 2011). Energy inputs are required throughout the food production chain: farm, storage, processing, handling, distribution and preparation for consumption. Biomass energy is a classic example of a direct link between agriculture (food) and energy production (Tashtoush et al. 2019). The energy mix in Kenya comprises 68% biomass, 22% petroleum, and 9% electricity (Sarkodie and Adom 2018). Charcoal and firewood are major sources of energy in most rural communities in Kenya.

3 Synergies and Trade-Offs in the Water-Energy-Food Nexus

Natural resources, in this case food, water, and energy, do not operate in isolation but are interrelated where one influences the other. Countries aim to achieve sustainable development through food, energy and water security with sustainable resource use being the core (Fig. 1). However, the relationships among food, energy and water are dynamic where an action in one sector may affect the other sector, which results in social, economic and environmental implications (Rasul and Sharma 2016). Provision of water and energy services and ensuring sufficient food for the population are often met with competing forces, and thus conflicts are inherent.

Understanding trade-offs and synergies or complementarities in the WEF nexus can provide new insights for developing effective adaptation (Rasul and Sharma 2016). It is imperative to minimize the trade-offs that exist in the WEF nexus in order to provide a framework for action to reduce the negative consequences, and maximize the synergies, which provide opportunities for adaptation in the nexus (Table 1). For example, the construction of a multi-purpose dam for hydropower generation increases water supply for domestic, industrial and agricultural use. Water for irrigation from the dam is useful for expanding crop production, and thus enhancing food security. The operation of the reservoir should, however, be carried out sustainably to avoid negative effects on the ecosystem, especially the downstream communities.

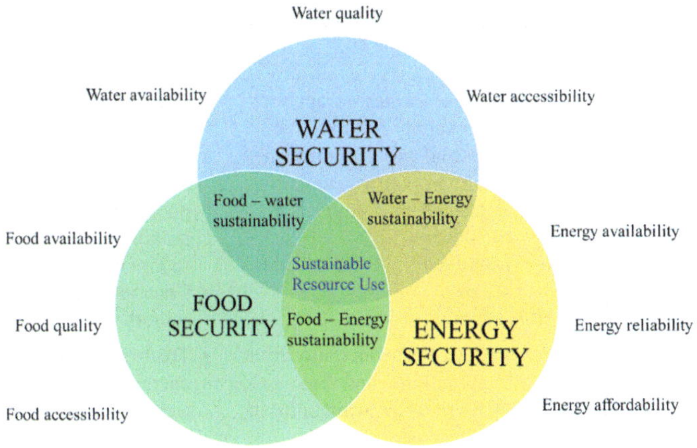

Fig. 1 Elements of the WEF nexus. *Source* Developed by the authors

Sustainable agriculture, based on best management practices, will reduce the overuse of the water resource and the negative impact on its quality.

Water used for the production of biomass energy (in this case biofuels) produces trade-offs in terms of increased water allocation to agriculture, and thus reducing the allocation to other sectors. However, proper planning of biofuel crops, e.g., using rainfed conditions coupled with appropriate variety selection, will help reduce the pressure on existing water sources.

4 Impacts of Climate Change on WEF Nexus in Kenya

Kenya's demographic and economic trends are some of the demand-side drivers affecting the WEF security nexus. The ever-growing population is estimated to reach 65 million by 2030 and 96 million by 2050 (UNDESA 2015) with the majority of the population dwelling in the urban centres. This has resulted in a high rate of urbanization, which in effect has led to the development of slums and increased pressure on limited available WEF resources (Wakeford 2017).

The main supply side of the WEF nexus is rainfall, which is also a major key driver affecting the security and stability of nexus. Rainfall variability is in itself affected by both climate change and El Nino Southern Oscillation. The availability and quality of water resources impact food production, energy generation, water for drinking, agriculture and other development activities.

Kenya's average precipitation is estimated to be 630 mm per annum (FAO 2016) with uneven distribution across the country. A vast part of Kenya is classified as arid or semi-arid land (ASAL) with greater rainfall variation every year (Mekonnen and Hoekstra 2014). Kenya is classified as a water-scarce country with less than 500 m^3 per capita per year, the severest category of water stress (WWAP 2014). Due to

Table 1 Examples of synergies and trade-offs in the WEF nexus

Development	Synergies	Trade-offs
Multi-purpose reservoir	• Increased water storage for water supply for domestic, industrial and agricultural use, and hydropower generation • Enhanced crop production for biomass energy • Improved food and nutritional security	• Decreased water for downstream users • Agricultural intensification leads to overuse of water and poor water quality through pollution from fertilizers and sediments • Chances of enhanced soil salinity
Recycling and re-use of wastewater programmes	• Reduce freshwater allocation to agriculture, and thus increase water availability in other sectors • Re-used wastewater minimizes fertilizer use, thus indirectly less energy input in the food production cycle	• Treatment techniques use energy depending on the technology used • Uncertainty on food safety • Wastewater re-use increases soil salinity (Gao et al. 2021)
Desalinization	• Improved water availability for domestic and industrial (food processing) use	• Desalinization techniques use a substantial amount of energy
Coal power plant	• Diversification of energy sources, hence reducing climate vulnerability • The use of treated wastewater for cooling reduces pressure on freshwater resources available to other sectors	• Increased air pollution and emission of GHGs • Coal power plant uses a lot of water for cooling, thus threatening water availability to other key sectors

regional heterogeneous rainfall patterns, climate change tends to reduce or increase rainfall availability. After the 1998 El Nino rainfall, the rainfall pattern in Kenya has greatly changed. The northern parts of Kenya, which had previously been known to experience low rainfalls, have been in recent times experiencing extreme rainfall events leading to unexpected flooding, which has resulted in human and animal casualties. The Rift Valley lakes have continuously risen in water levels and flood events experienced with the latest scenario leading to the submergence of the Lake Nakuru National Parks facilities and the surrounding areas (GOK and UNDP 2021).

During certain periods of the year, climate change has resulted in increasing temperatures, thereby decreasing precipitation and increased evapotranspiration (Anderson et al. 2008), thereby affecting the soil moisture and increasing irrigation water demand (Wang et al. 2016). The food production sector heavily depends on water resources, energy inputs for production, processing, storage and food preparation. With climate change, Kenya's agriculturally productive regions are susceptible to drought caused by ENSO (Mogaka 2006). The higher temperatures have led to increased demand for cooling systems in the houses, thereby increasing the electricity demand through frequent and extensive usage of air-conditioning.

Climate change adaptation is best achieved through holistic approaches under the WEF nexus instead of the silo (sectoral) approach, which is often practiced in Kenya. This is because adaptation requires addressing the full spectrum of challenges, including the underlying causes of vulnerability, managing climate risks, and building response capacity (Rasul and Sharma 2016). The WEF nexus approach shows the interdependence of systems in the critical areas/sectors of food, water and energy which are vulnerable to climate change impacts. Therefore, adaptation from the nexus perspective helps in identifying and recognising the relationships and interdependencies among the WEF areas. For example, in Kenya, it is critical to understand that adaptation and resilience strategies can be addressed by developing a working framework/mechanism that incorporates the key ministries of energy, agriculture, water and environment.

6 Conclusion and Recommendations

The understanding of relationships existing within the WEF nexus is important for planners and policymakers since there exists competing interests/demands, which are inherently evident in these critical sectors. These trade-offs are compounded by climate change, and thus adaptation mechanisms and strategies are required for Kenya to achieve SDGs. The current practice of treating each sector in the nexus in isolation is not sustainable since the WEF nexus sectors are interlinked and the action in one sector will undermine the others. The main adaptation mechanisms include policy changes to incorporate re-use of wastewater, prudent agricultural water management strategies, and techniques such as the adoption of water-saving irrigation technologies, conservation agriculture, and CSA practices, among others.

It is imperative to address sustainably and holistically the three key sectors through policy, legal and institutional frameworks and initiatives. Sectoral coordination is required in the development of WEF indicators and targets. Therefore, assessment models and tools should be developed to monitor the achievement of the WEF targets.

Nutritional security is often hidden in food security, and thus future studies in the WEF area need to incorporate nutritional security. The latter is related to health aspects, which imply that the nexus can expand to include the water-energy-food-health components. The WEF at a regional level, such as the East-African region, need to be explored in future since some of the components in WEF sectors are trans-boundary like rivers, and thus country-specific projects affect the regional countries in meeting their WEF targets.

References

AbdelHady RS, Fahmy HS, Pacini N (2017) Valuing of Wadi El-Rayan ecosystem through water–food–energy nexus approach. Ecohydrol Hydrobiol 17(4):247–253

Anderson J, Chung F, Anderson M, Brekke L, Easton D, Ejeta M, Peterson R, Snyder R (2008) Progress on incorporating climate change into management of California's water resources. Clim Change 87(1):91–108

Bellfield H (2015) Water, energy and food security nexus in Latin America and the Caribbean. Global Canopy Programme

Bijl DL, Bogaart PW, Dekker SC, van Vuuren DP (2018) Unpacking the nexus: different spatial scales for water, food and energy. Glob Environ Chang 48:22–31

Bizikova L, Roy D, Swanson D, Venema HD, McCandless M (2013) The water-energy-food security nexus: towards a practical planning and decision-support framework for landscape investment and risk management. The International Institute for Sustainable Development

Boulle M (2019) The hazy rise of coal in Kenya: the actors, interests, and discursive contradictions shaping Kenya's electricity future. Energy Res Soc Sci 56:101205

Chivenge P, Mabhaudhi T, Modi AT, Mafongoya P (2015) The potential role of neglected and underutilised crop species as future crops under water scarce conditions in sub-Saharan Africa. Int J Environ Res Public Health 12:5685–5711

Falchetta G, Gernaat DE, Hunt J, Sterl S (2019) Hydropower dependency and climate change in sub-Saharan Africa: a nexus framework and evidence-based review. J Clean Prod 231:1399–1417

FAO (2011) Energy-smart food for people and climate. Food and Agriculture Organization, Rome, Italy

FAO (2016) AQUASTAT main database. Food and Agriculture Organization, Rome

FAO (2017) The future of food and agriculture–trends and challenges. Annual report. Food and Agriculture Organization, Rome

Finley JW, Seiber JN (2014) The nexus of food, energy, and water. J Agric Food Chem 62(27):6255–6262

Gao Y, Shao G, Wu S, Xiaojun W, Lu J, Cui J (2021) Changes in soil salinity under treated wastewater irrigation: a meta-analysis. 255:106986

Garcia DJ, You F (2016) The water-energy-food nexus and process systems engineering: a new focus. Comput Chem Eng 91:49–67

GCF (2019) Water, energy and food Nexus to address climate change impacts in Central Tunisa. Green climate fund

GOK (2007) Kenya vision 2030. Government of the Republic of Kenya, Nairobi

GOK (2010) National climate change response strategy. Government of Kenya, Nairobi

GOK (2016a) The climate change act, 2016. Government of Kenya, Nairobi

GOK (2016b) Kenya national adaptation plan: 2015–2030. Government of Kenya, Nairobi

GOK (2016c) National climate change framework policy. Government of Kenya, Nairobi

GOK (2016d) The water act. Government Printer, Nairobi

GOK (2017) Kenya climate smart agriculture strategy-2017–2026. Ministry of Agriculture, Livestock and Fisheries, Nairobi

GOK and UNDP (2021) Rising water levels in Kenya's Rift valley lakes, Turkwel Gorge dam and Lake Victoria: a scoping report. In: Resources MoEaN (ed) Governmnent of Kenya and United Nations Development Programme Nairobi

Heard BR, Miller SA, Liang S, Xu M (2017) Emerging challenges and opportunities for the food–energy–water nexus in urban systems. Curr Opin Chem Eng 17:48–53

Hoff H (2011) Understanding the Nexus. Background paper for the Bonn 2011 Nexus conference: the water, energy and food security Nexus. Stockholm Environment Institute, Stockholm

IEA (2008) World energy outlook. International Energy Agency, Paris

Khalid AM, Sharma S, Dubey AK (2021) Concerns of developing countries and the sustainable development goals: case for India. Int J Sust Dev World 28(4):303–315

Kibria G, Haroon AY, Nugegoda D (2016) Climate change and water security: impacts, future scenarios, adaptations and mitigations. New India Publishing Agency, New Delhi

Mahlknecht J, González-Bravo R, Loge FJ (2020) Water-energy-food security: a Nexus perspective of the current situation in Latin America and the Caribbean. Energy 194:116824

Mekonnen MM, Hoekstra AY (2014) Water conservation through trade: the case of Kenya. Water Int 39(4):451–468
Mogaka H (2006) Climate variability and water resources degradation in Kenya: improving water resources development and management. World Bank Washington D.C.
Mpandeli S, Nhamo L, Naidoo D, Mabhaudhi T, Nhemachena C, Nhamo L, Liphadzi S, Hlahla S, Modi AT (2018) Climate change adaptation through the water-energy-food Nexus in Southern Africa. Int J Environ Res Public Health 15(10):1–19
Mwendwa P, Giliba RA (2012) Climate change impacts and adaptation strategies in Kenya. Chin J Popul Resour Environ 10(4):22–29
Niang I, Ruppel OC, Abdrabo MA, Essel A, Lennard C, Padgham J, Urquhart P (2014) Impacts, adaptation, and vulnerability. Part B: regional aspects. Contribution of working group II to the fifth assessment report of the intergovernmental panel on climate change. Cambridge University Press, Cambridge, United Kingdom
Pandey S, Bhandari H, Hardy B (2007) Economic costs of drought and rice farmers' coping mechanisms: a cross-country comparative analysis. International Rice Research Institute, New Delhi
Rasul G (2014) Food, water, and energy security in South Asia: a nexus perspective from the Hindu Kush Himalayan region. Environ Sci Policy 39:35–48
Rasul G, Sharma B (2016) The nexus approach to water–energy–food security: an option for adaptation to climate change. Climate Policy 16(6):682–702
Ringler C, Bhaduri A, Lawford R (2013) The nexus across water, energy, land and food (WELF): potential for improved resource use efficiency? Curr Opin Environ Sustain 5(6):617–624
Ringler C, Willenbockel D, Perez N, Rosegrant M, Zhu T, Matthews N (2016) Global linkages among energy, food and water: an economic assessment. J Environ Stud Sci 6(1):161–171
Sarkodie SA, Adom PK (2018) Determinants of energy consumption in Kenya: a NIPALS approach. Energy 159:696–705
Tashtoush F, Al-Zubari W, Shah A (2019) A review of the water–energy–food nexus measurement and management approach. Int J Energy Water Resour 3(4):361–374
UNDESA (2015) World population prospects: the 2015 revision, key findings and advance tables. United Nations Department of Economic and Social Affairs, New York
UNESCO (2014) The United Nations world water development report. United Nations Educational, Scientific and Cultural Organization, Paris
UNU (2014) World water day 2014: water and energy. United Nations University, Hamilton, United States
Wakeford JJ (2017) The water–energy–food Nexus in a climate vulnerable, frontier economy: the case of Kenya. Quantum Global Research Lab
Wang X-J, Zhang J-Y, Shahid S, Guan E-H, Wu Y-X, Gao J, He R-M (2016) Adaptation to climate change impacts on water demand. Miti Adapt Strat Global Change 21(1):81–99
White DJ, Hubacek K, Feng K, Sun L, Meng B (2018) The water-energy-food Nexus in East Asia: a tele-connected value chain analysis using inter-regional input-output analysis. Appl Energy 210:550–567
Wichelns D (2017) The water-energy-food nexus: is the increasing attention warranted, from either a research or policy perspective? Environ Sci Policy 69:113–123
WRI (2020) Mainstreaming CCA in Kenya: lessons from Makueni and Wajir Counties. Water Resource Institute
WWAP (2014) The United Nations world water development report 2014: water and energy. UNESCO, Paris

Willis Awandu is currently undertaking his Ph.D. studies at the Technical University of Darmstadt, Germany with a focus on Hydrokinetic River Turbines development using local materials for electricity production and use in off-grid rural areas. He holds an MSc. Water Resources Engineering and Management from the University of Stuttgart, and a BSc. Civil Engineering from Jomo Kenyatta University of Science and Technology in Kenya. His research interest is in Water and Sanitation (WASH), Urban Water Management, Hydraulic Structures, Rive Engineering and Environmental Engineering, with an extension to Water-Energy-Food nexus and associated interactions.

Edwin Kimutai Kanda holds a Doctorate degree in Philosophy in Agricultural Engineering specializing in Irrigation Engineering from University of KwaZulu-Natal, South Africa. Dr. Kanda also holds MSc degrees in Water Engineering and Project Management, and a Bachelor of Technology in Civil and Structural Engineering. His research interests include Water and Environmental engineering, Irrigation Engineering, Hydrology, Water, Sanitation and Hygiene (WASH), and Climate Change. Dr. Kanda's research focuses also on the energy-food and water nexus. He is also interested in emerging contaminants in river basins in Kenya with special reference to antibiotic pollution.

Susan Namaemba Kimokoti holds a Doctorate degree in Diplomacy and International Relations. Her areas of academic and research interest are in the fields of diplomacy, peace studies, resource-use conflict management, disaster management and International Relations. She is also keen on the sustainable resource use and amicable exploitation of the natural resources.

Chapter 5
The Impact of Climate Change on Groundwater Resources in Northwestern Morocco

Mohammed Bahir

Abstract Climate change poses a global challenge, intricately linked to the intricate, long-term interplay among environmental factors and the dynamics of economics, society, technology, and politics. These interactions lead to notable repercussions at the regional level. This investigation utilizes data from nine campaigns conducted in 1990, 1995, 1997, 2009, 2015, 2016, 2017, 2018, and 2019 to evaluate groundwater quality in the Essaouira region within the context of a changing climate. The hydrogeochemical analysis reveals that the Cenomanian–Turonian aquifer's groundwater exhibits mix facies such as Cl–Ca–Mg, Cl–Ca, Cl–Na, and HCO_3–Ca, with the Cl–Ca–Mg mix facies dominating along with Cl–Ca. Examination of correlations between major element concentrations indicates that groundwater mineralization is influenced by the dissolution of evaporitic minerals (halite, gypsum, and/or anhydrites) and carbonates (dolomite). Additionally, reverse ion exchange and marine intrusion, particularly in the Plio-Quaternary aquifer, contribute to the overall mineralization. The spatio-temporal analysis of groundwater quality in the study area demonstrates a gradual deterioration over time and space. Notably, the Essaouira basin emerges as particularly susceptible to climate change due to its reliance solely on meteoric waters for recharge.

Keywords Climate change · Coastal aquifers · Groundwater quality · Hydrogeochemical characteristics · Essaouira

1 Introduction

Climate change is a global challenge that entails intricate, long-term interactions among environmental factors and the dynamics of economics, society, technology, and politics. These interactions result in significant effects at the regional level, as

M. Bahir (✉)
High Energy and Astrophysics Laboratory, Faculty of Sciences Semlalia, Cadi Ayyad University, Marrakesh, Morocco
e-mail: bahir@uca.ac.ma

M. Behnassi et al. (eds.), *The Water, Climate, and Food Nexus*,
https://doi.org/10.1007/978-3-031-50962-9_5

documented by Lebel and Vischel (2005), Alpert et al. (2008), and Misra (2014), particularly in the Middle East and North Africa (MENA) region, as highlighted by Bahir et al. (2000, 2008, 2013a, b, 2016, 2021a, b), Bahir and Mennani (2002), El Mountassir 2023, El-Kharraz et al. (2012), and Ouhamdouch et al. (2018, 2020), among others.

In arid and semi-arid regions, rainfall stands out as a pivotal factor in climatic characterization. Examining recent climate evolution is crucial for identifying optimal solutions to problems arising from the interplay between water requirements and availability. This, in turn, facilitates the improved management of water resources, as emphasized by Bahir et al. (2018, 2019, 2020, 2021c, d), Bahir and Ouhamdouch (2020), Carreira et al. (2018), and El Mountassir et al. (2021a, b, 2022a, b), along with the insights from Ragab and Prudhomme (2002).

Research on climate change indicates that global warming in the MENA region exceeds the global average. While the global increase is estimated at 0.74 °C in the twentieth century, MENA countries experience fluctuations between 1 and 2 °C, as highlighted by Solomon et al. (2007), Green et al. (2011), and Ouhamdouch et al. (2018) in alignment with Ragab and Prudhomme (2002). Furthermore, precipitation has declined in the Mediterranean region, the Sahel, Southern Africa, and specific parts of South Asia across various temporal and spatial scales, as reported by Alpert et al. (2008) and IPCC (2013).

Morocco, akin to other MENA countries, has faced recurrent drought periods, as documented by Bahir et al. (2000, 2001, 2014), Bahir and Mennani (2002), Bahir and Ouhamdouch (2020), Driouech (2010), Babqiqi (2014), and Vicente-Serrano (2006). The country contends with limited water resources, estimated at 20 billion cubic meters, translating to an average of 700 m^3/year/inhabitant, indicative of a situation characterized by fairly high water stress (Babqiqi 2014). The prevalence of years with rainfall deficits exceeds those with ample rainfall, particularly during the cycles of 1980–1985, 1990–1995, and 2007–2010, as observed by Driouech (2010), Stour and Agoumi (2008), and Sinan et al. (2009). Babqiqi (2014) notes that a comparison of average annual temperatures between the periods 1971–1980 and 1998–2007 reveals an increasing trend (ranging from 0.3 to 2.5 °C, depending on the region). This reduction in precipitation and temperature increase, attributed to climate change, is anticipated to adversely affect water resources, particularly in arid and semi-arid regions (Bougharlou et al. 2014, Chenaker et al. 2017, El Mountassir et Bahir 2023).

Although limited studies have addressed the impact of climate change on water resources in these areas, they have primarily focused on surface water (Abutaleb et al. 2018; Hallouz et al. 2019; Xu et al. 2004). Some studies have assessed the impact of global warming on groundwater, concentrating on piezometric and/or hydrochemical approaches (Al-Maktoumi et al. 2018; Berhail 2019; Lachaal et al. 2018). In contrast, our study evaluates the influence of climate change on the water resources of the Essaouira basin in Morocco by integrating various approaches, including hydroclimatology, piezometry, hydrochemistry, and isotopy.

The study area, designated as the Essaouira syncline basin, encompasses 6000 km^2 and is situated within the Atlantic Atlas, the westernmost section of the southwestern

Moroccan basin (Dresh 1962; Duffaud 1960). The basin is bordered to the north by the Hadid anticline, to the south by Tidzi wadi, to the east by the Bouabout region, and to the west by the Atlantic Ocean. It is subdivided into two sections: the "Bouabout unit" (upstream part) and the "coastal zone" (downstream part) (Fig. 1).

From a morphological perspective, the study area comprises a series of synclinal basins filled with formations ranging from the Triassic to the Quaternary (see Fig. 2). Elevations vary between 400 and 1600 m in the upstream section and are less than 400 m in the downstream part (Bahir et al. 2012). Hydrographically, the study area is characterized by a less-developed network, primarily represented by the Ouazzi

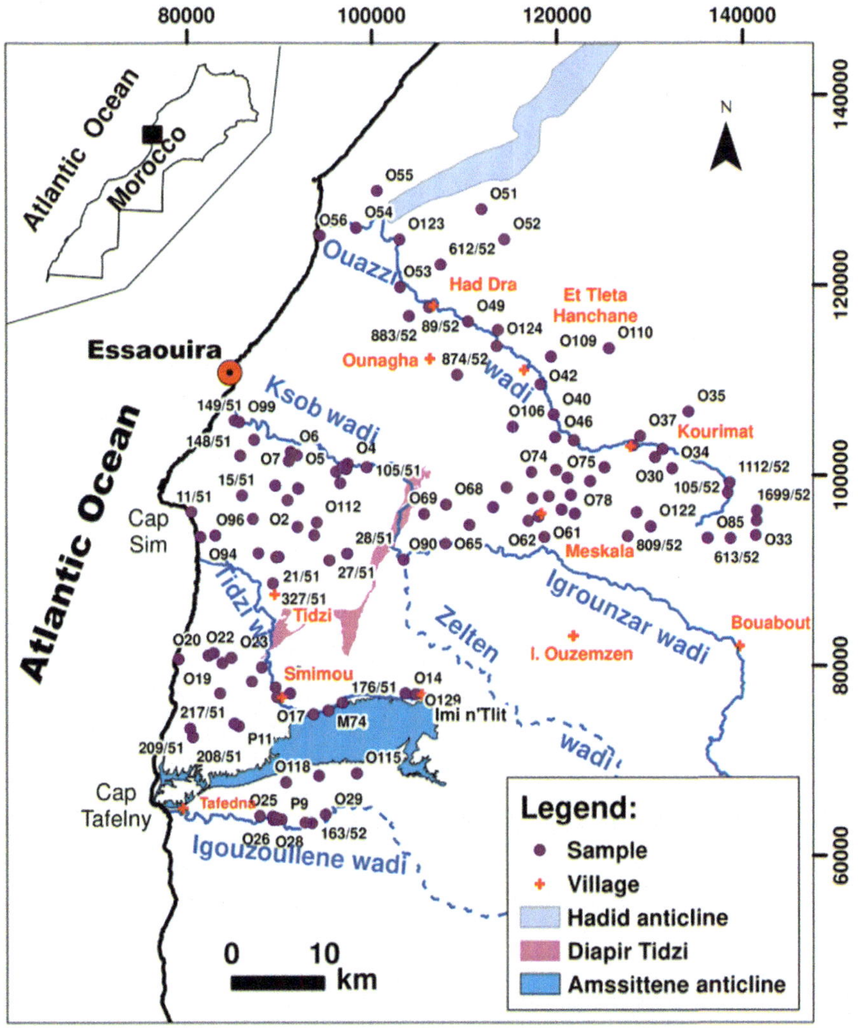

Fig. 1 Location of study area. *Source* Developed by the author

wadi in the north and the Igouzoullene wadi in the south. This network also includes the Ksob wadi, formed by the confluence of the Igrounzar and Zelten wadis, and the Tidzi wadi. All these wadis ultimately flow into the Atlantic Ocean (refer to Fig. 2).

From a geological and hydrogeological standpoint, the upper section of the study area is characterized by the exposure of Middle and Upper Cretaceous formations, specifically Albian-Vraconian, Cenomanian, and Turonian formations, as documented by Duffaud (1960) and Amghar (1989) (refer to Fig. 3). These formations consist of limestone and dolomitic strata interspersed with marl and sandstone.

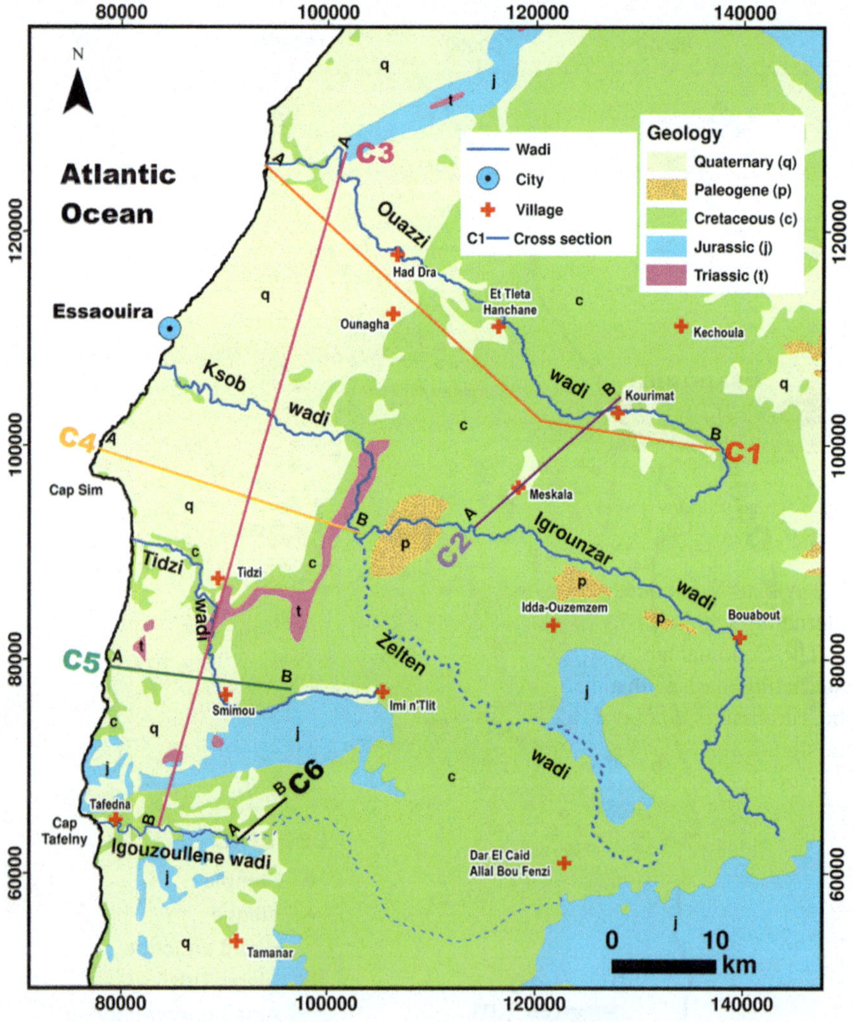

Fig. 2 Geological map of the study area and cross section's location. *Source* Developed by the author

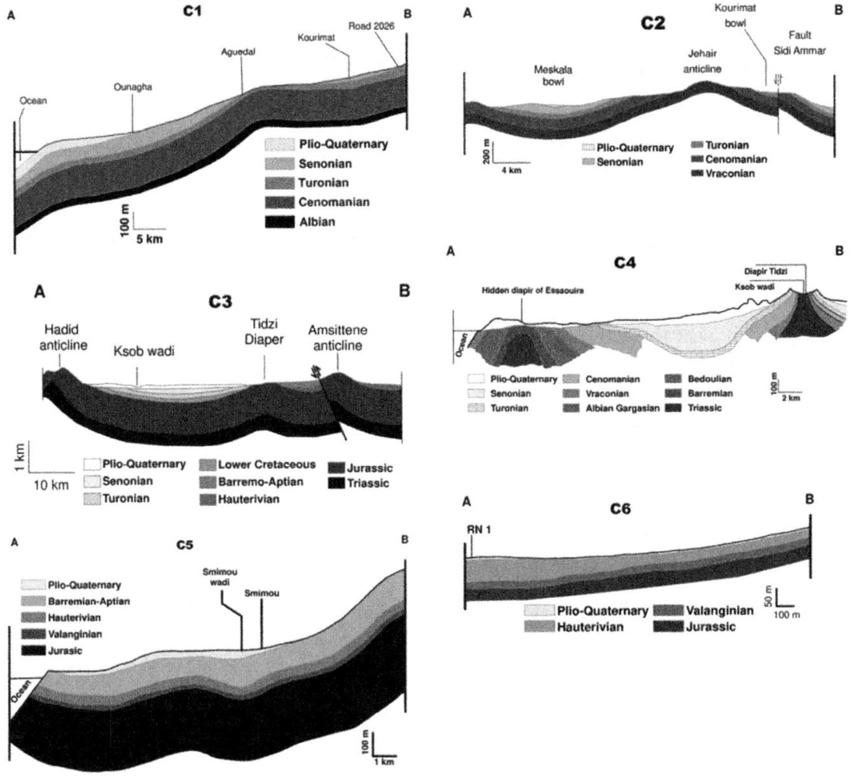

Fig. 3 Cross sections (cf Fig. 2). *Source* Developed by the author

The Albian-Vraconian formations encompass sandstone and limestone dolomites, alternating with sandstone banks and sandy clays.

The Cenomanian, with an approximate thickness of 200 m, is characterized by alternating marls with anhydrite, lumachellic, and dolomitic limestones. In contrast, the Turonian is composed of limestones abundant in silica. Within these synclines lie significant water reservoirs, notably the Cenomanian–Turonian aquifer, which remains the most crucial in the region. According to Jalal et al. (2001), this aquifer exhibits transmissivities ranging between 2.2×10^{-4} and 2.7×10^{-1} m^2/s.

The downstream section comprises four aquifers: (i) the Plio-quaternary; (ii) the Turonian in the northern part, between Ksob wadi and Tidzi wadi; (iii) the Barremian-Aptian located between Tidzi wadi and Amssittene anticline; and (iv) the Hauterivian, the southern limit of the study area, positioned between the Amssittene anticline and Igouzoullene wadi (refer to Fig. 2).

The Plio-quaternary is characterized by a matrix of limestone sandstone, housing a significant water table formed within the synclinal structure by the Senonian marls (see Fig. 3). Mennani (2001) reports transmissivities for this water table ranging between 6.1×10^{-2} and 4.5×10^{-5} m^2/s.

Regarding the Turonian, represented by limestones, it contains a captive aquifer beneath the Senonian marls within the synclinal structure, possibly in direct contact with the Plio-Quaternary at the structure's boundaries (refer to Fig. 3). The transmissivity of this aquifer is reported to range between 0.8×10^{-4} and 2.7×10^{-2} m^2/s (Mennani 2001).

The Barremo-Aptian aquifer consists of Barremian formations (about 70 m thick) characterized by alternating gray marl (with traces of gypsum), fractured fossiliferous limestones, and sandstone, as well as Aptian formations (about 100 m thick) composed of red clays and sandstone with intercalations of dolomitic sandstones or bioclastic limestones (Duffaud 1960; Duffaud et al. 1966) (see Fig. 3). The reported transmissivity for this aquifer is approximately 1.5×10^{-3} m^2/s (Mennani 2001).

The Hauterivian aquifer, with a thickness of about 200 m, is composed of marly clays and fractured siliceous limestones, along with marly and dolomitic limestones that are more or less fractured (see Fig. 3) (Duffaud 1960; Duffaud et al. 1966). According to Mennani (2001), transmissivities for this aquifer vary between 1.6×10^{-5} and 6.7×10^{-5} m^2/s.

2 Materials and Methods

In this study, data from nine campaigns conducted in the years 1990, 1995, 1997, 2009, 2015, 2016, 2017, 2018, and 2019 were utilized to evaluate groundwater quality in the Essaouira region within the framework of climate change (Bahir et al. 2012; EL Mountassir et al. 2020, 2021c, d, e, 2022a, b; El Mountassir 2023; El Mountassir et Bahir 2023; Ouhamdouch et al. 2018, 2020). Field measurements, including electrical conductivities, temperatures, pH, and nitrates, were collected using a portable conductivity meter (HI-9829 multiparametric instrument), and water level depths were measured employing a 200 m piezometric probe.

The chemical analyses were conducted at the Laboratory of Hydrogeology at the Faculty of Sciences Semlalia in Marrakech, Morocco, for the campaigns spanning from 1990 to 2009. Subsequently, analyses for the period from 2015 to 2019 were carried out at the Laboratory of Geosciences and Environment-ENS at the Ecole Normale Supérieure in Marrakech, Morocco. The determination of SO_4^{2-} anion content employed the nephelometric method (Rodier et al. 2009), while concentrations of Ca^{2+} and Mg^{2+} cations were measured using the complexometry method (EDTA), and those of Cl^- were determined by the Mohr method (Rodier et al. 2009). The Na^+ and K^+ contents were ascertained through flame photometry (Rodier et al. 2009), and HCO_3^- contents were determined by titration using a sulfuric acid solution. All samples exhibited an ion balance of less than 10%, validating the obtained results. A Geographic Information System (GIS) was employed for the spatial mapping of electrical conductivity and physicochemical elements.

3 Results and Discussion

An hydrogeochemical methodology proves to be an invaluable instrument in delineating the chemical composition of groundwater. This composition is significantly shaped by the properties of the geological formations, the hydrodynamic behaviors of the aquifers, and is further influenced by climatic and exploitation parameters.

3.1 Electrical Conductivity

Electrical conductivity is mainly governed by the concentration of dissolved ions (WHO 2011). It is closely linked to the lithological nature of the soil, the speed and direction of groundwater flow as well as the groundwater residence time. This parameter is proportional to the temperature. The results obtained are shown in Fig. 4.

The electrical conductivity of Cenomano-Turonian waters (upstream part), analyzed in 1995, does not exceed 3 mS/cm, with more than 80% falling into classes below 1.5 mS/cm. In 2007, the values of the EC knew an increase where some samples present values higher than 2.5 mS/cm. Thereafter, the EC values are increasing to reach values of 5 mS/cm in certain places, since 2016.

This increase may be the result of the decrease in precipitation experienced by the study area in recent decades. As for the Plio-Quaternary aquifer, the EC values vary between 1 and 6 mS/cm for 1990, 1995, 2004, and 2009 campaigns. The values become higher and higher towards the west and the south. This can be explained in the south by the influence of the Triassic terrains rich in halite, and the west by the combined action of the hidden diapir of Essaouira as well as that of aerosols and sea spray near the ocean. While the low values are observed in the north-east and east (aquifer recharge zone).

From 2015, we see the appearance of outliers reaching 12 mS/cm in 2019, and this towards the west. EC values measured for the Plio-Quaternary aquifer are greater than those obtained for the Cenomano-Turonian aquifer. This could be due to the effect of the Triassic formations (Tidzi diapir) rich in halite and the leaching of sea spray and aerosols by the rains feeding the aquifer. The outliers measured since 2015, and which increase to reach 12 mS/cm in 2019, are due to the intrusion of seawater.

The Turonian aquifer has known the EC measurement from 2004 to 2019. The obtained results show an oscillation of the values of this parameter around 2 mS/cm with a slight increase exceeding 2.5 mS/cm in 2018 and 2019. This stability of the EC within this aquifer can be explained by its captive nature and by the homogeneity of their facies consisting of limestones and dolomitic limestones. For the Barremian-Aptian aquifer, the majority of the samples have EC values that range over the class 1–2 mS/cm, while the rest fall within the class 2–4 mS/cm. The highest values are observed towards the north-west and west, under the influence of the Triassic salt formations as well as the effect of sea spray and aerosols leached by the rain infiltrating towards the aquifer. Towards the east, the influence of these factors decreases

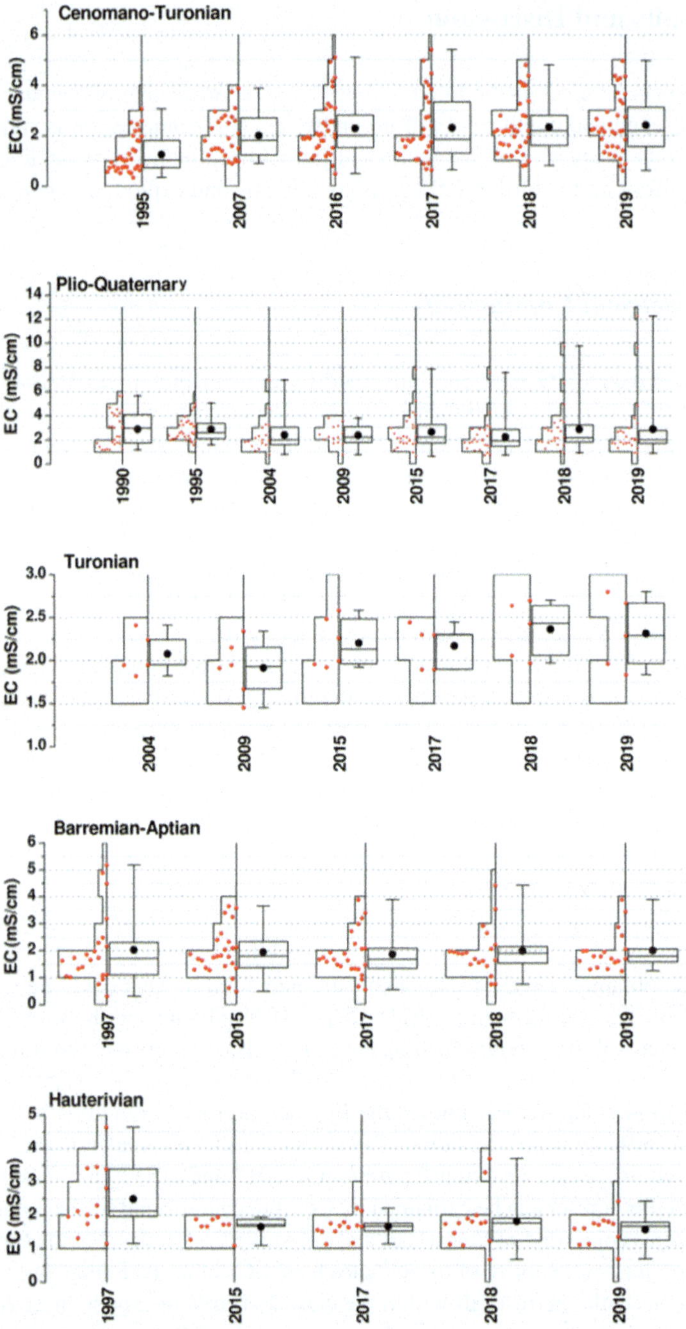

Fig. 4 Temporal evolution of the electrical conductivity of groundwater. *Source* Developed by the author

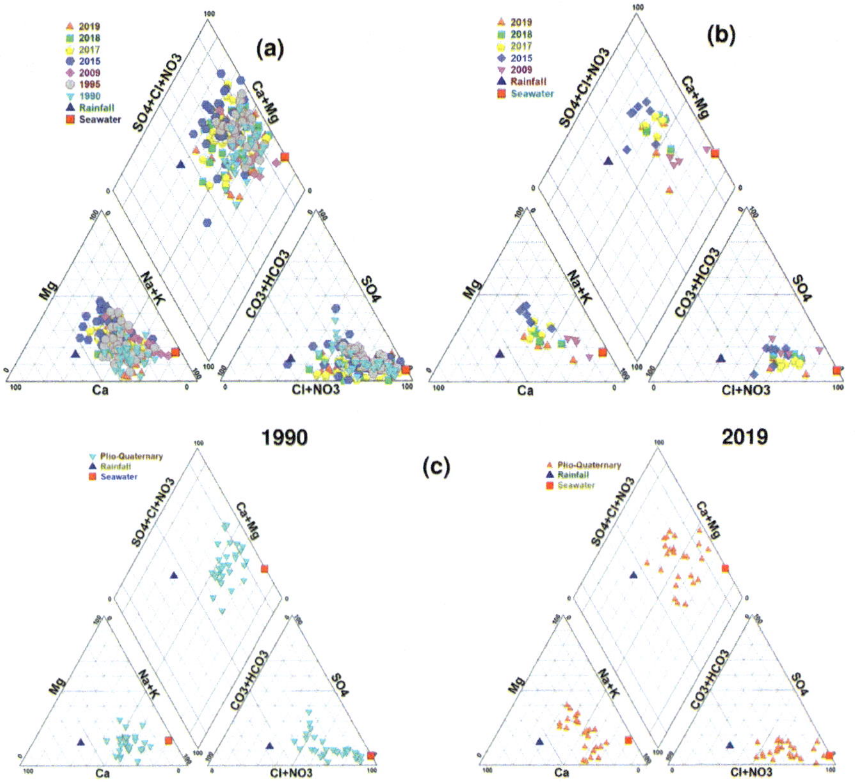

Fig. 6 Piper diagram of analysed samples of **a** Plio-Quaternary from 1990 to 2019 and of **b** Turonian from 2009 to 2019, and **c** comparison between samples of 1990 and 2019 for the Plio-Quaternary aquifer. *Source* Developed by the author

3.3 Groundwater Mineralization

To ascertain the origin and primary processes responsible for groundwater mineralization in the study area, we conducted a comprehensive examination of correlations among the principal major elements.

Chloride, recognized as a conservative ion consistently present in natural waters at highly variable concentrations (Fetter 1993), is commonly associated with sodium. In the groundwater of the upper segment, chloride concentrations range widely from 113 to 1818 mg/l, with an average of 574 mg/l. Sodium concentrations within this segment exhibit variability from 12 to 541 mg/l, with an average of 167 mg/l. The Piper diagram (Fig. 5) highlights the predominance of Cl^- ions in the waters. In the lower segment, chloride contents fluctuate between 120 and 4800 mg/l, averaging 620 mg/l, while sodium concentrations range from 28 to 1950 mg/l, with an average of 261 mg/l. The Plio-Quaternary aquifer displays the highest Na^+ and Cl^- concentrations.

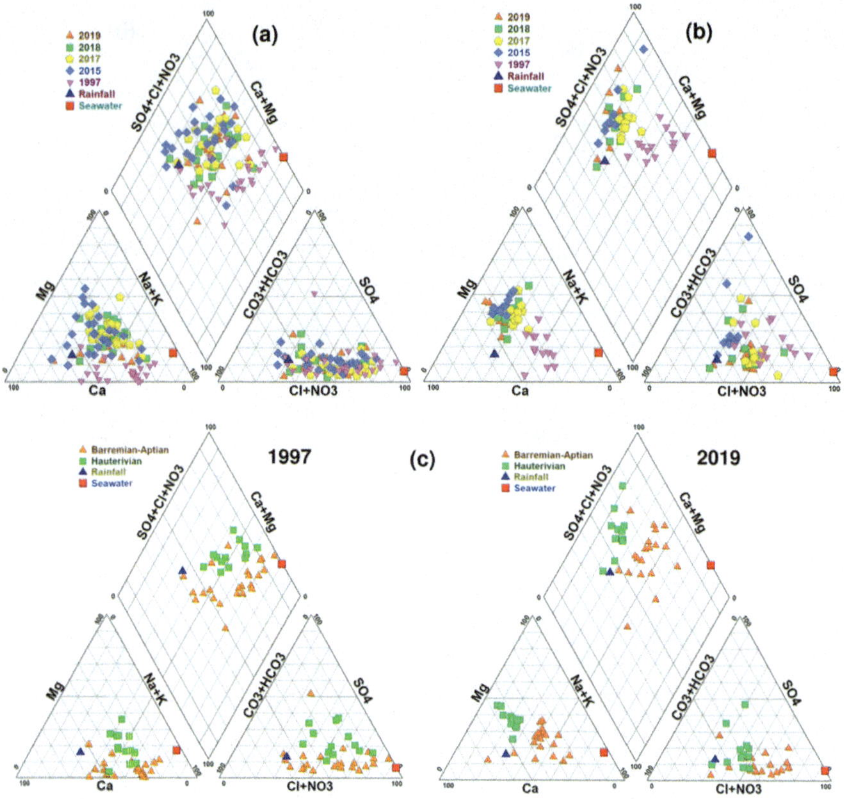

Fig. 7 Piper diagram of analysed samples of **a** Barremian-Aptian and **b** Hauterivian aquifers from 1997 to 2019, and **c** comparison between samples of 1997 and 2019 for the two aquifers. *Source* Developed by the author

The Na^+ versus Cl^- correlation diagram (refer to Fig. 8a) reveals a notable positive correlation between these two ions, suggesting a shared origin. Some data points are scattered along the halite dissolution line (line 1:1), indicative of the contribution of halite to groundwater mineralization in the study area. Conversely, other samples are situated below and parallel to the line 1:1, indicating a Na^+ deficit. This discrepancy implies the involvement of a phenomenon other than halite dissolution in groundwater mineralization.

The observed Na^+ deficit relative to Cl^- may be associated with ion exchange reactions within the aquifer matrix. As depicted in Fig. 8f, Na^+ ions are released from the complex and replaced by Ca^{2+} ions, as per Eq. (1) (Capaccioni et al. 2005):

$$Na^+ + \frac{1}{2}Ca - X_2 \rightarrow Na - X + \frac{1}{2}Ca^{2+} \tag{1}$$

With X being the natural exchanger.

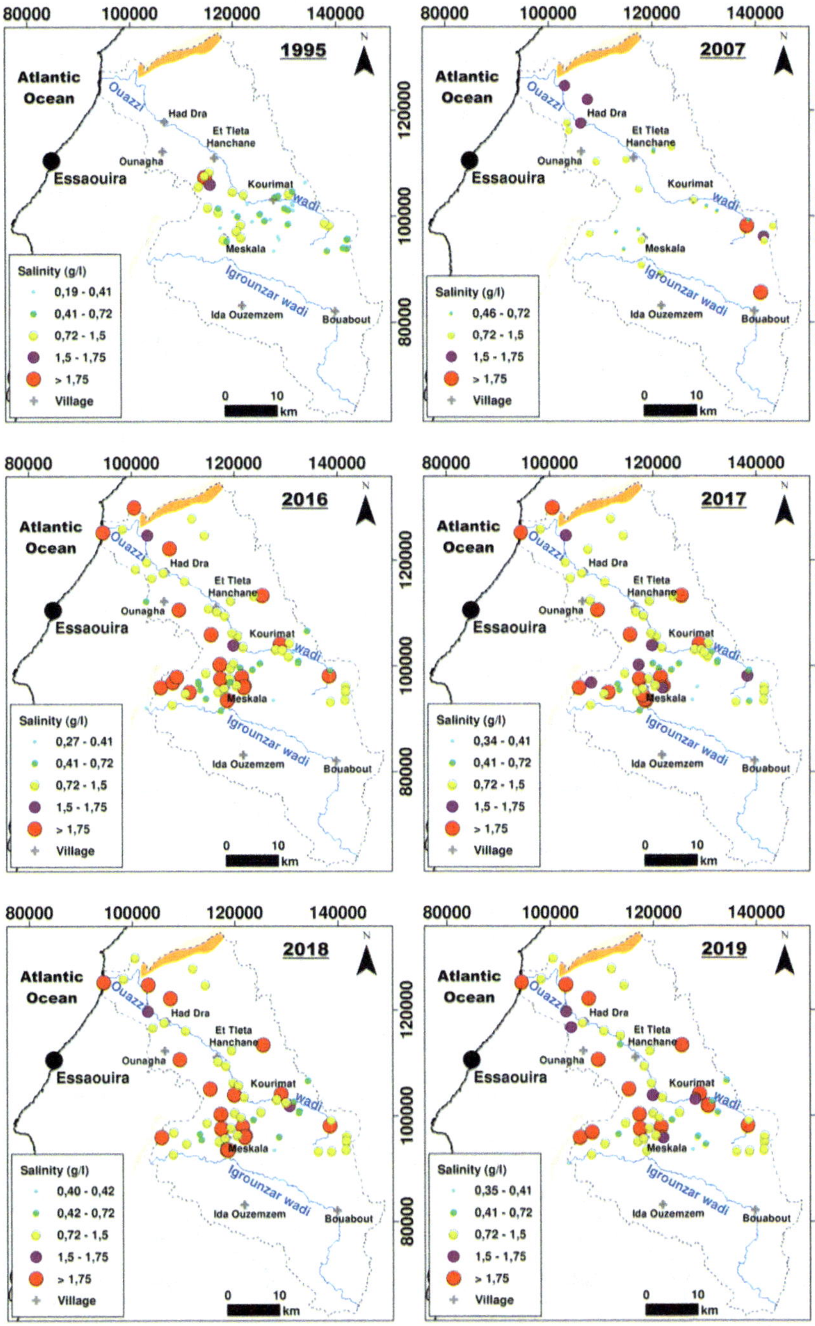

Fig. 9 Spatial distribution of salinity in Cenomanian–Turonian aquifer. *Source* Developed by the author

Analyzing the maps in Fig. 9 reveals that salinity values become more pronounced over time and advance from east to west during the six campaigns. For instance, in the Sebt Kourimat region, a recharge area for the Cenomanian–Turonian aquifer, salinity values fluctuated around 0.46 g/l in 1995 and reached 2.9 g/l in 2019. The overall spatio-temporal evolution of salinity depicts an upward trend.

In the downstream section, the Plio-Quaternary aquifer exhibits variations in groundwater salinity, with values ranging between 0.6 and 3.4 g/l and an average of 1.7 g/l in 1990. Subsequent years demonstrate a fluctuating pattern: between 0.9 and 3 g/l with an average of 1.6 g/l in 1995, from 0.4 to 4.1 g/l with an average of 1.3 g/l in 2004, ranging from 0.9 to 2.2 g/l with an average of 1.4 g/l in 2009, between 0.3 to 4.7 g/l with an average of 1.5 g/l in 2015, from 0.4 to 4.8 g/l with an average of 1.53 g/l in 2017, between 0.5 and 6.5 g/l with an average of 1.6 g/l in 2018, and ranging between 0.46 and 8.4 g/l with an average of 1.7 g/l in 2019 (refer to Fig. 10).

The spatial distribution, as depicted in Fig. 10, highlights higher salinity values in the southern and western parts, attributed to factors such as distance from recharge zones, residence time, influence of Triassic terrains, and impact from the sea (marine intrusion, particularly noted in well 11/51). Conversely, lower salinity values prevail in the north (along the Ksob wadi) and east of the Plio-Quaternary aquifer, representing recharge zones. This trend aligns with the temporal evolution of groundwater salinity in the Plio-Quaternary aquifer, demonstrating a consistent upward trajectory over the years, indicative of deteriorating groundwater quality.

Regarding the Turonian aquifer, minimal salinity values hover around 0.8 g/l, while maximum values reach around 1.3 g/l, with an average of 1.1 g/l for the campaigns spanning 2004, 2009, 2015, 2017, 2018, and 2019 (see Fig. 10). Unlike the Plio-Quaternary aquifer, the temporal evolution of groundwater salinity in the Turonian aquifer does not exhibit a significant trend, potentially attributed to its considerable depth and captive nature.

In the case of the Barremian-Aptian aquifer, salinity values fluctuate between 0.2 and 3.2 g/l, averaging 1.1 g/l for the 1997 campaign. Subsequent campaigns in 2015, 2017, 2018, and 2019 showcase variations ranging from 0.3 to 2.8 g/l, with an average of 1.1 g/l, and between 0.7 and 2.4 g/l, averaging 1.2 g/l, respectively (refer to Fig. 11).

The spatiotemporal variation in groundwater salinity within the Barremian-Aptian aquifer, as illustrated in Fig. 11, reveals a subtle upward trajectory in the minimum salinity values, while the maximum values exhibit marginal stability. In the case of the Hauterivian aquifer, salinity levels range between 0.6 to 2.6 g/l for the 1997 campaign, 0.5 to 1.1 g/l (with an average of 0.8 g/l) for the 2015 samples, 0.6 to 1.2 g/l (average of 0.9 g/l) for the 2017 campaign, 0.4 to 1.1 g/l (average of 0.8 g/l) for the 2018 campaign, and 0.4 to 1.3 g/l (average of 0.8 g/l) for the 2019 samples. The spatiotemporal pattern of Hauterivian aquifer salinity indicates a minor dilution in the analyzed waters for the years 2015, 2017, 2018, and 2019 when compared to the samples from the 1997 campaign. This shift could be attributed to the installation of the Igouzoullene dam in 2004, upstream, facilitating aquifer recharge.

Given the semi-arid climate prevalent in the study area, with recent trends toward aridity, coupled with reduced precipitation and heightened temperatures, leading to

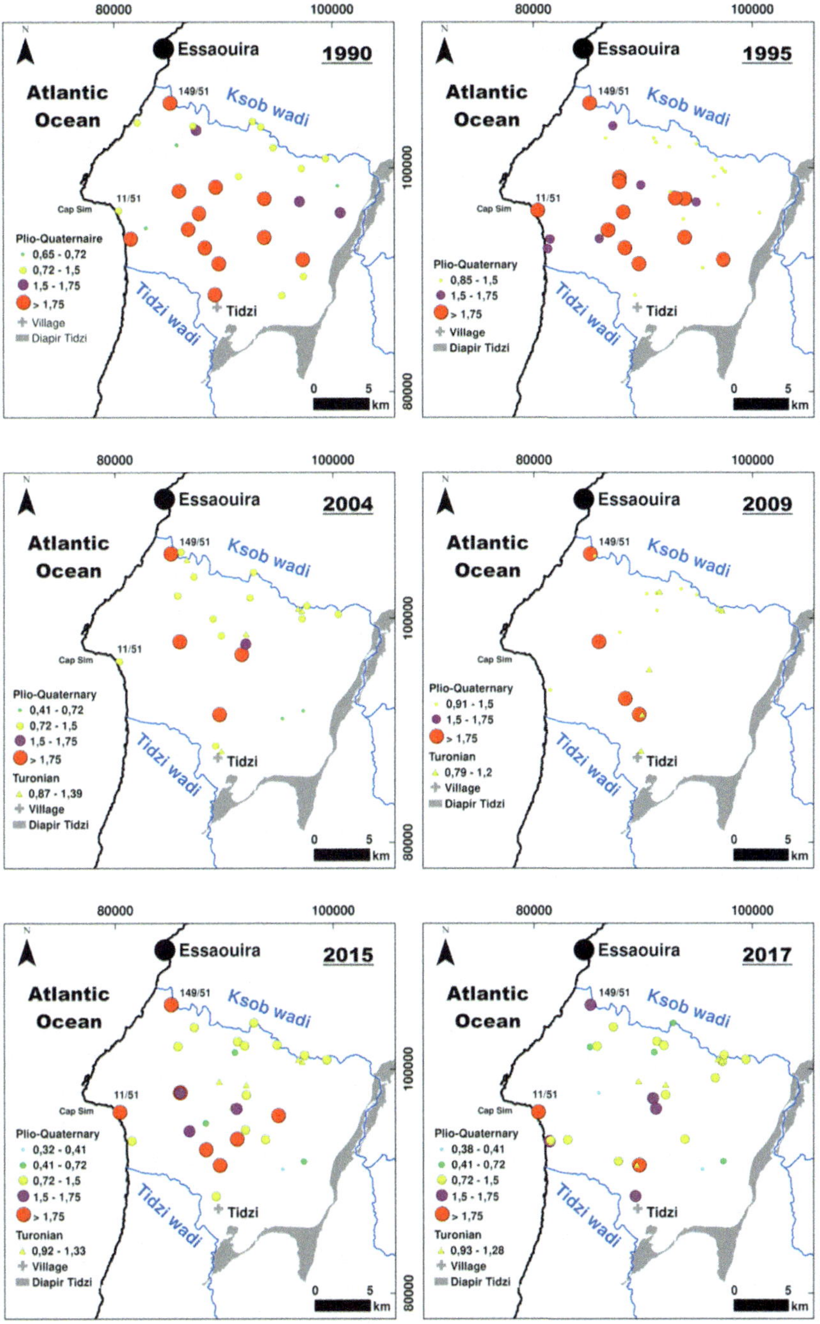

Fig. 10 Spatial distribution of salinity in Plio-Quaternary and Turonian aquifers. *Source* Developed by the author

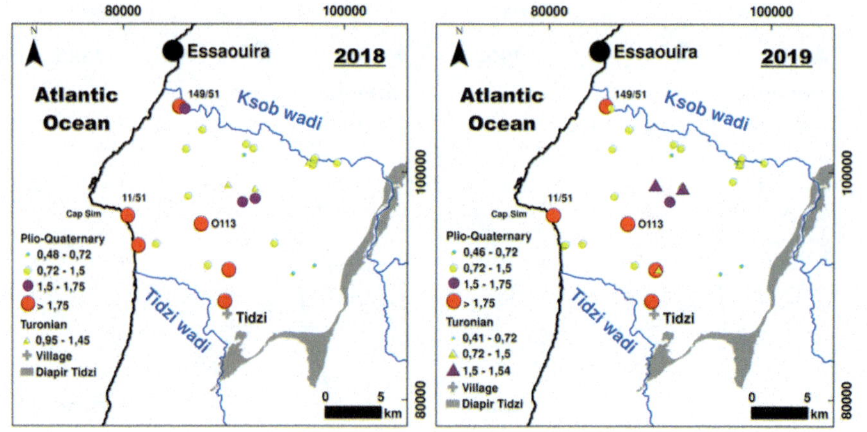

Fig. 10 (continued)

extended drought periods causing evaporation impacts on both surface and groundwater, particularly in shallow waters, the degradation of groundwater quality is primarily attributed to these climatic conditions. The decline in piezometric levels is a consequential effect of climate change.

4 Conclusion

The water resource within the Essaouira basin is constrained and exhibits spatial and temporal disparities. This limitation poses a potential threat to water availability, exacerbated by the dwindling resource attributed to the discernible impacts of climate change. The hydrogeochemical investigation revealed that the Cenomanian–Turonian aquifer's groundwater manifests diverse facies, including Cl–Ca–Mg, Cl–Ca, Cl–Na, and HCO_3–Ca, with the prevalence of the Cl–Ca–Mg facies and Cl–Ca. Examination of the temporal evolution of these facies indicates a lack of significant changes. Groundwater from the Plio-Quaternary and Turonian aquifers exhibits mixed characteristics of Cl–Na and Cl–Ca–Mg. The chemical facies experienced a subtle transformation from the Cl–Na facies to a combination of Cl–Na and Cl–Ca–Mg for the Plio-Quaternary aquifer, and from the Cl–Na facies to the Cl–Ca–Mg facies for the Turonian aquifer. In contrast, the Barremian-Aptian and Hauterivian aquifers generally display three chemical facies: Cl–Na, Cl–Ca–Mg, and HCO_3–Ca–Mg, with the Cl–Ca–Mg facies dominating. Notably, there was a noteworthy shift in facies during the study period, transitioning from the Cl–Na facies to the Cl–Ca–Mg facies. Correlation analyses of major element concentrations indicate that groundwater mineralization is influenced by the dissolution of evaporitic minerals (halite, gypsum, and/or anhydrites) and carbonates (dolomite), reverse ion exchange, and

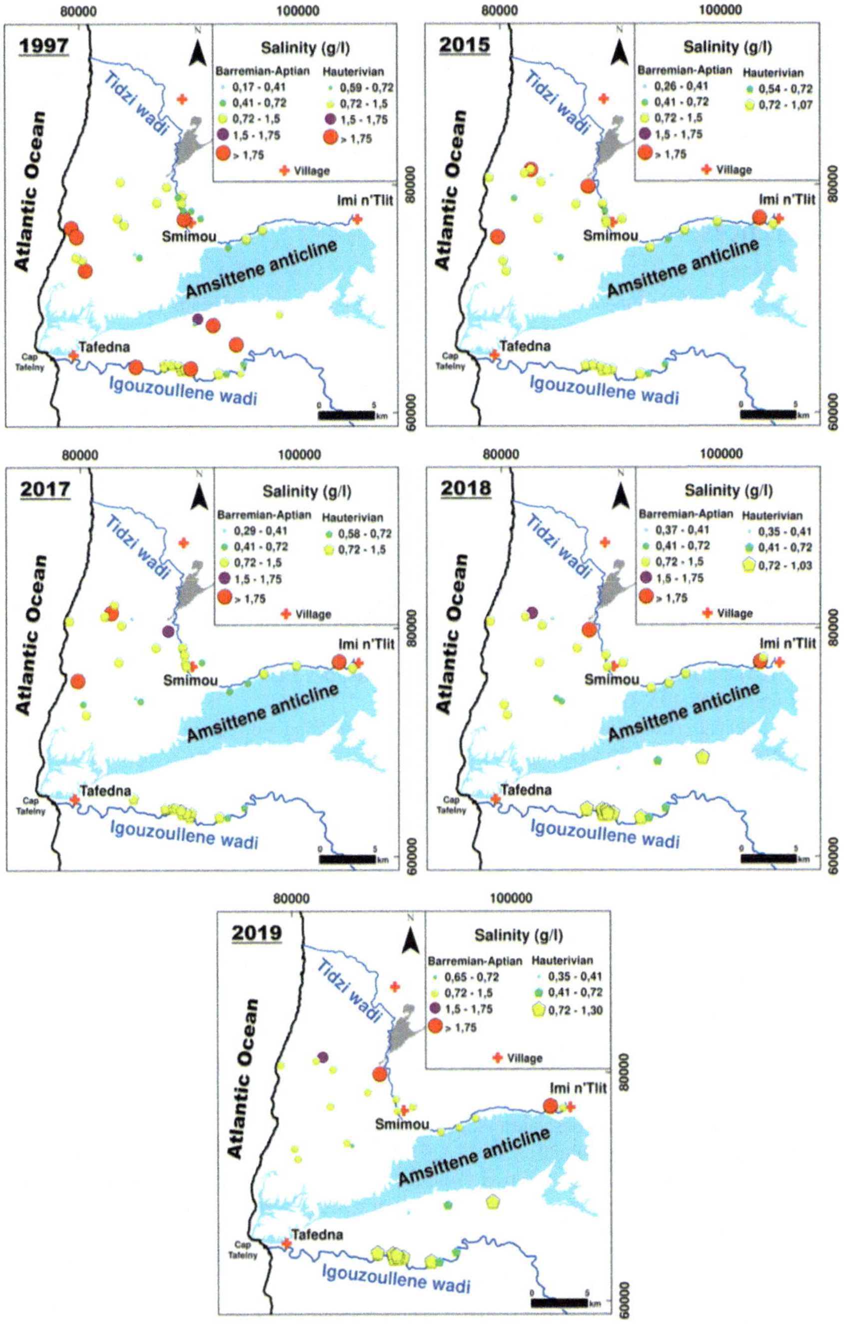

Fig. 11 Spatial distribution of salinity in Barremian-Aptian and Hauterivian aquifers. *Source* Developed by the author

marine intrusion, particularly evident in the Plio-Quaternary aquifer. The spatiotemporal assessment of groundwater quality in the study area reveals a gradual deterioration over time and space. However, the Essaouira basin is particularly susceptible to climate change due to its exclusive reliance on meteoric waters for recharge.

References

Abutaleb KAA, Mohammed AHE, Ahmed MHM (2018) Climate change impacts, vulnerabilities and adaption measures for Egypt's Nile Delta. Earth Syst Environ 2:183–192

Al-Maktoumi A, Zekri S, El-Rawy M, Abdalla O, Al-Wardy M, Al-Rawas R, Charabi Y (2018) Assessment of the impact of climate change on coastal aquifers in Oman. Arab J Geosci 11:501. https://doi.org/10.1007/s12517-018-3858-y

Alpert P, Krichak SO, Shafir H, Haim D, Osetinsky I (2008) Climatic trends to extremes employing regional modeling and statistical interpretation over the E. Mediterranean. Glob Planet Change 63:163–170

Amghar M (1989) Apports des méthodes d'analyses de la tectonique cassante à la connaissance de l'histoire alpine du Haute Atlas Occidental. L'exemple du versant nord du bloc ancien et de l'Atlas d'Agadir (Haut Atlas, Maroc). Thèse 3ème cycle, Université Cadi Ayyad, Maroc

Babqiqi A (2014) Changements Climatiques au Maroc: Etude du cas de la Région de Marrakech Tensift Al Haouz et implications sur l'agriculture à l'horizon 2030. Thèse de doctorat, Université Cadi Ayyad

Bahir M, Mennani A (2002) Problématique de la gestion des eaux souterraines au Maroc. Estud Geológicos 58:103–108

Bahir M, Mennani A, Jalal M, Youbi N (2000) Ressources hydriques du bassin synclinal d'Essaouira (Maroc). Estud Geol 56:185–195

Bahir M, Jalal M, Blavoux (2001) Apport de l'hydrochimie isotopique à la connaissance des zones de recharge en zones aride et semi-arides: Cas du bassin de Meskala-Kourimat. J Environ Hydrol

Bahir M, Carreira, P Silva, MO, Fernandes P (2008) Caractérisation hydrodynamique, hydrochimique et isotopique du système aquifère de Kourimat (Bassin d'Essaouira, Maroc). Estud Geol 64:61–73. https://doi.org/10.3989/egeol.08641433

Bahir M, Chkir N, Trabelsi R, Ammar H, Friha, Zouari K, Chamchati H (2012) Hydro-geochemical behaviour of two coastal aquifers under severe climatic and human constraints: comparative study between Essaouira basin in Morocco and Jeffara basin in Tunisia. Int J Hydrol Sci Technol 2:75–100

Bahir M, El Moukhayar R, Youbi N, Chamchati H, Chkir Ben Jemaa N (2013a) Management and protection of groundwater resources in semi-arid zones: application of hydrochemical methodologies to Essaouira Synclinal Basin, Morocco. Int J Hydrol Sci Technol 3:160–175

Bahir M, El Moukhayar R, Chkir N, Chamchati H, Fernandes PG, Carreira PM (2013b) Groundwater chemical evolution in the Essaouira Aquifer Basin—NW Morocco. Open J Mod Hydrol 3:8. https://doi.org/10.4236/ojmh.2013.33017

Bahir M, El Moukhayar R, Chamchati H, Youbi N, Carreira PM, Chkir N (2014) Recharge and hydrogeochemical evolution groundwater in semi-arid zone (Essaouira Basin, Morocco). Comun Geol 3 :651–653. http://www.lneg.pt/iedt/unidades/16/paginas/26/30/185

Bahir M, Ouhamdouch S (2020) Groundwater quality in semi-arid environments (Essaouira Basin, Morocco) Carbonates Evaporites 35:1–16. https://doi.org/10.1007/s13146-020-00576-7

Bahir M, Ouhamdouch S, Carreira PM (2016) La ressource en eau au Maroc face aux changements climatiques; cas de la nappe Plio-Quaternaire du bassin synclinale d'Essaouira. Comun Geol 103(1):35–44

Bahir M, Ouhamdouch S, Carreira PM (2018) Isotopic and geochemical methods for studying water–rock interaction and recharge mode: application to the Cenomanian-Turonian and Plio-Quaternary aquifers of Essaouira Basin, Morocco. Mar Freshwater Res 69:1290–1300

Bahir M, Ouazar D, Ouhamdouch S (2019) Hydrogeochemical investigation and groundwater quality in Essaouira region, Morocco. Mar Freshwater Res 70:1317–1332

Bahir M, Ouhamdouch S, Ouazar D, El Moçayd N (2020) Climate change effect on groundwater characteristics within semi-arid zones from western Morocco. Groundw Sustain Dev 11:1–14. https://doi.org/10.1016/j.gsd.2020.100380

Bahir M, Ouhamdouch S, Ouazar D, Chehbouni A, Ouarani M, El Mountassir O (2021a) Groundwater quality of the alluvial and carbonate aquifers of Essaouira basin (Morocco). Carbonates Evaporites 36:23. https://doi.org/10.1007/s13146-021-00697-7

Bahir M, El Mountassir O, Ouazar D, Chehbouni A, Carreira PM (2021b) Stable isotope and quality of groundwater around ksob sub-basin, Essaouira, Morocco. Sustain Water Resour Manag. https://doi.org/10.1007/s40899-021-00553-5

Bahir M, El Mountassir O, Ouazar D, Carreira PM (2021c) Use of WQI and isotopes to assess groundwater quality of coastal aquifers (Essaouira, Morocco). In: Abrunhosa M, Chambel A, Peppoloni S, Chaminé HI (eds) Advances in geoethics and groundwater management: theory and practice for a sustainable development. Advances in science, technology and innovation (IEREK Interdisciplinary Series for Sustainable Development). Springer, Cham. https://doi.org/10.1007/978-3-030-59320-9_51

Bahir M, El Mountassir O, Ouazar D, Carreira PM (2021d) Hydrochemical analysis and evaluation of groundwater quality in Ouazi Basin (Essaouira, Morocco). In: Abrunhosa M, Chambel A, Peppoloni S, Chaminé HI (eds) Advances in geoethics and groundwater management: theory and practice for a sustainable development. Advances in science, technology and innovation (IEREK Interdisciplinary Series for Sustainable Development). Springer, Cham. https://doi.org/10.1007/978-3-030-59320-9_50

Berhail S (2019) The impact of climate change on groundwater resources in northwestern Algeria. Arab J Geosci 12:770. https://doi.org/10.1007/s12517-019-4776-3

Bougheriou E, Bouri S, Khanfir H, Zarhloule Y (2014) Impacts of climate change on water resources in arid and semi-arid regions: Chaffar Sector, Eastern Tunisia. Desalin Water Treat 52:2082–2093

Capaccioni B, Didero M, Paletta C, Didero L (2005) Saline intrusion and refreshening in a multilayer coastal aquifer in the Catania Plain (Sicily, Southern Italy): dynamics of degradation processes according to hydrochemical characteristics of groundwaters. J Hydrol 307:1–16

Carreira PM, Bahir M, Ouhamdouch S, Fernandes PG, Nunes D (2018) Tracing salinization processes in coastal aquifers using an isotopic and geochemical approach: comparative studies in western Morocco and southwest Portugal. Hydrogeol J 26:2595–2615

Chenaker H, Houha B, Valles V (2017) Isotope studies and chemical investigations of hot springs from North-Eastern Algeria. J Mater Environ Sci 8:4253–4263

Dresh J (1962) Le Haut Atlas Occidental. Dans "Aspect de la géomorphologie du Maroc." Notes Et Mémoire Du Service Géologique-Maroc 96:107–121

Driouech F (2010) Distribution des précipitations hivernales sur le Maroc dans le cadre d'un changement climatique: descente d'échelle et incertitudes. Thèse de doctorat, Université. Toulouse, France

Duffaud F (1960) Contribution à l'étude stratigraphique du bassin secondaire du Haut Atlas Occidental (Sud-Ouest du Maroc). Bull Soc Géol Fr 7:728–734

Duffaud F, Brun L, Planchot B (1966) Bassin du Sud-Ouest Marocain (SW Morocco Basin). In: Reyre (ed) Bassin Sédimentaire du Littoral Africain, partie I, Paris, pp 5–12

El Kharraz J, El-Sadekb A, Ghaffourc N, Mino E (2012) Water scarcity and drought in WANA countries. Procedia Eng 33:14–29

El Mountassir O (2023) Hydrochemistry and vulnerability of water resources in a semi-arid environment: A multidisciplinary approach and the use of GIS tools (Case of the Essaouira Basin, Morocco) "Hydrochimie et Vulnérabilité des Ressources en Eau dans un milieu semi-aride :

Approche pluridisciplinaire et utilisation des outils SIG (Cas du bassin d'Essaouira, Maroc)" (Doctoral dissertation, Order No. 713). Cadi Ayyad University, Morocco
El Mountassir O, Bahir M (2023) The assessment of the groundwater quality in the coastal aquifers of the Essaouira Basin, Southwestern Morocco, using hydrogeochemistry and isotopic signatures. Water 15(9):1769. https://doi.org/10.3390/w15091769
El Mountassir O, Bahir M, Ouazar D, Ouhamdouch S, Chehbouni A, Ouarani M (2020) The use of GIS and water quality index to assess groundwater quality of Krimat aquifer (Essaouira; Morocco). SN Appl Sci 2:871. https://doi.org/10.1007/s42452-020-2653-z
El Mountassir O, Bahir M, Ouazar D, Chehbouni A, Carreira PM (2021a) Temporal and spatial assessment of groundwater contamination with nitrate using nitrate pollution index (NPI), groundwater pollution index (GPI), and GIS (case study: Essaouira basin, Morocco). Environ Sci Pollut Res. https://doi.org/10.1007/s11356-021-16922-8
El Mountassir O, Bahir M, Ouazar D, Chehbouni A, Carreira PM (2021b) Geochemical and isotopic evidence of groundwater salinization processes in the Essaouira region, north-west coast, Morocco. SN Appl Sci. https://doi.org/10.1007/s42452-021-04623-3
El Mountassir O, Ouazar D, Bahir M, Chehbouni A, Carreira PM (2021c) GIS-based assessment of aquifer vulnerability using DRASTIC model and stable isotope: a case study on Essaouira basin. Arab J Geosci 14:321. https://doi.org/10.1007/s12517-021-06540-6
El Mountassir O, Bahir M, Ouazar D, Carreira PM (2021d) Nitrate pollution in groundwater of the Ouazi basin: case of Essaouira (Southwestern Morocco). In: Abrunhosa M, Chambel A, Peppoloni S, Chaminé HI (eds) Advances in geoethics and groundwater management: theory and practice for a sustainable development. Advances in science, technology and innovation (IEREK Interdisciplinary Series for Sustainable Development). Springer, Cham. https://doi.org/10.1007/978-3-030-59320-9_49
El Mountassir O, Bahir M, Ouazar D, Carreira PM (2021e) For a better understanding of recharge and salinization mechanism of a Cenomanian–Turonian aquifer. In: Abrunhosa M, Chambel A, Peppoloni S, Chaminé HI (eds) Advances in geoethics and groundwater management: theory and practice for a sustainable development. Advances in science, technology and innovation (IEREK Interdisciplinary Series for Sustainable Development). Springer, Cham. https://doi.org/10.1007/978-3-030-59320-9_42
El Mountassir O, Bahir M, Ouazar D, Chehbouni A, Carreira PM (2022a) Evaluation of nitrate source and its distribution in the groundwater of Essaouira basin. Sustain Water Resour Manag 8:28. https://doi.org/10.1007/s40899-022-00609-0
El Mountassir O, Bahir M, Chehbouni A, Dhiba D, El Jiar H (2022b) Assessment of groundwater quality and the main controls on its hydrochemistry in a changing climate in Morocco (Essaouira Basin). Sustainability 14:8012. https://doi.org/10.3390/su14138012
Fetter CW (1993) Contaminant hydrogeology. Macmillan Publishing Company, New York
Green TR, Makoto T, Henk K, Gurdak JJ, Allen DM, Hiscock KM, Holger T, Alice A (2011) Beneath the surface of global change: impacts of climate change on groundwater. J Hydrol 405:532–560
Hallouz F, Meddi M, Mahe G, Karahacane H, Rahmani SEA (2019) Tendance des précipitations et évolution des écoulements dans un cadre de changement climatique : bassin versant de l'oued Mina en Algérie. RSE 32:83–114
IPCC (2013) In Stocker TF, Qin D, Plattner G-K, Tignor M, Allen SK, Boschung J, Nauels A, Xia Y, Bex V, Midgley PM (eds) Climate change 2013: the physical science basis. Contribution of working group I to the fifth assessment report of the intergovernmental panel on climate change. Cambridge University
Jalal M, Blavoux B, Bahir M, Bellion Y, Laftouhi N, Puig JM, Mennani A, Daniel M (2001) Etude du fonctionnement du système aquifère karstique Cenomano-Turonian de l'oued Igrounzar (Bassin d'Essaouira, Maroc). J Afr Earth Sci 32:803–817
Lachaal F, Chargui S, Messaoud RB, Chekirbane A, Tsujimura M, Mlayah A, Massuel S, Leduc C (2018) Impacts of global changes on groundwater resources in north-east Tunisia: the case of the Grombalia phreatic aquifer. In: Calvache ML et al (eds) Groundwater and global change in the

Western Mediterranean Area, Environmental Earth Sciences. Springer International Publishing Switzerland, pp 179–188
Lebel T, Vischel T (2005) Climat et cycle de l'eau en zone tropicale: un problème d'échelle. CR Geoscience 337:29–38
Marfia AM, Krishnamurthy RV, Atekwana EA, Panton WF (2004) Isotopic and geochemical evolution of ground and surface waters in a karst dominated geological setting: a case study from Belize, Central America. Appl Geochem 19:937–946
Mennani A (2001) Apport de l'hydrochimie et de l'isotopie à la connaissance du fonctionnement des aquifères de la zone côtière d'Essaouira (Maroc Occidental). Thèse de doctorat, Université Cadi Ayyad, Maroc
Misra AK (2014) Climate change and challenges of water and food security. Int J Sustain Built Environ 3:153–165
Ouhamdouch S, Bahir M, Carreira P (2018) Impact du changement climatique sur la ressource en eau en milieu semi-aride: exemple du bassin d'Essaouira (Maroc). Rev Des Sci De L'eau 31:13–27
Ouhamdouch S, Bahir M, Ouazar D, Goumih A (2020) Zouari K (2020) Assessment of the climate change impact on the future evapotranspiration and flows from a semi-arid environment. Arab J Geosci 13:82. https://doi.org/10.1007/s12517-020-5065-x
Parkhurst DL, Appelo CAJ (1999) User's guide to PHREEQC (version 2)—a computer program for speciation, batch–reaction, one–dimensional transport, and inverse geochemical calculations. USGS Water–Resources Investigation Report 99-4259, 312 pp
Piper AM (1944) A graphic procedure in the geochemical interpretation of water analyses. Trans Am Geophys Union 25:914–923
Ragab R, Prudhomme C (2002) Climate change and water resources management in arid and semi-arid regions: prospective and challenges for the 21st century. Biosystems Eng 81:3–34
Rodier J, Legube B, Merlet N, Mialocq JC, Leroy P, Houssin M, Lavison G, Bechemin C, Vincent M, Rebouillon P, Moulin L, Chomodé P, Dujardin P, Gosselin S, Seux R, Al Mardini F (2009) L'Analyse de l'eau. 9e édition, Dunod, Paris
Solomon S, Qin D, Manning M, Chen Z, Marquis M, Averyt KB, Tignor M, Miller HL (eds) (2007) Climate change 2007: "e physical science basis; contribution of working group I to the fourth assessment report of the intergovernmental panel on climate change. Cambridge University Press, Cambridge, 996 p
Sinan M, Boussetta M, El Rherari A (2009) Changements climatiques: causes et conséquences sur le climat et les ressources en eau. Revue HTE 142:21–30
Stour L, Agoumi A (2008) Sécheresse climatique au Maroc durant les dernières décennies. Hydroécol Appl 16:215–232
Vicente-Serrano SM (2006) Spatial and temporal analysis of droughts in the Iberian Peninsula (1910–2000). Hydrolog Sci J 51:83–97
WHO (2011) Guidelines for drinking water quality, 4th ed. Geneva
Xu ZX, Chen YN, Li JY (2004) Impact of climate change on water resources in the Tarim River Basin. Water Resour Manag 18:439–458

Prof. Dr. Eng. Mohammed Bahir, Ph.D., Hydrogeological Engineer, has published over 200 research/professional papers and reports. He co-authored more than a dozen books and book chapters. He supervised and/or co-supervised more than 20 PhD students. He is a reviewer of many international scientific journals with high impact factors, including Environment, Development, and Sustainability Journal, Environmental Earth Sciences Journal, Hydrological Sciences Journal, Sécheresse, Gaia, Communicaçoés International, Science of the Total Environment, Marine and Freshwater Research Journal, Proceedings of the Indian National Science Academy, and Groundwater for Sustainable Development. He also coordinates more than a dozen research projects.

The Asia Research Awards, in association with Times of Research and the World Research Council, announce that Professor Mohammed Bahir has been selected as the recipient of the prestigious ASTRA 2023 (Asia's Science, Technology, and Research Awards).

Chapter 6
Assessment of the Climate Change Impact on the Past and Future Evapotranspiration and Flows from a Semi-arid Environment

Mohammed Bahir

Abstract Ensuring access to potable water for a substantial portion of the population, effectively managing and conserving this frequently overutilized resource, regulating agricultural and industrial water consumption, and safeguarding the natural environment constitute significant challenges, particularly in the context of developing nations. The primary objective of this investigation was to assess the influence of climate change on the hydrological patterns within the Essaouira basin during the time frame of 2020–2050. To accomplish this, the Rural Genius GR2M model was employed to simulate streamflows based on precipitation and evapotranspiration data. The Mann–Kendall and Pettitt tests were employed to scrutinize the temporal series for both their consistency and directional trends. The chronological sequence of potential monthly evapotranspiration (ETP), covering the span from 1978 to 2005, demonstrates an escalating tendency, denoting a 4.2% increase. This trend in potential evapotranspiration remains evident under different Representative Concentration Pathways (RCPs) of CMIP5 (Coupled Model Intercomparison Project Phase 5) scenarios, specifically 2.6, 4.5, and 8.5 RCPs, where the trend rates are measured at 1.9%, 1.5%, and 1.6%, respectively. The investigation into the correlation between precipitation and streamflows, from 1978 to 2005 within the Essaouira basin, establishes the presence of a causative relationship between these two variables. This relationship holds true across all regions subject to arid and semi-arid climatic conditions. The insights derived from these findings can provide a foundation for safeguarding and managing water resources within the Essaouira watershed, including the construction of upland reservoirs along the Igrounzar, Zelten, and Ksob Wadi, enhancing the basin's overall water management strategy.

Keywords Climate change · Essaouira basin · Evapotranspiration · Flows · GR2M model

---

M. Bahir (✉)
High Energy and Astrophysics Laboratory, Faculty of Sciences Semlalia, Cadi Ayyad University, Marrakesh, Morocco
e-mail: bahir@uca.ac.ma

M. Behnassi et al. (eds.), *The Water, Climate, and Food Nexus*,
https://doi.org/10.1007/978-3-031-50962-9_6

1 Introduction

Ensuring widespread access to potable water, addressing the challenges of safeguarding a frequently overexploited and inadequately managed resource, regulating water usage in agriculture and industry, and preserving the environment are all pivotal concerns, particularly for developing nations (Bahir et al. 2016, 2018a, b, 2019, 2021a, b; El Mountassir et al. 2020, 2021a, b, d, e, 2022a). Among these considerations, one can highlight the potential transformations of increasingly human-influenced natural ecosystems, the collective management of environmental risks (such as floods, droughts, and groundwater salinization), and the specter of global warming, both in the present century and beyond (Bahir et al. 2013a, b, 2014, 2016, 2018c; El Gayar and Hamed 2018).

The issues surrounding water resources and their management are intricately tied to the future prospects of the Maghreb region (Agoumi 1999). Regardless of the effects of climate change, the pronounced susceptibility of watersheds to even minor fluctuations in climatic variables implies that the water volume available for utilization will be substantially impacted by declining runoff (El Mountassir et al. 2021c).

Similar to arid and semi-arid regions worldwide, Morocco, a country characterized by a Saharan climate, has grappled with recurrent drought periods dating back to the early 1980s (Bahir et al. 2000, 2001). These episodes have evidently had consequences for water availability (Bahir and Mennani 2002; Bahir et al. 2008, 2012, 2018a, b). In these settings, typified by the paucity or complete absence of surface water resources, the primary determinant in the water balance equation is rainfall.

In recent decades, extensive research has been dedicated to assessing the impact of climate change on water resources in arid and semi-arid regions (Bahir et al. 2016, 2020; Bahir and Ouhamdouch 2020; Belarbi et al. 2015; Ouhamdouch et al. 2018). These studies have consistently revealed a combination of quantitative and qualitative deterioration, as evidenced by a decline in both water quality and the levels of groundwater and surface water. This situation serves as a clear warning sign, necessitating proactive intervention, particularly by scientists, practitioners, and policymakers, to regulate, safeguard, and enhance these increasingly scarce resources. For such intervention to be effective, it is imperative to establish a comprehensive understanding of the mechanisms governing these resources and their interconnected ecosystems.

Over the past few decades, numerous hydrological models have been developed, many of which focus on the rainfall-runoff relationship, offering substantial flexibility and resource savings in terms of both time and material. Among these models, the Rural Genius (RG) model, developed by CEMAGREF, stands out as a tank model that can operate at different annual time steps, such as GR1A, monthly GR2M, and daily GR4J. This model has been successfully employed by various researchers, yielding highly satisfactory results (Dezetter et al. 2008; Belarbi et al. 2015; Rwasoka et al. 2014). A precipitation model holds significant importance, as it not only facilitates the estimation of available resources for development but also supports the long-term development of these resources by incorporating data and climatic scenarios.

To illustrate the application of these concepts, we turn our attention to the Essaouira basin. In this case, a rainfall-runoff model has been formulated for the period spanning 2020 to 2050. The aim is to characterize the sub-basins within the Essaouira region, including the Igrounzar, Zelten, and Ksob sub-basins, while simulating the future behavior of water flow under varying scenarios (RCPs 2.6, 4.5, and 8.5) in accordance with the CIMP5 model. The RG conceptual model, operational at monthly time steps and defined by two parameters (GR2M), has been chosen for its suitability in processing the data from the sub-basins under investigation.

The Ksob watershed (comprising the Igrounzar, Zelten, and Ksob Wadis) plays a central role in the Essaouira basin. It spans an area of approximately 1500 km^2 and is situated in the southeast of Essaouira city. Characterized by a semi-arid climate, this region receives annual rainfall of no more than 300 mm, while the average annual temperature hovers around 20 °C (El Mountassir et al. 2022b; Ouhamdouch et al. 2018). The landscape of the basin is typified by low hills with gently rounded summits, intersected by a network of shallow watercourses. This topographical arrangement divides the study area into three distinct sub-basins: the Igrounzar sub-basin in the north, the Zelten sub-basin in the south, and the Ksob sub-basin, which encompasses the entire watershed to the west. The studied basin is of sedimentary type containing formations ranging from Trias to Plio-Quaternary. The Triassic and Jurassic formations have only very small outcrops, and are located in the heart of anticlines (Hadid anticline, Amsittene anticline, and Tidzi diaper), while tertiary and quaternary formations meet in the synclines (Fig. 1).

The Triassic rock formations encompass saliferous red clays and doleritic basalts, while the Jurassic strata consist of alternating carbonate deposits, including limestones and dolomites, interspersed with marls rich in evaporite minerals such as gypsum and anhydrite (Peybernès et al. 1987). As proposed by Duffaud et al. (1966), the Lower Cretaceous period is primarily characterized by the presence of limestones and marls. Transitioning into the Middle Cretaceous, we observe marly-sandstone deposits during the Aptian phase, followed by green marl formations in the Albian stage, concluding with dolomitic marls and limestones (EL Mountassir et Bahir 2023).

The research area under investigation comprises a substantial synclinal region that is geographically open to the Atlantic Ocean. It has been shaped by various geological processes, resulting in the formation of multiple synclines. Notably, these include:

i. The Bouabout synclinal basin, which constitutes the upper part of the region and is traversed by the Oued Igrounzar.
ii. The Essaouira synclinal basin, located in the downstream area, separated from the upper basin by the Tidzi diapir.

The upper part of the study area is characterized by an aquifer that is situated within the Cenomanian–Turonian limestone and dolomitic limestone formations (El Mountassir et al. 2022a, b).

Conversely, the lower region of the study area contains groundwater resources within two primary reservoirs (El Mountassir et al. 2022c):

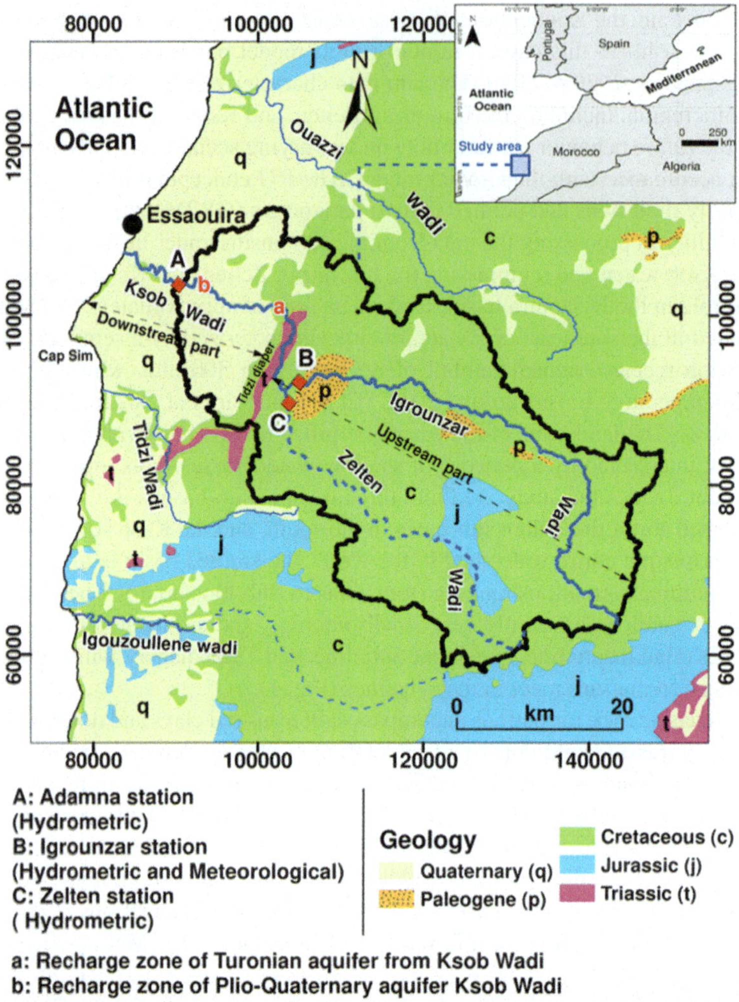

Fig. 1 Location and geological map of study area. *Source* Developed by the author

The first reservoir consists of Plio-Quaternary formations, encompassing marine calcareous sandstone or dune matrix and harboring a substantial shallow aquifer.

The second reservoir is associated with the Turonian formation, which is relatively confined beneath the Senonian marls within the synclinal structure. It likely interfaces with the Plio-Quaternary reservoir along the margins of this structure, notably to the north, toward the Ksob Wadi, to the west along the approach to the Atlantic Ocean, and to the east and south in the proximity of the Tidzi diapir (see Fig. 1).

Given the limited availability of climatic and hydrometric data, only information from three monitoring stations—namely, Igrounzar, Zelten, and Adamna—was utilized. The Igrounzar station served as the sole monitoring site for recording parameters such as rainfall, flow rates, temperature, humidity, and evaporation. Meanwhile, the Adamna station enabled the measurement of rainfall and flow rates, and the Zelten station facilitated the measurement of flow rates exclusively.

The dataset employed in this study comprises records of rainfall, potential evapotranspiration, and flow rates spanning a 28-year period (1978 to 2005). These datasets, supplied by the Tensift Basin Hydraulic Agency (ABHT), are based on monthly values collected at the three aforementioned stations. Importantly, there are no instances of missing data within the dataset, which covers the entire study period (1978–2005). The geographical locations of the three monitoring stations are depicted in Fig. 1.

The study area encompasses a basin situated within a semi-arid climatic zone, where influences from the ocean, continent, and mountainous terrain are distinctly manifested (Bahir et al. 2021a, 2022; Ouhamdouch et al. 2018). Notably, this region experiences pronounced aridity, particularly during the summer months. The aridity is notably exacerbated as one moves from the Atlantic Ocean inland. This gradient is attributed to the increasing distance from oceanic influences, leading to reduced precipitation and heightened thermal disparities (Ouhamdouch et al. 2018).

Annual precipitation levels exhibit noteworthy year-to-year variations, centering around an average of approximately 300 mm (Bahir et al. 2021c, d; El Mountassir et al. 2021d, e). According to Ouhamdouch et al. (2018), there is a discernible overall declining trend in annual precipitation, amounting to a 12% reduction. Concurrently, temperature patterns exhibit substantial seasonal fluctuations, with winter minimum temperatures plummeting to -11 °C and maximum temperatures reaching around 40 °C. The annual average temperature hovers around 20 °C, and a discernible upward trend is observed in the average annual temperatures, indicating a warming trend of approximately 2 °C (Ouhamdouch et al. 2018).

2 Methodology

The evaluation of the impact of climate change on future flows in the Essaouira basin was conducted through the integration of the GR2M model, climate parameter projections using the RCP 2.6, 4.5, and 8.5 scenarios, and the application of the Mann–Kendall and Pettitt tests.

The GR2M model, operating at monthly time steps, was originally developed by Makhlouf and Michel (1994) and has been utilized in various river basins in France, consistently demonstrating superior performance in comparison to other hydrological models in this domain. Subsequently, the GR2M model has proven to be effective in arid and semi-arid regions (Sakaa et al. 2015). Hence, considering data availability and the model's track record, it was selected for assessing the impact of climate change on future flows in the Essaouira basin.

The GR2M model structure employed in this study is the latest version, as developed by Mouelhi et al. (2006), who drew upon the experience gained during the

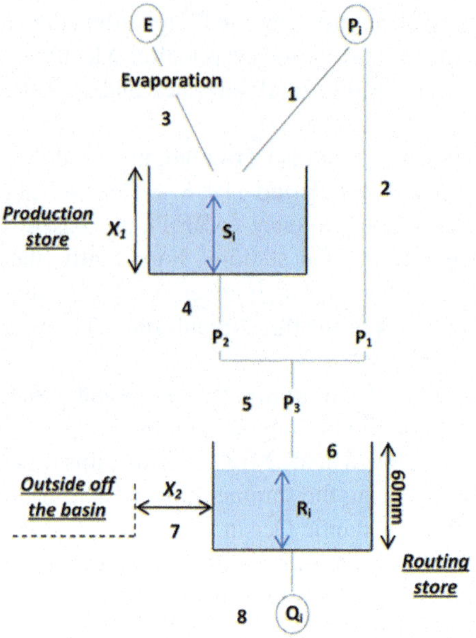

$$(1)\quad S_1 = \frac{S_i + X_1\varphi}{1!}, \text{ with } \varphi = \tanh\left(\frac{P_i}{X_1}\right)$$

$$(2)\quad P_1 = P_i + S_i - S_1$$

$$(3)\quad S_2 = \frac{S_1(1-\Psi)}{1+\Psi\left(1-\frac{S_1}{X_1}\right)}, \text{ with} \Psi = \tanh\left(\frac{E}{X_1}\right)$$

$$(4)\quad S_i = \frac{S_2}{\left[1+\left(\frac{S_2}{X_1}\right)^3\right]^{1/3}}, P_2 = S_2 - S_1$$

$$(5)\quad P_3 = P_1 + P_2$$

$$(6)\quad R_1 = R_i + P_3$$

$$(7)\quad R_2 = X_2 \times R_1$$

$$(8)\quad Q_i = \frac{R_2^2}{R_2+60}, \quad R_i = R_2 - Q_i$$

Fig. 2 Structure of GR2M model. *Source* Adapted from Mouelhi et al. (2006)

creation of the GR4J model, which operates at a daily time step (Perrin et al. 2003). The GR2M model conceptually divides each basin into two reservoirs: a soil reservoir represented as Si, controlling the production function with a maximum capacity of X1 (in millimeters, the model's first free parameter); and a routing reservoir, denoted as Ri, controlling the transfer function with a capacity of 60 mm (as illustrated in Fig. 2).

The model takes monthly inputs of rainfall (P) and potential evapotranspiration (E), and produces flow outputs (Q) in millimeters. Notably, this model version introduces a second free parameter, X2 (dimensionless). X2 is utilized to rectify any water balance discrepancies that may arise from potential biases in climate and flow time series. During each time step of the modeling process, rainfall is directed either for infiltration into the soil reservoir (1) or as surface runoff towards the routing reservoir (P1) (2). The soil reservoir Si reaches the level S1 (in millimeters), after which a portion is lost due to evaporation (3), resulting in a new level, S2 (in millimeters). A part of the soil moisture, denoted as P2 (in millimeters), is then transferred to the routing reservoir through percolation (4). The combined P1 and P2 (referred to as P3, (5)) enters the routing reservoir, reaching the level Ri (6). The routing reservoir undergoes water exchange with the external environment, which can lead to a net gain or loss of water (7). If X2 is greater than 1, external water sources contribute to the basin's supply; conversely, if X2 is less than 1, the routing reservoir supplies the flow Qi (8) (as shown in Fig. 2).

The calibration and validation of the GR2M model involved the application of the Nash criterion (Nash and Sutcliffe, 1970), considered the most robust evaluation criterion. This criterion hinges on the comparison between estimated and observed flow rates, as described by the following equation (Eq. 1):

$$\text{Nash}(Q) = 100 \times \left[1 - \frac{\sum_i (Q_{iobs} - Q_{ical})^2}{\sum_i \left(Q_{iobs} - \overline{Q}_{obs}\right)}\right] \tag{1}$$

With Q_{iobs} represents observed flow; Q_{ical} represents calculated flow and $\overline{Q}_{obs}$ is the observed flow rates average.

The GR2M model is considered efficient when the estimated flows are close to the observed flow rates, i.e. when the Nash criterion is close to 100%. According to Perrin et al. (2003), a performance $\geq 70\%$ is satisfactory.

The potential monthly evapotranspiration (ETP) was calculated using Thornthwaite formula (Eq. 2) (Thornthwaite 1948). The choice of the Thornthwaite method lies in the fact that the other methods (e.g. Pennman, Turc) use unavailable climatic parameters (e.g. insolation duration, hygrometric degree, wind speed …).

$$\text{ETP} = 16 \times \left(\frac{10 \times t}{I}\right)^a \times F(m, \lambda) \tag{2}$$

With

ETP: potential monthly evapotranspiration (mm).

t: Average monthly temperature in °C.

I: Annual thermal index is calculated using the equation (Eq. 3).

$$\text{I} = \sum i, \tag{3}$$

with $\text{i} = \left(\frac{t}{5}\right)^{1.514}$

F: Correlation coefficient that is based on month (m) and latitude (λ).

a: Complex function of the thermal index calculated via the formula (Eq. 4)

$$\text{a} = 0.49239 + \left(1792 \times 10^{-5} \times I\right) + (771 \times 10^{-7} \times I^2) + (675 \times 10^{-9} \times I^3) \tag{4}$$

For the climate projection, the results of the study realized by Ouhamdouch and Bahir (2017) were used. By the same approach, future rainfall from the Adamna station was calculated. The GR2M model, with parameters X_1 and X_2 obtained during calibration and validation phase, by using historical values of rainfall and potential monthly evapotranspiration, has been applied again to estimate future flows under new climatic conditions determined under the RCPs 2.6, 4.5 and 8.5 scenarios. In order to determine the trend direction of the flow series as well as the break dates of these series, the Pettitt test (break detection), and the Mann–Kendall test were applied using the XLSTAT® software (trial version).

Pettitt Test

The Pettitt (1979) test allows to examine the existence of a break in a time series from a formulation derived from that of the Mann–Whitney test. This test is based on the calculation of the variable $U_{t,T}$ defined by equation (Eq. 5):

$$U_{tT} = \sum_{i=1}^{T} \sum_{j=i+1}^{T} Dij \tag{5}$$

where $Dij = -1$ if $(xi - xj) > 0$, $Dij = 0$ if $(xi - xj) = 0$, $Dij = 1$ if $(xi - xj) > 0$.

Mann–Kendall's Test

The Mann–Kendall test (Mann 1945, Kendall 1975) is a test used to detect the presence or the absence of a linear trend within a time series. With the a_i series $(a_1, a_2, \ldots a_n)$, this method sets the standard U_{MK} multi-variable standard as follows (Eq. 6):

$$U_{MK} = \frac{S}{\sqrt{Var\ (s)}} \tag{6}$$

With, $S = \sum_{i=1}^{n-1} \sum_{j=i+1}^{n} sgn(a_j - a_i)$, $Var(s) = \frac{n(n-1)(2n+5)}{18}$, and n the number of data in the series.

The trend sense is defined by the statistical coefficient "U_{MK}". If $U > 0$, the trend is upward and if $U < 0$ the trend is downward.

3 GR2M Model Performance During Calibration and Validation Phase

The evaluation of the GR2M model revolves around the analysis of performance indicators, including the Nash criterion and the determination coefficient R^2.

During the model calibration process, we systematically adjusted the values of the parameters X1 and X2 until optimal values of the determination coefficient and Nash criterion were achieved (both exceeding 70%). The model defines a range for X1, spanning from 140 to 2640, and for X2, ranging from 0.21 to 1.31. For the validation phase, we utilized precipitation and evapotranspiration data spanning from January 1997 to December 2001, which were not used in the model's development. The decision to select this particular period is attributed to its consistent achievement of a Nash criterion exceeding 70% for both the calibration and validation phases.

The GR2M model's evaluation incorporates an analysis of performance parameters, notably the Nash criterion and R^2, alongside a visual comparison of observed and simulated flow hydrographs during the calibration and validation phases (Kouamé et al. 2013).

For the Igrounzar station, the Nash criterion values obtained during calibration (83%) and validation (78%) are notably higher than the 70% threshold. This demonstrates the robust modeling capability of the GR2M model concerning the data collected at the Igrounzar station. The calculated difference between the Nash criterion values during the calibration and validation periods stands at 5%. This differential serves as an indicator of the model's robustness. It is important to note that this 5% degradation in the Nash criterion is considered quite acceptable, as its absolute value remains below 10% (Kouamé et al. 2013). The hydrographs, illustrated in Fig. 3a and b, corresponding to the calibration and validation phases, exhibit high-quality characteristics. Furthermore, the comparison between simulated and observed flow rates during calibration and validation (as depicted in Fig. 3a and b) results in noteworthy R^2 values ($R^2 = 0.88$ for both phases). Indeed, the rain-flow modeling achieved through the GR2M model delivers highly satisfactory and promising outcomes at the Igrounzar station.

Regarding the Zelten station, both the calibration and validation phases yielded a Nash criterion value of 72%. This result underscores the continued effectiveness of the GR2M model in accurately modeling the data from the Zelten station. Notably, no difference between the Nash values obtained in the calibration and validation phases was observed for this station, indicative of the model's efficiency. The hydrographs presented in Fig. 4a and b exhibit a remarkable alignment between the curves representing observed and simulated flows. The correlation coefficient derived from the diagram depicting observed flow rates against simulated flows is calculated as 0.81 for both the calibration and validation phases. This further substantiates the robustness of the GR2M model when applied to the Zelten station.

Concerning the Adamna station, the Nash criterion and the determination coefficient R^2 results for this station are summarized in Table 1. The Nash criterion achieved during the calibration phase stands at 74%, surpassing the 70% threshold, thus validating the correctness of the GR2M model's calibration for the Adamna station. This calibration is subsequently affirmed with a Nash criterion of 76% during the validation phase. The strong alignment between the curves representing observed and simulated flows, as well as the substantial R^2 values displayed in Fig. 5a and b, further emphasize the excellent performance of the GR2M model at the Adamna station.

4 Future Evapotranspiration and Flows

As indicated by Ouhamdouch and Bahir (2017), the predictive model employed to estimate future temperature and precipitation levels within the Essaouira basin is CanESM2, a contributing model within the CMIP5 framework. Their integrated

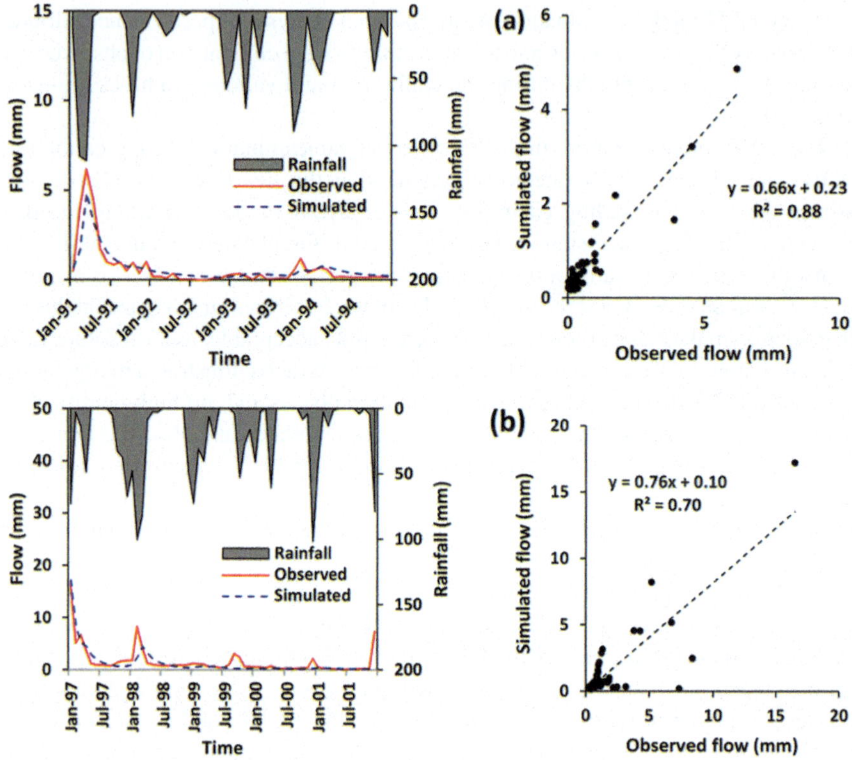

Fig. 3 Calibration **a** and validation **b** of GR2M model obtained at Igrounzar station. *Rainfall-flow observed and simulated (left). Correlation diagram between observed and simulated flow (right). Source* Developed by the author

approach, combining this model with statistical tests conducted using Pettitt and Mann–Kendall tests, reveals a projected increase in mean temperatures. Under the RCP 2.6 scenario, a rise of 0.72 °C is anticipated, while the RCP 4.5 and RCP 8.5 scenarios predict increases of 0.57 °C and 0.69 °C, respectively, by the year 2050. These results offer valuable insights into the climate outlook within the region (Ouhamdouch and Bahir 2017).

Given that the GR2M model utilizes data on rainfall and evapotranspiration to simulate flows, the calculation of future flow patterns requires forecasts of future rainfall and evapotranspiration. To estimate future evapotranspiration, future temperature data were employed in conjunction with the Thornthwaite formula (Eq. 2).

Table 2 presents the results of the statistical tests applied to the prospective monthly potential evapotranspiration time series at the Igrounzar station. The Mann–Kendall test under different scenarios, RCPs 2.6, 4.5, and 8.5, indicates a significant upward trend, supported by respective UMK values of + 0.59, + 0.78, and + 0.68. This trend is further corroborated by the Pettitt test (Fig. 6), revealing breakpoints in the series in April 2029, March 2036, and March 2034 for RCP 2.6, 4.5, and 8.5,

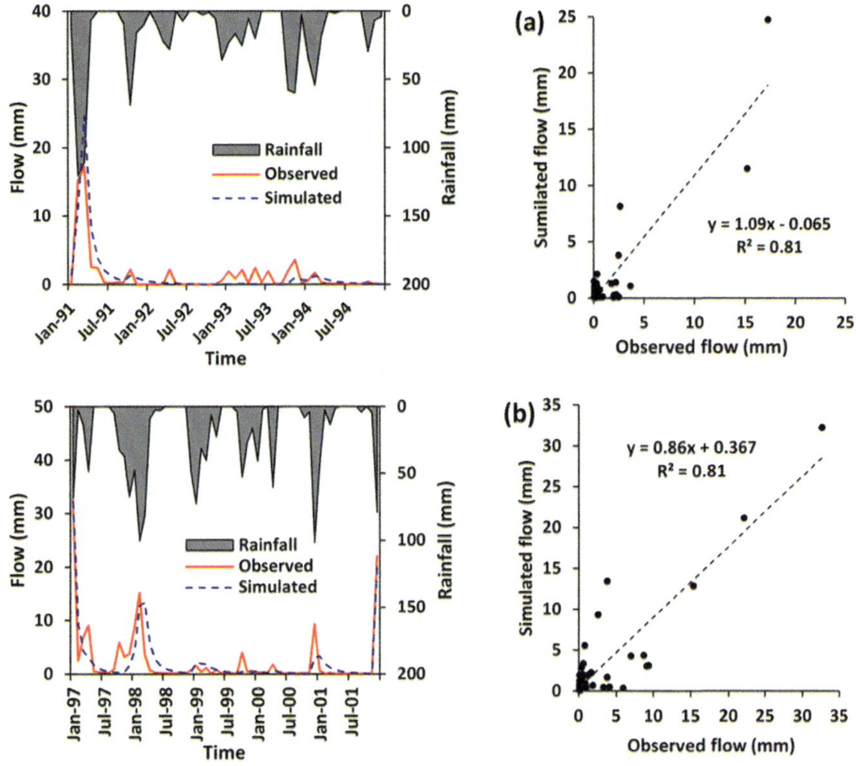

Fig. 4 Calibration **a** and validation **b** of GR2M model obtained at Zelten station. *Rainfall-flow observed and simulated (left). Correlation diagram between observed and simulated flow (right). Source* Developed by the author

respectively. For the study period extending from 2020 to 2050, the mean potential monthly evapotranspiration before and after the breakpoints exhibits varying increases. Under RCP 2.6, the means are 127.17 and 129.59 mm/month, reflecting a 1.9% surplus. In the case of RCP 4.5, the means stand at 125.75 and 127.63 mm/month, translating to a 1.5% excess. Lastly, under RCP 8.5, the means are 128.21 and 130.33 mm/month, indicating a 1.6% surplus (Fig. 6). The rising trend in potential monthly evapotranspiration is attributed to the parallel upward trajectory of future temperatures, a relationship also established by Ouhamdouch and Bahir (2017).

The findings from the simulation of future flow patterns under various scenarios (RCPs 2.6, 4.5, and 8.5) for the 2020–2050 period are presented in Fig. 7. In the context of the RCP 2.6 scenario, the Mann–Kendall test applied to the future flow series reveals a consistent upward trend across all three stations, namely Igrounzar, Zelten, and Adamna (refer to Table 2). This observed trend is further substantiated by the results of the Pettitt test (see Fig. 7a), indicating breakpoints in the series for January 2043, December 2042, and December 2042, corresponding to the Igrounzar, Zelten, and Adamna stations, respectively.

Table 1 Results of GR2M Calibration and validation for the Igrounzar, Zelten and Adamna stations: parameter values, Nash criterion and determination coefficient

Parameter	Units	Calibration (01/1991 to 12/1994)	Validation (01/1997 to 12/2001)
Igrounzar station			
X_1	mm	665.14	665.14
X_2	–	1.01	1.01
Nash	%	83	78
R^2	–	0.88	0.88
Zelten station			
X_1	mm	148.41	148.41
X_2	–	0.93	0.93
Nash	%	72	72
R^2	–	0.81	0.81
Adamna station			
X_1	mm	148.41	148.41
X_2	–	0.65	0.65
Nash	%	74	76
R^2	–	0.84	0.73

Specifically, for the Igrounzar station, the mean flow rates before and after the breakpoint amount to 0.45 m^3/s and 0.54 m^3/s, respectively, reflecting an estimated increase of 20% during the 2020–2050 period. In the case of the Zelten station, the average flow rates prior to and following the breakpoint are 0.59 m^3/s and 0.76 m^3/s, respectively, indicating an increase of 28.80% over the study period (as illustrated in Fig. 7a). Similarly, for the Adamna station, the flow series displays an average value before and after the breakpoint of 0.76 m^3/s and 0.91 m^3/s, respectively, representing an estimated surplus of 19.70% (as depicted in Fig. 7a).

In the context of the RCP 4.5 scenario, the results of both the Mann–Kendall and Pettitt tests applied to the future flow series for the three stations consistently indicate a downward trend, revealing deficits of 42.5%, 42.1%, and 40.6% for the Igrounzar, Zelten, and Adamna stations, respectively (as demonstrated in Fig. 7b). The identified breakpoints in the future flow series correspond to June 2040 for Igrounzar, May 2039 for Zelten, and May 2040 for Adamna stations.

In line with the RCP 8.5 scenario, a trend similar to that observed in the RCP 2.6 scenario emerges. The Mann–Kendall test results (as detailed in Table 2) and Pettitt test outcomes, with a 5% confidence interval (as represented in Fig. 7c), consistently demonstrate an upward trend in future flows. Specifically, this upward trend amounts to 44.4%, 53.8%, and 43.7% for the Igrounzar, Zelten, and Adamna series, respectively, projected by the year 2050.

Figure 8 presents a comparative analysis of observed and projected flows. It's worth noting that the observed flows at the Adamna station are approximately twice as large as those recorded at the Igrounzar and Zelten stations. This disparity arises from the fact that Wadi Ksob, where the Adamna station is located, results from

5 Conclusion

Similar to many semi-arid basins, the Essaouira Basin is confronting a reduction in precipitation resulting from the impacts of climate change. Given the interconnected nature of flow and rainfall patterns within these basins, this research was undertaken to assess the repercussions of climate change on the hydrological dynamics in the Essaouira basin during the period spanning from 2020 to 2050.

In pursuit of this objective, the Rural Genius GR2M model was employed to simulate flow patterns predicated on the interplay of rainfall and evapotranspiration. The Mann–Kendall and Pettitt tests served as valuable tools for scrutinizing the consistency and directionality of the time series analyzed. Notably, the time series of potential monthly evapotranspiration (ETP) for the period from 1978 to 2005 exhibited an upward trajectory, indicating a 4.2% increase. In contrast, when examined under the aegis of the 2.6, 4.5, and 8.5 RCPs of CIMP5, the rate of trend evolution stood at 1.9%, 1.5%, and 1.6%, respectively.

The examination of the interrelationship between rainfall and flows in the Essaouira basin during the 1978–2005 period underscored a discernible cause-and-effect connection between these two parameters. Remarkably, this causal association retained its validity across all zones characterized by arid and semi-arid climates.

A glimpse into the future flows at the Igrounzar station, as modeled using the GR2M framework under the RCP 2.6 and 8.5 scenarios, reveals a burgeoning upward trend of 20% and 44.4%, respectively, by the year 2050. In stark contrast, the envisaged future flow series during the same temporal scope but under the aegis of the RCP 4.5 scenario points to a distinct downward trajectory of 42.5%.

Similarly, the projections for the Zelten station illuminate future flow dynamics marked by a pronounced upward trend of 28.8% in the RCP 2.6 scenario, 53.8% in the RCP 8.5 scenario, and a contrary downward trend of 42.1% in the RCP 4.5 scenario. Looking at the future flows anticipated at the Adamna station by 2050, an upward shift of 19.70% and 43.7% is anticipated under the RCPs 2.6 and 8.5 scenarios, respectively, while a descending trend of 40.6% is projected under the RCP 4.5 scenario.

These findings have significant implications for water resource conservation and management within the Essaouira watershed, suggesting the potential for the construction of hill reservoirs along the Igrounzar, Zelten, and Ksob Wadi. These hydraulic structures hold promise in moderating water flow during periods of flooding and, consequently, replenishing aquifers. The adoption of hill dams remains a pertinent strategy for watersheds grappling with arid, semi-arid, and Saharan climatic conditions.

References

Agoumi A (1999) Introduction à la problématique des changement climatiques. Projet Magrébin sur les changements climatiques RAB/94/G31: SOMIGRAF mai 1999, 43p

Bahir M, Mennani A (2002) Problématique de la gestion des eaux souterraines au Maroc. Estud Geológicos 58:103–108
Bahir M, Ouhamdouch S (2020) Groundwater quality in semi-arid environments (Essaouira Basin, Morocco). Carbonates Evaporites 35:1–16. https://doi.org/10.1007/s13146-020-00576-7
Bahir M, Mennani A, Jalal M, Youbi N (2000) Contribution à l'étude des ressources hydriques du bassin synclinal d'Essaouira (Maroc). Estud Geol 56:185–195
Bahir M, Jalal M, Blavoux (2001) Apport de l'hydrochimie isotopique à la connaissance des zones de recharge en zones aride et semi-arides: Cas du bassin de Meskala-Kourimat. J Environ Hydrol
Bahir M, Carreira, P Silva, MO, Fernandes P (2008) Caractérisation hydrodynamique, hydrochimique et isotopique du système aquifère de Kourimat (Bassin d'Essaouira, Maroc). Estud Geol 64:61–73. https://doi.org/10.3989/egeol.08641433
Bahir M, Chkir N, Trabelsi R, Ammar H, Friha, Zouari K, Chamchati H (2012) Hydro-geochemical behaviour of two coastal aquifers under severe climatic and human constraints: comparative study between Essaouira basin in Morocco and Jeffara basin in Tunisia. Int J Hydrol Sci Technol 2:75–100. https://doi.org/10.1504/IJHST.2012.045940
Bahir M, El Moukhayar R, Youbi N, Chamchati H, Chkir Ben Jemaa N (2013) Management and protection of groundwater resources in semi-arid zones: application of hydrochemical methodologies to Essaouira Synclinal Basin, Morocco. Int J Hydrol Sci Technol 3:160–175
Bahir M, El Moukhayar R, Chkir N, Chamchati H, Fernandes PG, Carreira PM (2013b) Groundwater chemical evolution in the Essaouira Aquifer Basin—NW Morocco. Open J Mod Hydrol 3:8. https://doi.org/10.4236/ojmh.2013.33017
Bahir M, El Moukhayar R, Chamchati H, Youbi N, Carreira, PM, Chkir N (2014) Recharge and hydrogeochemical evolution groundwater in semi-arid zone (Essaouira Basin, Morocco). Comun Geol 3:651–653. http://www.lneg.pt/iedt/unidades/16/paginas/26/30/185
Bahir M, Ouhamdouch S, Carreira PM (2016) La ressource en eau au Maroc face aux changements climatiques; cas de la nappe Plio-Quaternaire du bassin synclinale d'Essaouira. Comun Geol 103(1):35–44
Bahir M, Ouhamdouch S, Carreira PM (2018a) Geochemical and isotopic approach to decrypt the groundwater salinization origin of coastal aquifers from semi-arid areas (Essaouira basin, Western Morocco). Environ Earth Sci 77:485. https://doi.org/10.1007/s12665-018-7663-4
Bahir M, Ouhamdouch S, Carreira PM (2018b) Isotopic and geochemical methods for studying water–rock interaction and recharge mode: application to the Cenomanian-Turonian and Plio-Quaternary aquifers of Essaouira Basin, Morocco. Mar Freshwater Res 69:1290–1300
Bahir M, Ouazar D, Ouhamdouch S (2018c) Characterization of mechanisms and processes controlling groundwater salinization in coastal semi-arid area using hydrochemical and isotopic investigations (Essaouira basin, Morocco). Environ Sci Pollut Res 25:24992–25004
Bahir M, Ouazar D, Ouhamdouch S (2019) Hydrogeochemical investigation and groundwater quality in Essaouira region, Morocco. Mar Freshwater Res 70:1317–1332
Bahir M, Ouhamdouch S, Ouazar D, El Moçayd N (2020) Climate change effect on groundwater characteristics within semi-arid zones from western Morocco. Groundw Sustain Dev 11:1–14. https://doi.org/10.1016/j.gsd.2020.100380
Bahir M, Ouhamdouch S, Ouazar D, Chehbouni A, Ouarani M, El Mountassir O (2021a) Groundwater quality of the alluvial and carbonate aquifers of Essaouira basin (Morocco). Carbonates Evaporites 36:23. https://doi.org/10.1007/s13146-021-00697-7
Bahir M, El Mountassir O, Ouazar D, Chehbouni A, Carreira PM (2021b) Stable isotope and quality of groundwater around ksob sub-basin, Essaouira, Morocco. Sustain Water Resour Manag. https://doi.org/10.1007/s40899-021-00553-5
Bahir M, El Mountassir O, Ouazar D, Carreira PM (2021c) Use of WQI and isotopes to assess groundwater quality of coastal aquifers (Essaouira, Morocco). In: Abrunhosa M, Chambel A, Peppoloni S, Chaminé HI (eds) Advances in geoethics and groundwater management: theory and practice for a sustainable development. Advances in science, technology & innovation (IEREK Interdisciplinary Series for Sustainable Development). Springer, Cham. https://doi.org/10.1007/978-3-030-59320-9_51

Bahir M, El Mountassir O, Ouazar D, Carreira PM (2021d) Hydrochemical analysis and evaluation of groundwater quality in Ouazi basin (Essaouira, Morocco). In: Abrunhosa M, Chambel A, Peppoloni S, Chaminé HI (eds) Advances in geoethics and groundwater management: theory and practice for a sustainable development. Advances in science, technology & innovation (IEREK Interdisciplinary Series for Sustainable Development). Springer, Cham. https://doi.org/10.1007/978-3-030-59320-9_50

Bahir M, El Mountassir O, Chehbouni A, Dhiba D, El Jiar H, Carreira PM (2022) Hydrogeochemical and isotopic assessment for characterizing groundwater quality and recharge processes in the Essaouira Basin, Northwestern Morocco. Arab J Geosci 15(7):1–21. https://doi.org/10.1007/s12517-022-09817-6

Belarbi H, Touaibia B, Boumechra N, Amiar S, Baghli N (2015) Sécheresse et modification de la relation pluie-débit: Cas du bassin versant de l'Oued Sebdou (Algérie Occidentale). Hydrolog Sci J 62:124–136

Dezetter A, Girard S, Paturel JE, Mahé G, Ardoin-Bardin S, Servat E (2008) Simulation of runoff in West Africa: Is there a single data-model combination that produces the best simulation results? J Hydrol 354:203–212

Driouech F, Mahé G, Déqué M, Dieulin C, El Heirech T, Milano M, Benabdelfadel A, Rouche N (2010) Evaluation d'impacts potentiels de changements climatiques sur l'hydrologie du bassin versant de la Moulouya au Maroc. In: Global change: facing risks and threats to water resources (Proceedings 6th FRIEND world conference, Fes, Morocco, October 2010), 561–567. IAHS Publ. 340. IAHS Press

Duffaud F, Brun L, Planchot B (1966) Bassin du Sud-Ouest Marocain (SW Morocco Basin). In: Reyre (ed) Bassin Sédimentaire du Littoral Africain, partie I, Paris, pp 5–12

El Gayar A, Hamed Y (2018) Climate change and water resources management in Arab countries. In: Kallel A, Ksibi M, Ben Dhia H, Khélifi N (eds) Recent advances in environmental science from the Euro-Mediterranean and surrounding regions. EMCEI 2017. Advances in science, technology & innovation (IEREK Interdisciplinary Series for Sustainable Development). Springer, Cham

El Mountassir O, Bahir M, Ouazar D, Ouhamdouch S, Chehbouni A, Ouarani M (2020) The use of GIS and water quality index to assess groundwater quality of krimat aquifer (Essaouira; Morocco). SN Appl Sci 2:871. https://doi.org/10.1007/s42452-020-2653-z

El Mountassir O, Bahir M, Ouazar D, Chehbouni A, Carreira PM (2021a) Temporal and spatial assessment of groundwater contamination with nitrate using nitrate pollution index (NPI), groundwater pollution index (GPI), and GIS (case study: Essaouira basin, Morocco). Environ Sci Pollut Res 29:17132–17149

El Mountassir O, Bahir M, Ouazar D, Chehbouni A, Carreira PM (2021b) Geochemical and isotopic evidence of groundwater salinization processes in the Essaouira region, north-west coast, Morocco. SN Appl Sci. https://doi.org/10.1007/s42452-021-04623-3

El Mountassir O, Ouazar D, Bahir M, Chehbouni A, Carreira PM (2021c) GIS-based assessment of aquifer vulnerability using DRASTIC model and stable isotope: a case study on Essaouira basin. Arab J Geosci 14:321. https://doi.org/10.1007/s12517-021-06540-6

El Mountassir O, Bahir M, Ouazar D, Carreira PM (2021d) Nitrate pollution in groundwater of the Ouazi basin: case of Essaouira (Southwestern Morocco). In: Abrunhosa M, Chambel A, Peppoloni S, Chaminé HI (eds) Advances in geoethics and groundwater management: theory and practice for a sustainable development. Advances in science, technology & innovation (IEREK Interdisciplinary Series for Sustainable Development). Springer, Cham. https://doi.org/10.1007/978-3-030-59320-9_49

El Mountassir O, Bahir M, Ouazar D, Carreira PM (2021e) For a better understanding of recharge and salinization mechanism of a Cenomanian–Turonian aquifer. In: Abrunhosa M, Chambel A, Peppoloni S, Chaminé HI (eds) Advances in geoethics and groundwater management: theory and practice for a sustainable development. Advances in science, technology & innovation (IEREK Interdisciplinary Series for Sustainable Development). Springer, Cham. https://doi.org/10.1007/978-3-030-59320-9_42

El Mountassir O, Bahir M, Ouazar D, Chehbouni A, Carreira PM (2022a) Evaluation of nitrate source and its distribution in the groundwater of Essaouira basin. Sustain Water Resour Manag 8:28. https://doi.org/10.1007/s40899-022-00609-0

El Mountassir O, Bahir M, Chehbouni A, Dhiba D, El Jiar H (2022b) Assessment of groundwater quality and the main controls on its hydrochemistry in a changing climate in Morocco (Essaouira Basin). Sustainability 14:8012. https://doi.org/10.3390/su14138012

El Mountassir O, Bahir M, Chehbouni A, Dhiba D, El Jiar H (2022c) Assessment of groundwater quality and the main controls on its hydrochemistry in a changing climate in Morocco (Essaouira Basin). Sustainability 14(13):8012. https://doi.org/10.3390/su14138012

El Mountassir O, Bahir M (2023) The Assessment of the groundwater quality in the coastal aquifers of the essaouira basin, Southwestern Morocco, using hydrogeochemistry and isotopic signatures. Water 15:1769.

Kendall MG (1975) Multivariate nonparametric tests for trend in water quality. Water Resour Bull 24:505–512

Kouamė KF, Kouassi AM, N'guessan BTM, Kouao JM, Lasm T, Saley MB (2013) Analyse de tendances dans la relation pluie-débit dans un contexte de changements climatiques: cas du bassin versant du N'zo-Sassandra (Ouest de la Côte d'Ivoire). IJIAS 2:92–103

Makhlouf Z, Michel C (1994) A two-parameter monthly water balance model for French watersheds. J Hydrol 162:299–318

Mann HB (1945) Nonparametric tests against trend. Econometrical 13:245–259

Mouelhi S, Michel C, Perrin C, Andréassian V (2006) Stepwise development of a two-parameter monthly water balance model. J Hydrol 318:200–214

Nash JE, Sutcliffe JV (1970) River discharge forecasting through conceptual models. Part I - a discussion of principles. J Hydrol 10:282–290

Ouhamdouch S, Bahir M (2017) Climate change impact on future rainfall and temperature in semi-arid areas (Essaouira Basin, Morocco). Environ Process 4:975–990

Ouhamdouch S, Bahir M, Carreira P (2018) Impact du changement climatique sur la ressource en eau en milieu semi-aride: exemple du bassin d'Essaouira (Maroc). Rev Des Sci De L'Eau 31:13–27

Perrin C, Michel C, Andréassian V (2003) Improvement of a parsimonious model for stream discharge simulation. J Hydrol 279:275–289

Pettitt AN (1979) A non-parametric approach to the change-point problem. Appl Statist 28(2):126–135

Peybernès B, Bouaouda MS, Almeras Y, Ruget CH, Cugny P (1987) Stratigraphy of the Lias-Dogger deposits from Essaouira coastal basin (Morocco) before and during the beginning of oceanic accretion in the central Atlantic; comparisons with the Agadir basin. CR Acad Sci 305:1449–1555

Rwasoka DT, Madamombe CE, Gumindoga W, Kabobah AT (2014) Calibration, validation, parameter identifiability and uncertainty analysis of a 2-parameter parsimonious monthly rainfall-runoff model in two catchments in Zimbabwe. Phys Chem Earth 67–69:36–46

Sakaa B, Boulghobra N, Chaffai H, Hani A, Djabri L (2015) Application of GR2M for rainfall-runoff modeling in Kébir Rhumel watershed, North East of Algeria. World Appl Sci J 33(10):1623–1630

Solomon S, Qin D, Manning M, Chen Z, Marquis M, Averyt KB, Tignor M, Miller HL (eds) (2007) Climate change 2007: "e physical science basis; contribution of working group I to the fourth assessment report of the intergovernmental panel on climate change. Cambridge University Press, Cambridge, 996 p

Thornthwaite CW (1948) An approach toward a rational classification of climate. Geor Review 38:55–94

Zeroual A, Meddi M, Bensaad S (2013) The impact of climate change on river flow in arid and semi-arid rivers in Algeria. In: Proceedings of H01, IAHS-IAPSO-IASPEI Assembly, Gothenburg, Sweden, July 2013 (IAHS Publ. 359, 2013)

Prof. Dr. Eng. Mohammed Bahir, Ph.D., Hydrogeological Engineer, has published over 200 research/professional papers and reports. He co-authored more than a dozen books and book chapters. He supervised and/or co-supervised more than 20 PhD students. He is a reviewer of many international scientific journals with high impact factors, including Environment, Development, and Sustainability Journal, Environmental Earth Sciences Journal, Hydrological Sciences Journal, Sécheresse, Gaia, Communicaçoés International, Science of the Total Environment, Marine and Freshwater Research Journal, Proceedings of the Indian National Science Academy, and Groundwater for Sustainable Development. He also coordinates more than a dozen research projects.

The Asia Research Awards, in association with Times of Research and the World Research Council, announce that Professor Mohammed Bahir has been selected as the recipient of the prestigious ASTRA 2023 (Asia's Science, Technology, and Research Awards).

Chapter 7
Evolution of Historical and Future Precipitations and Temperatures Within Essaouira Basin Under Climate Change Effect

Mohammed Bahir, Otman El Mountassir, and Mohamed Behnassi

Abstract In recent decades, the specter of global warming has loomed large, emerging as a substantial and impending threat to the well-being of the global populace. Within the context of nations particularly susceptible to the repercussions of climate change, Morocco, notably its regions characterized by arid and semi-arid climatic conditions, assumes a prominent position as one of the most exposed to vulnerability. To gain deeper insights into the intricate interplay between climate change and the availability of water resources, a comprehensive approach has been adopted. This approach encompasses a climatological analysis encompassing factors such as rainfall and temperature, alongside an exploration of groundwater levels, salinity, and isotopic methodologies. A meticulous examination of the annual precipitation patterns has been conducted, drawing upon the graphical representation offered by the Nicholson rainfall index. Furthermore, this investigation has employed statistical tools, notably the Pettitt and Mann–Kendall tests, to probe the nuances of these climatic shifts. The results that have emerged indicate an overarching negative trend in the basin, signifying a reduction in annual precipitation ranging from 12 to 16% over the span from 1978 to 2015. This decline in the average annual precipitation levels is juxtaposed against a concomitant escalation in temperatures, manifesting as a highly substantial warming effect. Specifically, this warming amounts to 1.2 °C in the downstream segment of the study region and 2.3 °C in its upstream counterpart, a

M. Bahir (✉) · O. El Mountassir
High Energy and Astrophysics Laboratory, Faculty of Sciences Semlalia, Cadi Ayyad University, Marrakech, Morocco
e-mail: bahir@uca.ac.ma

O. El Mountassir
e-mail: otman.elmountassir@ced.uca.ma

M. Behnassi (✉)
Economic, Social and Environmental Council (ESEC), Rabat, Morocco
e-mail: behnassi@gmail.com

Center for Environment, Human Security & Governance (CERES), Agadir, Morocco

Ministry of Higher Education and Innovation, Rabat, Morocco

M. Behnassi et al. (eds.), *The Water, Climate, and Food Nexus*,
https://doi.org/10.1007/978-3-031-50962-9_7

phenomenon that underscores the pronounced continentality of temperature patterns in this geographical area. In light of these findings, there exists a compelling rationale for the inclusion of stringent local and global policies aimed at mitigating the emission of greenhouse gases, fostering the utilization of clean and renewable energy sources, and engendering awareness and education among the populace. These endeavors are of paramount importance, serving to expedite the adoption of adaptive behaviors capable of addressing the challenges posed by a warmer and drier climate. This is particularly pertinent for individuals inhabiting regions characterized by semi-arid, arid, and Saharan climates.

Keywords Climate change · Water resources · Morocco · Pettitt and Mann–Kendall · Essaouira basin

1 Introduction

In recent decades, the specter of global warming has assumed a palpable presence, emerging as a formidable threat to humanity's well-being (Kaid Rassou et al. 2006; Chkir et al. 2008; Bouya et al. 2011). Of particular significance is Morocco, a nation characterized by its arid and semi-arid climates, and notably, it stands out as one of the most vulnerable in the face of climate change's impacts (Boko et al. 2007; Kuriqi 2014; Kuriqi et al. 2016; Bahir et al. 2000, 2020a, b). This vulnerability is compounded by the spatial and temporal variability in rainfall, with a discernible decrease observed toward the southern reaches of the country (Driouech et al. 2013; Babqiqi 2014).

The Fourth Climate Change Report by the Intergovernmental Panel on Climate Change (IPCC 2007) has illuminated the likelihood of Morocco experiencing a precipitous decline in precipitation, which could culminate in a substantial reduction of up to 20%. This adverse shift is coupled with a prospective rise in temperatures, projected to reach an increment of 2.5–5.5 °C by the end of the century. However, the global climate models hitherto employed by the IPCC suffer from limited resolution, with a notable absence of precise topographical, vegetative, and soil information. In light of the imperative need for robust climate change assessments, particularly concerning socio-economic sectors like agriculture and water resources, there is an exigent requirement for high-resolution climate scenarios.

Several investigations have probed the patterns of rainfall in Morocco, revealing a preponderance of years marked by rainfall deficits over surplus years. These studies have collectively identified a pervasive downward trajectory, marked by reductions ranging from 12 to 23%, contingent on the region in question (Stour and Agoumi 2009; Driouech 2010; Driouech et al. 2013; Ouhamdouch et al. 2016; Ouhamdouch and Bahir 2017; Ouhamdouch 2020). Furthermore, comparisons of mean annual temperatures over two distinct periods, 1971–1980 and 1998–2007, have unveiled an upward shift, exemplified by an increase spanning from 0.3 to 2.5 °C, depending on the specific geographical region under scrutiny (Babqiqi 2014; Blunden and Derek

2015; Bahir et al. 2016). This contemporary scenario portends the stark reality of limited water resources, which currently hover at an average of 606 m^3 per person per year, indicative of a state of acute water stress.

The principal objective of this study revolves around the assessment of groundwater conditions within a semi-arid Moroccan environment and their interplay with global warming, with the Essaouira Basin serving as an illustrative example. To this end, a comprehensive analysis has been executed, encompassing the evolution of climatic parameters such as temperature and precipitation, groundwater levels, salinity, and the isotopic composition of groundwater.

Situated on the Atlantic Coast of Morocco, the Essaouira Basin occupies a spatial expanse of 6000 km^2 and comprises several distinct aquifer systems belonging to two synclinal units: the Bouabout unit in the eastern domain and the Essaouira synclinal unit in the western territory. The boundary demarcating these two units is delineated by the Triassic formations' outcrop, notably the Tidzi diapir (Fig. 1).

Within the geological and structural context, the Triassic and Jurassic formations exhibit minimal outcrops and are concentrated within the central regions of anticlines, particularly notable in the Jbel Hadid NW, Jbel Amsitene SW, and the diapir Tidzi, as illustrated in Fig. 1.

The Triassic formations encompass components characterized by saliferous red clays, doleritic basalts, and sandstone pelites. In contrast, the Jurassic formations are primarily constituted of carbonate deposits, encompassing limestones and dolomites, interspersed with marls rich in gypsum and anhydrite. The lower Cretaceous stratum is predominantly composed of marls and limestones, boasting an average thickness of 200 m (Duffaud et al. 1966). The middle Cretaceous epoch initiates with deposits of Martian sandstone from the Aptian, averaging 60 m in thickness, followed by pyritic marl of the Albian, which presents an approximate thickness of 100 m. Marls continue to dominate through the Cenomanian, characterized by a thickness of around 200 m and a significant presence of anhydrites, interspersed with some limestone units. These marls form the foundational bedrock of the Turonian aquifer, marked by fractured flint and averaging 60 m in thickness. The Cretaceous geological sequence culminates in dolomitic marls and limestones, surmounted by gypsiferous and siliceous gray marls, occasionally interspersed with Senonian sandstone formations. These stratigraphic layers demarcate the boundaries of the two aquifers, namely the Turonian and Plio-Quaternary aquifers, situated within the synclinal zone of Essaouira (Duffaud et al. 1966).

The architectural configuration of the study area is attributed to tectonic processes, with Tertiary age compression (Souid 1983; Amghar 1995) and distasic tectonics stemming from the Triassic and Jurassic eras (Duffaud et al. 1966; Souid 1983; Broghton and Trepanier 1993) culminating in the formation of anticlines and synclines. This structural organization facilitates the existence of multiple aquifer systems distributed as follows: In the upstream region, aquifers are nestled within the limestone and dolomitic limestones of the Cenomanian–Turonian. The impervious base and roof of this system are respectively secured by the gray clays of the Lower Cenomanian and the Senonian white marls (Bahir et al. 2007). Conversely, in the

Fig. 1 Isohyets map for the period 1978–2004 within Essaouira basin. *Source* Ouhamdouch et al. (2019) and Bahir et al. (2020c)

downstream sector, groundwater resources are harbored within two primary reservoirs: the Plio-quaternary, characterized by a marine calcareous sandstone matrix housing phreatic aquifers, and the Turonian, accommodating a water table that is swiftly confined beneath the Senonian marls, likely in direct proximity to the Plio-quaternary strata at the fringes of this geological structure (Bahir et al. 2007; Jalal et al. 2001; El Mountassir et al. 2022a, b, c).

2 Data and Methods

The climatological data, encompassing precipitation and temperature records, employed in this research were generously provided by the Hydraulic Basin Agency of Tensift (ABHT), as detailed in Annex I. Multiple datasets pertaining to precipitation measurements were also made accessible for analysis. In the climatic examination, data originating from the Adamna, Igrounzar, and Essaouira stations were utilized to scrutinize precipitation trends, while the evaluation of temperature patterns exclusively relied upon the last two stations. These two stations were the sole sources that provided concurrent data on both precipitation and temperature, thus rendering them conducive for a comprehensive climatic assessment.

To ensure the suitability of a dataset for the study of climate change, it is imperative that the observation period be of sufficient duration to accurately reflect the climatic conditions of the region. The World Meteorological Organization (WMO) has stipulated that an observation period exceeding 30 years is necessary to establish climatic trends. In our specific case, the three stations mentioned herein readily satisfy this criterion, with observation periods spanning over 30 years for both precipitation and temperature data.

For this study, the predictands were derived from the maximum daily temperatures, daily minimum temperatures, and daily precipitation measurements acquired during the periods 1987–2004 and 1978–2011 from the Igrounzar station. The model employed for this investigation is the second-generation Canadian Earth System Model (CanESM2), developed by the Canadian Centre for Climate Modeling and Analysis (CCCma) under the purview of Environment Canada. The rationale behind selecting this model resides in its unique capability to furnish daily predictor variables, facilitating direct integration into the Statistical DownScaling Model (SDSM). SDSM is an instrumental decision support tool for evaluating local climate change impacts, harnessing a robust statistical downscaling methodology. CanESM2 outputs were procured for three distinct climate scenarios, namely RCP 2.6, RCP 4.5, and RCP 8.5, and were subsequently employed within the ambit of this study.

The RCP 2.6 scenario was crafted by the IMAGE modeling team affiliated with the PBL Netherlands Environmental Assessment Agency (Thomson et al. 2011; Wayne 2013). This particular RCP delineates a mitigation scenario aimed at curtailing the escalation of global mean temperatures to a maximum of 2 °C. A pivotal assumption underpinning this scenario pertains to the rapid worldwide dissemination and immediate implementation of new, energy-efficient technologies.

The RCP 4.5, on the other hand, was formulated by the GCAM modeling group within the Pacific Northwest National Laboratory's Joint Global Change Research Institute (JGCRI) in the United States. This scenario represents a stabilization framework (Wayne 2013), with radiative forcing stabilizing at 4.5 W/m^2 in the year 2100, never exceeding this threshold (Thomson et al. 2011). Key postulates within this scenario encompass a global population peak surpassing 9 billion by 2065, subsequently declining to 8.7 billion by 2100, reductions in energy consumption, heightened fossil fuel utilization, substantial increases in renewable energy sources, and a considerable expansion of forested areas as a mitigation strategy.

Finally, the RCP 8.5 scenario was developed through the Integrated Assessment Framework by the International Institute for Applied System Analysis (IIASA), employing the MESSAGE model (Thomson et al. 2011; Wayne 2013). This scenario characterizes a high-emission trajectory typified by escalating greenhouse gas (GHG) emissions over time (Riahi et al. 2011). Salient assumptions integral to this pathway involve a continuous global population increase, reaching 12 billion by 2100, sluggish income growth with modest technological advancements, prolonged high energy demand, a shift towards coal-intensive technologies, and high emissions in the absence of comprehensive climate change policies (Riahi et al. 2011).

In addition to the aforementioned data sources, we incorporated large-scale atmospheric variables from the National Center for Environmental Prediction (NCEP) and the National Center for Atmospheric Research (NCAR) reanalysis project to establish a statistical linkage with the observed station data. These NCEP/NCAR predictor variables, generously provided by CCCma, were carefully selected to ensure temporal consistency and span the timeframe from 1961 to 2005. The utilization of identical variables from NCEP/NCAR aligns with the atmospheric parameters available from CanESM2 outputs, thereby facilitating a unified analytical approach. Both CanESM2 outputs and NCEP/NCAR reanalysis project data, specifically the Box_126x_44y dataset, furnished an identical set of 26 predictor variables (as delineated in Table 1). These datasets were sourced from the Canadian Climate Data and Scenarios website and constituted the basis for our study. To derive predictions regarding climate change within the Essaouira basin for the upcoming period spanning from 2018 to 2050, these datasets were meticulously processed by means of the SDSM and XLSTAT® software.

The SDSM model, drawing on previous research (Gagnon et al. 2005; Rashid and Mukand 2012; Parvaze et al. 2016), emerges as the preferred model for projecting local climate change.

The selection of predictors for this study involved a rigorous screening process, incorporating the evaluation of correlation coefficients (R), coefficients of determination (R^2), and corresponding *p*-values with respect to the predictands (as delineated in Table 2). Following this comprehensive selection procedure, we proceeded to the calibration and validation phases. The dataset spanning 28 years (1978–2005) for precipitation observations and the dataset spanning 18 years (1987–2004) for temperature observations were partitioned into two distinct groups. In the context of precipitation, the initial 22 years (1978–1999) served as the calibration dataset, while the subsequent 6 years (2000–2005) were reserved for validation. For temperature data, calibration encompassed the initial 13 years (1987–1999), with validation conducted over the remaining 5 years (2000–2004). The SDSM model was fine-tuned for this study by incorporating selected NCEP predictors through both annual and monthly sub-models.

Table 1 Predictor variables of the Global Climate Model (GCM) and the National Center for Environmental Prediction (NCEP)

#	Predictor	Description	#	Predictor	Description
1	mslp	Mean sea level pressure	14	p8_f	850 hPa air flow strength
2	p1_f	Surface air flow strength	15	p8_u	850 hPa zonal velocity
3	p1_u	Surface zonal velocity	16	p8_v	850 hPa meridional velocity
4	p1_v	Surface meridional velocity	17	p8_z	850 hPa vorticity
5	p1_z	Surface vorticity	18	p8_th	850 hPa wind direction
6	p1_th	Surface wind direction	19	p8_zh	850 hPa divergence
7	p1_zh	Surface divergence	20	p500	500 hPa geopotential height
8	p5_f	500 hPa air flow strength	21	p850	850 hPa geopotential height
9	p5_u	500 hPa zonal velocity	22	Prcp	Surface precipitation
10	p5_v	500 hPa meridional velocity	23	s500	Specific humidity at 500 hPa height
11	p5_z	500 hPa vorticity	24	s850	Specific humidity at 850 hPa height
12	p5-th	500 hPa wind direction	25	shum	Surface-specific humidity
13	p5_zh	500 hPa divergence	26	temp	Surface mean temperature

Source Ouhamdouch and Bahir (2017) and Bahir et al. (2020c)

To depict the temporal series of historical and future local climatic parameters, we employed two distinct methods: a graphical approach predicated on the analysis of future weighted rainfall indices, and a statistical approach involving the application of two tests—Pettitt's test for detecting structural breaks and the Mann–Kendall test for trend detection. The calculation of the rainfall index adhered to the methodology established by Nicholson, and the statistical tests were executed utilizing the XLSTAT® software.

2.1 *Rainfall Index (RI)*

The rainfall index measures the deviation from the average of the future rainfall series obtained from the SDSM model of the Igrounzar station. The annual rainfall index 'RI' is the reduced centered variable of the annual rainfall. It is obtained by the Eq. (1):

$$RI = \frac{X_i - X_m}{\sigma} \quad (1)$$

where x_i is the rainfall of the year i; x_m is the series average; and σ is the standard deviation.

Table 2 Selected set of predictor variables

Predictand	Predictor	R	R^2	*p*-value
Precipitation	p1_v	0.606	0.367	< 0.0001
	p8_f	0.633	0.4	< 0.0001
	p8_4	0.518	0.268	< 0.0001
	prcp	0.786	0.619	< 0.0001
Maximum temperature	p1_th	0.605	0.366	< 0.0001
	p5_v	0.628	0.395	< 0.0001
	p500	0.743	0.552	< 0.0001
	shum	0.757	0.573	< 0.0001
	temp	0.832	0.692	< 0.0001
Mean temperature	p1_th	0.604	0.365	< 0.0001
	p5_v	0.669	0.448	< 0.0001
	p500	0.831	0.691	< 0.0001
	shum	0.84	0.706	< 0.0001
	temp	0.903	0.816	< 0.0001
Minimum temperature	p1_th	0.593	0.351	< 0.0001
	p5_v	0.641	0.411	< 0.0001
	p500	0.76	0.577	< 0.0001
	shum	0.788	0.62	< 0.0001
	temp	0.835	0.696	< 0.0001

Note R = correlation coefficient; R^2 = coefficient of determination
Source Ouhamdouch and Bahir (2017) and Bahir et al. (2020c)

To better visualize the interannual fluctuation in the pluviometry, the seasonal variations are eliminated by weighting the annual pluviometric totals through Eqs. (2 to 6) recommended by Assani (1999):

$$x_t = 0.06x_{(t-2)} + 0.25x_{(t-1)} + 0.38x_{(t)} + 0.25x_{(t+1)} + 0.06x_{(t+1)} \quad (2)$$

where $3 \leq t \leq (n-2)$; x_t is the total weighted rainfall of the t term; $x_{(t-2)}$ and $x_{(t-1)}$ are the observed rainfall totals immediately preceding the t term; and $x_{(t+2)}$ and $x_{(t+1)}$ are the observed rainfall totals of two terms immediately following the t term.

The weighted rainfall totals of the first two ($x_{(1)}$ and $x_{(2)}$) and the last two ($x_{(n-1)}$ and $x_{(n)}$) terms in the series are calculated as follows:

$$x_{(1)} = 0.54x_{(1)} + 0.46x_{(2)} \quad (3)$$

$$x_{(2)} = 0.25x_{(1)} + 0.50x_{(2)} + 0.25x_{(3)} \quad (4)$$

to be 2039. The calculated UMK is negative, indicative of this declining trend (see Table 7).

The climatic data encompassing temperature and rainfall, acquired through the diligent efforts of the Tensift Hydraulic Basin Agency (ABHT), pertaining to the Igrounzar station, extend from the year 1988 to 2004. In tandem with this historical dataset, projections under the RCPs 2.6, 4.5, and 8.5 scenarios for the period spanning from 2018 to 2050 have facilitated the creation of ombrothermic diagrams (as depicted in Fig. 12). A meticulous comparison between these two diagrams underscores the prevailing temperature increase and concurrent precipitation decline. Evidently, the forecasts indicate a substantial reduction in precipitation, particularly in the months of October, November, December, and January, alongside an elevation in temperature.

The study area, characterized by a semi-arid climate, anticipates the return of rainfall during the period from 2018 to 2050, predominantly in April, May, June, July, August, and September. These months constitute the dry season within semi-arid regions and are likely to witness rainfall in the form of sporadic storms. An intriguing shift can be observed under all three scenarios (RCP 2.6, 4.5, and 8.5)

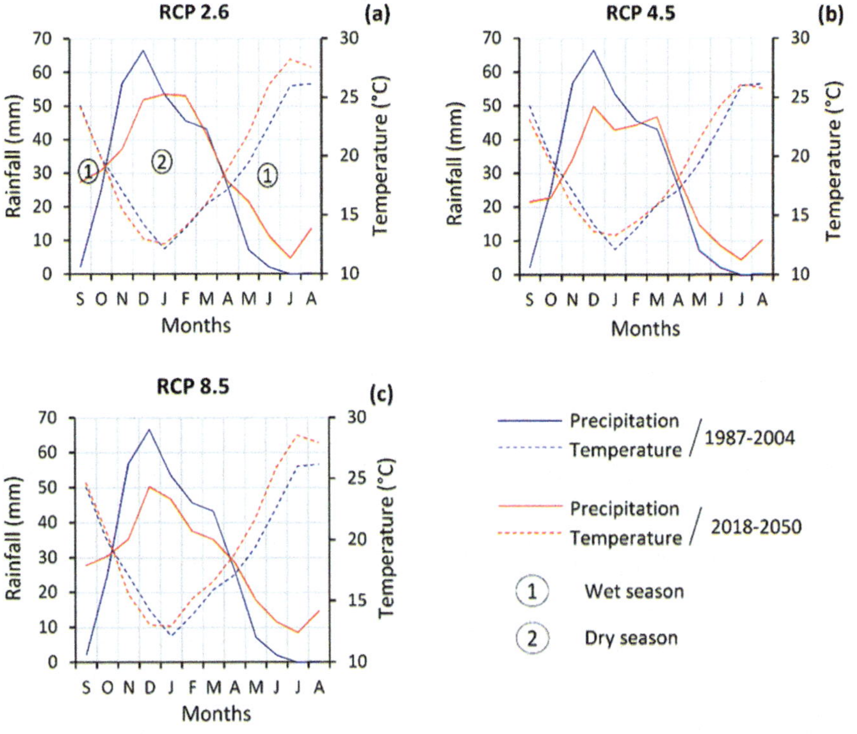

Fig. 12 Ombrothermic diagram of Igrounzar station for the period 1987–2004 (in blue) and the period 2018–2050 (in red) according to **a** RCP 2.6, **b** RCP 4.5, and **c** RCP 8.5 scenarios. *Source* Ouhamdouch and Bahir (2017) and Bahir et al. (2020c)

wherein November, which formerly belonged to the wet season during the 1988–2004 timeframe, is expected to transition into the dry season throughout 2018–2050. Furthermore, November and December, during the 2018–2050 period under all three scenarios, will exhibit a cooler climate compared to the corresponding months during 1987–2004.

The months ranging from January to August are expected to experience higher temperatures under the RCP 8.5 scenario, as opposed to the 1987–2004 period, while September and October are projected to maintain a similarity to their historical counterparts. Within the RCP 4.5 scenario, the months of April through July are expected to be warmer in comparison to the 1987–2004 period, while August, September, and October are foreseen to resemble the conditions of the 1987–2004 timeframe. For the RCP 2.6 scenario, February, March, September, and October are poised to remain consistent with the 1987–2004 period, while April through August is predicted to exhibit warmer temperatures relative to the historical data. In summation, this analysis suggests a contraction of the wet season from five months (November to March) to four months (December to March) for the Essaouira basin. This transformation is expected to be emblematic of analogous semi-arid zones across the globe due to the shared climatic characteristics.

5 Conclusion

The findings of our analysis, employing the Nicholson rainfall index graphical method and statistical tests, specifically Pettitt and Mann–Kendall tests, have unveiled a discernible negative trend in annual precipitation across the Essaouira basin. Over the period from 1978 to 2015, a substantial decline of 12–16% has been observed. This pronounced decrease in the average annual precipitation is concomitant with a noteworthy rise in temperatures, signifying a considerable warming of 1.2 °C in the downstream region of the study area and an even more substantial increase of 2.3 °C in its upstream vicinity. This temperature escalation is attributed to the influence of continental conditions on the temperature patterns within the study area.

To downscale future temperature variables and precipitation, statistical downscaling models have been employed. These models establish a statistical relationship between extensive atmospheric variables, referred to as predictor variables, and the observed temperature and precipitation data, termed predictands. Within the context of a semi-arid locale, exemplified by the Essaouira basin in western Morocco, our investigation has revealed that both maximum and minimum temperatures are notably responsive to surface wind direction, meridional velocity at 500 hPa, geopotential height at 500 hPa, specific near-surface humidity, and mean temperature at 2 m, serving as key predictors. In contrast, precipitation exhibits a dependency on predictor variables such as surface meridional velocity, air flow strength at 850 hPa, zonal velocity at 850 hPa, and surface precipitation.

By integrating these models and complementing them with the Nicholson rainfall index, Pettitt's and Mann–Kendall statistical tests, we have discerned a prominent upward trajectory in temperature variables. These projections indicate the potential for a mean temperature increase of up to 0.72 °C by the mid-century, marking the conclusion of our study period, under the RCP 2.6 scenario. Under the RCP 4.5 and RCP 8.5 scenarios, the increase is projected to reach 0.57 °C and 0.69 °C, respectively.

The outlook for precipitation varies across different time intervals and under the auspices of the three scenarios. In the context of RCP 2.6 and 8.5 scenarios, a significant increase in precipitation of 12.50% and 21.33%, respectively, is anticipated. This predicted resurgence of precipitation, under these scenarios, is likely to manifest in the form of storms, a prevailing meteorological phenomenon characterizing precipitation in contemporary semi-arid regions. The return of precipitation is anticipated during the summer months, as elucidated by the presented ombrothermic diagrams. Conversely, the RCP 4.5 scenario foresees a reduction in mean annual precipitation by approximately 17.29%. The ombrothermic diagrams underscore a contraction of the wet season from five months, spanning November to March, to a condensed four-month period, encompassing December to March.

Consequently, the climate within semi-arid regions, analogous to our study area, seems to be veering towards increased aridity, characterized by heightened temperatures and decreased moisture levels. This necessitates immediate action on both national and global fronts. It is imperative that policies encompass endeavors to curtail greenhouse gas emissions, foster the utilization of clean and renewable energy sources, and impart educational initiatives to expedite the adoption of adaptive behaviors, particularly among the residents of semi-arid, arid, and Saharan climates, to contend with the looming challenges of a warmer and drier climate.

References

Ait Brahim Y, El Mehdi SM, Kouraiss K, Sifeddine A, Bouchaoua L (2017) Analysis of observed climate trends and high resolution scenarios for the 21st century in Morocco. JMES 8:1375–1384

Amghar M (1995) Tectonique et inversion géodynamiquepost-rift dans le Haut Atlas Occidental: Structures, instabilité tectonique et magmatisme liés à l'ouverture de l'Atlantique central et la collision Afrique-Europe. Thèse d'Etat es-Sciences, Université Cadi Ayyad, Maroc

Assani AA (1999) Analyse de la variabilité temporelle des précipitations (1916–1996) à Lubumbashi (Congo-Kinshasa) en relation avec certains indicateurs de la circulation atmosphérique (oscillation australe) et océanique (El Niño/La Niña). Sécheresse 10(4):245–252

Babqiqi A (2014) Changements Climatiques au Maroc : Etude du cas de la Région de Marrakech Tensift Al Haouz et implications sur l'agriculture à l'horizon 2030. Thèse de doctorat, Université Cadi Ayyad, Marrakech, Maroc

Bahir M, Mennani A, Jalal M, Youbi N (2000) Contribution à l'étude des ressources hydriques du bassin synclinal d'Essaouira (Maroc). Estud Geol 56:185–195. https://doi.org/10.3989/egeol.00563-4150

Bahir M, Silva MO, Carreira P (2007) Ressources en eau en zones semi-aride, état des lieux: cas du bassin d'Essaouira (Maroc Atlantique). In: Bosch APulido, López Geta JA y Ramos G,González (eds) Los acuíferos costeros: retos y soluciones. Coastal aquifers: challenges and solutions. Instituto Geológico y Minero de España, Madrid, pp 529–540

Bahir M, Ouhamdouch S, Carreira PM (2016) La ressource en eau au Maroc face aux changements climatiques; cas de la nappe Plio-Quaternaire du bassin synclinale d'Essaouira. Comun Geol 103:35–44

Bahir M, Ouhamdouch S, Ouazar D, El Moçayd N (2020a) Climate change effect on groundwater characteristics within semi-arid zones from western Morocco. Groundwater Sustain Dev J 11:1–15

Bahir M, Ouhamdouch S, Ouazar D, Abdelghani C (2020b) Global warming and groundwater from semi-arid areas: Essaouira region (Morocco) as an example. SN Appl Sci 2:1245

Bahir M, Ouhamdouch S, Ouazar D, Chehbouni Abdelghani (2020c) Assessment of groundwater quality from semi-arid area for drinking purpose using statistical, water quality index (WQI) and GIS technique. Carbonates Evaporites 35:27. https://doi.org/10.1007/s13146-020-00564-x

Boko M, Niang I, Nyong A, Vogel C, Githeko A, Medany M, OsmanElasha B, Tabo R, Yanda P (2007) Africa. Climate change 2007: impacts, adaptation and vulnerability. In: Parry ML, Canziani OF, Palutikof JP, van der Linden PJ, Hanson CE (eds) Contribution of working group II to the fourth assessment report of the intergovernmental panel on climate change. Cambridge University Press, Cambridge, pp 433–467

Bouya B, Faouzi M, Ben Abbou M, Essahlaoui A, Bahir M, Youbi N, Hessane M (2011) The coastal aquifer of Mnasra (Gharb, Morocco): hydrogeology and hydrodynamic modelling. Comun Geológica 98:73–81

Blunden J, Derek SA (2015) State of the climate in 2014. Bull Am Meteorol Soc 96:S1–S267

Broghton P, Trepanier A (1993) Hydrocarbon generation in the Essaouira Basin of western Morocco. Am Assoc Pet Geol Bull 77:999–1015

Chen H, Xu CY, Guo S (2012) Comparison and evaluation of multiple GCMs, statistical downscaling and hydrological models in the study of climate change impacts on runoff. J Hydrol 434–435:36–45

Chkir N, Trabelsi R, Bahir M, Friha HA, Zouari K, Chamchati H, Monteiro JP (2008) Vulnérabilité des ressources en eaux des aquifères côtiers en zones semi-arides—Etude comparative entre les bassins d'Essaouira (Maroc) et de la jeffara (Tunisie). Commun Geol 2008(95):107–121

Driouech F (2010) Distribution des précipitations hivernales surle Maroc dans le cadre d'un changement climatique: descented'échelle et incertitudes. Thèse de Doctorat, Université de Toulouse, Toulouse, France

Driouech F, Ben Rached S, El Hairech T (2013) Climate variability and change in north African countries. In: Sivakumar MVK, Lal R, Selvaraju R, Hamdan I (eds) Climate change and food security in West Asia and North Africa. Springer, New York, pp 161–172

Dorji S, Herath S, Mishra BK (2017) Future climate of Colombo downscaled with SDSM-neural network. Climate 5(1):24. https://doi.org/10.3390/cli5010024

Duffaud F, Brun L, Planchot B (1966) Bassin du Sud-Ouest Marocain (SW Morocco Basin). In: Reyre D (ed) Bassin Sédimentaire du Littoral Africain, partie I, Paris, pp 5–12

El Mountassir O, Bahir M, Ouazar D, Chehbouni A, Carreira PM (2022a) Temporal and spatial assessment of groundwater contamination with nitrate using nitrate pollution index (NPI), groundwater pollution index (GPI), and GIS (case study: Essaouira basin, Morocco). Environ Sci Pollut Res 17:1–8

El Mountassir O, Bahir M, Ouazar D, Chehbouni A, Carreira PM (2022b) Evaluation of nitrate source and its distribution in the groundwater of Essaouira basin. Water Resour Manag 8:28. https://doi.org/10.1007/s40899-022-00609-0

El Mountassir O, Bahir M, Chehbouni A, Dhiba D, El Jiar H (2022c) Assessment of groundwater quality and the main controls on its hydrochemistry in a changing climate in Morocco (Essaouira Basin). Sustainability 14:8012. https://doi.org/10.3390/su14138012

Fiseha BM, Melesse AM, Romano E, Volpi E, Fiori A (2012) Statistical downscaling of precipitation and temperature for the Upper Tiber Basin in Central Italy. Int J Water Sci 1(3):1–14

Gagnon S, Singh B, Rousselle J, Roy L (2005) An application of the statistical downscaling model (SDSM) to simulate climatic data for streamflow modelling in Québec. Can Water Resour J 30:297–314

Chapter 8
Examining the Climate Change Phenomenon Using Temperature and Precipitation Observations: The Case of Erzurum

Yasemin Kuslu and Kenan Barik

Abstract Climate change is a reality today. The most important indicators of the climate change are the long-term changes in the spatial and temporal precipitation and temperature patterns. It is being felt more and more over time by extreme weather events and changes in temperature and precipitation trends. Data collected about the temperature and precipitation patterns provide important clues about climate change. For this purpose, different statistical analysis methods are used. Since hydrometeorological observations are data covering a series of time, they may be intermittent, short-term, irregular, and skewed. For this reason, homogeneity tests were applied before the statistical analysis. In this study, the Standard Normal Homogeneity test, the Mann–Kendall and the Spearman Rho tests from the non-parametric test groups, and the Linear Trend test from the parametric test groups was applied to all data. Applications were made on the temperature and precipitation values obtained from the observation stations that show slight differences from each other and represent three sub-climates in the city of Erzurum, Turkey, which has a semi-arid main climate. It is seen that there is an increase in the temperature trend and a decrease in the precipitation trend for all three stations. The statistical significance of the change in temperature and precipitation trends requires urgent action to be taken against climate change. It also gives an idea about the dimensions of adaptation studies.

Keywords Climate change · Temperature · Precipitation · Mann–Kendall test · Spearman Rho test · Anatolia · Erzurum

Y. Kuslu (✉)
Department of Agricultural Structures and Irrigation, University of Atatürk, Erzurum, Turkey
e-mail: ykuslu@atauni.edu.tr

K. Barik
Department of Soil Science and Plant Nutrition, University of Atatürk, Erzurum, Turkey
e-mail: kbarik@atauni.edu.tr

M. Behnassi et al. (eds.), *The Water, Climate, and Food Nexus*,
https://doi.org/10.1007/978-3-031-50962-9_8

1 Introduction

Our world has experienced many climatic changes over its existence. These changes generally continued in a natural motion until industrial processes became prevalent in the 1800s. However, after the industrial revolution, the contribution of human beings to climate change has been greater. Pollutants released into the atmosphere are increasing day by day. According to researchers, while 2 billion tons of carbon dioxide gas was released in the early 1900s in the world, this value increased 16 times and reached 35.2 billion tons in 2012 (IPCC 2013). The greenhouse effect, which occurs as a result of polluting factors, causes changes in the annual temperature trend. Climate change, as a result of the increase in the concentrations of greenhouse gases, causes temperature increases and the duration and frequency of extreme weather events, as well as changes in precipitation patterns and hydrological cycles. Generally, temperatures are trending higher, and affecting the hydrological cycle. In other words, climate change noticeably changes the temperature and precipitation trends. Information about temperature and precipitation, which are the main elements of the Earth's climate, is of great importance in determining the characteristics of global climate variations. These two events vary greatly on both spatial and temporal scales. The oscillations in these two parameters reveal important clues for understanding the general structure of the climate. Therefore, recent studies on climate change have focused on trend analyses of these two parameters.

Global climate change is a reality. The most important indicators of this are the variation seen in the spatial and temporal precipitation and temperature patterns. The changing character of precipitation occurs in the form of seasonality becoming evident in some regions—the seasonality of precipitation is a measure of the deviation of the precipitation falling from 25% in the fall, winter, spring, and summer seasons. As the amount of deviation increases, seasonality becomes stronger—less precipitation in some areas, and an increase in precipitation intensity in other areas (Er and Kuşlu 2019). The high natural variability of precipitation (over very short distances and varying greatly from year to year) raises the need for a detailed analysis of precipitation. While the lack of precipitation causes problems such as drought and, therefore, water scarcity, the increase in precipitation intensity causes natural disasters such as floods and overflows, and a change in water quality (Oskay et al. 2023).

According to the 6th Assessment Report of the Intergovernmental Panel on Climate Change, every 10 years over the last 30 years has been increasingly warmer than the last, and warmer than any 10-year period recorded since 1850. It has been noticed that there has been an increase in the amount of precipitation observed in the Northern Hemisphere since 1901 and changes in extreme weather events since 1950. Globally mean surface temperature showed a linear increase of 0.89 °C over the period from 1901 to 2012. It has been noticed that there has been an increase in the amount of precipitation observed in the Northern Hemisphere since 1901 and changes in extreme weather events since 1950. Globally, there has been a decrease in the number of cold days and nights, while the number of warm days and nights has increased. There has been an increase in the frequency and intensity of extreme precipitation events in many regions. The number of regions where extreme precipitation events are increasing is greater than the number of regions seeing decreases. The report also states that there is an increase in the frequency of heatwaves (IPCC 2021).

All these results should serve as a warning to countries to be prepared for the problems they will face. Combating climate change is possible with a good understanding of its current effects. To determine such effects, which vary according to features such as surface structure and vegetation, regional and more local studies (Karabulut and Cosun 2009; İstanbulluoglu et al. 2013; Doğan Demir et al. 2017; Topuz 2017; Ercan and Yüce 2018; Çeribaşı 2019; Dalkılıç 2019; Palta et al. 2019; Coşkun 2020; Tokgöz and Partal 2020; Er et al. 2021; Çelikyurt Uzuner et al. 2021; Topçu and Karaçor 2021; Yılmaz et al. 2021) are carried out in addition to the studies carried out across the world (Philandras et al. 2011; Jain and Kumar 2012; Adarsh and Janga Reddy 2015; Chowdhury et al. 2015; Larabi et al. 2020). In these studies, the trends of temporal series of hydro-meteorological parameters are examined to determine climate change.

Climate change is expected to have many possible effects on Turkey as well. It is among the risk group of countries in terms of the potential effects of global warming. In particular, it will be affected by predicted negative aspects such as the weakening and deterioration of water resources, forest fires, drought and desertification, and related ecological deterioration (Oskay et al. 2021, 2022). Determining trends is very important in terms of being able to see and interpret the situations caused by past trends, and making predictions about future trends. Precipitation regimes irregularity and variability in precipitation do not follow a significant course in Turkey. Therefore, it causes drought in some regions and excessive water problems in other regions. Drought can cause significant effects on hydrological systems, flora, and fauna, and negatively affect animal and plant production (Bölükbaşı Aktaş et al. 2022). In similar situations, it necessitates the emergence of new approaches and agricultural systems related to agricultural irrigation (Er and Kuşlu 2021; Kaya et al. 2017).

The adaptation of the agricultural sector to climate change is very important for the world food security. Adaptation is possible with determination and an understanding of climate change parameters. Such studies have not reached a sufficient level for the Erzurum region, which has cold and snowy winters and cool and dry summers, and regions with similar semi-arid climates. In this study, the temperature and precipitation data obtained from three different meteorology stations within the borders of Erzurum province were evaluated and information on climatic change was revealed.

2 Study Area

The province of Erzurum, which is considered a research area, is located in the east of Turkey and is one of the largest provinces in the Eastern Anatolia region in terms of population and area. Erzurum is located between 40°–15′ and 42°–35′ east longitudes and 40°–57′ and 39°–55′ north latitudes. It is surrounded by the Dumlu Mountains in the north and Palandöken Mountains in the south. Meadow-pasture land comprises 64% of the total agricultural land of the province. The amount of agricultural land is 460,252 ha, while 305,636 ha of this area is suitable for irrigation. Erzurum province has a voice, especially in animal husbandry which is the main livelihood of the provincial rural region. The widespread availability of pasture areas throughout the province and the current climatic conditions have been effective in the development of animal husbandry compared to other branches of agriculture.

Erzurum city center is at an altitude of 1890 m above sea level, and the altitude of agricultural lands varies between 1500–1800 m above sea level. The research area is at the intersection of the Aras, Fırat, and Çoruh basins and is a region with high agricultural potential. The Erzurum Plain (825 km^2) and the Pasinler Plain (540 km^2) are the most important plains. Erzurum is rich in water resources. The most important water resources are the Aras River, the Çoruh River, the Murat River, the Karasu Stream, the Dumlu Stream, the Teke Stream, the Serçeme Stream, and Erzurum—Pasinler groundwater reserves (Kuslu 2008).

Erzurum province, which is located in the terrestrial climate zone, is mostly dry and cool in summers, and usually cold and snowy in winters. The temperature difference between night and day is significant. The average annual temperature for many years (1929–2021) has been 5.3 °C. The coldest month is January with an average temperature of – 10.2 °C, and the hottest month is August at 19.5 °C. The annual average precipitation is 395.6 mm, the month with the highest precipitation is 70.7 mm in May and the least amount of precipitation occurs in January with 16.4 mm. Forty-four percent of the annual precipitation (174.2 mm) falls during the vegetation period of May–September. The relative humidity is the lowest in August with a value of 47% and the highest in December and January with a value of 76%, while the annual average relative humidity is 64% (MGM 2022).

The Erzurum airport station (7 km away from the city centre—39°57′N, 41°11′E) was chosen to represent dry and warm summers and cold and snowing winter climatic features. In the northern part of the city, where the transition zone to the rainy Black Sea climate is located, and the Tortum station (40°18′N, 41°32′E), is in a suitable location in terms of reflecting this climate. In the east of the city, a semi-arid climate is observed where summer heat is more effective. Horasan station (40°02′N, 42°10′E) was chosen to represent these features (Fig. 1).

Data on temperature and precipitation obtained from the General Directorate of Meteorology from 2000 to 2021 were taken into account as the observation period. Monthly total precipitation and average temperature values were used as meteorological data types.

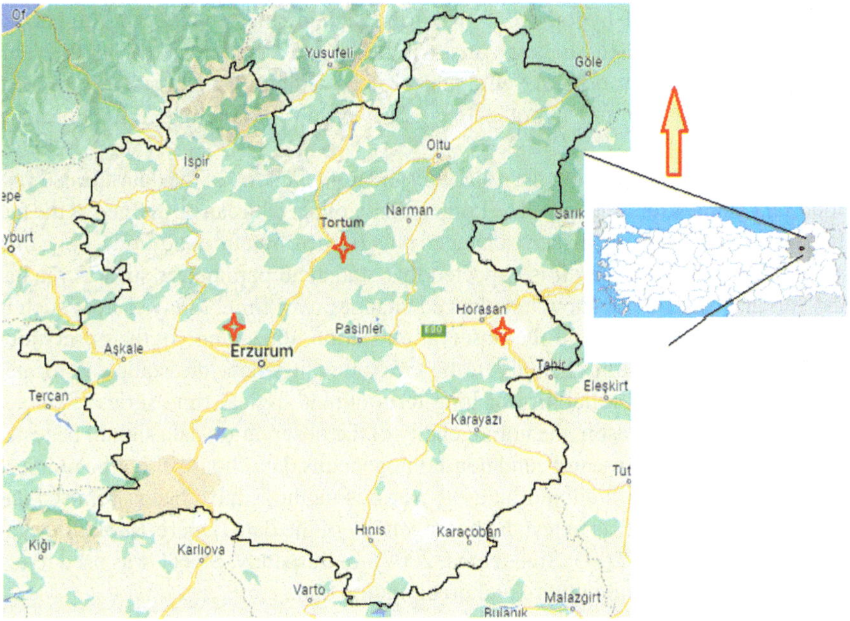

Fig. 1 Study area and the locations of the observation stations. *Source* Taken from Google Maps and developed by the authors

3 Statistical Analysis Methods

Statistical analyses in the study were made using the XLSTAT program. To test the homogeneity of the data to be used, the Standard Normal Homogeneity Test (SNHT) was performed. Linear trend from parametric tests and Mann–Kendall and Spearman Rho Tests from non-parametric tests were used to determine the trend in climatological and hydrological time series. The reason why non-parametric tests are preferred for analysis is to eliminate the negative effects of observations such as being short-term, intermittent, irregular, and skewed (Sneyers 1990).

3.1 Hypothesis Tests

First, we selected the appropriate method for hypothesis testing, and then the null hypothesis (H_0) and alternative hypothesis (H_1) were determined. The H_0 hypothesis expresses the researcher's counter-argument that he did not expect anything out of the ordinary (e.g. climate change changes the temperature trend, etc.). An alternative hypothesis is put forward against the null hypothesis (for example, climate change does not change the temperature trend, etc.). Then the significance level (α) is decided. The significance level expresses the probability of rejection as a result of hypothesis testing when the null hypothesis is true (the smaller the α value, the higher the reliability [1 − α]). Then, the test statistic is calculated and compared with the value read from the relevant critical value tables. If the calculated test statistic exceeds the critical value, the null hypothesis is rejected and the alternative hypothesis is accepted.

In climate analysis studies, the reliability and homogeneity of the data are also very important along with its length. The homogeneity of the data can be defined as the fact that the measurement results have changed only due to climatic conditions. For a time series to be homogeneous, it is expected that statistical parameters such as median and variance should not show unbalanced changes over time. Many factors such as the change in the location of the observation station, the change in the observation mode, and the structural changes in the environment where the station is located affect the quality and reliability of long-term climatological time series (Peterson et al.1998). For this reason, the homogeneity of the observation data should be tested before using it in any research, and non-homogeneous data should either be removed or homogenized, then it should be used. Various methods have been proposed and used by climate scientists to test the homogeneity of the data (Ducre-Rubiatille et al. 2003; Tomozeiu et al. 2005; Staudt et al. 2007). In this study, SNHT was used to test the homogeneity of temperature and precipitation data.

3.1.1 Standard Normal Homogeneity Test

SNHT was developed by Alexandersson (1986) to detect breaks in precipitation data. The break represents abrupt changes in the mean of a time series. In this test, the null hypothesis is that the time series is homogeneous, while the alternative hypothesis argues that there is a break in the mean and the data is not homogeneous. With this test, the time of breaking is also obtained.

Alexandersson defines a new T(k) series and uses the following equation to compare the mean of the first k years of observation data with the mean of the last n-k years of observation data.

$$T(k) = k\overline{z}_1^2 + (n-k)\overline{z}_2^2$$
$$k = 1, \ldots \mathrm{n},$$

$$\overline{z}_1 = \frac{1}{k}\sum_{1=1}^{k}(Y_i - \overline{Y})/s$$

$$\overline{z}_2 = \frac{1}{n-k}\sum_{1=k+1}^{n}(Y_i - \overline{Y})/s$$

Y_i: (i to show the year from 1 to n) the tested series, $\overline{Y}$: the mean of the series, and s: shows the standard deviation of the series.

Table 1 T_0 critical values for SNHT Test

Years (n)		20	30	40	50	70	100
Significance level	%1	9.56	10.4	11.0	11.3	11.8	12.3
	%5	6.95	7.65	8.1	8.45	8.8	9.15

Source Alexandersson (1986)

For the statistical significance test of the refraction, the T_0 test statistic is determined as follows:

$$T_0 = \max_{1 \le k < n} T(k)$$

If the T_0 value does not exceed the critical value read from Table 1, the null hypothesis is accepted, that is, the series is considered to be homogeneous.

3.1.2 Trend Analysis

Parametric and non-parametric statistical methods are used to determine the presence and severity of a statistically significant trend in a climatological time series (Duhan and Pandey 2013). For the unbiased estimates of the parametric methods, the data used should mostly comply with the normal distribution and meet many conditions such as the independence and stationarity of the data. However, it is often very difficult to verify whether climatological data meet these requirements.

Nonparametric methods, on the other hand, are independent of the distribution and characteristics of the data. With these methods, time series can be analysed easily (Van Belle and Hughes 1984). Non-parametric methods can also be easily applied to skewed data where there are missing data and do not fit the normal distribution (Cunderlik and Burn 2004).

In this study, both parametric (linear trend) and non-parametric (Mann–Kendall and Spearman Rho) methods were used.

3.1.3 Linear Trend Analysis

Regression analysis is a test to determine the relationship between two or more variables that have a cause-effect relationship between them. It is also a technique characterized by a mathematical model created to make predictions about the subject. The Linear Regression test is also a parametric test that assumes the data are normally distributed. It tests the relationship between X and Y variables and whether there is a linear trend. The equation used in linear regression is as follows:

$$y = ax + b$$

In the equation, the constant a represents the direction and amount of change, a positive change indicates an increasing change, and a negative indicates a decreasing change. If it is close to zero, it indicates there is no significant change.

3.1.4 Mann–Kendall Test

The Mann–Kendall test is a widely used statistical test to analyze trends in climatological and hydrological time series. It is a variation of the test known as Kendall's Tau. In particular, it is a suitable test for data with extreme values, non-linear tendencies, and non-normal distribution (Mann 1945; Kendall 1975). With this method, the order of the data in the series is used rather than the size. In this test, according to the null hypothesis of H_0, the data are distributed similarly regardless of time, that is, there is no trend, but according to the H_1 hypothesis, there is a trend in the series. The Mann–Kendall test statistic is defined as:

$$S = \sum_{i=1}^{n-1} \sum_{j=i+1}^{n} sgn\left(x_j - x_i\right)$$

In this equation, x_i and x_j represent the length of data and n data in years i and j, respectively.

$$\mathrm{sgn}\left(x_j - x_i\right) = \begin{cases} 1, \left(x_j - x_i\right) > 0 \\ 0, \left(x_j - x_i\right) = 0 \\ -1, \left(x_j - x_i\right) < 0 \end{cases}$$

In cases where $n \geq 10$, the distribution of the S statistic is quite suitable for the normal distribution. In this case, the mean of the statistic is zero and its variance is found with the following equation:

$$\mathrm{var}(S) = \frac{1}{18}\left[n(n-1)(2n+5) - \sum_{i=1}^{k} t_i(t_i - 1)(2t_i + 5)\right]$$

Whether the trend obtained as a result of the test is statistically significant or not is determined by comparing the normalized z value with the critical z values by calculating the following equation:

$$z = \begin{cases} \frac{S-1}{\sqrt{\mathrm{var}(S)}}, & S > 0 \\ 0, & S = 0 \\ \frac{S-1}{\sqrt{\mathrm{var}(S)}}, & S < 0 \end{cases}$$

The z values obtained are compared with the $z\,\alpha/2$ value obtained from the standard normal distribution table for the α significance level. If $|z| > z\,\alpha/2$, the H_0 hypothesis is rejected. In other words, there is a statistically significant trend in the series. In addition, there is a decreasing trend if the z value is negative and an increasing trend if it is positive.

3.1.5 Spearman Rho Test

This test is a non-parametric test used for trend analysis and is applied in comparison with the Mann–Kendall test. In this test, time-series data is assumed to be independent and evenly distributed. H_0 (zero hypotheses) means there is no trend over time, while the alternative hypothesis H_1 indicates an increasing or decreasing trend. This test statistic is expressed as R_{sp} and its standardized statistic is Z_{sp} as follows.

$$R_{sp} = 1 - \frac{6\sum_{i=1}^{n}(D_i - i)}{n\left(n^2 - 1\right)}$$

$$Z_{sp} = R_{sp}\sqrt{\frac{n-2}{1-R_{sp}}}$$

In these equations; D_i = sequence number of i observations, n = total size of time series data, i = order of observations of data, Z_{sp} = $(n - 2)$ Student's t distribution with degree of independence. Positive values of Z_{sp} indicate an increasing trend in hydrological time series, while negative values of Z_{sp} indicate a decreasing trend. If the $|z|$ value is greater than the $z\alpha$ value determined from the standard normal distribution table at the chosen α significance level ($|z| > z\ \alpha$), the H_0 hypothesis, which is based on the fact that the observation values do not change over time, is rejected and it is concluded that there is a certain trend (Spearman 1904).

4 Results

The analysis results of the SNHT performed to determine whether the data are homogeneous are shown in Table 2.

According to the homogeneity test results, none of the stations in the study area have extraordinary precipitation data. T_0 values of all stations remain below

Table 2 Standard normal homogeneity test analysis results of the data of the stations

Station	T_0		T_0 hypothesis
	Temperature	Precipitation	
Erzurum	3.65	3.09	+
Tortum	4.29	4.79	+
Horasan	2.51	4.06	+

the critical value of 6.95 (at a 5% significance level) and show homogeneous characteristics.

Linear Regression analysis was performed on the temperature and precipitation data taken from the stations. The temperature change over the years is shown in Fig. 2 and the annual precipitation trend is shown in Fig. 3. The equations obtained as a result of statistical analysis are given in Table 3 collectively as well as in the graphs.

When Fig. 2 is examined, it is seen that the average annual temperature trend in all Horasan, Tortum, and Erzurum stations increased, and the increase rates are close to each other in all three stations. The temperature increase trend is more pronounced in Tortum, which is a microclimatic region. This region is followed by Erzurum and Horasan regions, respectively.

When Fig. 3 is examined, it is observed that the annual total precipitation tends to decrease at the Erzurum, Tortum, and Horasan stations. It can be said that this trend is

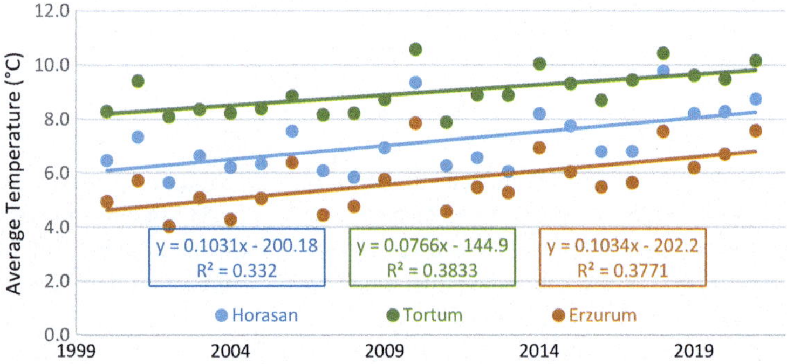

Fig. 2 Distribution of annual average temperature trend by stations. *Source* Developed by the authors

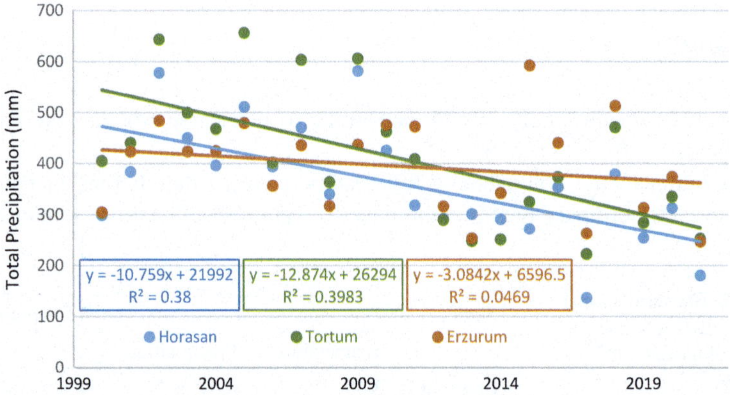

Fig. 3 Distribution of total annual precipitation change by stations. *Source* Developed by the authors

Table 3 Summary for statistics on Linear Regression analysis

		Stations		
		Erzurum	Tortum	Horasan
Temperature	Equation	y = 0.1034x – 202.2	y = 0.0766x – 144.9	y = 0.1031x – 200.18
	R Sq.	$R^2 = 0.3771$	$R^2 = 0.3833$	$R^2 = 0.3320$
	Maximum (°C)	7.9	10.6	9.8
	Minimum (°C)	4.0	7.9	5.7
	Average (°C)	5.7	9.0	7.2
Precipitation	Equation	y = – 3.0842x + 6596.5	y = – 12.874x + 26,294	y = – 10.759x + 21,992
	R Sq.	$R^2 = 0.0469$	$R^2 = 0.3983$	$R^2 = 0.3800$
	Maximum (mm)	593.3	656.9	582.2
	Minimum (mm)	247.5	222.6	137.2
	Average (mm)	395.6	410.1	360.6

highest at the Tortum station, followed by the Horasan and Erzurum stations. When the station observations are examined monthly, the highest temperature increase in Erzurum and Tortum stations occurred in June, and in May at the Horasan station (Fig. 4). In terms of precipitation decrease, April saw the highest at the Erzurum and Tortum stations. At Horasan station, the highest decrease occurred in September (Fig. 5). In the top three rankings, the highest temperature increase, and the highest precipitation decrease were observed mostly (83.3%) during the vegetation period (April-September).

The results of the Mann–Kendall Test are shown in Table 4. According to the table results, there is an increase in the annual temperature trend and a decrease in the precipitation trend. According to the stations, the highest temperature increase occurred at the Tortum station, followed by the Erzurum and the Horasan. Annual precipitation decrease is observed mostly at the Tortum station, followed by the Horasan station.

According to the analysis results, the temperature trend is positive at all three stations. The highest increase is seen in the observations measured at the Tortum station. Being statistically significant (95% confidence interval at Erzurum and Tortum stations, 90% confidence interval at Horasan station), there is an increasing trend in temperature in May and June.

According to Table 4, in the October precipitation trend at the Horasan station, in the 95% confidence interval, a decreasing trend is observed in the 90% confidence interval in April and July. At the Tortum station, there is a decreasing trend in the 90% confidence interval in April, June, July, August, and October. At the Erzurum station, the precipitation trend in April decreases in the 90% confidence interval. The station where precipitation decreases were observed most was the Tortum station, followed by the Horasan and Erzurum stations. At the Erzurum station in May, June, August,

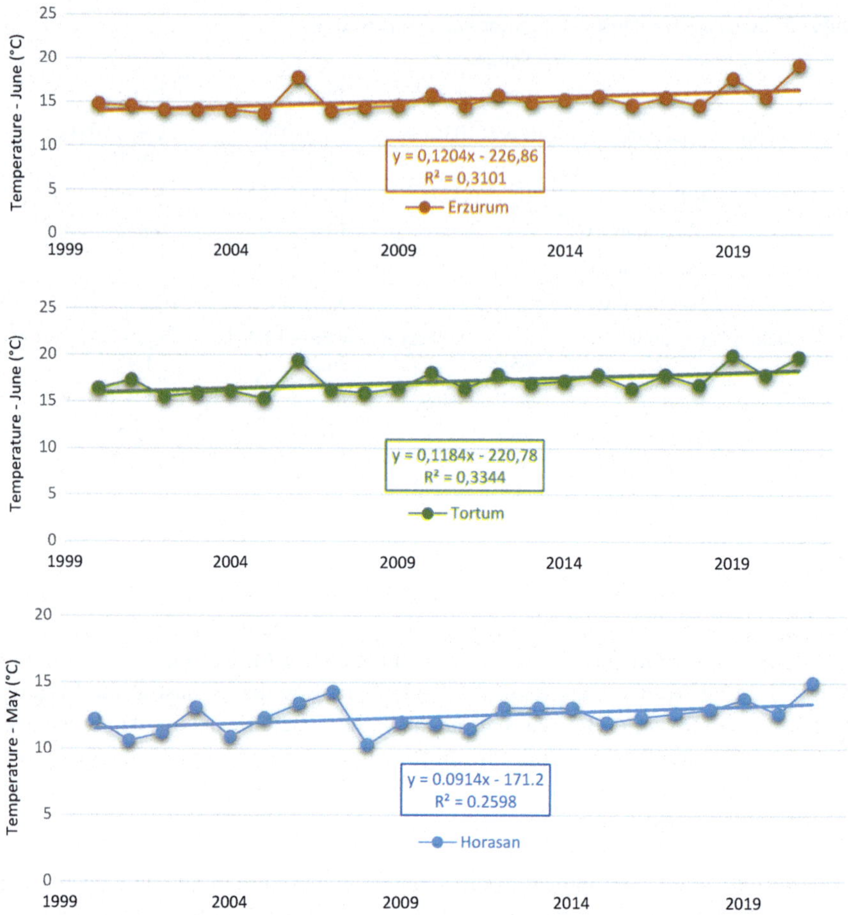

Fig. 4 Months with the highest monthly temperature increases by stations. *Source* Developed by the authors

and September; at the Tortum station in May and September; and at the Horasan station, although the precipitation trend is positive in March, it is not statistically significant.

The results of the Spearman Rho analysis are shown in Table 5. According to the table, the temperature trend for all three stations increased, and it is statistically significant (90% confidence interval). While there was a decrease in the precipitation trend at all three stations, the decreasing trend at the Tortum and Horasan stations was found to be statistically significant (90% confidence interval).

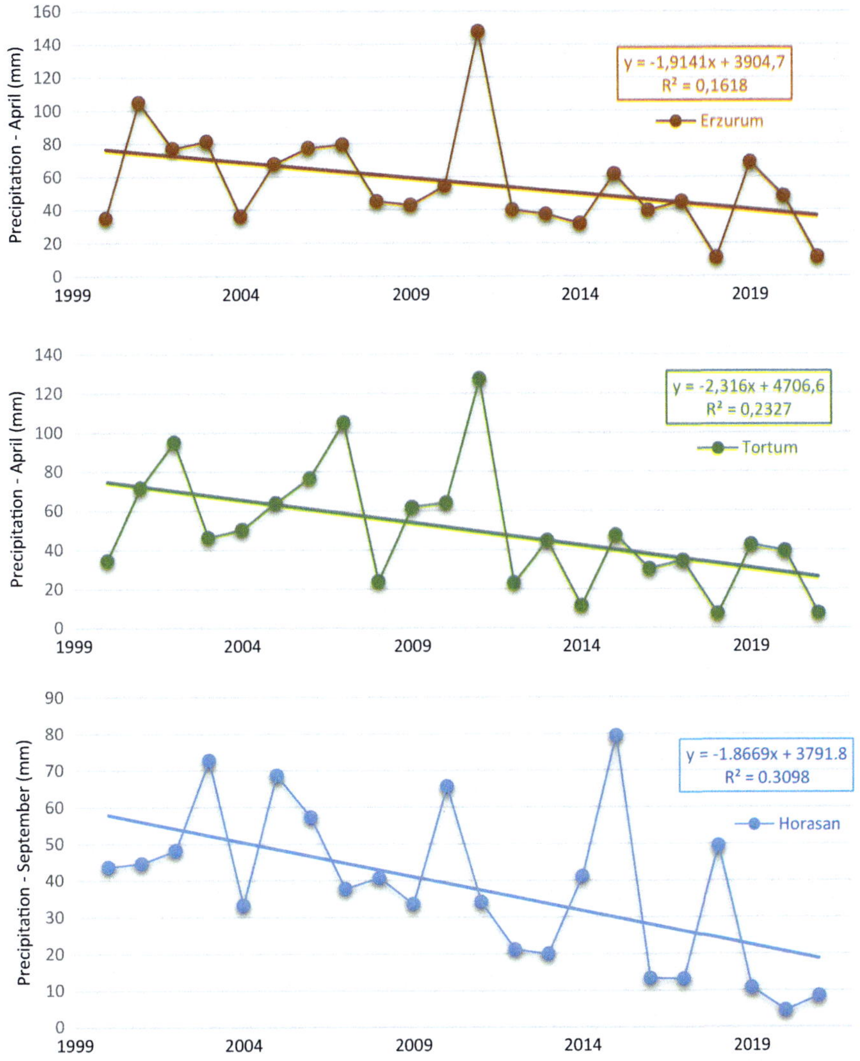

Fig. 5 Months with the highest monthly precipitation decrease by stations. *Source* Developed by the authors

Table 4 Results of the Mann–Kendall test by the stations

Months	Temperature trend values						Precipitation trend values					
	Erzurum		Tortum		Horasan		Erzurum		Tortum		Horasan	
January	0.82		0.81		**1.18**		– 0.11		– 0.62		– **1.29**	
February	0.61		0.45		0.90		– 0.18		– 0.60		– 0.36	
March	0.92		**1.01**		0.11		– 0.79		– **1.36**		**0.01**	
April	0.49		0.29		0.58		– **1.66***	▽	– **2.38***	▽	– **1.71***	▽
May	**3.13****	▲	**3.16****	▲	**2.66***	△	**0.43**		**0.06**		– 0.25	
June	**3.17****	▲	**3.42****	▲	**2.06***	△	**0.01**		– **1.91***	▽	– **1.43**	
July	0.89		**1.49**		0.46		– 0.36		– **2.19***	▽	– **2.39***	▽
August	0.80		**1.39**		0.43		**0.15**		– **2.18***	▽	– 0.95	
September	**1.42**		**1.03**		**1.38**		**0.40**		**0.01**		– 0.76	
October	0.64		0.60		0.01		– 0.11		– **1.86***	▽	– 3.17**	▼
November	0.64		0.18		0.49		– 0.59		– 0.54		– 0.46	
December	0.58		0.68		0.85		– 0.05		– 0.99		– 0.09	

Bold values indicate that a trend exists, but this trend is not statistically significant
*indicates the level of significance at the 90% and **95% confidence intervals

Table 5 Spearman Rho Test results for observation stations

Stations	Temperature statistics			Precipitation statistics		
	r_s		p (2-tailed)	r_s		p (2-tailed)
Erzurum	**0.62974***	△	0.00169	– 0.19486	▽	0.38485
Tortum	**0.64062***	△	0.00132	– **0.63410***	▽	0.00153
Horasan	**0.58152***	△	0.00453	– **0.63862***		0.00138

Bold values indicate that a trend exists, but this trend is not statistically significant
*indicates statistical significance level at 90% confidence interval

5 Conclusion

In this study, temperature and precipitation observations made from 2000 to the end of 2021 at the Erzurum-Airport (Erzurum), Tortum, and Horasan stations in Erzurum province were evaluated. Temperature and precipitation trends were determined in the region represented by the stations selected from the temperature and precipitation data. For this purpose, the Linear Regression Test from parametric tests and Mann–Kendall and Spearman Rho tests from non-parametric tests were performed. As a result of all three tests, it was seen that the direction of the temperature trend was positive at all stations. It can be said that the upward trend in the region represented by the Tortum station is positive and more effective. Significant increase in temperature were observed from March to September in the vegetation period. This situation stresses about the microclimate feature of the region. However, at the same station, it was observed that the precipitation trend is decreasing, and that trend is more pronounced in the March–October period. Increasing temperature and decreasing precipitation trends reveal the necessity of reviewing the applied agricultural systems. In this region, where vineyard and garden cultivation and greenhouse cultivation are intense due to their microclimate feature, it is necessary to select especially drought-resistant product varieties and to take measures to protect water resources.

The Horasan is the other station with a positive temperature increase trend and a decrease in precipitation. In the region represented by this station, the said trend is generally effective during the vegetation period. It will be very important to improve the practices regarding irrigation systems so that the agricultural potential of the region, where field agriculture is intensely practiced, is not affected by this trend. Surface irrigation methods in the region are the methods that farmers use a lot, pressurized irrigation methods have not yet become widespread.

In the region represented by the Erzurum station, there was a statistically significant increase in temperature, especially in May and June, which is the vegetation start period. Increases in temperature cause rapid evaporation of water from the soil and loss of moisture necessary for seed germination. This means that agricultural irrigation practices are brought forward, and the number of irrigation applications and the amount of water required increase. It is important to use water resources effectively and to choose drought-resistant varieties. The widespread use of pressurized irrigation systems, in which irrigation water is used more efficiently, and especially an emphasis on a drip and sub-surface drip irrigation systems, and digital technologies should be considered in preparation for future climate change scenarios.

In the future, studies on the observation and analysis of drought symptoms other than precipitation and temperature should increase. The scope of such studies can be expanded. It is clear that basin-based studies will show local changes more realistically. In addition, future vision studies are needed in terms of the effective use of natural resources and ensuring food safety regions, such as precipitation waters harvesting, water recovery, and recycled wastewater use in agriculture for all three stations.

References

Adarsh S, Janga Reddy M (2015) Trend analysis of rainfall in four meteorological subdivisions of Southern India using nonparametric methods and discrete wavelet transforms. Int J Climatol 35(6):1107–1124

Alexandersson H (1986) A homogeneity test applied to precipitation data. J Climatol 6:661–675

Bölükbaşı Aktaş ŞC, Er H, Kuşlu Y (2022) Meteorolojik faktörlerdeki değişme eğiliminin hayvansal üretim ve ürün kalitesine etkisi, In: Kökten, İnci (eds) Tarım ve Hayvancılığın Sürdürülebilirlik Dinamikleri Üzerine Akademik Çalışmalar, İKSAD Publishing

Chowdhury RK, Beecham S, Boland J, Piantadosi J (2015) Understanding South Australian rainfall trends and step changes. Int J Climatol 35(3):348–360

Coşkun S (2020) Tuz Gölü- Konya Kapalı Havzalarının yaz mevsimi ortalama sıcaklık, yağış, buharlaşma ve akim verilerindeki değişimlerin karşılaştırmalı trend analizi. J Soc Sci 46:123–138. ISSN: 2149-0821. https://doi.org/10.29228/SOBIDER.43355

Cunderlik JM, Burn DH (2004) Linkages between regional trends in monthly maximum flows and selected climatic variables. J Hydrol Eng 9(4):246–256

Çelikyurt Uzuner G, Gökcan M, Karateke V, Güven E, Öztopal A (2021) Trend analysis of annual maximum rainfall in Balıkesir. Bull Turkish Water Found No:001

Çeribaşı G (2019) Analyzing rainfall datas' of Eastern Black Sea Basin by using Sen method and trend methods. J Inst Sci Technol 9(1):254–264

Dalkılıç HY (2019) Trend analysis of precipitation. Erzincan Univ J Sci Technol 12(3):1537–1549. https://doi.org/10.18185/erzifbed.587610

Doğan Demir A, Demir Y, Şahin Ü, Meral R (2017) Trend analyses of temperature and precipitation and effect on agricultural in Bingol Province. Turk J Agric Nat Sci 4(3):284–291

Ducre-Rubiatille J, Vincent A, Boulet G (2003) Comparison of techniques for detection of discontinuities in temperature series. Int J Climatol 23:1087–1101

Duhan D, Pandey A (2013) Statistical analysis of long term spatial and temporal trends of precipitation during 1901 2002 at Madhya Pradesh, India. Atmos Res 122:136–149

Er H, Kuşlu Y (2019) The importance of rainwater harvesting, 3rd edn. International Congress of Applied Sciences, Diyarbakır

Er H, Kuşlu Y (2021) Smart irrigation systems practices in agriculture. In: Proceedings of abstract book of 4th international congress on agriculture, environment and health, 20–22 May, p 236

Er H, Koçyiğit S, Kuşlu Y (2021) Meteorolojik faktörlerin ve temiz su kaynaklarının arı ve arı ürünleri üzerine etkisi. In: Uygulamalarında T (ed) Kökten and İnci. İKSAD Publishing, Yenilikçi Yaklaşımlar, pp 295–311

Ercan B, Yüce MI (2018) Monthly temperature and precipitation trend analysis of Kilis Province DÜMF Mühendislik Dergisi 9(2):947–953

IPCC (2021) Climate change 2021: the physical science basis. In: Masson-Delmotte V, Zhai P, Pirani A, Connors SL, Péan C, Berger S, Caud N, Chen Y, Goldfarb L, Gomis MI, Huang M, Leitzell K, Lonnoy E, Matthews JBR, Maycock TK, Waterfield T, Yelekçi O, Yu R, Zhou B (eds) Contribution of Working Group I to the sixth assessment report of the intergovernmental panel on climate change. Cambridge University Press (In Press)

IPCC (2013) Climate change 2013: the physical science basis. In: Stocker TF, Qin D, Plattner G-K, Tignor M, Allen SK, Boschung J, Nauels A, Xia Y, Bexand V, Midgley PM (eds) Contribution of working group I to the fifth assessment report of the intergovernmental panel on climate change. Cambridge University Press, Cambridge, UK and New York, USA

İstanbulluoglu A, Bagdatli MC, Arslan C (2013) To evaluated with trend analysis of long-annual rainfall: Tekirdağ–Çorlu district application. J Tekirdağ Agric Fac 10(2):70–77

Jain SK, Kumar V (2012) Trend analysis of rainfall and temperature data for India. Curr Sci 102(1):37–49

Karabulut M, Cosun F (2009) Precipitation trend analyses in Kahramanmaraş. Coğrafi Bilimler Dergisi CBD 7(1):65–83. https://doi.org/10.1501/Cogbil_0000000095

practices and recharge scenarios. Similarly, studies in Libya, as conducted by El Hassadi (2008) using a hydrogeochemical approach and a two-stage finite element simulation algorithm, revealed a gradual increase in intrusion from 1960 to 2005, with intrusion extending up to 10 km from the coast in the Gefara plain near Tripoli. On the Algerian coast, research by Morsli et al. (2007) and Belkhiri et al. (2012) employed multivariable statistical and geochemical modeling techniques to identify marine intrusion in the region.

Moroccan coastal aquifers have also been affected by saltwater intrusion. The Saïdia sandy aquifer in the northeastern part of the country, for instance, demonstrated intrusion, as evidenced by electrical and logging soundings (El Halimi et al. 1999). In the Temara-Rabat area, which encompasses an aquifer system containing marine deposits, saltwater intrusion was confirmed (Pulido-Bosch 1999). In the southern regions of Morocco, such as the Sahel of Doukkala Abda, groundwater quality deteriorated due to marine intrusion (Fadili et al. 2012; Kaid Rassou et al. 2005).

This study focuses on the shallow Plio-Quaternary aquifer in the Essaouira basin along the Moroccan Atlantic coast to further explore the phenomenon of marine intrusion. The research employs hydrogeochemical and isotopic methodologies to achieve its objectives.

Geographically, the aquifer in question is situated on the Moroccan Atlantic coast, spanning from latitude 31° 24′ to 31° 49′N and longitude 9° 52′ to 9° 85′W. Its eastern boundary is marked by Triassic outcrops (Tidzi diapir), while its western limit is defined by the Atlantic Ocean (Fig. 1). The study region is characterized by a semi-arid climate, featuring an average annual rainfall of 300 mm and an average annual temperature of 20 °C (Bahir et al. 2020a, b; El Mountassir et al. 2022b). Ouhamdouch et al. (2020a) reported a decline in annual precipitation within the study area by 12–16%, along with a notable temperature increase ranging from 1.2 to 2.3 °C. This reduction in precipitation and temperature rise has led to declining piezometric levels and a deterioration of groundwater quality in the study area.

Geologically, the shallow aquifer primarily comprises sandstones, conglomerates, and is interspersed with sandy marl, limestone, and dolomite (Fig. 1). The aquifer, with a thickness ranging from 5 to 60 m, is underlain by impermeable gypsiferous and siliceous marls of the Senonian period (200 m) (Bahir et al. 2000). According to Mennani (2001), the shallow aquifer exhibits variable permeability, ranging from 0.27 to 132 m/d, and transmissivity values between 4.5×10^{-5} and 6.02×10^{-2} m^2/s (Mennani et al. 2001), with the highest values occurring near the Ksob wadi, a prominent recharge area. The piezometric map illustrates a flow direction from the southeast to the northwest, characterized by a hydraulic gradient ranging from 1.2% in the downstream segment to 2.5% in the upstream region. The direction of groundwater flow is largely influenced by the subsurface aquifer geometry.

Groundwater in North Africa

Groundwater serves as the primary source of water supply for various purposes, particularly in arid and semi-arid regions of numerous countries (Bahir et al. 2021c,

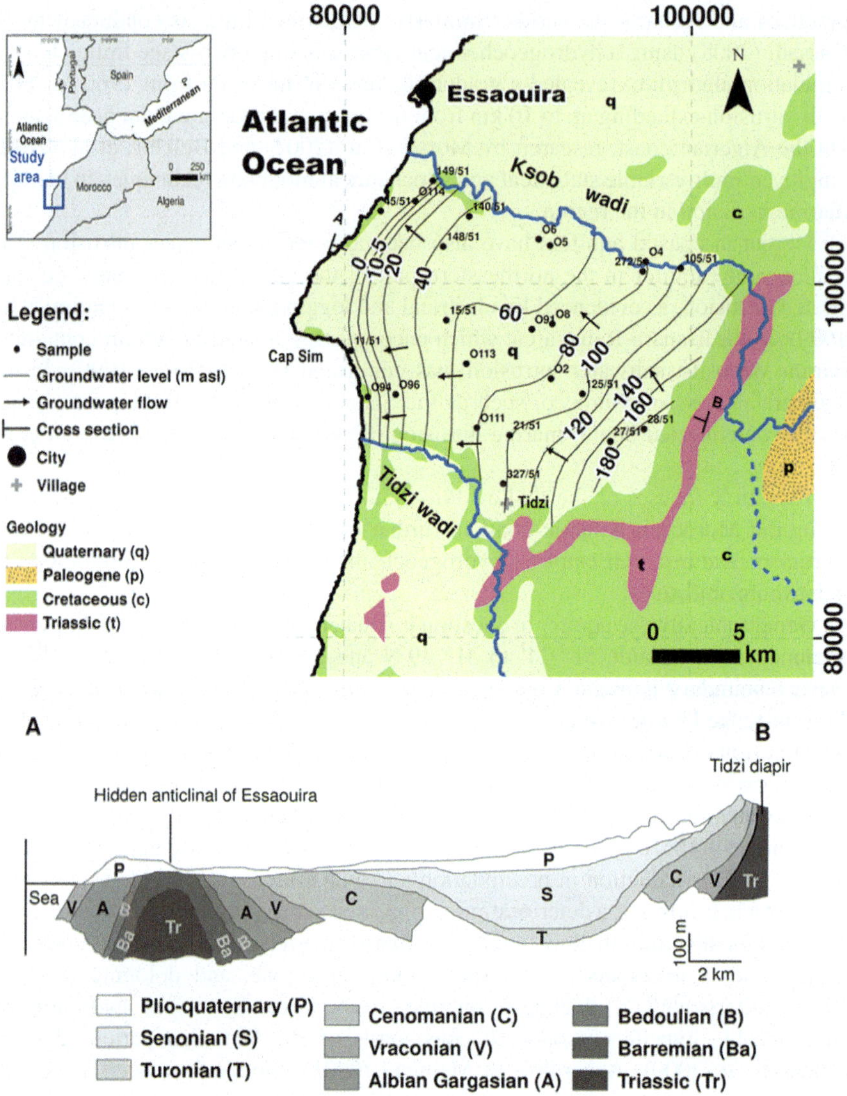

Fig. 1 Location and geological map of study area. *Source* Developed by the author

2022a; El Mountassir et al. 2022a). On a global scale, approximately 65% of groundwater is utilized for drinking, 20% for irrigation and livestock, and 15% for industrial and mining applications. Groundwater offers several advantages over surface water, including wide distribution, inherent stability, natural regulation, good water quality, and resistance to contamination. However, its utilization is not without challenges, as it is often buried underground and necessitates a comprehensive understanding of its distribution patterns before effective usage.

In recent decades, as urbanization and living standards have improved, the demand for groundwater resources has escalated, subjecting these resources to unprecedented pressures. A recent study published in Science by researchers from the University of California, Santa Barbara indicates that groundwater sustaining 2.5 billion people worldwide is at risk of depletion. Debra Perrone, co-author of the study, highlights that up to 20% of the world's wells are located within a depth of merely five meters below the water table. This precarious situation implies that millions of these wells may run dry due to factors such as global warming or excessive extraction (Podgorski and Berg 2020).

In the African context, the continent features around eighty transboundary river and lake basins, as well as approximately forty transboundary aquifer basins. The Africa Water Vision 2025 underscores that groundwater represents the primary, and often the sole, source of potable water for over 75% of the African population. Furthermore, groundwater accounts for more than 95% of Africa's freshwater resources, and the pollution and salinization of this resource are often irreversible on a human timescale.

In arid and desert regions of North Africa, groundwater from depths exceeding 500 m constitutes the sole water source for fulfilling local needs, encompassing agriculture, industry, tourism, and domestic consumption. Similar to many countries worldwide, the populations in southern Mediterranean nations grapple with water stress conditions, defined as those consuming more than 20% of their renewable water resources. Water withdrawal exceeding 40–50% signifies severe water stress. Presently, the United Nations projects that by 2025, 25 African countries will face water scarcity or water stress. According to UNEP (2006, 2010, 2011), an estimated 1.1 billion individuals lack access to clean drinking water, whether from surface or groundwater sources. Contaminated water directly contributes to five million deaths annually, with the majority occurring in sub-desertic Africa.

Water-related challenges, particularly the salinization of groundwater, have become a focal point for both scientific researchers and policymakers on a national level. These concerns have been addressed in numerous studies across Morocco. The country, akin to many Mediterranean nations, has faced recurrent periods of drought since the early 1980s. Its water resources are limited, estimated at 20 billion cubic meters, translating to an average of 700 m^3 per year per inhabitant. This situation indicates a notable level of water stress. The number of years experiencing a rainfall deficit has surpassed those with adequate rainfall, with a prevailing downward trend of 23%.

One of Morocco's pivotal aquifers, notable for its extent and geographical location, is the Plio-quaternary aquifer in the Essaouira basin, situated in the western part of the country. To evaluate the evolution of water quality, particularly concerning groundwater, in regions characterized by semi-arid climates and influenced by climate change, the Essaouira basin is employed as an illustrative case. In recent decades, this basin has encountered a sequence of drought periods, leading to the degradation of groundwater quality. The Essaouira basin holds strategic importance within Morocco, accounting for 10% of the total national aquifer reserves, encompassing eight aquifers out of a total of 80. Hence, comprehending the primary processes governing groundwater mineralization (Bahir et al. 2018b, 2021a, b) is essential for enhancing water resource management in these regions.

Coastal aquifers, particularly shallow ones, often confront vulnerabilities linked to seawater intrusion. This intrusion not only contributes to water resource deterioration but also contributes to the elevated salinity levels resulting from the dissolution of minerals due to evaporation (e.g., halite, gypsum, anhydrite), agricultural and industrial activities, as well as climate change. The present study focuses on the geochemical examination of the Essaouira basin, which is situated along the Atlantic coast of Morocco. The inhabitants of this basin primarily rely on groundwater as their principal water source for both domestic use and irrigation. The outcomes of this research are anticipated to substantially enhance management strategies aimed at safeguarding the scarce groundwater resources in the region. Furthermore, they contribute to a deeper understanding of the status of coastal aquifers within the Atlantic basin.

In recent decades, Morocco has confronted prolonged and unusual drought periods, leading to widespread reductions in free piezometric levels, the depletion of numerous water sources, and diminished wadi flows. The resource mobilization has now approached its limits, and optimal dam locations have already been exhausted. In this context, resource mobilization is becoming increasingly challenging, given the ever-growing demand. This marks the conclusion of the system known as "OFFERTA (OFFER)," as described by our Spanish counterparts.

One of the latest recommendations from the Higher Council for Water and Climate underscores that groundwater, primarily reserved for drinking water supply, will become progressively harder to attain in the future. This difficulty stems from deteriorating water quality due to water stress and the discharge of wastewater into the vicinity of major urban and rural settlements (Niemczynowicz 1999).

The potential of groundwater resources is estimated at 4×10^9 m^3, distributed across approximately 80 shallow and deep aquifers within the country's nine basins. These resources are subject to regular monitoring by the Hydraulics Directorate and the National Water Office. Cadi Ayyad University of Marrakesh, Morocco, collaborates with public authorities to study several basins, including that of Essaouira.

Groundwater is an integral component of the hydrological cycle and the subsurface environment. It exhibits interconnections with surface water and continuously interacts in both space and time through processes such as infiltration and drainage. Groundwater is replenished by a portion of infiltrated rainwater, subject to various atmospheric effects, including direct or indirect evaporation.

Moreover, it is intricately linked to the subsurface, where it serves as a vital and active constituent. Groundwater cannot be regarded as a distinct, isolated resource, nor can it be compared to other natural soil resources. Instead, it represents reservoirs of water and dynamic flows, akin to surface water. The renewal of groundwater is influenced by rainwater infiltration, a process that sustains its flow rates.

The spatial distribution of groundwater is notably more continuous than that of surface water, and it is heavily influenced by the geological formations that structure its dynamics. As a result, groundwater exhibits a slow flow rate, characterized by

- Inherent uncertainties stemming from measurement inaccuracies required for quantitatively understanding flow mechanisms and subsurface storage.
- Influence of climatic variability and potential hazards.
- Incomplete knowledge of the various actors involved and their behavioral patterns.
- Unidentified water demands in terms of both quantity and quality that may emerge urgently.

The selection of specific groundwater exploitation sites is often constrained by various factors. These limitations may be tied to the environmental structure, characteristics of the operational infrastructure, and the configuration of water use. Hydrogeological factors certainly play a role, encompassing the nature and structure of aquifers, their productivities, and water quality distributions. However, economic considerations also impose constraints, such as:

- The maximum allowable depth for resource extraction, whether it is to access the aquifer and initiate production (borehole depth) or to pump water, resulting in production costs that users must accept.
- Restrictions on exploitation to minimize impacts on spring flow rates, the regimes of other catchment areas, or even neighboring aquifers.
- Varying site-specific requirements for operational structures, contingent on their type and the specific water demands they serve.
- For groundwater, management is a matter of applying overarching principles across the resource system, treating it as a common good shaped by natural conditions. This approach establishes a balance between the objectives of exploitation and the goals of resource allocation and conservation, all driven by the broader public interest.

As groundwater is interdependent with surface water, its management is integrated into water management as a whole. It presupposes objectives, stakeholders, and instruments. Given their scale, the objectives in themselves constitute constraints on the exploitation of groundwater. They are multiple and must be reconciled to:

- satisfy the demands for underground water in quantity and quality;
- allocate the resource according to priorities;
- keep the potential resource in quantity and quality;
- keep the productivity and accessibility of groundwater, especially when it is intended for collective use;
- intensify the use of the resource deemed to be underexploited;
- avoid conflict interests between stakeholders.

Thus, these management objectives are in the common interest of both users and the community. The objectives listed above lead to the following question: “the management of groundwater is under whose responsibility?”.

It is the business of managers but still the business of everyone since groundwater is a common good; it is also the business of the state since groundwater is part of the hydraulic domain (2015 Morocco Water Law). It is of interest to those to whom it offers an accessible resource from which they benefit as well as to those who represent and reconcile multiple and often conflicting interests.

Actions are associated with the money and investigation aspects. The activities consist of investments for investigations only. The direct effects on groundwater are of interest to the operators, which are numerous and diversified because of the following:

- of the extension of aquifers;
- that the operation does not require public utility development like surface reservoirs;
- that the occupants of the land see themselves as 'rights holders' over groundwater.

At the level of farmers, microeconomic objectives predominate. They individually manage their production according to their criteria, but not the resource itself. They are normally united since their actions interfere, but paradoxically do not care much about the effects of their actions which accumulate within the same aquifer system.

On the other hand, the authority emanating from public power or management authority has no means of direct action on groundwater before the promulgation of the water law of 1995. It does not have the appropriate power to intervene on the behavior of operators. It only intervenes through various traditional indirect management instruments.

This responsibility lies with the Ministry of Equipment and Water, certain powers of which are delegated concerning agricultural water to the Regional Offices for Agricultural Development. Note that Thermo mineral waters are under the responsibility of the Ministry of Energy and Mines.

As for the water law, nine years after its promulgation, stipulating among other things the creation of hydraulic basin agencies, its effective application is still progressing slowly.

In general, three major constraints affect water resources in Morocco: The scarcity of resources, their irregularity in time and space (Nadifi 1998).

Concerning groundwater resources, there are more exactly 32 deepwater tables (depth ranging from 200 m to more than 1000 m) and 48 surface water tables (shallow water level). The former is difficult to access with a high economic cost, the latter is more accessible, but also more vulnerable to pollution and drought, such as the case of the Plio-Quaternary aquifer in the Essaouira basin (Bahir et al. 2001) ensuring the supply of drinking water to the city (198,400 inhabitants in 2014) and the surrounding rural agglomerations.

During the last decades, the exploitation of groundwater has continued to grow under the combined effects of demographic pressure, the search for satisfactory food self-sufficiency, industrialization, and the political will for balanced regional development. This growth is due not only to new drilling techniques and ever more efficient means of dewatering, but also an unfavorable climatic situation leading to an acceleration of groundwater operations to compensate for deficits in surface flow and rainwater. The combined effects of intensive operating conditions and unfavorable climatic conditions have led to hydrodynamic imbalance regimes, which have almost generally caused alarming reductions in reserves (Haouz, Souss, Angads,

Tafilalt) and, consequently, productivity losses of collective or individual collection works, dewatering of traditional works (*Khettaras*, shallow wells), and reduction in emergence flows.

But one can wonder if these worrying reductions are the result of a provisional state of groundwater and its capacity to regulate and if the more favorable climatic conditions were correcting the broken equilibria. Should we curb and advise against an intensification of these exploitations by freezing this precious natural 'capital'? On the contrary, should we encourage, guide, organize the exploitation of groundwater and take actions accordingly for a better valuation of this natural resource while avoiding irreversible disruption of the balance?

It is legitimate for those responsible of the planning and management of water resources, academics, and those responsible for economic development to ask these questions and seek appropriate answers. Be that as it may, and given the level of exploitation of groundwater, we will agree that a period of rigorous management of these resources, whether renewable or not, is now required, to best adapt quantity, quality, and users in a context where all the natural, socio-economic parameters are not completely mastered, nor controllable. This is what motivated the selection of the Essaouira basin as a case study for both its complexity and fragility.

In arid and semi-arid environments, the precipitations and temperatures are determining parameters for climatic characterization. Studying the evolution of recent climate variability, which is essential for better management of water resources, remains an essential tool for overcoming the problems resulting from the relationship between water needs and their availability.

Climate change is a global phenomenon. In the long term, it involves complex interactions between environmental factors and economic and social conditions, leading to significant regional effects (Filho 2012; Misra 2014; Vennetier et al. 2005), including the Maghreb (El Kharraz et al. 2012).

Morocco, like Algeria (Elouissi et al. 2017) and other Mediterranean countries (Nassopoulos 2012; Taabni and El Jihad 2012) have suffered from several periods of drought accompanied by a water shortage. Over the past decades, numerous studies have shown a downward trend in precipitations and an upward trend in temperatures across North Africa (Meddi and Meddi, 2007; Meddi et al. 2009; Sebbar 2013). Morocco, in particular, has experienced a drop in the average annual precipitation and remarkable warming that began since the late 1970s (Babqiqi 2014; Driouech 2010; Driouech et al. 2010; Sebbar 2013). The changes in climate parameters mentioned above are not the same for all areas and the intensity should be quantified locally to manage natural resources, in particular water.

Due to its geographic location and context, Morocco remains one of the countries most vulnerable to the effects of climate change. The first signs of climate change are already manifesting in changing temperatures and precipitations. The precipitations show an overall fluctuating decrease depending on the region, between 3 and 30% (Babqiqi 2014). Temperatures display an average increase of + 0.6 to + 1.4 °C depending on the region (DMN 2007). ESCAP (2017) forecasts a reduction in precipitation of 8–10 mm/month depending on the scenarios, by the end of the century over Morocco. This reduction is accompanied by an increase in temperature expected to reach 4 °C.

These changes will undoubtedly have harmful effects on the water resource already characterized by scarcity and a spatio-temporal irregularity. HCP (2013) indicates that Morocco would be exposed to move from a situation of water stress (< 1000 m^3/inhabitant/year) to a situation of water scarcity (< 500 m^3/inhabitant/year) by the 2030 horizon. This situation requires the rational exploitation of available water resources, the recycling of wastewater, and the use of water-saving—and production—techniques. Existing studies on the assessment of the climate change impact on water resources in Morocco remain very rare, general, and regional. It is therefore essential to move towards local studies, otherwise, on the 'hydrological and hydrogeological basins' scale. This makes it possible to specify the effects on this vital resource and to recommend the necessary adaptation measures.

2 Material and Methods

A total of 26 water samples representing the groundwater of the shallow aquifer within the Essaouira basin underwent comprehensive chemical and isotopic analysis (El Mountassir et al. 2022c). Of these samples, 24 were specifically dedicated to groundwater, with one sample reserved for rainfall and another for seawater. The on-site measurements of parameters including electrical conductivity (EC), pH, and temperature (T) were conducted using the HI-9829 Multiparametric Instrument. A 200-m probe was employed to gauge the depth of the water table. In order to capture values representative of the natural aquifer conditions, water samples were collected following a pumping duration of 15–20 min. These samples were meticulously preserved in 500 mL clean polyethylene bottles and stored at a temperature below 6 °C before being transported to the laboratory for further analysis.

The chemical element analysis was carried out at the Laboratory of Geosciences and Environment within the Ecole Normale Supérieure of Marrakech, Morocco. The concentrations of chloride (Cl) and sulfate (SO_4) anions were determined using the Mohr technique and the nephelometric technique, respectively. Calcium (Ca) and magnesium (Mg) concentrations were ascertained through the EDTA titrimetric method. Sodium (Na) and potassium (K) levels were measured via flame spectrometry. The contents of bicarbonate (HCO_3) and carbonate (CO_3) were analyzed through titration using 0.1 M HCl acid. Bromide concentration was determined using the Mettler Toledo SevenCompact meter. The ionic balance for all samples fell within the acceptable range of ± 10%.

The stable isotope composition of the groundwater samples was determined at the Laboratory of Radio Analyses and Environment within the National School of Engineers of Sfax, Tunisia. This analysis was conducted using the laser absorption spectrometer LGR DLT 100 (Penna et al. 2010), with a measurement uncertainty of ± 0.1‰ for $\delta^{18}O$ and ± 1‰ for $\delta^{2}H$. The results are presented in delta values,

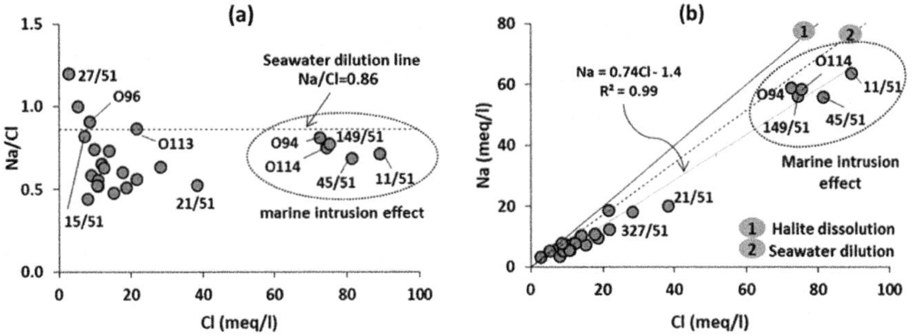

Fig. 6 Correlation diagram **a** Na/Cl versus Cl and **b** Na versus Cl of analysed samples. *Source* Developed by the author

the dissolution of evaporitic formations, which can result in elevated Na and Cl concentrations, remains challenging.

Nonetheless, for this coastal aquifer, the proximity of the sea to permeable sandy formations suggests the possibility of saltwater intrusion in specific regions of the aquifer. This intrusion could provide an explanation for the heightened concentrations of these elements. Moreover, the escalation in Cl levels might also be attributed to the dissolution of marine aerosols transported by the wind.

The average Na/Cl molar ratio in this context stands at 0.7, with a range spanning from 0.4 to 1.2. This average closely approximates the ratio found in seawater (0.86) (Jones et al. 1999), indicating that the Plio-Quaternary aquifer may indeed be subject to seawater contamination in select areas. This is particularly evident in instances where Cl concentrations exceed 70 meq/l, accompanied by a molar ratio of 0.7, as observed in samples 11/51, 45/51, 149/51, O94, and O114.

3.2.2 Ca/Mg Couple

One of the prominent characteristics associated with seawater intrusion is the elevated concentration of calcium in comparison to its content in seawater. The Mg/Ca ratio serves as a natural tracer for discerning the phenomenon of marine intrusion into coastal aquifers (Bouderbala 2015; Pulido-Leboeuf et al. 2003; Telahigue et al. 2018, 2020). This ratio displays an increment corresponding to the proportion of marine water within the mixture, with salinity represented by chloride concentrations (Pulido-Leboeuf et al. 2003). The rationale behind this behavior is that seawater exhibits a Ca/Mg ratio of 0.2, while freshwater typically features a ratio greater than 1. It is worth noting that saline water with a high calcium concentration may arise from various mechanisms, not necessarily linked to the cation exchange phenomenon (Jones et al. 1999). The Ca/Mg ratio diminishes as the proportion of seawater introduced into the mixture increases.

In the present case study, the ratio also experiences a decline in response to the chloride content. For instance, upon comparing point 27/51 (the least mineralized

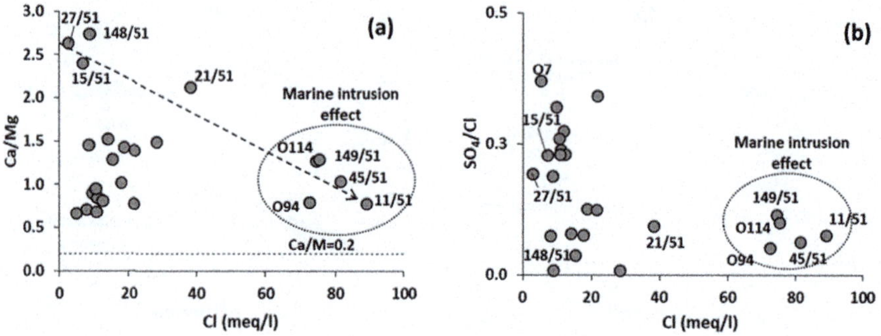

Fig. 7 Correlation diagram **a** Ca/Mg versus Cl and **b** SO_4/Cl versus Cl of analysed samples. *Source* Developed by the author

sample) and point 11/51 (the most mineralized sample), a distinct reduction in the Ca/Mg ratio is evident, coupled with an upsurge in chloride levels (Fig. 7a). This observation suggests a possible marine origin of the mineralization, particularly at well sites such as 11/51, 45/51, 149/51, O94, and O114, where the ratio closely resembles that of seawater and is associated with high chloride levels. In cases where the Ca/Mg ratio exceeds 1, seawater intrusion is unlikely to be implicated in the mineralization. Instead, another hydrochemical process, likely related to water–rock interaction, should be considered. This may involve phenomena such as cation exchange, which results in the release of calcium from the aquifer matrix and the adsorption of sodium from the solution (Gimenez et al. 2010).

3.2.3 *SO_4/Cl Couple*

As per Pulido-Lebouef et al. (2003), the SO_4/Cl ratio can be employed as a natural tracer for identifying the phenomenon of marine intrusion in coastal aquifers. This ratio tends to decrease as the proportion of seawater in the mixture increases (Bouderbala 2015; Pulido-Leboeuf et al. 2003; Telahigue et al. 2018, 2020; Tellam and Lloyd 1986).

For the shallow aquifer within the Essaouira basin, all the examined samples exhibit a SO_4/Cl ratio of less than 1, signifying the prevalence of chlorides over sulfates (Fig. 7b). The SO_4/Cl ratio among the sampled specimens ranges from 0.02 to 0.37, with the majority of them displaying a higher SO_4/Cl ratio than that of seawater (0.1) (Fig. 7b). This observation implies a blending of seawater with freshwater. The elevated sulfate content in these samples suggests the presence of additional sources, such as the dissolution of gypsum and anhydrite (Table 1). In particular, samples from wells 11/51, 45/51, 149/51, O94, and O114 exhibit a notably low SO_4/Cl ratio, coupled with high chloride contents. This indicates that the increase in salinity in these wells is primarily attributed to seawater intrusion, corroborating the results obtained through the Na/Cl and Ca/Mg ratios.

3.2.4 Br/Cl Couple

Bromide is recognized as a dependable indicator of the marine intrusion phenomenon (De Montety et al. 2008; Kim et al. 2003; Telahigue et al. 2018, 2020). Similar to chloride, bromide behaves as a conservative element and does not engage in reactions with the aquifer matrix, unless substantial amounts of organic matter are present (Davis et al. 1998). These two conservative elements offer insights into the origin of solutions and help identify potential contributions of marine water, as their concentrations remain unaffected by redox processes and are independent of low solubility minerals (Fedrigoni et al. 2001).

Given the extensive residence time of bromides and chlorides in oceanic masses, the Br/Cl ratio in contemporary seawater remains relatively constant, typically ranging between 1.5 and 1.7 × 10–3 (De Montety et al. 2008; Kim et al. 2003). This stability is preserved when the two elements share a common source. However, seawater can be distinguished from remnants of evaporated seawater or hypersaline waters (Starinsky et al. 1983), which arise from the dissolution of evaporite formations, as well as from anthropogenic sources such as wastewater effluents (Vengosh et al. 1998) or the return of irrigation water.

As seawater evaporates, the Br/Cl ratio remains constant until the onset of halite precipitation. During halite precipitation, the solution becomes enriched in bromide, leading to an increase in the Br/Cl ratio (Ben Hamouda et al. 2011). Notably, the Br/Cl ratio of residual brine continues to rise with increasing quantities of precipitated halite. Consequently, a solution originating from the concentration of seawater before halite saturation exhibits a Br/Cl ratio identical to that of seawater. Conversely, more concentrated brine, which has surpassed the phase of halite precipitation, demonstrates a higher Br/Cl ratio than seawater. Consequently, freshwater dissolving halite to saturation will exhibit a Br/Cl ratio lower than that of seawater, considering that primary halite is the sole chlorinated salt with a Br/Cl ratio lower than that of seawater. Simultaneously, a mixture of freshwater and brine that has undergone halite precipitation will possess a higher Br/Cl ratio than the marine ratio.

The correlation diagram between Br and Cl (Fig. 8a) reveals a robust positive correlation ($r^2 = 0.99$) between these two ions, suggesting a shared origin for bromides and chlorides. In the Br/Cl diagram as a function of Cl (Fig. 8b), samples 11/51, 45/51, 149/51, O94, and O114 fall within the seawater dilution field, displaying a Br/Cl ratio ranging from 1.5 to 1.7. This finding underscores their marine origin of salinity. The elevated bromide concentrations in certain wells, associated with a Br/Cl molar ratio exceeding that of seawater, can be attributed to the abandonment of these wells more than a decade ago and their substantial distance from the recharge zone, facilitating the formation of brines that have undergone the halite precipitation phase. Other data points exhibit lower molar ratios than seawater, indicating that these points are unaffected by the marine intrusion phenomenon and therefore suggest the existence of alternative sources of salinization, such as the dissolution of salts.

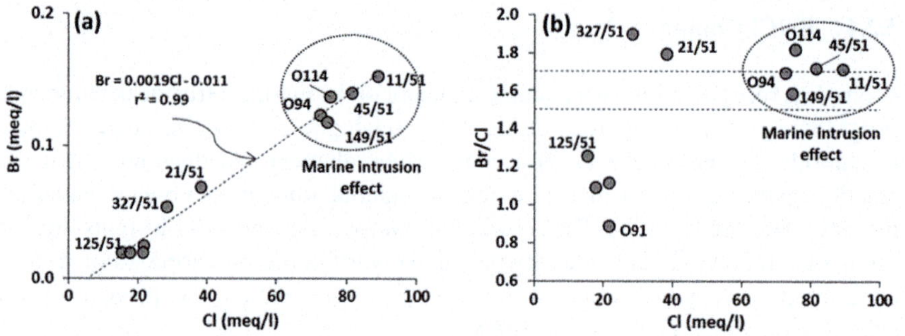

Fig. 8 Correlation diagram **a** Br versus Cl and **b** Br/Cl versus Cl of analysed samples. *Source* Developed by the author

3.2.5 *Cl/HCO$_3$ Versus Cl Plot*

The Cl/HCO$_3$ versus Cl graph (Fig. 9a, b) serves as a means to categorize water types, distinguishing between freshwater and seawater. Since seawater typically exhibits higher chloride (Cl^-) concentrations, while freshwater is characterized by higher bicarbonate (HCO_3^-) ion levels, this plot aids in the differentiation. Revelle (1941) and Todd (1959) established a classification system for water salinization based on Cl/HCO$_3$ ratios. This system distinguishes between water types unaffected by seawater intrusion or freshwater (< 0.5), those slightly affected by seawater intrusion (0.5–1.3), those moderately affected (1.3–2.8), those adversely affected (2.8–6.6), and those severely impacted by seawater intrusion (> 6.6). The graph essentially classifies water as freshwater, mixed water, or seawater.

The majority of data points in this study exhibit contamination resulting from marine intrusions, with none representing freshwater sources. This pattern is consistent with point 27/51, which displays contamination despite its aquifer recharge (Fig. 9a, 9b).

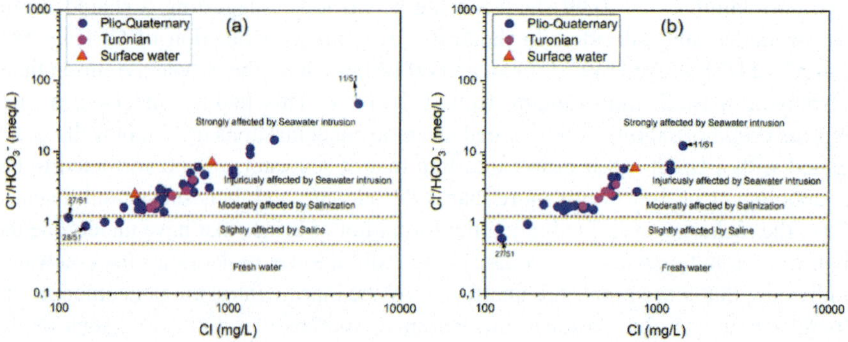

Fig. 9 Cl^-/HCO_3^- versus Cl plot for the campaign 2019 (**a**), and 2020 (**b**) (Bahir et al. 2022b)

3.3 Estimate of Mixture with Seawater

The most elevated chloride (Cl) concentrations are prominently observed in specific samples, with sample 11/51 exhibiting the highest recorded value at 89.1 meq/l, located at Cap Sim. Similarly, point 45/51, situated to the north, demonstrates notable Cl content of 81.5 meq/l. Additional samples display Cl concentrations within the range of 70–80 meq/l, exemplified by O114 (75.3 meq/l), 149/51 (74.4 meq/l), and O94 (72.5 meq/l). Conversely, the remaining samples encompass a spectrum of Cl concentrations that vary between 2.8 and 38.3 meq/l.

To quantify the degree of intrusion in each affected sample, a mixing rate (F) with seawater can be computed through a mass balance analysis of chlorides. This mixing rate can be estimated employing Eq. (2) as presented by Abou Zakhem and Hafez in 2007:

$$F(\%) = \frac{\left[Cl_{\text{sample}}\right] - \left[Cl_{\text{freash}}\right]}{\left[Cl_{sea}\right] - \left[Cl_{\text{freash}}\right]} \times 100 \quad (2)$$

With:

F corresponds to the seawater fraction.

Cl_{sample} corresponds to the concentration of chlorides in the water sampled.

Cl_{freash} corresponds to the concentration of chlorides in fresh groundwater. For our case, the average chloride concentrations of wells with electrical conductivity values less than 1000 μs/cm were used as Cl concentrations. Cl_{sea} corresponds to the concentration of chlorides in seawater.

The calculated mixing rates exhibit a range between 0% and 15.9% (as illustrated in Fig. 14). The most substantial values are observed in specific samples, such as 11/51 with a rate of 15.9%, 45/51 at 14.5%, 149/51 at 13.2%, O114 at 13.3%, and O94 at 12.8%. In contrast, the remaining samples depict mixing rates varying from 0% to 6.5%.

The elevated values are distributed primarily along the coastline, indicative of the seawater intrusion extending into the aquifer. The average width of this intrusion is estimated to be approximately 2 km. This contamination is likely attributed to a decline in the piezometric level due to decreased precipitation patterns observed over the previous decades within the study area (Ouhamdouch et al. 2018), as well as the impact of rising sea levels induced by global warming (IPCC, 2013).

4 Conclusion

The integration of findings derived from both the hydrogeochemical and isotopic approaches, encompassing parameters such as (Na, Cl), (Ca, Mg), (Br, Cl), (δ^2H, δ^{18}O), and (δ^{18}O, Cl), offers a more comprehensive understanding of the origins of groundwater salinization within the study area.

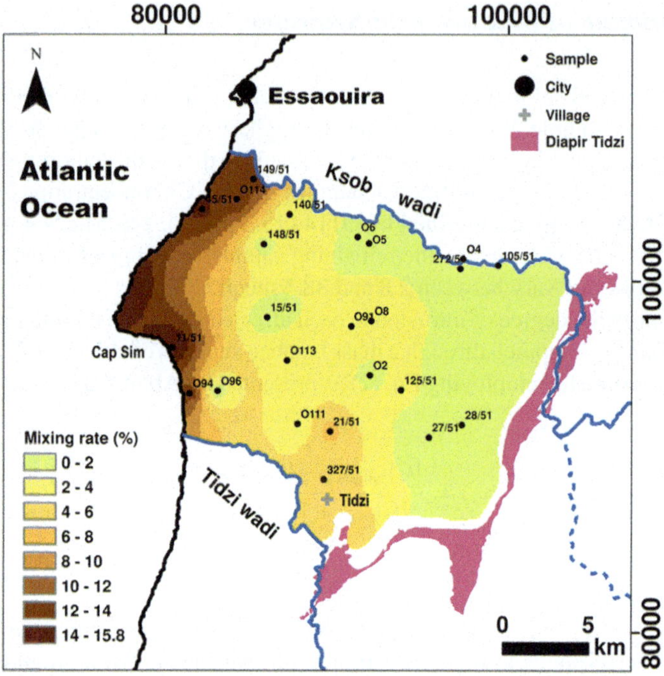

Fig. 14 Seawater fraction of analysed samples. *Source* Developed by the author

Ionic ratios such as Br/Cl approximating 1.5 to 1.7‰, Na/Cl approaching 0.86, as well as the presence of weak Mg/Ca and SO_4/Cl ratios collectively indicate the incursion of seawater into the fresh groundwater of the Plio-Quaternary aquifer in the Essaouira basin. This intrusion, supported by ionic ratios, is further substantiated by stable isotopes. The joint application of oxygen-18 content and chloride levels reveal a mixing rate of 15.9% at well 11/51, 14.5% at sample 45/51, 13.2% at well 149/51, 13.3% at point O114, and 12.8% at well O94. These elevated rates are distributed along the coastline, signifying the advancement of the saline front by approximately 2 km. The principal drivers behind this intrusion likely include a decline in the piezometric level due to reduced precipitation over recent decades in the study area and the impact of rising sea levels associated with climate change.

The case of the Essaouira region, located on Morocco's Atlantic coast, serves as an illustrative example of the multifaceted challenges encountered in semi-arid zones. Beyond its regional significance in water resource management, the methodology employed can be readily adapted and applied to other regions grappling with analogous climatic and anthropogenic constraints:

- Implementation of hill dams to mitigate wadi bed erosion and combat siltation by enhancing the longevity and mobilization of water resources via the dams.
- Establishment of artificial recharge mechanisms in areas characterized by deficits and vulnerability, especially those susceptible to seawater intrusion.

- Incorporation of desalinated seawater into available water resources to enhance their quality.
- Adoption of purified wastewater for irrigation purposes in cases where conditions permit, including golf courses, green spaces, and agriculture.
- Transition from aquifer overexploitation to surface water withdrawals, as a means to address water resource sustainability concerns.

References

Abou Zakhem B, Hafez R (2007) Environmental isotope study of seawater intrusion in the coastal aquifer (Syria). Environ Geol 51:1329–1339

Agoumi A (1999) Introduction à la problématique des changement climatiques. Projet Magrébin sur les changements climatiques RAB/94/G31 : SOMIGRAF mai 1999 ; 43p

Babqiqi A (2014) Changements Climatiques au Maroc: Etude du cas de la Région de Marrakech Tensift Al Haouz et implications sur l'agriculture à l'horizon 2030. Université cadi Ayyad, Thèse de doctorat

Bahir M, Mennani A, Jalal M, Youbi N (2000) Contribution à l'étude des ressources hydriques du bassin synclinal d'Essaouira (Maroc). Estud Geol 56:185–195

Bahir M, Jalal M, Blavoux (2001) Apport de l'hydrochimie isotopique à la connaissance des zones de recharge en zones aride et semi-arides : Cas du bassin de Meskala-Kourimat. J Environ Hydrol

Bahir M, Ouhamdouch S, Carreira PM (2016) La ressource en eau au Maroc face aux changements climatiques; cas de la nappe Plio-Quaternaire du bassin synclinale d'Essaouira. Comun Geol 103(1):35-44

Bahir M, Ouhamdouch S, Carreira PM (2018a) Geochemical and isotopic approach to decrypt the groundwater salinization origin of coastal aquifers from semi-arid areas (Essaouira basin, Western Morocco). Environ Earth Sci 77:485. https://doi.org/10.1007/s12665-018-7663-4

Bahir M, Ouhamdouch S, Carreira PM (2018b) Isotopic and geochemical methods for studying water–rock interaction and recharge mode: application to the Cenomanian-Turonian and Plio-Quaternary aquifers of Essaouira Basin, Morocco. Mar Freshwater Res 69:1290–1300

Bahir M, Ouazar D, Ouhamdouch S (2018c) Characterization of mechanisms and processes controlling groundwater salinization in coastal semi-arid area using hydrochemical and isotopic investigations (Essaouira basin, Morocco). Environ Sci Pollut Res 25:24992–25004

Bahir M, Ouazar D, Ouhamdouch S (2019) Hydrogeochemical investigation and groundwater quality in Essaouira region, Morocco. Mar Freshwater Res 70:1317–1332

Bahir M, Ouhamdouch S, Ouazar D, El Moçayd N (2020) Climate change effect on groundwater characteristics within semi-arid zones from western Morocco. Groundw Sustain Dev 11:1–14. https://doi.org/10.1016/j.gsd.2020.100380

Bahir M, Ouhamdouch S (2020b) Groundwater quality in semi-arid environments (Essaouira Basin, Morocco) Carbonates Evaporites 35:1–16. https://doi.org/10.1007/s13146-020-00576-7

Bahir M, Ouhamdouch S, Ouazar D, Chehbouni A, Ouarani M, El Mountassir O (2021a) Groundwater quality of the alluvial and carbonate aquifers of Essaouira basin (Morocco). Carbonates Evaporites 36:23. https://doi.org/10.1007/s13146-021-00697-7

Bahir M, El Mountassir O, Ouazar D, Carreira P.M (2021b) Use of WQI and isotopes to assess groundwater quality of coastal aquifers (Essaouira, Morocco). In: Abrunhosa M, Chambel A, Peppoloni S, Chaminé HI (eds) Advances in geoethics and groundwater management: theory and practice for a sustainable development. Advances in science, technology & innovation (IEREK interdisciplinary series for sustainable development). Springer, Cham. https://doi.org/10.1007/978-3-030-59320-9_51

Bahir M, El Mountassir O, Ouazar D, Carreira P.M (2021c) Hydrochemical analysis and evaluation of groundwater quality in Ouazi Basin (Essaouira, Morocco). In: Abrunhosa M, Chambel A, Peppoloni S, Chaminé HI (eds) Advances in geoethics and groundwater management: theory and practice for a sustainable development. Advances in science, technology & innovation (IEREK interdisciplinary series for sustainable development). Springer, Cham. https://doi.org/10.1007/978-3-030-59320-9_50

Bahir M, Mountassir EL, O, Chehbouni A, El Jiar H, Carreira P.M, (2022a) Hydrogeochemical and isotopic assessment for characterizing groundwater quality and recharge processes in the Essaouira Basin, Northwestern Morocco. Arab J Geosci 15(7):1–21. https://doi.org/10.1007/s12517-022-09817-6

Bahir B, El Mountassir B, Ouazar D, Chehbouni A, Dhiba D, Carreira PM (2022b) Assessment of seawater intrusion in the coastal plain aquifers of Essaouira basin using geochemical approaches. In: Ksibi M et al (eds) Recent advances in environmental science from the Euro-Mediterranean and surrounding regions (3nd Edition). Environmental science and engineering. Springer, Cham

Bear JJ, Cheng HDA (2010) Modeling groundwater flow and contaminant transport. In Hassanizadeh SM (ed) Theory and applications of transport in porous media. Springer Nature, Switzerland, pp 593–636

Bear J, Cheng AHD, Sorek S, Ouazar D, Herrera I (1999) Seawater intrusion in coastal aquifers-concepts, methods, and practices. In: Bear J, Cheng AHD, Sorek S, Ouazar D, Herrera I (eds) Theory and applications of transport in porous media. Kluwer Academic Publishers, Dordrecht/Boston/London, pp 51–71

Behnassi M, McGlade K (eds) (2017) Environmental change and human security in Africa and the Middle East. Springer International Publishing, Cham. Springer, Dordrecht, 336p

Behnassi M, Muteng'e MS, Ramachandran G, Shelat KN (2014) Vulnerability of agriculture, water and fisheries to climate change. Towards sustainable adaptation strategies. Springer, Dordrecht, 336p

Belkhiri L, Mouni L, Boudoukha A (2012) Geochemical evolution of groundwater in an alluvial aquifer: Case of El Eulma aquifer, East Algeria. J Afr Earth Sci 66:46–55

Ben Hamouda MF, Tarhouni J, Leduc C, Zouari K (2011) Understanding the origin of salinization of the Plio-quaternary eastern coastal aquifer of Cap Bon (Tunisia) using geochemical and isotope investigations. Environ Earth Sci 63:889–901

Bennouna A, El Hebil C (2016) Energy needs for Morocco 2030, as obtained from GDP-energy and GDP-energy intensity correlations. Energy Policy 1(88):45–55

Bouderbala A (2015) Groundwater salinization in semi-arid zones: an example from Nador plain (Tipaza, Algeria). Environ Earth Sci 73:5479–5496

Carreira PM, Marques JM, Nunes D (2014) Source of groundwater salinity in coastline aquifers based on environmental isotopes (Portugal): Natural vs. human interference. A review and reinterpretation. Appl Geochem 41:163–175

Carreira PM, Bahir M, Ouhamdouch S, Fernandes PG, Nunes D (2018) Tracing salinization processes in coastal aquifers using an isotopic and geochemical approach: comparative studies in western Morocco and southwest Portugal. Hydrogeol J 26:2595–2615

Castany G (1982) Principes et méthodes de l'hydrogéologie. Edition Dunod, Paris, p 236p

Craig H (1961) Isotopic variations in meteoric waters. Science 133:1702–1703

Custodio E (1997) Seawater intrusion in coastal aquifers. Guidelines for study, monitoring and control, water report N°11. Food and Agriculture Organization of the United Nation, Rome, Italie, p 152

Custodio E (2002) Coastal aquifers as important natural hydrogeological structures. In: Bocanegra M, Massone H (eds) Groundwater and human development, pp 1905–1918

Davis SN, Whittemore DO, Fabryka-Martin J (1998) Uses of chloride/bromide ratios in studies of potable water. Groundwater 36:338–350

De Montety V, Radakovitch O, Vallet-Coulomb C, Blavoux B, Hermitte D, Vallès V (2008) Origin of groundwater salinity and hydrogeochemical processes in a confined coastal aquifer: case of the Rhône delta (Southern France). Appl Geochem 23:2337–2349

Direction de la Météorologie Nationale (DMN) (2007) Les changements climatiques au Maroc: observations et projections, Maroc-Météo, Casablanca
Driouech F (2010) Distribution des précipitations hivernales sur le Maroc dans le cadre d'un changement climatique: descente d'échelle et incertitudes. Thèse de doctorat, Université, Toulouse, France
Driouech F, Mahé G, Déqué M, Dieulin C, El Heirech T, Milano M, Benabdelfadel A, Rouche N (2010) Evaluation d'impacts potentiels de changements climatiques sur l'hydrologie du bassin versant de la Moulouya au Maroc. In: Global change: facing risks and threats to water resources (Proc. VIth FRIEND World Conf., Fes, Morocco, October 2010). IAHS Publ. 340. IAHS Press, pp 561–567
Edmunds WM, Milne CJE (2001) Palaeowaters in coastal Europe: evolution of groundwater since the late Pleistocene. Geol Soc, Special Publications, London 189:289–311
El Hassadi A (2008) Seawater intrusion in Derna located in the Green Mountain region, Libya: a threatening recurrent phenomenon calling for desalination. Desalination 220:189–193
El Kharraz J, El-Sadek A, Ghaffour N, Mino E (2012) Water scarcity and drought in WANA countries. Procedia Eng 1(33):14–29
El Mountassir O, Bahir M (2023) The assessment of the groundwater quality in the coastal aquifers of the Essaouira Basin, Southwestern Morocco, using hydrogeochemistry and isotopic signatures. Water 15(9):1769. https://doi.org/10.3390/w15091769
El Mountassir O, Bahir M, Ouazar D, Chehbouni A, Carreira PM (2021a) Geochemical and isotopic evidence of groundwater salinization processes in the Essaouira region, north-west coast, Morocco. SN Appl Sci. https://doi.org/10.1007/s42452-021-04623-3
El Mountassir O, Bahir M, Ouazar D, Carreira P.M (2021b) Nitrate pollution in groundwater of the Ouazi Basin: case of Essaouira (Southwestern Morocco). In: Abrunhosa M, Chambel A, Peppoloni S, Chaminé HI (eds) Advances in geoethics and groundwater management: theory and practice for a sustainable development. Advances in science, technology & innovation (IEREK interdisciplinary series for sustainable development). Springer, Cham. https://doi.org/10.1007/978-3-030-59320-9_49
El Mountassir O, Bahir M, Ouazar D, Carreira PM (2021c) For a better understanding of recharge and salinization mechanism of a cenomanian–turonian aquifer. In: Abrunhosa M, Chambel A, Peppoloni S, Chaminé HI (eds) Advances in geoethics and groundwater management: theory and practice for a sustainable development. Advances in science, technology & innovation (IEREK interdisciplinary series for sustainable development). Springer, Cham. https://doi.org/10.1007/978-3-030-59320-9_42
El Mountassir O, Driss O, Mohammed B, Abdelghani C, Carreira Paula M (2021d) GIS-based assessment of aquifer vulnerability using DRASTIC model and stable isotope: a case study on Essaouira basin. Arab J Geosci 14:321 (2021).https://doi.org/10.1007/s12517-021-06540-6
El Mountassir O, Bahir M, Ouazar D, Chehbouni A, Carreira PM (2021e) Geochemical and isotopic evidence of groundwater salinization processes in the Essaouira region, North-West Coast, Morocco. SN Appl Sci. https://doi.org/10.1007/s42452-021-04623-3
El Mountassir O, Bahir M, Ouazar D, Chehbouni A, Carreira PM (2022a) Temporal and spatial assessment of groundwater contamination with nitrate using nitrate pollution index (NPI), groundwater pollution index (GPI), and GIS (case study: Essaouira basin, Morocco). Environ Sci Pollut Res. https://doi.org/10.1007/s11356-021-16922-8
El Mountassir O, Bahir M, Ouazar D, Chehbouni A, Carreira PM (2022b) Evaluation of nitrate source and its distribution in the groundwater of Essaouira basin. Sustain Water Resour Manag 8:28. https://doi.org/10.1007/s40899-022-00609-0
El Mountassir O, Bahir M, Chehbouni A, Dhiba D, El Jiar H (2022c) Assessment of groundwater quality and the main controls on its hydrochemistry in a changing climate in Morocco (Essaouira Basin). Sustainability 14(13):8012. https://doi.org/10.3390/su14138012
Elouissi A, Habi M, Benaricha B, Boualem SA (2017) Climate change impact on rainfall spatio-temporal variability (Macta watershed case, Algeria). Arab J Geosci 10:496
ESCAP U, ECA U, ECE U, ESCWA U (2017) World economic situation and prospects

Fadili A, Mehdi K, Malaurent P, Riss J, Boutayeb K, Guessir H (2012) Influence de la marée océanique sur la variation du niveau piézométrique de l'aquifère karstique côtier de Oualidia (Maroc). Afr Geosci Rev 19:135–150
Farid I, Trabelsi R, Zouari K, Beji R (2013) Geochemical and isotopic study of surface and groundwaters in Ain Bou Mourra basin, central Tunisia. Quatern Int 303:210–227
Fedrigoni L, Krimissa M, Zouari K, Maliki A, Zuppi GM (2001) Origine de la minéralisation et comportement hydrogeochimique d'une nappe phréatique soumise à des contraintes naturelles et anthropiques sévères: exemple de la nappe de Djebeniana (Tunisie). C R Acad Sci 332:665–671
Filho WL (2012) Climate change and the sustainable use of water resources, climate change management. Springer, 801 p. ISBN 978-3-642-22265-8
Geyh MA (2000) An overview of 14C analysis in the study of the groundwater. Radiocarbon 42:99–114
Gibbs RJ (1970) Mechanisms controlling world water chemistry. Science 170:1081–1090
GIEC (2007) Résumé à l'intention des décideurs. In: Solomon, S, Qin D, Manning M, Chen Z, Marquis M, Averyt KB, Tignor M and Miller HL (eds) Bilan 2007 des changements climatiques: Les bases scientifiques physiques. Contribution du Groupe de travail I au quatrième Rapport d'évaluation du Groupe d'experts intergouvernemental sur l'évolution du climat. Cambridge University Press, Cambridge, UK and New York, NY, USA
Gimenez E, Bencini A, Pranzini G (2010) Hydrogeochemical considerations about the origin of groundwater salinization in some coastal plains of Elba Island (Tuscany, Italy). Environ Geochem Health 32:243–257
El Halimi N, Alia AC, Lebbe L, Beeuwsaert E, Walraevens K (1999) The fresh-/salt-water flow and distribution in a cross-section through the dunes of Saidia Plain (Northeastern Morocco). In 15th Salt Water Intrusion meeting, Ghent 1998. Proceedings, pp 205–211
Hamed Y, Hadji R, Redhaounia B, Zighmi K, Bâali F, El Gayar A (2018) Climate impact on surface and groundwater in North Africa: a global synthesis of findings and recommendations. Euro-Mediterr J Environ Integr 3:25. https://doi.org/10.1007/s41207-018-0067-8
Hassani My D, Mokssit A, Henia L (1998) Changements climatiques et ressources en eau dans les pays du Maghreb. Juin 1998:55p
Haut-Commissariat au Plan (HCP) (2013) Séminaire sur: "La comptabilité des actifs naturels pour une croissance verte dans la région MENA" 26 et 27 mars 2013, Marseille, France. https://www.hcp.ma
IPCC (2013) In: Stocker TF, Qin D, Plattner GK, Tignor M, Allen SK, Boschung J, Nauels A, Xia Y, Bex V and Midgley PM (eds) Climate change 2013: the physical science basis. Contribution of working group I to the fifth assessment report of the intergovernmental panel on climate change. Cambridge University Press, Cambridge, UK and New York, NY, USA
Jones BF, Vengosh A, Rosenthal E, Yechieli Y (1999) Geochemical investigations. In Bear J et al (eds) Seawater intrusion in coastal aquifers. Kluwer Academic Publisher, pp 51–71
Kaid Rassou K, Fakir Y, Bahir M, Zouari K, Marah M, Monteiro JP (2005) Apports des analyses isotopiques à la compréhension du fonctionnement des aquifères côtiers du bassin hydrologique de la lagune d'Oualidia. Comun Geol 92:129–142
Kim Y, Lee K, Koh D, Lee D, Lee S, Park W, Koh G, Woo N (2003) Hydrogeochemical and isotopic evidence of groundwater salinization in a coastal aquifer: a case study in Jeju volcanic island, Korea. J Hydrol 270:282–294
Loucks DP, Van Beek E (2017) Water resource systems planning and management: an introduction to methods, models, and applications. Springer
Meddi H, Meddi M (2007) Variabilité spatiale et temporelle des précipitations du nord-ouest de l'Algérie. Geographia Technica 2:49–55
Meddi M, Talia M, Martin C (2009) Recent evolution of weather conditions and flows on the basin of Macta (Northwest of Algeria). PhysioGeo 23:61–84
Mennani A (2001) Apport de l'hydrochimie et de l'isotopie à la connaissance du fonctionnement des aquifères de la zone côtière d'Essaouira (Maroc Occidental). Université Cadi Ayyad, Maroc, Thèse de doctorat

Mennani A, Blavoux B, Bahir M, Bellion Y, Jalal M, Daniel M (2001) Apports des analyses chimiques et isotopiques à la connaissance du fonctionnement des aquifères plio-quaternaire et turonien de la zone synclinale d'‘Essaouira (Maroc occidental). J Afr Earth Sci 32:819–835
Misra AK (2014) Climate change and challenges of water and food security. Int J Sustain Built Environ 3:153–165
Morsli B, Habi M, Hamoudi A (2007) Contraintes et perspectives des aménagements hydroagricoles et antiérosifs en Algérie. Actes des JSIRAUF, Hanoi, 6–9 November 2007.
Nadifi K (1998) Ressource en eau du Maroc, Séminaire « échelle et modèles de gestion de l'eau du bassin du Haouz au Bassin occidental de la Méditerranée, Faculté des sciences Semlalia de Marrakech
Nassopoulos H (2012) Les impacts du changement climatique sur les ressources en eaux en Méditerranée. Université Paris-Est, France, Thèse de doctorat, p 153
Niemczynowicz J (1999) Urban hydrology and water management–present and future challenges. Urban Water 1;1(1):1–4
Ouhamdouch S, Bahir M, Carreira P (2018) Impact du changement climatique sur la ressource en eau en milieu semi-aride: exemple du bassin d'Essaouira (Maroc). Rev Des Sci De L'eau 31:13–27
Ouhamdouch S, Bahir M, Ouzar D, Carreira PM, Zouari K (2019) Evaluation of climate change impact on groundwater from semi-arid environment (Essaouira Basin, Morocco) using integrated approaches. Environ Earth Sci 78:449. https://doi.org/10.1007/s12665-019-8470-2
Ouhamdouch S, Bahir M, Ouazar D, Goumih A (2020) Zouari K Assessment of the climate change impact on the future evapotranspiration and flows from a semi-arid environment. Arab J Geosci 13:82. https://doi.org/10.1007/s12517-020-5065-x
Ouhamdouch S, Bahir M, Ouazar D (2020b) Climate change impact assessment on a carbonate aquifer under semi-arid climate; example of the Cenomanian-Turonian aquifer within Meskala-Ouazzi region (Essaouira Basin, Morocco). Arab J Geosci (2020) 13:173.https://doi.org/10.1007/s12517-020-5180-8
Paniconi C, Khlaifi I, Lecca G, Giacomelli A, Tarhouniy J (2001) Modeling and analysis of seawater intrusion in the Coastal Aquifer of Eastern Cap-Bon, Tunisia. Transport Porous Med 43:3–28
Parkhurst DL, Appelo CAJ (1999) User's guide to PHREEQC (Version 2) a computer program for speciation, batch-reaction, one-dimensional transport, and inverse geochemical calculations: U.S geological survey. Water-Resour Invest Rep 99(4259):312
Penna D, Stenni B, Wrede S, Bogaard TA, Gobbi A, Borga M, Fischer BMC, Bonazza M, Charova Z (2010) On the reproducibility and repeatability of laser absorption spectroscopy measurements for [delta] 2H and [delta] 18O isotopic analysis. Hydrol Earth Syst Sc 14:1551–1566
Piper AM (1944) A graphic procedure in the geochemical interpretation of water analyses. Trans Am Geophys Union 25:914–923
Podgorski J, Berg M (2020) Global threat of arsenic in groundwater. Science 22:368(6493):845–850
Post VEA (2004) Groundwater salinization processes in the coastal area of the Netherlands due to transgressions during the Holocene. PhD thesis, Vrije University, Amsterdam, 138 p
Pulido-Bosch A, Tahiri A, Vallejos A (1999) Hydrogeochemical characteristics of processes in the Temara aquifer in northwestern Morocco. Water Air Soil Poll 114(3):323–337
Pulido-Leboeuf P, Pulido-Bosch A, Ml C, Vallejos A, Andreu JM (2003) Strontium, SO_4/Cl and Mg/Ca ratios as tracers for the evolution of seawater into coastal aquifers: the example of Castell de Ferro-aquifer. CR Geosci 335:1039–1048
Rabeiy RE (2018) Assessment and modeling of groundwater quality using WQI and GIS in Upper Egypt area. Environ Sci Pollut Res 25(31):30808–30817. https://doi.org/10.1007/s11356-017-8617-1
Revelle R (1941) Criteria for recognition of the sea water in ground-waters. EOS Trans Am Geophys Union 22(3):593–597
Sadeg SA, Karahanoglu N (2001) Numerical assessment of seawater intrusion in the Tripoli region, Libya. Environ Geol 40:1151–1168

Sebbar A (2013) Etude de la variabilité et de l'évolution de la pluviométrie au Maroc (1935–2005): Réactualisation de la carte des précipitations. Université Hassan II, Thèse de doctorat, p 155
Selvakumar S, Chandrasekar N, Kumar G (2017) Hydrogeochemical characteristics and groundwater contamination in the rapid urban development areas of Coimbatore, India. Water Resour Ind 17:26–33. https://doi.org/10.1016/j.wri.2017.02.002
Sherif MM (1999) The Nile Delta aquifer in Egypt, chap 17. In: Cheng A, Sorek S, Ouazar D, Herrera A (eds) Bear J. Theory and application of transport in porous media, Kluwer Academic Publishers, Dordrecht/Boston/London, pp 559–590
Starinsky A, Bielski M, Lazar B, Steinitz G, Raab M (1983) Strontium isotope evidence on the history of oilfield brines, Mediterranean Coastal Plain, Israel. Geochim Cosmochim Acta 47:687–695
Taabni M, El Jihad MD (2012) Eau et changement climatique au Maghreb : quelles stratégies d'adaptation ? Les Cahiers D'outre-Mer 260:493–518
Telahigue F, Agoubi B, Souid F, Kharroubi A (2018) Groundwater chemistry and radon-222 distribution in Jerba Island, Tunisia. J Environ Radioact 182:74–84
Telahigue F, Mejri H, Mansouri B, Souid F, Agoubi B, Chahlaoui A, Kharroubi A (2020) Assessing seawater intrusion in arid and semi-arid Mediterranean coastal aquifers using geochemical approaches. Phys Chem Earth 115.https://doi.org/10.1016/j.pce.2019.102811
Tellam JH, Lloyd JW (1986) Problems in the recognition of seawater intrusion by chemical means: an example of apparent equivalence. Q J Eng Geol Hydroge 19:389–398
Teshome F (2020) Seasonal water quality index and suitability of the water body to designated uses at the eastern catment of lake Hawassa. Environ Sci Pollut Res 27(1):279–290. https://doi.org/10.1007/s11356-019-06794-4
Todd DK (1959) Annotated bibliography on artificial recharge of ground water through 1954. US Government Printing Office
UNEP (2006) Africa's lakes: atlas of our changing environment, Nairobi
UNEP (2010) "Africa water Atlas". Division of early warning and assessment (DEWA). United Nations environment programmed (UNEP), Nariobi
UNEP (2011) Vital climate graphics Africa
United Nations Economic and Social Commission for Western Asia (ESCWA) et al (2017) Arab climate change assessment report—main report. Beirut, E/ESCWA/SDPD/2017/RICCAR/Report
Vella C, Fleury TJ, Raccasi G, Provansal M, Sabatier F, Bourcier M (2005) Evolution of the Rhone delta plain in the Holocene. Mar Geol 222–223:235–265
Vengosh A, Gert JD, Starinsky A (1998) Boron isotope and geochemical evidence for the origin of Urania and Bannock brines at the eastern Mediterranean: effect of water-rock interactions. Geochim Cosmochim Acta 62:3221–3228
Vennetier M, Vila B, Liang EY, Guibal F, Ripert C, Chandioux O (2005) Impact du changement climatique sur la productivité forestière et le déplacement d'une limite bioclimatique en région méditerranéenne française. Ingénieries-EAT 49:49–61

Dr. Otman El Mountassir is a hydrogeologist specializing in water and environmental engineering. He holds a Master's degree in Water and Environmental Engineering from the Faculty of Sciences Kenitra, and a Ph.D. in Geosciences Geressources and Environment from Cadi Ayyad University under the supervision of Professor Mohammed Bahir. He has published articles in leading international journals, indexed in databases such as Scopus and Thomson. He has also acted as a reviewer for many international journals. His expertise includes Hydrogeology, Hydrology, Stable Isotopes, Hydrochemistry and Modelling. He is affiliated with the High Energy and Astrophysics Laboratory of the Semlalia Faculty of Sciences, Cadi Ayyad University, Morocco.

help predict, test, and compare various reasonable alternatives of any hydrogeologic system. The approximations made by the models about physical systems and their closeness with the real systems help to decide about their usefulness and their reliability. This also requires a close understanding of the physical system and assumptions used to derive numerical equations (Cooley 1997; Lin and Anderson 2003; Torak 1993). The assumptions made during the modeling process have a key role to play, which could be about the flow direction, geometry of aquifer, its heterogeneity and anisotropy of bed material, transport mechanism and chemical reactions, etc. For the solution of problems related to groundwater management and to make decisions many professionals still rely on various models in spite of the fact that such models just provide estimations and their predicted values are not much reliable due to various simplifying assumptions which are embedded in the numerical equations (Anderson et al. 2015).

A number of groundwater modeling codes are available in the market nowadays; however, they all have their limitations, different operational characteristics merits, and demerits. Every modeling project has its own objectives, therefore it is very important to select a specific modeling code depending upon the requirements of the project, because it is not necessary that every code will fulfill all the requirements of any project. To understand the limitations, capabilities, and operational capabilities of any code available in the market it requires expertise because it is hard to understand it merely from its documentation (Dagan 1982). For any particular project, the code's systematic and comprehensive description and classified information help in its selection process.

5 Characteristics of Regional Groundwater Model

The regional groundwater models would be useful if they should preferably cover the entire basin for proper management and assessment of the basin. The regional-scale models must include the spatial distribution of pumping and recharge characteristics. Some distinct features of such a regional-scale model are (1) The topographic elevation difference provides the principal driving force to determine the flow direction of groundwater. Seepage from surface water bodies and precipitation are the major sources of recharge to the system. Whereas evapotranspiration, springs, and pumping are the major discharging components. (2) The model layers must include information about aquitards because of their importance in determining leakages, especially in a regional scale model. (3) As a single source the interaction between ground and surface water interaction is very important and must be included in the model setup. Whereas interaction among surface water bodies e.g. rivers, canals, etc. with groundwater could be treated as line source/sink; however constant head boundary or head-dependent boundaries may be used to represent stationary water bodies like lakes, etc. specifically for steady-state models (Zhou and Li 2011).

The grid size of the regional scale models can fluctuate a lot due to large area variations e.g. these models could be used to simulate the area which could extend from few km^2 to thousands of km^2. If the simulated area is large enough some special care should be adopted i.e.: (a) appropriate upscaling techniques should be employed for the aggregation of cell hydraulic conductivity values with proper care; (b) the predicted head for a cell should not be treated as a single point value because it gives an average value that corresponds to several square kilometers area; (c) Such models could be calibrated using contour maps having observed Vs modeled head values but it is preferred that observed discharge data of pumping and evapotranspiration should be used for comparison purposes; (d) During the conversion of a regional-scale model to local scale grid refinement is required that could be achieved by utilizing telescopic refinement technique rather than adopting local grid refinement technique, this could also help to increase the resolution of several points of interest e.g. wells and contaminant transport, etc. (Zhou and Li 2011).

6 Hydrogeological Conceptual Model Approach

The conceptual model approach is very common for the last few years in developing complex flow and transport models. The points, arcs, lines, polygons, etc. are used to describe features in the conceptual modeling approach. The availability of data and its complexity drive the grid features of a three-dimensional model grid employing the traditional modeling approach. There are possibilities of errors, and it could be a tedious job because a large number of data has to be incorporated into the model. Therefore it is more convenient to construct a conceptual model having a grid first that could be transformed into a numerical model later on along with the input data (Zhou and Li 2011).

This approach is very useful especially if the scale of the model is regional. The conceptual model approach can have a lot of benefits e.g. defining a model grid resolution is not a prerequisite. Valuable time can be saved by this technique by automatically computing the conductance value for river cells which could consume a lot of time in its manual computations. The transient variables like pumping rates could be assigned exclusively by adding as a curve between time and stress data. During the process of conversion of a conceptual model to a numerical model, the data regarding stress periods is also assigned to the model automatically. This approach has another big advantage i.e. in case there is any change in the model then the transformation process from conceptual to numerical is also very fast because the conceptual model is constructed without taking into consideration the space and time factors of a numerical model. Therefore more reliable and accurate model results could be anticipated due to the modeler's liberty to construct and compare a variety of conceptual models side by side in a short duration (Zhou and Li 2011).

Likewise, to reduce the inherent instability in some finite difference models e.g. MODFLOW to some extent featured objects are used to assign boundary conditions to the grid cells. To calculate the elevation of a river cell along its length and the entry of heads and elevation values for each cell was also a time-consuming and tedious job therefore modeler often tends to choose small cells to relate minor values to the same group which created the problem of non-convergence of the model. In the conceptual model approach along the river or canal's length interpolated values are provided which is a good solution to the problem. It also prevents any abrupt change in the cell value and a more realistic representation of the boundary conditions is made possible which are in close agreement with real-world natural conditions (Zhou and Li 2011).

7 Application of Groundwater Flow Modeling

The models represent the simplified form of real systems which are very complex and cumbersome in nature. The models can help in understanding the complexities of the real world and could provide the best possible solutions in solving the problems (Kumar 2015). The description of the complexity in natural systems and the processes that occur in the system naturally is best linked up by the conceptual models. Therefore the construction of a reasonable conceptual model is the first step towards the development of a successful model (Zhou and Herath 2017). The conceptual models help in identifying the aquifers and their areal extent, the associated water bodies, by analyzing the stresses in the system i.e. pumping and recharge, etc., and the movement of water through a porous medium (Anderson and Woessner 1992).

The numerical model quantifies the conceptual model that encompasses different governing equations which represent physical properties of the aquifer, stress periods and system geometry, etc. (Sadeghi-Tabas et al. 2017). The hydraulic heads and flow rates along with hydraulic conductivity values are used to represent the response of the aquifer when it is subjected to different stresses when it is transformed from conceptual to numerical model (Sahoo and Jha 2017). Computer programming can help in easily transforming the involved complex equations. To best represent systems and their physical properties several codes and computer programs are widely in use nowadays. The code or computer program is only written once although it can be changed however models can vary according to the study objectives hence the input parameters and code for a specified groundwater system are collectively termed as a model (Anderson and Woessner 1992).

8 Global Climate Models

To study the behavior and variations in the global climate various tools are used and climate models are one of those tools. The climate models are of different types and can be classified as simple climate models (SCMs), having intermediate complex nature, like (EMICs) and three-dimensional global climatic models or general circulation models (GCMs), etc. (Green et al. 2011).

Among these, the (GCMs) are considered the most efficient tools for the projection of future scenarios and the current global climate simulation. Recently, GCMs, provided by the Coupled Model Intercomparison Project Phase 5 (CMIP5), have been widely applied in climate change analysis (Hosseinzadehtalaei et al. 2017). The interaction of flows in the hydrosphere, cryosphere, biosphere, and atmosphere within the climate system are usually accounted for in the formation of these models. The research in the climate field is quite complex, and, with the advancement in knowledge about the climatic processes, climate models are also developing rapidly, although the development of early GCMs started in the 1960s and 1970s. The GCMs spatial resolution has increased enormously during the last decade; moreover, the incorporation of physical processes like simple rainfall and carbon dioxide emission to water vapor feedback, has also improved tremendously (Jones and Wigley 2010).

The current climate models incorporate the prevalent continental processes, which could have an impact or could affect the climate on a large scale for the next coming decades. However, the deficiency exists as some potential processes—e.g. glaciation on a global scale—which could have impacts on longer time scales are still missing. Some newly developed GCMs have oceanic components, and the process is going on. The modeling of sea ice, freshwater fluxes, and mixing schemes of rivers and estuary are possible with these new developments (Randall et al. 2007). It is now possible to simulate frost days, cold air bursts, and intense hot temperatures with the GCMs quite reasonably. But the extreme precipitation event simulation is still dependent on the selected threshold, resolution and parameterization, etc. Generally, it is quite possible with these models that they may produce too many days with weak precipitation i.e. less than 10 mm per day, and maybe overall too little precipitation for intense events, i.e. more than 10 mm per (Fant et al. 2016; Randall et al. 2007).

9 Downscaling

The discrepancy among the data obtained from the local meteorological observatory and GCMs coarse spatial scales are addressed by the downscaling techniques (De Caceres et al. 2018; Hewitson and Crane 2006; Wilby and Wigley 1997). Two variables, i.e. temperature and precipitation are of utmost importance in meteorological science and their future projection is also necessary over the global scale. The local prediction of climate, especially temperature and precipitation are not so accurate, however, the estimation of ratios and differences from previous to projected scenarios for these variables is fairly consistent (Loáiciga 2009).

The development of regional and GCMs for the improved climatic projections, that could couple both the atmospheric processes and groundwater, is likely to come in the future (Cohen et al. 2006; Gutowski et al. 2002). The large scale of GCMs and small/local scale of surface or groundwater models are a real challenge for scientists because the local scale models primarily require daily input data of few km^2 and that also of higher resolution that is normally not possible for climate models (Bouraoui et al. 1999; Loáiciga 2009). There are two major types of downscaling techniques

(i) dynamic downscaling (ii) Statistical downscaling. Nowadays downscaling techniques are widely used because of the maturity in methods since the inception of the third assessment report IPCC-2001 (Griggs and Noguer 2002). Though, the large-scale downscaling for multi-model climate change simulations is still not common and is only available for some areas (Christensen et al. 2007).

10 Dynamic Downscaling

In the dynamic downscaling technique, the coarse resolution global climate models (GCM) are nested with higher resolution (RCM). The transient atmospheric boundary conditions are defined by the RCMs around a finite domain by utilizing the GCM data set. Which helps to model the atmosphere's physical dynamics ranging from 20 to 50 km or maybe less by utilizing horizontal grid spacing. RCMs also has some limitations just like GCMs they also require large computations. This limits the simulation period, model domain size, and the capability of the model to perform the number of experiments (Guo and Wang 2016).

11 Statistical Downscaling

In this technique, the present and past empirical data are collectively used in order to minimize the difference between coarse resolution GCMs and point climatic observations. The statistical downscaling models are used to establish a relation between the predictors and predictands. Which helps in estimating the regional or local characteristics of the climate. There are several statistical models available which can be classified as synoptic weather models, stochastic weather generators and regression models, etc. The statistical downscaling models are comparatively inexpensive and can be applied to various GCMs outputs very conveniently. The climate change impact studies often require local-level information which could be easily inferred from these models (Stoll et al. 2011). However, the validity of these models' results remains questionable due to their basic data assumption mechanisms which require model calibration with high-quality data set.

12 Future Scenario Analysis

The groundwater demand in the semi-arid to arid regions is always on the rise due to mounting pressure of growing more food and fiber for the accelerated population demand. The surface water supplies are decreasing day by day due to a reduction in canal water supplies and river flows. Moreover, the spatial and temporal uncertainties and variability in intensity and duration of rainfall are also one of the major causes.

Therefore, future scenarios are required to be developed using three-dimensional hydrogeological models to simulate the future response of the aquifer under climate change scenarios. The model simulations could help to analyze the hydrodynamics of the aquifer system.

13 Climate Change Scenarios

There are several variables due to which changes in the natural ecosystem are induced. Climate change is one of the most important potential variables which is responsible for this change (IPCC 2007). To analyze the changes in aquifer's future response to stresses caused by the impacts of climate change a comprehensive analysis is required. Climate change can cause changes in various parameters among them evapotranspiration and precipitation have large implications, especially in the regions where irrigated agriculture is predominant. The hydrodynamics of the groundwater can be studied by adopting any statistical downscaling technique in these areas. The IPCC emission scenarios H3A2 and H3B2 could be synthesized for future climate change projections in these kinds of regions for the assessment of future outcomes and associated uncertainties due to climate change these scenarios could act as a convenient tool in analyzing responses and various driving forces.

14 Impact of Climate Change on Future Groundwater Fluctuations

Due to the regular and chaotic behavior of climate, it is usually considered a metastable system like all other adaptive systems. There is an improvement in the representation of climatic behavior over the past few decades (Soldatenko 2019) due to advancements within the global climate models. This has helped to simulate short-term variation like the chaotic behavior of the climatic variables. To examine the impacts of climate change on groundwater generally, two approaches are used:

- GCMs results could be used to formulate various scenarios encompassing groundwater demand for the agriculture sector and other competing users like industry and domestic by coupling it with the precipitations amount and pattern changes (Zhou et al. 2010);
- Or the regional climate models (RCM) data could be utilized for the development of more detailed and explicit models predicting hydrological impacts over the entire region understudies such as an aquifer or river basin etc. (Zhou et al. 2010).

The available information and its quality plus quantity play a pivotal role in the use of any particular model. In some cases, such as research studies more precise and comprehensive models may be required whereas simpler straightforward approach-based models could be more appropriate in investigating the initial stages of any problem. However, the element of uncertainty within these climatic scenarios remains pertinent in predicting these impacts. That is why still the researchers are in the process of developing a consensus about the timing and magnitude of uncertainty reduction in the model's predictions (Soden et al. 2018).

The observed spatial and temporal data of climate variables e.g., temperature, humidity, winds, precipitation, etc. is used extensively for the verification of climate models that can provide a level of confidence in the model's climatic projections. The climate system is governed by several biochemical and geophysical processes which can influence the ability of a model to simulate interactions within the climate system (Flato et al. 2014).

The groundwater recharge is very much dependent on the precipitation amount and its distribution, any changes in the pattern of precipitation would highly likely alter the recharge pattern resultantly some areas will receive more recharge whereas others may suffer the reductions (Thomas et al. 2016).

For the assessment of climate change impacts on groundwater resources, it is imperative to understand the extent and magnitude of groundwater vulnerability to droughts, over-exploitation as well as deterioration in its quality (Thomas et al. 2016). To understand the spatial and temporal availability of groundwater a better quantification of the regional water budget is required. Which must include all the components of the hydrologic cycle involving moisture present in the atmosphere, soil, and vegetation cover. This information can be a good source of evidence for groundwater managers dealing with present and future climate change regimes.

Many factors influence the estimation of recharge. As its estimation is a complex process not only the metrological conditions but also the physiographic conditions of the aquifer, its geological parameters, the topography of the area, and even the type of vegetation and its areal extent can influence the estimation process (Li and Merchant 2013; Scibek and Allen 2006). Therefore, several parameters e.g. aquifer properties, spatial and temporal changes in groundwater withdrawals, groundwater level fluctuations and even assessment of any historical drought event needs to be linked together for a better understanding of groundwater problems. To understand the complex groundwater regimes and several interrelated processes which could influence the system Numerical groundwater models are by far the most comprehensive tools used by scientists. The assessment of groundwater level changes in real-time in any aquifer can be done by simulation studies by means of incorporating various stresses such as groundwater withdrawals, recharge, precipitation, droughts, etc. (Sulzbacher et al. 2012). The future sustainable availability of groundwater is highly dependent on long-term climate trends and their possible implications on groundwater recharge and water level fluctuations.

15 Conclusions and Recommendations

To understand the characteristics of any region, it is imperative to investigate the relationship between climate change and loss of fresh groundwater resources. Especially for the developing countries where agriculture is contributing at large to its economy and population increase is putting immense stress on the natural resources. Although the quantification of climate change and its impacts is still difficult, particularly at a regional scale, mitigating the adversaries of climate change still requires serious efforts. As groundwater is a precious resource and its depletion is not only damaging agriculture, rather it also has impacts on ecological and social functions.

Numerical groundwater models coupled with global and regional climate models data can provide useful information on the impacts of climate change on the groundwater resources of any region. Although there are uncertainties in the predictions of climate models which need to be addressed. Moreover, not only do hydrological processes contribute to groundwater recharge, but rather physical features of the land surface and soil profile may also have an effect on it which needs further exploration. Likewise, the reliability and authenticity of downscaled values of climate variables such as precipitation are still a question mark although it is a very important parameter, that is widely downscaled in climate change studies. The regional models are often unable to represent the mesoscale processes occurring at the local scale due to their representative spatial and temporal sizes in comparison to larger-scale regional precipitation. Resultantly, the summer precipitation is often underestimated by the global scale models which require further characterization. Nevertheless, climate change and its impacts are a reality, and to prevent water resources problems in the future, particularly groundwater problems physical-based groundwater system models, with possible climate change scenarios, are required.

References

Adane GW (2014) Groundwater modeling and optimization of irrigation water use efficiency to sustain irrigation in Kobo Valley, Ethiopia UNESCO-IHE]

Anderson MP, Woessner WW (1992) Applied modeling of groundwater flow-Simulation of flow and advective transport. Academic Press, New York

Anderson MP, Woessner WW, Hunt RJ (2015) Applied groundwater modeling: simulation of flow and advective transport. Academic press

Arora NK (2018) Environmental sustainability—necessary for survival. Springer

Benz SA, Bayer P, Blum P (2017) Global patterns of shallow groundwater temperatures. Environ Res Lett 12(3):034005

Bogardi JJ, Bharati L, Foster S, Dhaubanjar S (2021) Water and its management: dependence, linkages and challenges. In: Handbook of water resources management: discourses, concepts and examples. Springer, pp 41–85

Bouraoui F, Vachaud G, Li L, Le Treut H, Chen T (1999) Evaluation of the impact of climate changes on water storage and groundwater recharge at the watershed scale. Clim Dyn 15(2):153–161

Change C (2007) IPCC fourth assessment report. Phys Sci Basis 2:580–595

Chapter 11
Hydrogeochemical Processes Regulating the Groundwater Quality and Its Suitability for Drinking and Irrigation Purpose in a Changing Climate in Essaouira, Southwestern Morocco

Mohammed Bahir, Otman El Mountassir, and Mohamed Behnassi

Abstract Groundwater is vital for water supply and environmental protection, especially in semi-arid and desert regions. An integrated assessment, focused on the combined use of Water Quality Index (WQI), Irrigation Water Quality Index (IWQI), geochemical and isotopic ($\delta^{18}O$, $\delta^{2}H$ and $\delta^{3}H$) tools, was performed in the Cenomanian–Turonian aquifer during the campaigns 2017, 2018, 2019, and 2020. Hydrogeochemical analysis reveals that the groundwater is of mixed Ca–Mg–Cl, Ca–HCO_3, Ca–Cl, and Ca–SO_4 types, with a dominance of the first type. The analysis of the abundance of the main cations and anions shows the dominance of $Ca^{2+} > Na^{+} > Mg^{2+} > K^{+}$ for cations and the dominance of $Cl^{-} > HCO_3{}^{2-} > SO_4{}^{-} > NO_3{}^{-}$ for anions. The WQI in the Cenomanian–Turonian aquifer was divided into 32.8% (good areas), 44.3% (poor areas), 21.3% (very poor areas) and 1.6% (areas unsuitable for consumption). The IWQI showed that 43% of the studied samples have a high to severe restriction level, while 57% of the studied samples were placed in the low to moderate restrictions for irrigation use. The piezometry of the study area showed that the water flows generally from northeast to northwest and northeast to southwest. Groundwater mineralization in the Cenomanian–Turonian aquifer system is

M. Bahir (✉) · O. El Mountassir
High Energy and Astrophysics Laboratory, Faculty of Sciences Semlalia, Cadi Ayyad University, Marrakesh, Morocco
e-mail: bahir@uca.ac.ma

O. El Mountassir
e-mail: otman.elmountassir@ced.uca.ma

M. Behnassi
Economic, Social and Environmental Council (ESEC), Rabat, Morocco

Center for Environment, Human Security and Governance (CERES), Rabat, Morocco

Ministry of Higher Education and Innovation, Rabat, Morocco

M. Behnassi
e-mail: behnassi@gmail.com

M. Behnassi et al. (eds.), *The Water, Climate, and Food Nexus*,
https://doi.org/10.1007/978-3-031-50962-9_11

controlled by dissolution, including dissolution of evaporite minerals and reverse cation exchange. Stable isotope signatures ($\delta^{18}O$, $\delta^{2}H$) indicate that the groundwater samples are of meteoric origin without significant evaporation. The most of spring has tritium < 1TU. This could be explained by a reduction in annual precipitation recorded in the study area. The recharge altitude of the aquifer was estimated between 375 to 1275 m, following an altitudinal gradient of 0.26% per 100 m.

Keywords Groundwater quality · IWQI · WQI · Isotope stable · Tritium · Hydrogeochemical processes

1 Introduction

Groundwater is the main commodity/source of water supply for drinking, agricultural, and industrial purposes, exclusively in the arid and semi-arid regions of many countries (El Mountassir et al. 2021a, b; Bahir et al. 2021a, b, c). Globally, 65% of groundwater is used for drinking purposes, 20% for irrigation and livestock, and 15% for industry and mining (Bahir and Mennani 2002; Bahir et al. 2013b; Bahir et al. 2022a, b). As an important water resource, groundwater has a series of advantages over surface water, i.e. having a wide distribution, good stability, natural regulation, good water quality, and not easy to be contaminated. However, groundwater also has some disadvantages for exploitation, such as being more difficult to be accessed than surface water due to being buried underground, and requiring full understanding of the distribution rules prior to use (Bahir et al. 2013b; El Mountassir et Bahir 2023). In recent decades, however, with the continuous development of human activities, urbanization, and the continuous improvement of life quality, in a context of a changing climate and degradation of water by pollution, the demand for groundwater resources is increasing and groundwater resources are facing greater pressure than ever (EL Mountassir et al. 2022a; Bahir et al. 2007). The groundwater that feeds 2.5 billion people on Earth is in danger of drying up, according to a study recently published in 'Science' by two researchers at the University of California, Santa Barbara. "Up to 20 percent of the world's wells are no more than five meters below the water table. This means that millions of them could dry up if groundwater levels drop due to global warming or over-abstraction" (Cribb 2010).

In Africa, there are about eighty transboundary river and lake basins, and at least forty transboundary aquifer basins. The Africa Water Vision 2025 points out that groundwater is the main, and often the only, source of drinking water for more than 75% of the African population. Groundwater constitutes more than 95% of Africa's freshwater resources, and the pollution and salinization of this resource is often irreversible on a human scale (El Mountassir and Bahir 2023).

In the study area, the knowledge of hydrogeochemical characteristics of groundwater is an essential issue for the sustainable use of water resources and to mitigate the deterioration of the ecological environment. Considering this, using hydrochemical parameters, hydrogen and oxygen isotopes, and ^3H radioactivity, we first investigated: (a) hydrogeochemical characteristics and their changes along the flow path of groundwater in the Cenomanian–Turonian aquifer; and (b) the origin of aquifer recharge.

2 Study Area

2.1 Location and Climate

The study area is a set of synclinal structures located between Igrounzar wadi in the south and Hadid anticline in the north (Fig. 1). The Meskala-Ouazzi sub-basins are a part of the Essaouira coastal region, where the climate is a semi-arid type, with a mean annual rainfall of approximately 300 mm year^{-1} and a temperature of 20 °C (EL Mountassir et al. 2021c, 2021d, 2022b; El Mountassir 2023).

2.2 Geologic and Hydrogeologic Setting

The study area is marked by the outcrop of formations of Middle and Upper Cretaceous age, in particular, Albian-Vraconian, Cenomanian and Turonian (Duffaud 1960; Amghar 1989, Bahir et al. 2002) (Fig. 1). These formations are composed of limestone and dolomitic benches interspersed with marl and sandstone. The Albian-Vraconian formations contain sandstone and limestone dolomites alternating with sandstone banks and sandy clays (Bahir et al. 2022a, 2022b).

The Cenomanian (about 200 m thickness) is represented by alternating marls with anhydrite, lumachellic and dolomitic limestones (Bahir et al. 2000a, 2000b, 2013a, 2013b). As for the Turonian, it is composed of limestones with an abundance of silica (Bahir et al. 2007). These synclines contain important water reservoirs, notably the Cenomanian–Turonian aquifer, which remains the most important in the region. According to Jalal et al. (2001), this aquifer has transmissivities varying between 2.2×10^{-4} and 2.7×10^{-1} m^2/s.

3 Materials and Methods

3.1 Analytical Techniques and Groundwater Sampling

In this investigation, the results of four campaigns 2017, 2018, 2019, and 2020 (El Mountassir 2023) were used to assess the quality of groundwater in the Cenemano-Turonian aquifer. Electrical conductivities, temperatures, pH and nitrates were

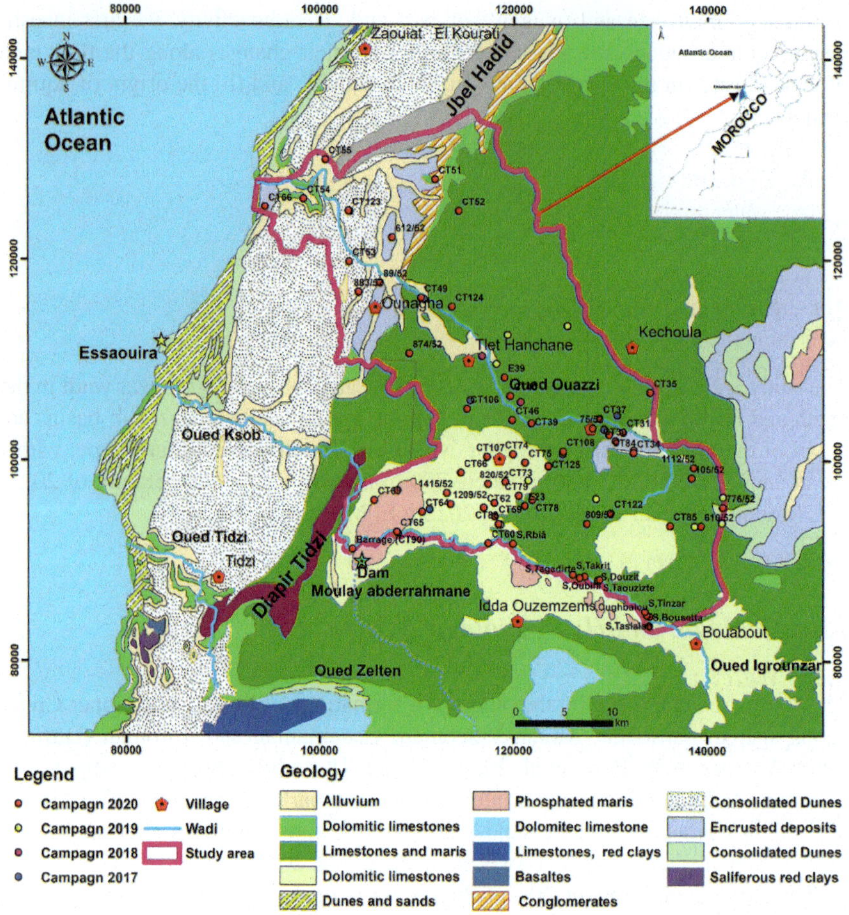

Fig. 1 Location of the study area and distribution of the sampling points (El Mountassir 2023)

measured in situ with a portable conductivity meter (HI-9829 multiparametric instrument), and the depth of the water level was measured using a 200 m piezometric probe.

The analyses of the chemical elements were carried out at the Laboratory of Geosciences and Environment-ENS at the Ecole Normale Superieure of Marrakech, Morocco for the campaigns 2017 to 2019 (El Mountassir 2023). For the 2020 campaign (El Mountassir 2023), the chemical analyses were conducted in the labs of Mohamed VI Polytechnic University of Benguerir (UM6P), Morocco. The samples analyses for Ca^{2+}, Mg^{2+}, Cl^{-}, SO_4^{2-}, NO_3^{-} and HCO_3^{-} were performed using a SKALAR San^{++} Continuous Flow Analyzer (CFA), while Na and K concentrations were measured using Atomic Absorption Spectrophotometry (AAS). The SO_4^{2-} anion contents were determined by the nephelometric method (Rodier et al. 2009). Concentrations of Ca^{2+} and Mg^{2+} cations were measured by the complexometry method (EDTA) and those of Cl^- by the Mohr method (Rodier et al. 2009). The Na^+ and K^+ contents were determined by flame photometry (Rodier et al. 2009). As for HCO_3^- contents, they were determined by titration using a sulfuric acid solution. All the samples display an ion balance of less than 10%, which allowed us to validate the obtained results. The obtained results are grouped in Table 1 (El Mountassir 2023).

Water samples were collected from various sources, including wells, springs, surface water, and a dam, as part of the 2020 campaign (El Mountassir 2023). These samples were analyzed at the laboratory located at the Center "Centro de Ciencias e Tecnologias Nucleares (C^2TN)", Campus Tecnológico e Nuclear in Lisbon, Portugal. The analysis involved the use of the SIRA 10 VG-ISOGAS mass spectrometer to determine the values of δ^2H and $\delta^{18}O$, which are stable isotopes of water.

To ensure precise results, each sample was subjected to three measurements following the procedures proposed by Friedman (1953) and Epstein and Mayeda (1953) for stable isotopes (δ^2H and $\delta^{18}O$). The standard method involved equilibrating the water samples with CO_2 and H2 to obtain $\delta^{18}O$ and δ^2H values, respectively. The instrument was then calibrated using the International Atomic Energy Agency (IAEA) standards, specifically the Vienna Standard Mean Ocean Water (VSMOW), with an accuracy range of $\pm$ 1.0‰ for δ^2H and $\pm$ 0.1‰ for $\delta^{18}O$.

3.2 Water Quality Index

The Water Quality Index (WQI) is a powerful method for assessing drinking water quality and regulating water sources (Brown et al. 1970). The WQI evaluation standard focuses on evaluating different chemical components and shows the effect of different chemical factors on the overall quality of drinking water (Xiao et al. 2019). Also, these indicators have been adapted to evaluations of water quality around the world, especially in the Essaouira Basin concurrent study area, with substantial impacts on the population and decision-makers for managing groundwater resources (Brown et al. 1970; Horton 1965; Xiao et al. 2019). The water quality index was determined according to the following equations (Eqs. 1 and 2):

$$WQI = \sum \left[W_i \times (\frac{C_i}{S_i}) \times 100 \right] \tag{1}$$

$$W_i = \frac{\omega_i}{\sum \omega_i} \tag{2}$$

The relative weight ωi is the weight of each chemical parameter calculated for drinking purposes by its relative effects on human health (Brown et al. 1970; Horton 1965; Xiao et al. 2019; El Mountassir et al. 2022b), as shown in Table 2. In every groundwater sample, the measured concentration of each chemical element is 'Ci,' and the corresponding standard value for each chemical parameter is 'Si' (WHO 2011). Based on WQI values, water quality can be divided into five distinct classes.

Table 1 Results of the physico-chemical parameters of all campaign (2017. 2018. 2019. and 2020) in the coastal zone of Essaouira basin in the Cenemano-Turonian aquifer (El Mountassir 2023)

Sample	H	pH	T	EC	Ca^{2+}	Mg^{2+}	Na^{+}	K^{+}	HCO_3^-	Cl^-	SO_4^{2-}	NO3$^-$	WQI	IWQI	IB
			°C	µS/cm	mg/L										%
Campagne Mai 2017															
54/43	–	7.1	22.2	3014	224.4	94.8	446.0	0.0	427.1	1134.4	76.9	37.2	180.0		– 3
75/52	445	7.8	18.6	2107	226.5	143.4	89.7	0.0	262.4	326.1	648.6	80.6	158.7	61.7	– 2
89/52	180	7.3	22.2	1962	110.2	75.4	183.9	0.0	421.0	340.3	163.4	43.4	123.5	54.8	– 2
105/52	603	7.3	18.8	2707	535.1	200.6	64.4	11.7	311.2	141.8	1441.4	6.2	209.5	67.1	8
612/52	177	7.4	21.9	3274	112.2	58.3	287.4	15.6	250.2	425.4	124.9	62.0	149.9	19.1	8
613/52	674	7.9	18.4	1200	80.2	60.8	142.5	3.9	323.4	283.6	76.9	62.0	102.2	64.5	– 2
648/52	485	7.3	22.5	1153	128.3	64.4	55.2	3.9	347.8	170.2	177.8	31.0	94.6	73.0	– 2
776/52	621	7.2	20	1421	156.3	87.5	59.8	3.9	366.1	198.5	331.5	24.8	109.9	69.2	– 3
824/52	341	7.6	20.4	3700	336.7	175.0	305.8	11.7	427.1	1120.2	648.6	31.0	230.5		– 8
874/52	270	7.4	21.5	3935	128.3	114.3	498.9	3.9	408.8	1148.6	216.2	68.2	208.2	1.9	– 9
883/52	167	7.1	23.4	2424	148.3	94.8	195.4	0.0	451.5	482.1	269.1	37.2	145.3	39.1	– 7
1112/52	556	7.7	21.5	1175	156.3	99.7	46.0	3.9	317.3	141.8	269.1	24.8	98.5	75.8	9
1699/52	600	7.2	20.7	1808	112.2	60.8	119.5	3.9	421.0	113.4	254.6	6.2	104.8	69.7	1
1871/52	–	7	25.1	1856	154.3	85.1	108.1	0.0	414.9	397.0	201.8	24.8	120.5	46.3	– 7
CT30	486	7.4	19.9	2413	224.4	149.5	115.0	0.0	360.0	467.9	547.7	62.0	165.6	43.6	– 5
CT31	489	7.7	17.9	1025	106.2	48.6	55.2	0.0	299.0	226.9	105.7	31.0	83.7	72.9	– 9
CT32	458	7.1	20.7	1589	82.2	40.1	112.7	0.0	396.6	156.0	144.1	24.8	96.0	69.2	– 8
CT33	641	7.6	21.2	1348	170.3	70.5	39.1	3.9	335.6	212.7	355.5	24.8	108.1	58.9	– 9
CT34	486	7.3	23.8	1110	116.2	68.1	46.0	3.9	323.4	156.0	187.4	24.8	89.2	75.4	– 2

(continued)

Table 1 (continued)

Sample	H	pH	T	EC	Ca^{2+}	Mg^{2+}	Na^{+}	K^{+}	HCO_3^-	Cl^-	SO_4^{2-}	NO3⁻	WQI	IWQI	IB
			°C	µS/cm	mg/L										%
CT36	455	8	19.9	2100	174.3	86.3	209.2	3.9	427.1	354.5	374.8	37.2	143.5	35.1	− 1
CT37	454	7.2	20.7	4659	328.7	306.3	273.6	0.0	512.6	808.3	1239.6	192.2	319.7	7.4	− 6
CT38	456	7.3	21.3	1007	100.2	54.7	50.6	3.9	384.4	156.0	96.1	31.0	87.8	74.2	− 6
CT39	375	7.4	20.7	2224	140.3	110.6	165.5	3.9	421.0	425.4	168.2	130.2	164.3	45.3	− 2
CT40	341	7.1	21.8	1876	166.3	119.1	94.3	0.0	543.1	340.3	321.9	49.6	144.0	48.5	− 8
CT41	393	7.7	22.7	2044	150.3	124.0	117.2	0.0	433.2	382.9	369.9	6.2	129.4	44.8	− 6
CT42	291	7.8	22.6	1910	184.4	128.8	71.3	3.9	549.2	382.9	288.3	18.6	139.3	39.4	− 6
CT45	438	7.6	18.7	2811	210.4	125.2	280.5	0.0	476.0	935.9	100.9	49.6	174.9	15.2	− 6
CT46	406	6.9	22.6	2907	464.9	175.0	66.7	3.9	353.9	198.5	1191.5	18.6	199.4	63.2	5
CT47	401	7.2	19.9	3961	336.7	151.9	386.2	11.7	408.8	1120.2	562.1	62.0	239.8		− 5
CT48	303	7.4	20.6	1640	112.2	48.6	246.0	0.0	274.6	467.9	48.0	49.6	109.3	33.2	2
CT49	194	–	–	–											–
CT51	174	7.4	23.1	2003	72.1	88.7	317.3	3.9	518.7	482.1	57.7	6.2	123.2	14.3	3
CT52	315	7.2	21.8	1908	56.1	86.3	338.0	0.0	506.5	439.6	43.2	31.0	123.6	21.8	6
CT53	180	7.4	22.1	2436	118.2	77.8	255.2	0.0	335.6	340.3	245.0	18.6	124.2	42.1	7
CT54	64	7.7	21.7	1806	108.2	72.9	280.5	0.0	445.4	411.2	158.5	18.6	120.1	28.4	2
CT55	38	7.2	21.8	3245	160.3	103.3	436.8	0.0	299.0	1106.0	62.5	37.2	168.5		− 4
CT56	1.6	7.3	20.9	3620	192.4	110.6	459.8	82.1	299.0	1006.8	153.7	167.4	259.2		2
CT82	189	7.2	23.6	2387											–

(continued)

Table 1 (continued)

Sample	H	pH	T	EC	Ca^{2+}	Mg^{2+}	Na^{+}	K^{+}	HCO_3^-	Cl^-	SO_4^{2-}	$NO3^-$	WQI	IWQI	IB
			°C	µS/cm	mg/L										%
CT83	300	7.1	22.3	1742	146.3	125.2	50.6	0.0	402.7	297.8	321.9	62.0	131.8	63.7	– 7
CT84	467	7.2	21.6	1578											–
CT85	702	7.6	19	1341	122.2	64.4	112.7	0.0	408.8	312.0	100.9	37.2	105.2	66.7	– 6
CT86	195	7.4	23	1962	142.3	99.7	209.2	0.0	610.2	510.5	158.5	37.2	145.5	21.2	– 7
CT7	564	7.6	20.6	635	76.2	55.9	25.3	3.9	360.0	113.4	33.6	18.6	69.5	66.9	– 4
72/52	389	7.3	19.9	1884	172.3	109.4	89.7	3.9	683.4	482.1	9.6	12.4	135.0	33.5	– 8
108/52	412	7.7	22.9	903	84.2	41.3	59.8	3.9	390.5	170.2	19.2	37.2	84.2	73.0	– 9
809/52	630	7.8	20.3	666	88.2	36.5	41.4	3.9	274.6	127.6	28.8	24.8	66.6	67.1	1
820/52	430	7.5	21	3836	232.5	166.5	285.1	0.0	360.0	1106.0	254.6	130.2	226.1	4.0	– 8
1209/52	–	7.2	25.7	1054	80.2	55.9	64.4	3.9	390.5	170.2	62.5	18.6	82.8	71.3	– 5
1415/52	332	7.3	24.1	881	96.2	42.5	48.3	3.9	439.3	141.8	24.0	18.6	80.8	74.9	– 7
CT59	382	7.8	20.7	3324	274.5	201.8	80.5	3.9	500.4	950.1	173.0	130.2	220.2	23.0	– 9
CT60	374	8.1	20.2	1043	104.2	79.0	59.8	7.8	384.4	212.7	168.2	12.4	92.6	70.9	– 5
CT61	–	7.4	29.3	4462	190.4	177.5	381.6	0.0	323.4	1191.1	326.7	93.0	228.4		– 7
CT62	384	8	19.1	1857	140.3	114.3	89.7	0.0	323.4	496.3	182.6	68.2	131.7	47.6	– 9
CT63	310	7.1	21.6	6776	432.9	294.2	761.0	11.7	366.1	2470.9	571.7	37.2	364.3		– 5
CT64	304	7.6	20.5	1765	194.4	75.4	147.1	7.8	384.4	382.9	336.3	12.4	126.7	50.7	– 4
CT65	262	7.2	22	1822	152.3	111.8	126.4	0.0	524.8	354.5	240.2	24.8	130.3	36.0	– 4
CT66	410	7.5	21.8	1553	96.2	93.6	92.0	0.0	353.9	340.3	134.5	43.4	108.2	61.0	– 7

(continued)

Table 1 (continued)

Sample	H	pH	T	EC	Ca^{2+}	Mg^{2+}	Na^{+}	K^{+}	HCO_3^{-}	Cl^{-}	SO_4^{2-}	$NO3^{-}$	WQI	IWQI	IB
			°C	µS/cm	mg/L										%
CT81		8.7	21.9	2194	194.4	114.3	85.1	3.9	390.5	382.9	446.8	6.2	133.1	47.9	– 8
CT90		9.6	19.5	601	74.1	35.2	27.6	3.9	189.2	127.6	100.9	6.2	56.5	70.7	– 6
Min	1.4	7.2	16.2	601	66.1	30.4	11.5	0.0	189.2	85.1	4.8	6.2	56.5	0.7	– 9
Max	701.5	9.6	24.6	6845	583.2	329.4	561.0	35.2	689.5	1705.1	2219.7	167.4	315.3	70.7	9
Mean	388.0	8.0	20.9	2481.8	177.2	115.6	152.9	5.4	382.2	495.5	283.5	37.0	144.1	44.2	– 3
Campagne Mars 2019															
75/52	445.2	8	18	2750	268.5	167.7	98.9	11.7	378.3	453.8	821.6	62.0	189.4	43.7	– 8
89/52	178.1	7.4	21.7	2203	188.4	113.0	131.0	7.8	512.6	482.1	278.7	43.4	150.4	39.4	– 7
105/52	602.6	7.1	17.3	3646	769.5	109.4	55.2	11.7	537.0	340.3	1941.0	12.4	277.7	39.8	– 8
612/52	176.2	7.4	20.8	3286	280.6	117.9	218.4	62.6	402.7	950.1	293.1	68.2	218.8	12.3	– 8
613/52	674.1	8	14.9	1843	146.3	91.2	98.9	23.5	476.0	411.2	76.9	80.6	145.4	45.5	– 6
776/52	622.7	7.3	18.8	2173	232.5	156.8	89.7	3.9	561.4	283.6	595.8	37.2	161.8	47.8	– 3
874/52	269.9	7.3	20.8	4349	196.4	183.5	452.9	7.8	500.4	1301.0	394.0	62.0	234.1		– 9
883/52	165.9	7.1	23	2804	202.4	155.6	183.9	3.9	585.8	666.5	374.8	37.2	176.0	26.5	– 8
1112/52	555.7	7.5	21.4	1411	134.3	100.9	75.9	3.9	500.4	269.4	254.6	12.4	111.5	60.9	– 8
1209/52	–	7.2	23.6	1075	118.2	96.0	27.6	3.9	488.2	326.1	19.2	18.6	95.6	54.7	– 9
1699/52	597.5	7.6	20.3	1543	182.4	176.2	96.6	15.6	433.2	368.7	384.4	12.4	135.2	48.8	5
CT30	485	7.6	18	3389	420.8	260.1	151.7	11.7	683.4	666.5	1297.2	43.4	262.0	12.6	– 8
CT31	485.5	7.5	17.5	1192	134.3	66.9	80.5	3.9	378.3	312.0	105.7	12.4	92.4	64.3	– 5

(continued)

Table 1 (continued)

Sample	H	pH	T	EC	Ca^{2+}	Mg^{2+}	Na^+	K^+	HCO_3^-	Cl^-	SO_4^{2-}	$NO3^-$	WQI	IWQI	IB
			°C	µS/cm	mg/L										%
CT33	640.3	7.4	19.6	1584	156.3	110.6	101.2	3.9	476.0	312.0	418.0	6.2	122.1	58.2	– 8
CT34	485.2	7.2	22.2	1261	124.2	97.2	55.2	3.9	402.7	297.8	216.2	24.8	103.5	65.6	– 8
CT35	518.4	7.8	21	862	88.2	31.6	57.5	0.0	305.1	113.4	28.8	31.0	70.2	77.1	1
CT37	453.5	7.2	20.5	4965	312.6	297.8	349.4	7.8	634.6	935.9	1172.3	173.6	323.3		– 7
CT39	374.3	7.5	20.7	2087	124.2	81.4	213.8	23.5	421.0	524.7	158.5	18.6	133.5	30.8	– 5
CT40	340.2	7.2	21	2149	192.4	131.3	131.0	3.9	720.0	397.0	317.1	18.6	154.7	30.4	– 7
CT42	291	7.7	22.1	2440	190.4	145.9	115.0	35.2	897.0	482.1	173.0	6.2	179.8	26.0	– 8
CT46	398.9	7	22.8	2939	551.1	130.1	50.6	7.8	634.6	312.0	1412.5	0.0	227.9	48.5	– 9
CT49	195.1	7.5	22.7	2148	154.3	122.8	144.8	3.9	512.6	538.8	187.4	24.8	138.1	44.0	– 7
CT51	173.7	7.2	22.2	2277	184.4	87.5	181.6	11.7	659.0	567.2	100.9	0.0	144.3	28.3	– 8
CT52	315.1	7.4	21	2140	108.2	89.9	193.1	3.9	524.8	467.9	105.7	31.0	130.5	31.1	– 7
CT53	179.2	7.3	20.8	2657	166.3	134.9	186.2	3.9	427.1	680.6	288.3	43.4	157.4	36.5	– 9
CT54	63.1	7.6	21.5	2266	124.2	81.4	190.8	3.9	421.0	482.1	187.4	24.8	128.1	40.3	– 8
CT55	37.4	7.6	21	2230	130.3	38.9	209.2	3.9	421.0	443.1	91.3	12.4	116.5	31.5	– 7
CT56	1	7.2	20.7	4277	246.5	74.1	388.5	168.1	451.5	1106.0	158.5	167.4	317.0		– 6
CT85	701.1	7.1	16.5	1574	102.2	91.2	80.5	3.9	408.8	301.3	129.7	31.0	105.2	61.4	– 6
CT90	–	8.4	17.8	615	82.2	41.3	32.2	3.9	244.1	184.3	38.4	6.2	59.3	67.0	– 6
CT105	503.9	7.8	21	796	104.2	80.2	11.5	0.0	500.4	184.3	14.4	12.4	81.8	61.9	– 6
CT106	357	7.3	19.2	4216	420.8	296.6	179.3	11.7	585.8	666.5	1667.2	6.2	274.4	9.9	– 8

(continued)

Table 1 (continued)

Sample	H	pH	T	EC	Ca^{2+}	Mg^{2+}	Na^{+}	K^{+}	HCO_3^-	Cl^-	SO_4^{2-}	$NO3^-$	WQI	IWQI	IB
			°C	µS/cm	mg/L										%
CT107	437.3	7.1	19.5	3690	258.5	165.3	298.9	3.9	573.6	1148.6	177.8	86.8	219.5		– 8
CT108	455	7.8	19.3	1336	136.3	94.8	64.4	11.7	439.3	297.8	264.2	18.6	113.0	63.0	– 9
CT109	323.3	7.4	20.8	1873	116.2	46.2	186.2	3.9	323.4	439.6	110.5	31.0	109.3	44.3	– 7
CT110	400.6	7.4	19.2	4370	314.6	255.3	278.2	86.0	598.0	1176.9	687.0	24.8	284.2		– 6
CT122	655.3	7.6	18.9	882	180.4	88.7	13.8	3.9	549.2	241.1	57.7	18.6	100.4	50.8	– 1
CT123	-	7.4	22.9	3220	260.5	167.7	305.8	7.8	598.0	950.1	326.7	37.2	201.4		– 4
CT124	225.6	7.3	21.5	2650	210.4	122.8	160.9	3.9	476.0	694.8	235.4	12.4	151.0	38.3	– 8
72/52	387.4	7.5	19.2	2428	140.3	104.5	181.6	19.6	598.0	581.4	14.4	6.2	125.2	28.3	– 5
809/52	629.7	7.8	20	736	140.3	55.9	18.4	0.0	292.9	187.9	33.6	24.8	72.5	65.7	5
820/52	429.7	7.5	21.1	4380	430.9	75.4	395.4	11.7	524.8	1350.6	177.8	117.8	257.8		– 7
1415/52	330.5	7.2	24	972	102.2	62.0	154.0	3.9	488.2	212.7	28.8	105.4	118.4	65.3	2
CT61	–	7.1	21.7	2179	188.4	141.0	126.4	7.8	573.6	638.1	96.1	55.8	156.4	28.0	– 6
CT62	383.4	7.8	18	2249	152.3	96.0	197.7	11.7	372.2	567.2	269.1	62.0	149.9	39.3	– 8
CT64	303.3	7.6	20.3	2199	262.5	89.9	158.6	7.8	360.0	538.8	283.5	6.2	135.1	45.8	1
CT65	257.8	7.7	20.8	2381	180.4	102.1	131.0	7.8	598.0	467.9	230.6	12.4	145.8	30.0	– 9
CT66	407.4	7.5	22	1888	134.3	98.5	119.5	3.9	427.1	453.8	110.5	31.0	120.1	43.7	– 6
CT68	390.3	7.5	20.2	3842	364.7	99.7	225.3	15.6	427.1	645.2	812.0	12.4	207.7	16.7	– 7
CT69	325.2	7.4	21.5	4530	292.6	229.7	395.4	43.0	500.4	1205.3	946.5	31.0	274.1		– 9
CT73	394.4	7.8	19.6	2123	264.5	139.8	131.0	3.9	720.0	623.9	76.9	12.4	155.9	47.7	– 1

(continued)

Table 1 (continued)

Sample	H	pH	T	EC	Ca^{2+}	Mg^{2+}	Na^{+}	K^{+}	HCO_3^-	Cl^-	SO_4^{2-}	NO3⁻	WQI	IWQI	IB
			°C	µS/cm	mg/L										%
CT74	428.1	7.4	20.8	1559	132.3	96.0	82.8	3.9	427.1	467.9	81.7	12.4	107.0	25.8	– 10
CT75	417.7	7.5	21.4	1510	160.3	110.6	80.5	3.9	488.2	411.2	148.9	18.6	116.8	48.0	– 5
CT78	437.9	7.6	20.9	3016	168.3	72.9	285.1	19.6	488.2	680.6	96.1	161.2	202.7	10.8	– 8
CT79	399.3	7.7	21.4	2500	150.3	93.6	292.0	15.6	476.0	850.8	14.4	6.2	145.8	16.3	– 6
CT80	411.7	7.2	20.3	5738	485.0	102.1	540.3	11.7	561.4	1818.6	269.1	68.2	294.3		– 9
CT125	408.2	7.8	21.3	1020	112.2	81.4	43.7	3.9	488.2	226.9	72.1	18.6	93.8	57.3	– 6
Min	1	7	14.9	615	82.2	31.6	11.5	0.0	244.1	113.4	14.4	0.0	59.3	9.9	– 10
Max	701.1	8.4	24	5738	769.5	297.8	540.3	168.1	897.0	1818.6	1941.0	173.6	323.3	77.1	5
Mean	387.3	7.5	20.5	2428.4	213.6	118.5	166.7	13.9	499.2	573.8	339.0	37.4	162.7	41.9	– 6
Campagne Juillet 2020															
S.Oubihi		7.03	20.5	1109	110.8	59.7	38.8	3	429.4	84.2	179	10.8	85.4	62.6	4
S.Douzit		7.04	19.6	1030	109.8	57.3	34.6	2.5	378.2	74.9	162	16.6	80.7	64.7	1
S.Taouzizte		7.37	23.14	1537	159.4	80.1	63.5	8.5	390.4	165.6	253	58.9	118.9	68.6	– 1
S.Tagadirte		7.23	21.8	1597	160	92.1	67	3.9	420	159.7	303	20.8	110.4	67.4	– 1
S.Takrit		7.12	20.8	1504	153.9	94.7	70.9	3.1	436	163.5	274	10.8	105.1	66.9	– 3
S.Bousetta		7.85	22	636	80.4	19.8	24	1.2	220	40.4	58	30.4	57.8	71.9	– 2
S.Taslalaft		7.15	21	541	102.6	41.6	23.9	2.7	292.8	54.1	119	27.5	66.7	70.6	– 2
S.Oughbalou		7.6	21	868	102.6	41.6	23.2	2.8	285.5	50.9	126	27.4	71.8	68.2	– 2
S.Tinzar		7.4	21	806	99.2	40.1	22	2.7	283	50.2	125	27.8	70.0	68.8	– 1

(continued)

Table 2 Weight and relative weight of each parameter used for the WQI calculation

Physico-Chemical parameters	WHO Standard (2011)	Weight (wi)	Relative weight (Wi) $W_i = \frac{w_i}{\sum_i^n wi}$
pH	8.5	4	0.114
EC (μS/cm)	1500	4	0.114
TDS (mg/L)	1000	5	0.142
Cl^- (mg/L)	250	3	0.086
SO_4^{2-} (mg/L)	250	4	0.114
NO_3^- (mg/L)	45	5	0.142
HCO_3^- (mg/L)	120	3	0.086
Na^+ (mg/L)	200	2	0.057
Ca^{2+} (mg/L)	75	2	0.057
Mg^{2+} (mg/L)	50	1	0.029
K^+ (mg/L)	12	2	0.057
		35	**0.998**

The quality of water is excellent when WQI is less than 50; the quality of water is good when WQI < 100 and above 50; the quality of water is poor when WQI is between 100 and 200; the quality of water is very poor when WQI range between 200 and 300; groundwater is unsuitable for drinking purposes if WQI above 300. Furthermore, using the IDW tool of ArcGis software version 10. 2 (ESRI 1999), the spatial distribution of the estimated WQI values in the Essaouira basin was obtained.

3.3 *Irrigation Water Quality Index*

Irrigation water quality index (IWQI) was calculated using a method developed in Brazil by Meireles et al. (2010), and the criteria and standards developed by Ayers and Westcot (1985). Before starting data analysis, the concentration units were transformed from [mg/L] to [meq/L] according to the conversion factors given by (Lesch and Suarez 2009).

Meireles et al. (2010) developed an IWQI that reflects the water toxicity, sodicity, soil salinity and miscellaneous risks impacts on sensitive crops. They developed this IWQI by conducting two steps:

- In the first step using the statistical method (principal component analysis), they identified the electrical conductivity (EC), sodium ion concentration (Na^+), bicarbonate ion concentration (HCO_3^-), chloride ion concentration (Cl^-), and sodium adsorption ratio (SAR) as the parameters that cause the most variability of water quality.

Table 3 Weights for the IWQI parameters according to Meireles et al. (2010)

Paramètres	Poids (wi)
EC	0.211
Na^+	0.204
HCO_3^-	0.202
Cl^-	0.194
SAR	0.189
Total	1

- The second step is defining the quality sub-index (Q_i) and the unit weightage (W_i) for each parameter. The values of (W_i) were defined according to the parameter's values in their studied area and the criteria of (Meireles et al. 2010) listed in Table 3, the W_i values were normalized such that their sum equals one. The values of Q_i were calculated using Eq. (3), based on the tolerance limits shown in Table 4 (Ayers and Westcot 1994), and the data of the collected samples from the studied area in Table 1.

$$Q = (X_{ij} - X_{\inf}) \times Q_{iamp} / Q_{i\max} \quad (3)$$

where;

Q_{imax} is the maximal value of qi in each category.
X_{ij} is the parameter marked value.
X_{inf} is the minimal value of the category to the parameter.
Q_{iamp} is the ampleness of the category.
X_{amp} is the ampleness of the category to each parameter.

The IWQI is dimensionless, four categories with the range of 0 to 100 and calculated using Eq. 4:

$$IWQI = \sum_{i=1}^{n} Q_i \times W_i \quad (4)$$

Table 4 Limiting values of (Q_i) calculations (Ayers and Westcot 1994)

HCO_3^- (meq/L)	Cl^-	Na^+	SAR (meq/L)	EC (µS/cm)	Qi
1–1.5	1–4	2–3	2–3	200–750	85–100
1.5–4.5	4–7	3–6	3–6	750–1500	60–85
4.5–8.5	7–10	6–9	6–12	1500–3000	35–60
< 1 or >= 8.5	< 1 or >= 10	< 2 or >= 9	< 2 or >= 12	< 200 or >= 3000	0–35

The IWQI values were classified into 5 types: No restriction (85–100), Low restriction (70–85), Moderate restriction (55–70), High restriction (40–55) and severe restriction (0–40) (Table 5) (Meireles et al. 2010).

4 Results and Discussion

4.1 Groundwater Characteristics and Hydrochemistry

The chemical composition of groundwater is influenced by geochemical processes that occur as water moves through the minerals in the aquifer, which influences its physical characteristics. The primary qualities are the physicochemical parameters to identify the quality, nature, and type of groundwater. This chapter used several factors to better present the large set of sampling data for different campaign like 2017, 2018, 2019, and 2020 (El Mountassir 2023). Table 1 reveals the minimum, maximum, and the mean, with associated box plot in Fig. 2 for the 2020 campaign. The summary of physico-chemical parameters and hydrochemical properties of groundwater were statistically evaluated and the results were compared to the World Health Organization (WHO) standards.

Electrical conductivity (EC) is a good indicator to determine the degree of overall mineralization of the water. The electrical conductivities obtained by the 2020 campaign range from 541 to 5285 μS/cm with an average of 2129 μS/cm (Table 1; Fig. 2). The water samples analyzed in this study showed considerable differences in their chemical composition. The Taslalaft spring, situated in the aquifer's recharge area, had the lowest electrical conductivity value at 541 μS/cm. On the other hand, the well labeled N° CT56, located close to the sea, recorded the highest value at 5285 μS/cm. It is important to note that these values exceed the Moroccan standards for evaluating the overall quality of groundwater, which state that the electrical conductivity should not exceed 2700 μS/cm (El Mountassir 2023). In general, the electrical conductivity of the Cenemano-Turonian aquifer is at the standards; with 46% of the water samples had good to excellent conductivity. The remaining 26% have conductivity between 2700 and 5285 μS/cm, indicating mineralized water and are unsuitable for human consumption. Salinity increases from the Bouabout area (northeast) to the Atlantic Ocean (southwest) in the direction of groundwater discharge. The reason for the high salinity of groundwater in these areas is due to the natural concentration of salts, when evaporation and interaction of water with rocks exceeds the precipitation in the area. In addition, high groundwater salinity also occurs in the shallow aquifer (Plio-Quaternary) that is close to the Atlantic Ocean, due to marine intrusion into the aquifer (Bahir et al. 2018b). In addition, another reason for water mineralization is due to leaching of the evaporated material-rich limestone-dolomitic matrix (Mennani et al. 2001).

pH is an important measure of water quality. It describes the acidic or basic nature of a substance. The groundwater in the study area is moderately alkaline with

Table 5 Irrigation water quality index characteristics (Meireles et al. 2010)

Recommandation		Water use restrictions	IWQI
Plant	Soil		
No toxicity risk for most plants	May be used for the majority of soils with low probability of causing salinity and sodicity problems. Leaching recommended within irrigation practices. except for in soils with extremely low permeability"	No restriction (NR)	85–100
Avoid salt sensitive plants	Recommended for use in irrigated soils with light texture or moderate permeability. Salt leaching recommended. Soil sodicity in heavy texture soils may occur. being recommended to avoid its use in soils with high clay	Low restriction (LR)	70–85
Plants with moderate tolerance to salts may be grown	May be used in soils with moderate to high permeability values. moderate leaching of salts suggested	Moderate restriction (MR)	55–70
Should be used for the irrigation of plants with moderate to high tolerance to salts with special salinity y control practices. except water with low Na. Cl and HCO^{-3} values	May be used in soils with high permeability without compact layers. High frequency irrigation schedule should be adopted for water with EC above 2000 μS cm^{-1} and SAR above 7.0	High restriction (HR)	40–55
Only plants with high salt tolerance. except for waters with extremely low values of Na^+. Cl^- and HCO_3^-	Should be avoided for irrigation under normal conditions. May be used occasionally in special cases. Water with low salt levels and high SAR require gypsum application. In high saline water, soils must have high permeability. and excessive water should be applied to avoid salt accumulation	Severe restriction (SR)	0–40

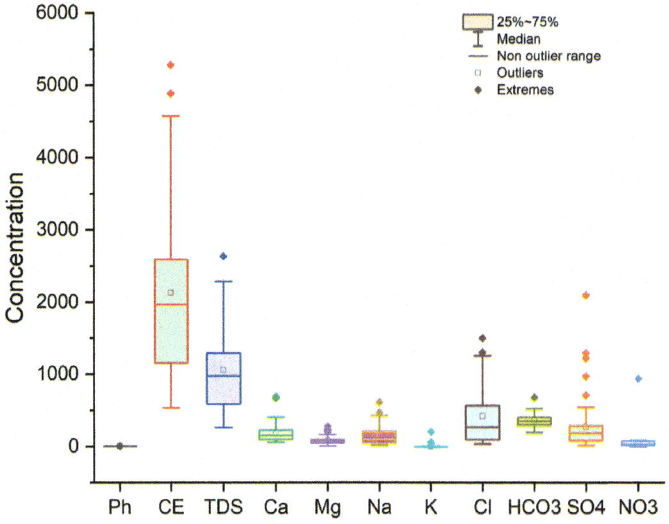

Fig. 2 Box plot showing ion concentrations (mg/L) except pH and EC (μS/cm) of the campaign 2020 (El Mountassir 2023)

a pH ranging from 7 to 8.2 with an average of 7.5 (Table 1; Fig. 2). The nature of the rocks and sediments surrounding the recharge water entering the groundwater determines the pH of the groundwater. It also depends on the length of time the current groundwater has been in contact with a specific rock. The greater the impact of the rock chemistry on the composition and pH of the groundwater, the longer the contact time. For this analysis, the pH of all groundwater samples is within the acceptable standard established by WHO (2011) (Table 2).

The dominant cation in the groundwater was Ca^{2+} with an average of 193.6 mg/L, while the dominant anions were Cl^- with an average of 424 mg/L. Generally, the order of abundance of cations was $Ca^{2+} > Na^+ > Mg^{2+} > Ca^{2+} > K^+$, while that of anions was $Cl^- > HCO_3^- > SO_4^{2-} > NO_3^-$ (Table 1).

4.2 Depth of Groundwater Level

The state of low and high groundwater is illustrated by the groundwater level map of the study area (Fig. 3). The state of the hydraulic gradient of the Cenemano-Turonian aquifer at different points plays an important role on the direction of groundwater. One of the main parameters for the transmission and movement of soluble materials is the hydraulic gradient, as a high hydraulic gradient contributes to faster movement of contaminants, and thus strong downstream pollution expansion (Chitsazan et al. 2017).

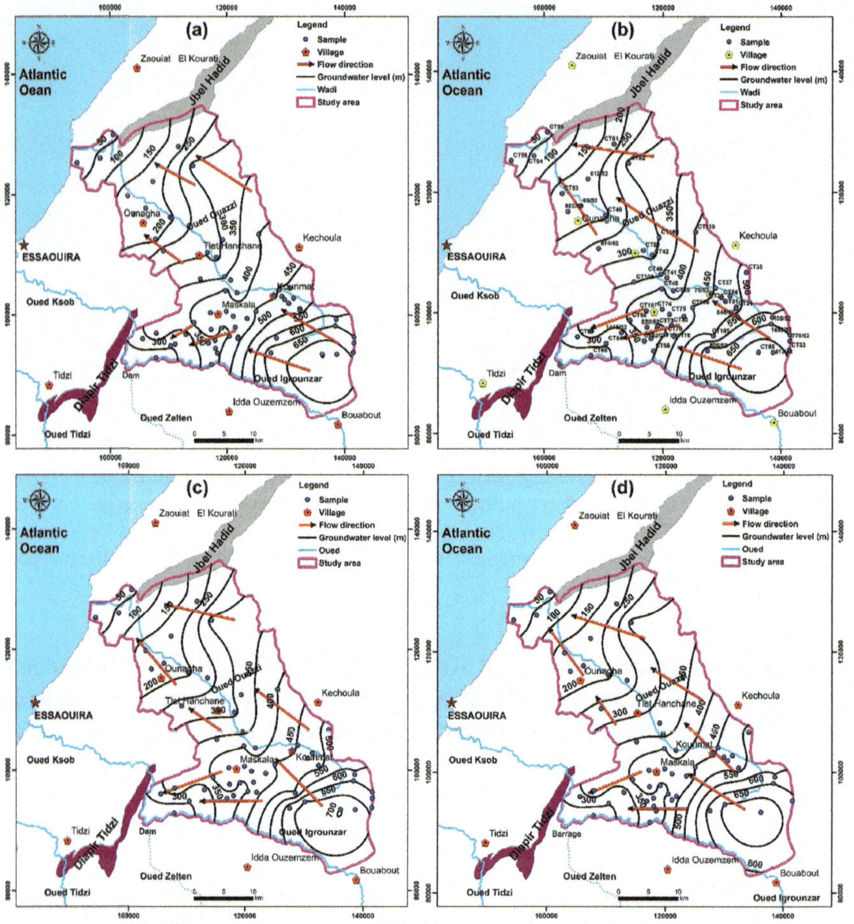

Fig. 3 Piezometric maps of the Cenemano-Turonian aquifer for all campaign 2017, 2018, 2019, and 2020 (El Mountassir 2023)

Water level fluctuation is controlled by variations in precipitation, topography, lithology, and water extraction in the study area. The depth of groundwater level varied from 1 to 50 m below ground level during 2017, from 2 to 50.4 m during 2018, from 1.4 m to 53.8 m during 2019, and from 2.5 m to 53.8 m during 2020 (Fig. 1). Groundwater levels are shallow near coastal wells (Plio-Quaternary aquifer), and deeper away from the coast (Cenemano-Turonian aquifer) (El Mountassir 2023).

The groundwater flow keeps the same direction during the four campaigns (2017, 2018, 2019, 2020), with a decrease of the water level due to the decrease of precipitation in the last years. The piezometry of the study area showed that the water flows generally from northeast to northwest and northeast to southwest of the Cenemano-Turonian aquifer (Fig. 3). Therefore, groundwater direction can detect and control

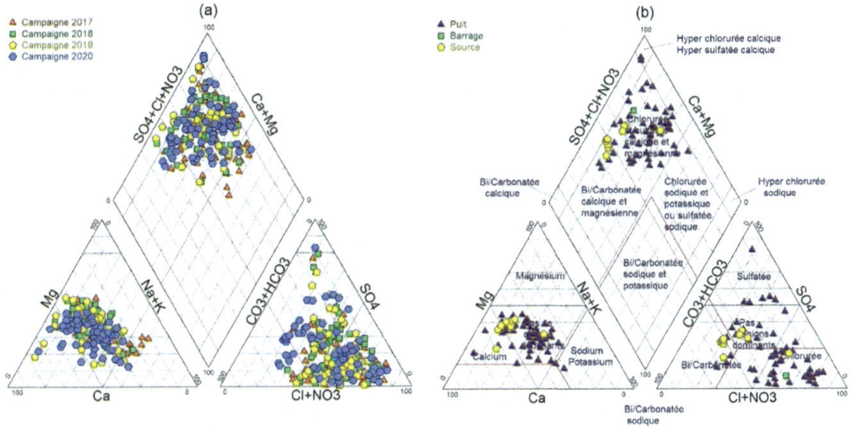

Fig. 4 Piper diagram of all campaign (2017, 2018, 2019, 2020) (**a**); piper diagram of the campaign 2020 (**b**) (El Mountassir 2023)

water pollution in the aquifers. The groundwater elevation in the study area is varied from 0.8 to 700.9 m (Table 1; Fig. 1).

The hydraulic gradient is 2.5% upstream, 1.1% in the middle and 1.6% downstream. To the south, in the Meskala area, the Igrounzar wadi drains the entire Cenomanian–Turonian aquifer (Fig. 1).

4.3 Hydrochemical Facies

Data from the chemical analysis of groundwater samples were used to investigate geochemical processes in the Cenomanian–Turonian aquifer and to evaluate factors affecting water quality changes. Data from each campaign were first plotted on Fig. 3 of Piper's trilinear diagrams (Piper 1944) to determine the chemical facies that was associated with the groundwater samples. Cations and anions were plotted in the left and right triangles as single points. These points were then projected into the central diamond-shaped area parallel to the upper edges of the central area. The points in the diamond-shaped area represent specific hydrochemical facies and water types. For each water sample, only one point was obtained in the diamond-shaped area, which represents a water type.

The representative points of the Cenomanian–Turonian aquifer waters on the Piper diagram (Fig. 3b) of the 2020 campaign constitute a cloud of mixed facies points Ca–Mg–Cl (66.15%), Ca–HCO_3 (6.15%), Na–Cl (3.07%), and Ca–SO_4 (24.61%) with a dominance of the first type (El Mountassir 2023).

The presence of different hydrochemical facies is related to the diversity of geology and geochemical processes in the Cenomanian–Turonian aquifer.

4.4 Origin of Groundwater Mineralization

Water–rock interactions result in progressive changes in groundwater chemistry along groundwater flow paths. The use of ion ratios can be useful in understanding the geochemical processes that contribute to groundwater mineralization.

The Na^+ versus Cl^- relationship has often been used to identify mechanisms of mineralization acquisition in arid and semi-arid regions (Dixon and Chiswell 1992). A strong correlation observed between Na^+ and Cl^- with $R^2 = 0.9$ (Fig. 5a) in the Cenemano-Turonian aquifer system indicates probable dissolution of halite (NaCl) by groundwater. This is reflected in the high Na^+ and Cl^- contents, and especially in a Na/Cl ratio close to 1. Halite dissolution is corroborated by negative saturation indices (Fig. 6). The excess of Cl^- over Na^+ is due to the phenomenon of reverse cation exchange by the fixation of Na^+ and the release of Ca^{2+} and Mg^{2+} from the aquifer matrix (Fig. 7a) (Hem 1985). In the discharge zone (downstream part of the study area), groundwater in the coastal zone is progressively affected by surface water recharge, human activities, evaporation, and seawater intrusion resulting in elevated Na^+ and Cl^- values (El Mountassir 2023).

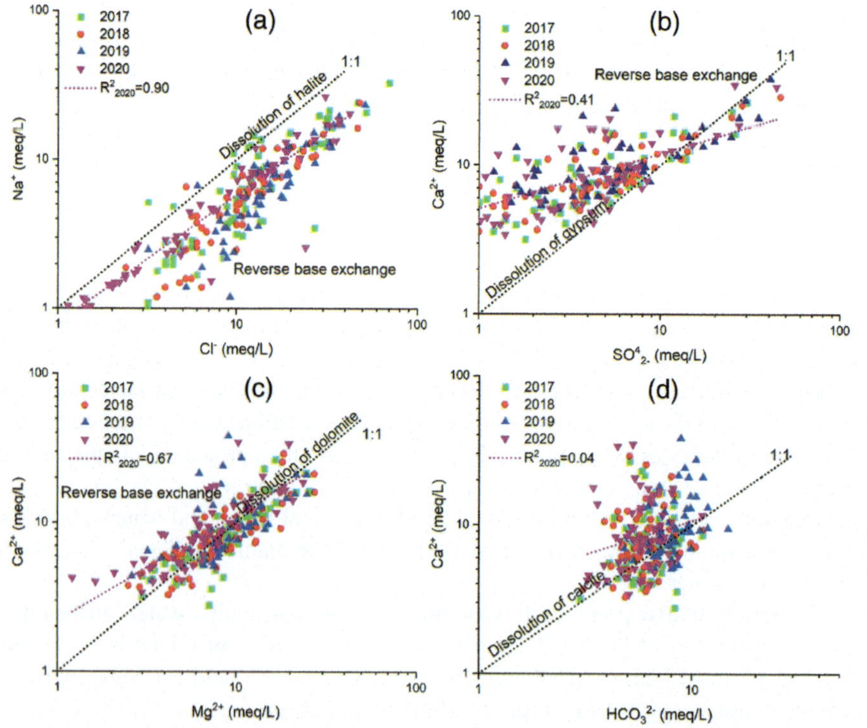

Fig. 5 Bivariate plots of Na^+ versus Cl^- (**a**); Ca^{2+} versus SO_4^- (**b**); Ca^{2+} versus Mg^{2+} (**c**); Ca^{2+} versus HCO_3^- (**d**) (El Mountassir 2023)

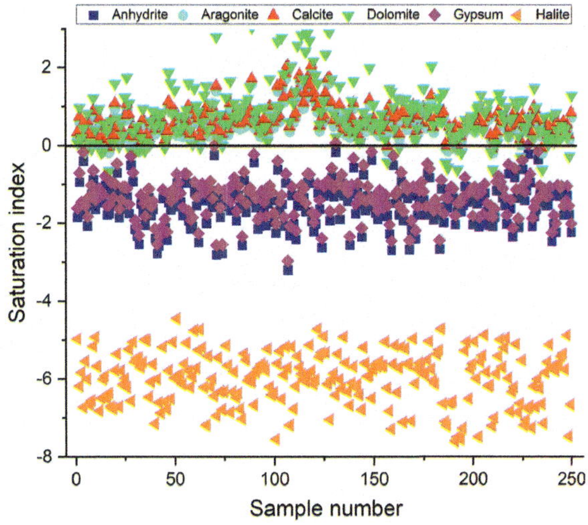

Fig. 6 Saturation index (SI) for relevant minerals of all groundwater campaigns (El Mountassir 2023)

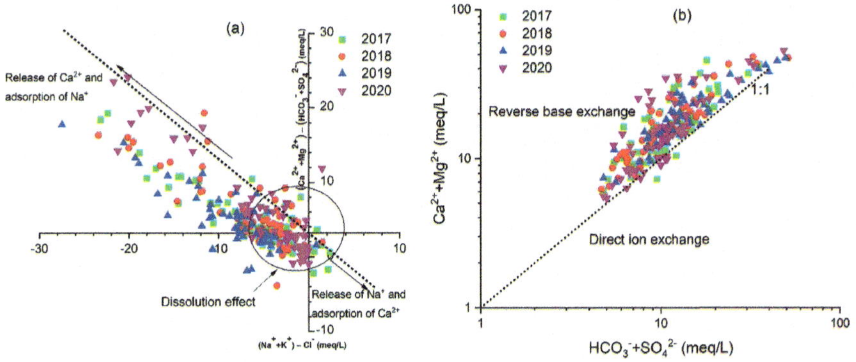

Fig. 7 Bivariate plot $((Ca^{2+} + Mg^{2+}) - (HCO_3^- - SO_4^{2-}))$ versus $((Na^+ + K^+) - Cl^-)$ (**a**); $(Ca^{2+} + Mg^{2+})$ versus $(HCO_3^- + SO_2^-)$ (**b**) (El Mountassir 2023)

The gypsum and anhydrite saturation indices (Fig. 6) show negative values. This reflects the contribution of gypsum and anhydrite dissolution to groundwater mineralization.

The Ca^{2+} versus SO_4^{2-} correlation plot shows that some points from the four campaigns (2017, 2018, 2019, and 2020) are aligned with the gypsum and/or anhydrite dissolution line, while the majority of samples are located above the 1:1 line reflecting an excess of Ca^{2+} over SO_4^{2-}. The excess Ca^{2+} could be explained by the process of reverse cation exchange (Fig. 7a).

The Ca^{2+} versus Mg^{2+} correlation shows a strong positive correlation (Fig. 5c). This reflects the fact that these two elements probably have the same origin. The calculated dolomite saturation indices are around zero, reflecting dolomite saturation of the groundwater.

The Ca^{2+} versus HCO_3^- correlation diagram (Fig. 5d) shows the contribution of calcite dissociation and reverse cation exchange to groundwater mineralization in the Cenomanian–Turonian aquifer system. Calcite saturation indices show values around zero (Fig. 6), reflecting an equilibrium state between calcite dissolution and precipitation. For all four campaigns (2017, 2018, 2019, and 2020), the interaction between Ca^{2+} and HCO_3^- (Fig. 5d) is weak with $R^2 = 0.04$. This illustrates the negligible involvement of calcite dissolution in groundwater mineralization in the study area.

The (Ca^{2+} + Mg2 +) versus ($HCO_3^- + SO_2^-$) diagram (Fig. 7b) shows that cation exchange is not the only process controlling groundwater composition. The diagram (Fig. 7b) can be divided into three groups. The samples lie along the 1:1 axis showing dissolution of gypsum, calcite, and dolomite (Fisher and Mullican 1997). The second group is represented by samples showing a slight predominance of $Ca^{2+} + Mg^{2+}$ over $HCO_3^- + SO_4^{2-}$, and the third group is represented by samples showing a marked increase in $Ca^{2+} + Mg^{2+}$ concentration. The high concentration of $Ca^{2+} + Mg^{2+}$ relative to $HCO_3^- + SO_2^-$ may be the result of reverse ion exchange (Rajmohan and Elango 2004). Similarly, increased calcium and bicarbonate concentrations in groundwater are thought to be responsible for the dissolution of carbonate minerals. This suggests that reverse ion exchange reactions and weathering are the dominant processes controlling groundwater chemistry due to excess HCO_3^- (Singh et al. 2015).

Groundwater mineralization in the Cenomanian–Turonian aquifer system is controlled by dissolution, including dissolution of evaporite minerals and reverse cation exchange (El Mountassir 2023).

4.5 *Mechanisms Controlling Groundwater Chemistry*

The Gibbs diagram is an important tool for identifying these geochemical processes (Gibbs 1970). It consists of a plot of the ratio of dominant cations Na/(Na + Ca) and anions Cl/(Cl + HCO_3) versus TDS to determine the general mechanisms controlling the chemical composition of the groundwater being studied. Essentially, groundwater chemistry is controlled by three dominant natural mechanisms, including rock weathering dominance, evaporation dominance, and precipitation dominance (Gibbs 1970). As shown in Fig. 8, possible causes of cation and anion dominance in groundwater samples are water–rock interaction and evaporative processes. The dominance of calcium and bicarbonate ions in groundwater samples is due to the weathering process of aquifer rocks. Evaporation significantly increases the concentrations of ions formed by chemical weathering, leading to higher salinity.

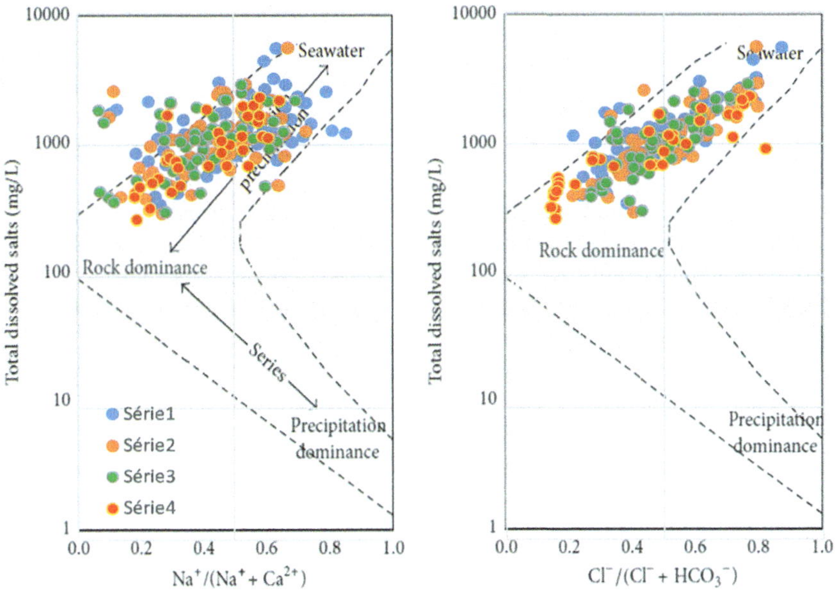

Fig. 8 Gibbs diagrams for major ion composition of groundwater in the Cenomanian–Turonian aquifer (El Mountassir 2023)

4.6 Assessment of Groundwater Quality for Consumption

In order to preserve human health, WHO has set limit values that should not be exceeded if they want to meet international consumption standards. Also, not all countries in the world follow the same standards regarding the quality of drinking water, each country has defined its standards of drinking water quality, some adopt their standards and others choose those recommended by WHO (2011). Morocco has set national standards determined by the Ministry of Land Management, Water and Environment (MATEE 2000) for the potability of water. The difference between Moroccan standards and WHO limits reflects the water management required in Morocco.

In order to assess the water quality of the Cenomanian–Turonian aquifer for domestic purposes, the electrical conductivity, chloride and nitrate contents were analyzed. The water classification was made on the basis of the simplified groundwater grid in Table 3. Evaluation was also done by comparing the levels of major elements (Ca^{2+}, Mg^{2+}, Na^{+}, K^{+}, HCO_3^{-}, Cl^{-}, SO_4^{2-}, and NO_3^{-}) with the standards set by the World Health Organization (WHO 2011; Fig. 2).

Cl^{-} and SO_4^{2-} ion concentrations in groundwater ranged from 40.4 to 1502 mg/L, and 26 to 2106 mg/L with a mean value of 424 and 274.6 mg/L, respectively (Table 1; Fig. 2) (El Mountassir 2023).

Precipitation and leaching from sedimentary rocks and soils, urban effluents, wind-borne salts in precipitation, domestic and industrial waste discharges, and salt-water infiltration are all potential sources of chloride in water (Gunn 1985). Naturally, this affects the taste of the water. The high concentration of chloride in water does not pose a health hazard (Jain et al. 2010). The acceptable and maximum limits of Cl^- in drinking water are 200 and 600 mg/L (WHO 2011). All 51 groundwater samples are within the acceptable limit, but 14 groundwater samples exceed this limit.

Sulfate (SO_4^{2-}) occurs as inorganic sulfate in the water. Eleven of the sample sites exceeded 400 mg/L, the acceptable level for drinking water (WHO 2011). In addition, the high concentration of sulfate in some of these samples could be due to the dissolution of gypsum. In drinking water, the appropriate value for sulfate (SO_4^{2-}) is 240 mg/L (WHO 2011). Almost 40% of the groundwater samples exceeded the bicarbonate level in the 65 groundwater samples from the 2020 campaign (Table 1).

HCO_3^- richness (197.5 to 680 mg/L with an average of 356.6 mg/L; Table 1, Fig. 2) results from CO_2, which is supplied to groundwater from the soil zone, where it is stored by decomposition of organic matter. It promotes the dissolution of minerals (Subba Rao et al. 2019). No direct negative effects on human health have yet been reported due to HCO_3^- (Berner and Berner 1987).

In the last decades, there has been an increasing trend of nitrate pollution in groundwater, especially in the Essaouira basin, due to rapid population growth, urbanization, and tourism, as well as massive application of nitrogen fertilizers (ammonium nitrate and urea), and agricultural practices that are also rich in nitrates. High NO_3^- concentration in drinking water can pose a risk to human health (WHO 2011). According to Berner and Berner (1987), the main sources of nitrogen in water are: biological nitrogen fixation, precipitation, and fertilizer activity. Nitrate concentration in groundwater in the Cenemano-Turonian aquifer ranges from 2 to 940.5 mg/L, with an average of 56.9 mg/L (Table 1; Fig. 2). The majority of people in the study area depend on agriculture for their survival and use various fertilizers to increase crop yields, which could be one of the reasons for the high nitrate levels in the area. The concentration of nitrate in 40% of the groundwater samples exceeded the WHO recommended limit of 50 mg/L for consumption.

For drinking water, the maximum acceptable concentration of calcium is 75 mg/L (WHO 2011). In addition, the majority of samples exceeded the maximum allowable limits prescribed by WHO (2011). According to WHO (2011), the maximum allowable concentration of magnesium (Mg^{2+}) in drinking water is 150 mg/L. Five groundwater samples were above this limit. The concentrations of Ca^{2+} and Mg^{2+} in groundwater are approximately below the maximum allowable limits prescribed by WHO (2011) for drinking purposes, and range from 68.7 to 699.1 mg/L and 14.7 to 285 mg/L, respectively (Table 1; Fig. 2).

The acceptable concentration level for sodium in drinking water is 200 mg/L (WHO 2011). This limit was exceeded by 17 groundwater samples (Table 2). High Na^+ intakes increase blood pressure, hyperosmolarity, arteriosclerosis, and edema (Nikpey et al. 2017). According to WHO (2011), the allowable potassium potability standard is 12 mg/L. Seven groundwaters have reported K+ concentrations above this limit (WHO 2011). However, due to the weather resistant nature of potassium

Table 8 Results of isotopic analysis of samples collected in the Cenemano-Turonian aquifer for the campaign 2020 (El Mountassir 2023)

Echantillons	X	Y	Z	pH	T	EC	Cl^-	$\delta^{18}O$	$\delta^{2}H$	$\delta^{3}H$
	m				°C	µS/cm	meq/L	‰ versus SMOW		UT
S. Oubihi	127,318	88,434	559	7.03	20.5	1109	2.38	− 5.87	− 34.5	0.13
S. Douzit	128,684	88,075	573	7.04	19.6	1030	2.11	− 6.01	− 34.9	0.49
S. Taouzizte	128,893	88,109	577	7.37	23.14	1537	4.67	− 5.85	− 33.8	0.3
S. Tagadirte	126,783	88,300	576	7.23	21.8	1597	4.50	− 5.19	− 31.3	0.43
S. Takrit	126,084	88,625	524	7.12	20.8	1504	4.61	− 5.36	− 32.5	0.62
S. Bousetta	134,028	83,475	752	7.85	22	636	1.14	− 5.12	− 32.8	0.49
S. Taslalaft	134,057	84,422	698	7.15	21	541	1.53	− 6.31	− 37.9	0.05
S. Oughbalou	133,793	84,537	710	7.6	21	868	1.44	− 6.16	− 39.1	0.96
S. Tinzar	133,584	84,920	677	7.4	21	806	1.42	− 6.07	− 39.2	0.83
S. Rbiâ	119,900	91,680	44	7.4	23	955	1.92	− 5.93	− 31.9	
CT123	102,959	124,849	95	7.3	23	2993	18.55	− 5.24	− 28.5	
CT88	118,419	93,635	404	7.45	23	3345	21.44	− 5.32	− 29.9	
CT60	117,368	91,734	389	7.5	21.6	1353	5.18	− 5.28	− 30.9	
CT66	114,500	98,776	422	7.7	23	1754	9.07	− 4.37	− 26	
809/52	127,541	93,680	638	8	21.5	666	1.39	− 5.87	− 31.2	
CT74	119,893	100,610	436	8	21.8	1390	7.21	− 4.54	− 28	
1209/52	113,440	95,628	355	7.3	25	985	2.81	− 5.4	− 34.6	
CT107	117,229	100,359	444	7.4	21.9	3270	26.80	− 4.21	− 24.3	
874/52	109,190	110,628	283	7.4	22.3	4323	36.67	− 4.8	− 24.9	
E39	119,017	108,246	332	7.7	21.2	1198	2.80	− 5.58	− 29.3	

(continued)

Table 8 (continued)

Echantillons	X	Y	Z	pH	T	EC	Cl^-	$\delta^{18}O$	δ^2H	δ^3H
	m				°C	μS/cm	meq/L	‰ versus SMOW		UT
CT39	121,773	103,695	390	7.6	21.2	1496	5.44	– 6.09	– 32.5	
75/52	128,124	103,181	461	7.9	20.7	2338	7.40	– 5.46	– 30.2	
CT35	134,110	106,742	543	7.9	21	850	2.69	– 5.21	– 27.2	
CT34	132,390	100,749	526	7.6	23.6	1148	2.20	– 5.92	– 32.1	
1112/52	138,581	99,269	591	7.5	24.5	1440	2.69	– 5.99	– 35.8	
776/52	141,554	95,298	637	7.4	23	2090	4.86	– 6.27	– 35.6	
CT40	119,621	106,390	347	7.4	22.8	1750	6.27	– 6.29	– 36.6	
CT108	125,042	100,877	469	7.1	20.2	1157	4.01	– 3.16	– 15.2	
CT125	123,518	99,423	458	7.5	25	1394	2.38	– 5.62	– 29.8	
CT49	110,383	116,191	222	7.3	23.8	2005	12.22	– 5.31	– 28.2	
O124	113,568	115,297	264	7.2	22.7	2417	16.11	– 5.12	– 28.4	
612/52	107,407	122,209	204	7.4	22	3800	33.85	– 4.5	– 22.7	
CT52	114,269	124,836	324	7.5	22.3	1970	12.20	– 4.92	– 24.2	
CT51	111,820	128,010	224	7.1	23	2126	14.41	– 4.45	– 22	
89/52	106,174	117,707	190	7.6	23	2106	11.73	– 4.7	– 24.5	
CT54	98,283	126,062	74	7.5	22.3	1891	10.66	– 4.89	– 23.6	
CT56	94,383	125,259	11	7.2	21.5	5285	30.76	– 4	– 19.6	
CT122	130,057	94,712	662	7.6	21	865	1.43	– 6.01	– 32.2	
Dam (CT90)	103,393	91,167	238	8.17	26.7	637	7.22	0.58	2.2	

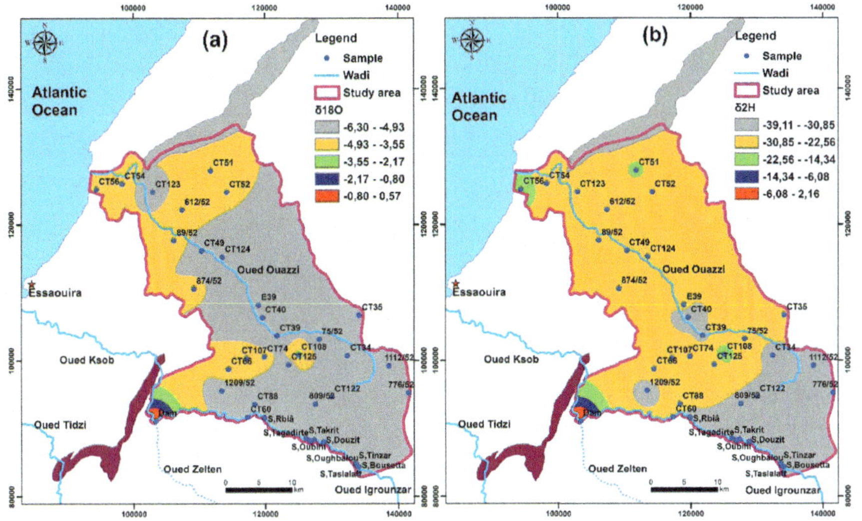

Fig. 11 Spatial distribution map. **a** $\delta^{18}O$ and **b** $\delta^{2}H$ for the 2020 campaign (El Mountassir 2023)

A total of 39 samples collected in 2020 were analyzed for stable isotopes ($\delta^{18}O$ and $\delta^{2}H$) with nine samples of tritium ($\delta^{3}H$). For the oxygen-18 samples range from—6.31‰ to 0.58‰, with an average of—5.17‰. For deuterium, the maximum value is 2.2‰ versus SMOW and the minimum value is equal to—39.2‰ versus SMOW with an average of—29.02‰ vs SMOW (Table 8).

For the isotopic characterization of the study area, two reference lines were used: the global meteoric water line (GMWL) Eq. (5) (Craig 1961)

$$\delta^{2}H = 8 * \delta^{18}O + 10 \tag{5}$$

And the local meteoric line of the Essaouira basin (LMWL) according to Eq. (6) (Mennani 2001)

$$\delta^{2}H = 7.95 * \delta^{18}O + 11.3 \tag{6}$$

The results of the isotopic analyses were plotted on the deuterium–oxygen 18 diagram; this diagram (Fig. 12a) shows that two groups of waters can be distinguished.

The most enriched samples characterize the first group of the study area, lying below the two reference lines (GMWL and LMWL), as do the samples (CT74, CT66, CT90, 1209/52). These samples could indicate aquifer recharge from evaporated meteoric water and return of evaporated irrigation water. Evaporation may occur either prior to recharge as precipitation passes through the atmosphere with low water content or after recharge by leaching of heavy isotopes accumulated during the dry season in the unsaturated zone (El Mountassir 2023).

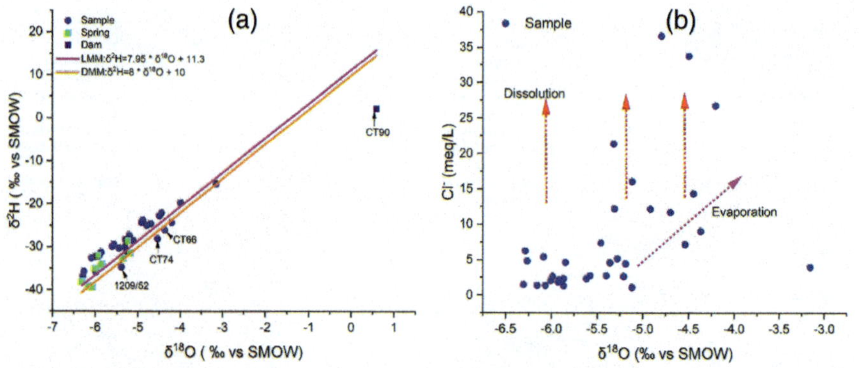

Fig. 12 Isotopic data in the study area, **a** $\delta^{18}O$ versus δ^2H, **b** $\delta^{18}O$ versus Cl^- for the 2020 campaign. *Source* Developed by the authors

The second group represents the majority of the samples, located near the LMWL and GMWL. These waters originate from rapid infiltration of meteoric water with no change in isotopic content. The δ^2H and $\delta^{18}O$ values at the intersection correspond to the isotopic values of (initial) recharge from precipitation to groundwater (Clark and Frjtz 1997).

The primary source of groundwater and surface water recharge in the region is rainwater, as shown in the regression plot (Fig. 12a). Groundwater samples show no close resemblance to standard isotopic valucs (0.0‰) of marine origin. The relatively high $\delta^{18}O$ (0‰) values as well as the higher TDS and Cl^- concentrations are characteristics of seawater origin, while the low $\delta^{18}O$ levels are indicative of fresh groundwater sources (Wang and Jiao 2012).

4.8.3 *Relationship Between Cl^- and $\Delta^{18}o$*

Environmental isotopes are used to find the source of salinity in groundwater, e.g., by leaching of salts, percolating water from evaporite deposits, saltwater intrusion from the sea/surface brackish water/brine etc., or by concentration of dissolved ions by evaporation. Being conservative, stable isotopes do not change composition during transport in a single groundwater system or in leach water by dissolution of salts. However, when salinity is caused by mixing a saline water source with fresh water, the resulting water will have a different isotopic composition and salinity. This property is used in this study to identify salinity sources. The $\delta^{18}O$, as well as the conservative chloride ion, is linearly correlated on a mixing line within limits given by the freshwater and saltwater components (Payne 1983).

The relationship between chloride and stable isotopes of water ($\delta^{18}O$) are used in this study to give reliable interest to the combination of chemical and isotopic data to confirm the major processes controlling groundwater salinity. A significant increase in Cl^- concentration coupled with $\delta^{18}O$ enrichment is observed in the diagram (Fig. 12b) could be the result of evaporation (either during recharge, from shallow water levels, or seepage of village wastewater due to lack of sewerage channels as the case of Meskala village) as in sample CT107 with $Cl^- = 950.1$ mg/L and $\delta^{18}O = -4.21$, as the evaporation process enriches both $\delta^{18}O$ values and Cl^- concentrations (Foster et al. 2018). Groundwater $\delta^{18}O$ values are enriched by surface water recharge (Oued Igrounzar) especially in springs in the upstream part of the study area (oubihi, douzit.), while the variation in Cl^- concentrations depends on the salinity of the infiltration water that is ultimately attached to the water table. The salinity of infiltration water comes from both surface water and soil along groundwater pathways (Foster et al. 2018). Furthermore, independent of isotopic exchange, high Cl^- concentrations with low $\delta^{18}O$ values are associated with mineral accumulation processes within the aquifer (mineral dissolution and transpiration) such as in samples 776/52 and CT123. Around the Bouabout area, the geographical distribution map shows a comparatively greater depletion of $\delta^{18}O$ values and low chloride concentration in groundwater, indicating potential recharge locations that confirms by low salinity and electrical conductivity around 1000 μS/cm (Fig. 11; Table 8) (El Mountassir 2023).

4.8.4 Recharge Altitude of the Aquifer

Locating the recharge zone is essential for assessing groundwater resources and defining the boundaries of an aquifer protection area. The stable isotope elevation effect solves this problem in areas with marked topography. Thus, it is possible to determine the average altitude of the water supply area of a spring by placing its average isotopic composition on a regional correlation line between the isotopic composition of oxygen or hydrogen and altitude. This line can be defined by the relationship of the $\delta^{18}O$ of precipitation to the altitude of the precipitation. The isotopic composition of precipitation varies with the altitude of the sample. Precipitation is increasingly rich in $\delta^{18}O$ and δ^2H with increasing altitude (Clark and Frjtz 1997).

To determine the recharge elevation of the Cenemano-Turonian aquifer, we projected the isotopic contents onto the altitudinal line graph of precipitation and water sources in Morocco (Abourida 2007), which has a gradient of 0.25‰ per 100 m for $\delta^{18}O$, this value is consistent with that defined for the High Atlas (0.27‰ per 100 m) (El Ouali 1999), and that of the Essaouira Basin our study area (− 0.26‰ per 100 m) (Bahir et al. 2000a, b); the projection (Fig. 13) reveals that the recharge altitudes of the Cenomano-Turonian aquifer range from 375 to 1275 m. The following are the main areas of recharge: (1) the lowest recharge zone is located at an altitude of 375 m; this is the Tlat Hanchane area, (2) The Krimat area is the intermediate recharge zone with altitudes between 450 and 650 m. (3) Another recharge zone of the studied aquifer is located at an altitude between 650 and 1275 m (the extreme east of the study area) (El Mountassir 2023).

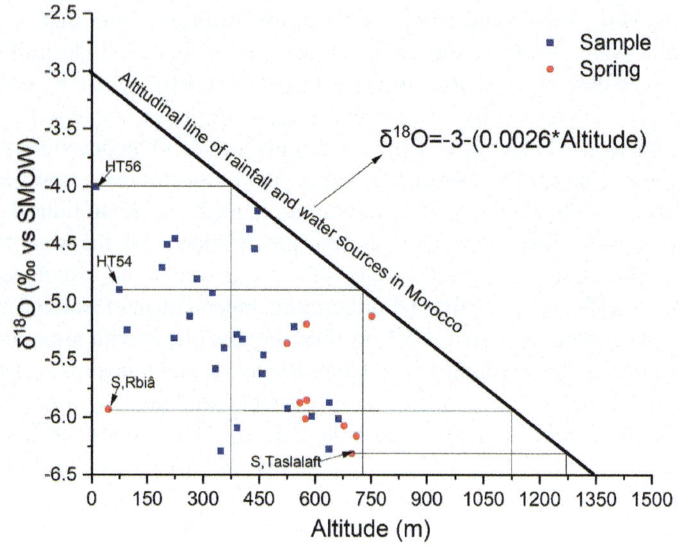

Fig. 13 Estimated recharge elevation of the Cenemano-Turonian aquifer for the year 2020 (El Mountassir 2023)

4.8.5 Age Dating of Groundwater in the Cenemano-Turonian Aquifer Using Tritium (δ3H)

After the nuclear weapons tests of the 1950s and 1960s, atmospheric tritium levels almost returned to natural levels (pre-1950 levels), which means that tritium is becoming more effective in determining the age of groundwater. It has also been shown that tritium is a non-reactive noble gas and in water is not subject to chemical reactions, absorption, or dissolution/precipitation processes; therefore, it is conservative of the geochemical fingerprint of the source (Telloli et al. 2022). According to Mazor (1991), a tritium content greater than 2 TU indicates recent water recharge, a tritium content greater than 1 TU indicates post-nuclear water recharge, while a content less than 1 TU indicates pre-nuclear water recharge or a mixture of modern and ancient waters.

In 2020, 9 samples of tritium in the Cenemano-Turonian aquifer were analyzed. They are presented in (Table 8; Fig. 14). The δ^3H contents range from 0.05 to 0.96 TU, with an average of 0.47 TU. The nine measurements are present sources of recharge to the Cenemano-Turonian aquifer. The groundwater with values less than or equal to 1 TU are considered old water because of overexploitation of the aquifer and a decrease in precipitation in recent years generating a low recharge of the aquifer. No values above 1 TU indicate recent recharge (El Mountassir 2023).

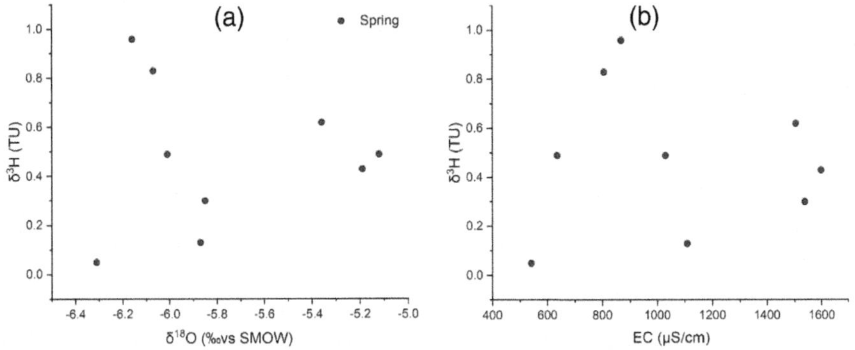

Fig. 14 Tritium versus $\delta^{18}O$ (**a**), δ^3H versus EC (**b**) for the campaign 2020 (El Mountassir 2023)

5 Conclusion

The combination of major element and stable isotope geochemistry allowed us to understand the hydrodynamic functioning of the Cenomanian–Turonian aquifer and to clarify the geologic factors that control its water chemistry and mineralization process. In this chapter, water quality and suitability were evaluated using GIS-based indexing methods (WQI and WQII).

The groundwater flow keeps the same direction during the four campaigns (2017, 2018, 2019, and 2020) with a decrease in water level due to the decrease in precipitation in recent years. The study of the piezometry of the study area showed that the water flows generally from northeast to northwest and northeast to southwest of the Cenomanian–Turonian aquifer.

Hydrogeochemical analysis reveals that the groundwater is of mixed Ca–Mg–Cl, Ca–HCO_3, Ca–Cl, and Ca–SO_4 types, with a dominance of the first type. The analysis of the abundance of the main cations and anions shows the dominance of $Ca^{2+} > Na^+ > Mg^{2+} > K^+$ for cations and the dominance of $Cl^- > HCO_3^- > SO_4^- > NO_3^-$ for anions. The interaction time of groundwater with rocks and anthropogenic pollutants play important roles in the observation of different water types in the Cenomanian–Turonian aquifer. The study of the origin of groundwater mineralization shows that two phenomena control this salinization: the dissolution of evaporite minerals (halite, gypsum and anhydrite) and the phenomenon of reverse cation exchange.

The spatial distribution of the WQI in the Cenomanian–Turonian aquifer was divided into 32.8% (good areas), 44.3% (poor areas), 21.3% (very poor areas) and 1.6% (areas unsuitable for consumption),

In addition, the combination of irrigation water quality parameters such as Irrigation Water Quality Index (IWQI), Sodium Adsorption Ratio (SAR), Electrical Conductivity (EC), Sodium Ratio (%Na), Chloride Concentration, and Permeability Index (PI) was used to assess the problems that could develop in the soil due to the current irrigation practices in the study area. Based on the SAR classification, all water samples are considered suitable for irrigation, and the possibility of hazardous salinization is ruled out, especially due to the high infiltration capacity of the soil. The IWQI showed that 43% of the studied samples have a high to severe restriction level, while 57% of the studied samples were placed in the low to moderate restrictions for irrigation use.

The spatial distribution of DWQI and WQI provided a clear visualization of the quality of groundwater for drinking and irrigation in the study area. This makes it easier for decision makers to assess the quality of groundwater used in the study area. In addition, the GIS maps can recommend the range of appropriate crops grown and used for the soil field in the study area.

Stable isotope signatures ($\delta^{18}O$, $\delta^{2}H$) indicate that the groundwater samples are of meteoric origin. These tracers reveal that direct infiltration provides recharge to the aquifer in precipitation without significant evaporation. The combination of chemical and isotopic parameters, particularly Cl^- versus $\delta^{18}O$, confirms the results obtained by the hydrochemical approach by confirming that dissolution is one of the main phenomena contributing to groundwater mineralization in the study area. The comparison of the $\delta^{3}H$ content of the samples collected in 2016 and 2020 shows a low value of $\delta^{3}H$, around zero and less than 1 TU, which translates by a low recharge rate. This could explain this situation following a reduction in annual precipitation recorded in the area. On the other hand, the recharge area of the Cenomanian–Turonian aquifer varies between 375 and 1275 m, with the lowest recharge located in the Bouzerktoun area and the highest in the Bouabout area.

Taking all this into consideration, continuous monitoring of groundwater, control of salinity through proper management, implementation of more efficient irrigation methods, a good leaching regime, maintenance of low groundwater levels, as well as awareness raising of agricultural workers are necessary to decrease the soil degradation process in the Essaouira region.

Acknowledgements The authors appreciate the help they have received from their colleagues throughout the data collection and the field survey. The editor and reviewers deserve special thanks for their insightful critiques and helpful suggestions. The authors gratefully acknowledge the technical staff at the Laboratory of Mohammed VI Polytechnic University in Benguerir city and those of the Centro de Ciencias e Tecnologias Nucleares (C^2TN) in Lisboa city for their constant help

Conflicts of Interest There are no conflicts of interest declared by the authors.

Funding There is no funding for this study.

References

Abd El-Aziz SH (2017) Evaluation of groundwater quality for drinking and irrigation purposes in the north-western area of Libya (Aligeelat). Environ Earth Sci 76:147. https://doi.org/10.1007/s12665-017-6421-3

Abourida A (2007) Approche hydrogeologique de la nappe du Haouz (Maroc) par teledetection, isotopie, SIG et modelisation (PhD thesis). Cadi Ayyad University, 146p

Amghar M (1989) Apports des méthodes d'analyses de la tectonique cassante à la connaissance de l'histoire alpine du Haute Atlas Occidental. L'exemple du versant nord du bloc ancien et de l'Atlas d'Agadir (Haut Atlas, Maroc). Thèse 3ème cycle, Université Cadi Ayyad, Maroc

Ayers RS, Westcot DW (1985) In: Water quality for agriculture, vol. 29. Food and Agriculture Organization of the United Nations, Rome. http://www.fao.org/3/T0234E/T0234E00.htm#TOC

Ayers RS, Westcot DW (1994) Water quality for agriculture, FAO Irrigation and Drainage Paper Water: Rome, pp 1–120

Bahir M, Mennani A (2002) Problématique de la gestion des eaux souterraines au Maroc. Estudios Geologicos 8(3–4):103–108

Bahir M, Mennani A, Jalal M, Youbi N (2000a) Ressources hydriques du bassin synclinal d'Essaouira (Maroc). Estud Geol 56:185–195

Bahir M, Mennani A, Jalal M, Youbi N (2000b) Contribution à l'étude des ressources hydriques du bassin synclinal d'Essaouira (Maroc). Estud Geol 56(3–4):185–195. https://doi.org/10.3989/egeol.00563-4150

Bahir M, Carreira P, Da Silva MO, Fernandes P (2007) Caractérisation hydrodynamique, hydrochimique et isotopique du système aquifère de Kourimat (Bassin d'Essaouira, Maroc). Estud Geol 64(1):61–73. https://doi.org/10.3989/egeol.08641433

Bahir M, El Moukhayar R, Youbi N, Chamchati H, Chkir Ben Jemaa N (2013a) Management and protection of groundwater resources in semi-arid zones: application of hydrochemical methodologies to Essaouira Synclinal Basin, Morocco. Int J Hydrol Sci Technol 3(2):160–175. https://doi.org/10.1504/IJHST.2013.057627

Bahir M, El Moukhayar R, Chkir N, Chamchati H, Fernandes PG, Carreira PM (2013b) Groundwater chemical evolution in the Essaouira Aquifer Basin—NW Morocco. 2013b, Open J Modern Hydrol 3:8. https://doi.org/10.4236/ojmh.2013.33017

Bahir M, El Mountassir O, Ouazar D, Carreira PM (2021a) Hydrochemical analysis and evaluation of groundwater quality in Ouazi Basin (Essaouira, Morocco). In: Abrunhosa M, Chambel A, Peppoloni S, Chaminé HI (eds) Advances in geoethics and groundwater management: theory and practice for a sustainable development. Advances in science, technology & innovation (IEREK Interdisciplinary Series for Sustainable Development). Springer, Cham. https://doi.org/10.1007/978-3-030-59320-9_50

Bahir M, El Mountassir O, Ouazar D, Carreira PM (2021b) Use of WQI and isotopes to assess groundwater quality of coastal aquifers (Essaouira, Morocco). In: Abrunhosa M, Chambel A, Peppoloni S, Chaminé HI (eds) Advances in geoethics and groundwater management: theory and practice for a sustainable development. Advances in science, technology & innovation. Springer, pp 251–255. https://doi.org/10.1007/978-3-030-59320-9_51

Bahir M, Mountassir EL, O, Ouazar D, Chehbouni A, Carreira P.M, (2021c) Stable isotope and quality of groundwater around Ksob sub-basin, Essaouira, Morocco. Sustain Water Resour Manag 7:73

Bahir M, El Mountassir O, Dhiba D, Chehbouni A, Carreira PM, Elbiar H (2022a) Combining stable isotope and WQI methods to study the groundwater quality: a case study in Essaouira city, Morocco. SN Appl Sci 4:317

Bahir M, El Mountassir O, Chehbouni A, El Jiar H, Carreira PM (2022b) Hydrogeochemical and isotopic assessment for characterizing groundwater quality and recharge processes in the Essaouira Basin, Northwestern Morocco. Arab J Geosci 15:1–21. https://doi.org/10.1007/s12517-022-09817-6

Berner EK, Berner RA (1987) The global water cycle: geochemistry and environment. Prentice Hall, Englewood Clifs

Brown RM, McClelland NI, Deininger RA, Tozer RG (1970) A water quality index-do we dare. Water Sewage Works 117(10)

Chitsazan M, Tabari MM, Eilbeigi M (2017) Analysis of temporal and spatial variations in groundwater nitrate and development of its pollution plume: a case study in Karaj aquifer. Environ Earth Sci 76(11):1–27. https://doi.org/10.1007/s12665-017-6677-7

Clark ID, Frjtz P (1997) Environmental isotopes in hydrogeology. Lewis Publishers, New York

Craig H (1961) Isotopic variations in meteoric waters. Science. 26;133(3465):1702–3
Cribb J (2010) The coming famine: the global food crisis and what we can do to avoid it. University of California Press
Dixon W, Chiswell B (1992) The use of hydrochemical sections to identify recharge areas and saline intrusions in alluvial aquifers, southeast Queensland, Australia. J Hydrol 130:299–338
Duffaud F (1960) Contribution à l'étude stratigraphique du bassin secondaire du Haut Atlas Occidental (Sud-Ouest du Maroc). Bull Soc Géol Fr 7:728–734
El Mountassir O (2023) Hydrohémie et vulnérabilité des ressources en eau dans un milieu semi-aride: Approche multidisciplinaire et utilisation des outils SIG (Cas du bassin d'Essaouira, Maroc) "Hydrochemistry and Vulnerability of Water Resources in a Semi-Arid Environment: A Multidisciplinary Approach and the Use of GIS Tools (Case of the Essaouira Basin, Morocco)". Thèse, N713, Université Cadi Ayyad
El Mountassir O, Bahir M (2023) The assessment of the groundwater quality in the coastal aquifers of the Essaouira basin, Southwestern Morocco, using hydrogeochemistry and isotopic signatures. Water 5;15(9):1769
El Mountassir O, Ouazar D, Bahir M, Chehbouni A, Carreira PM (2021a) GIS-based assessment of aquifer vulnerability using DRASTIC model and stable isotope: a case study on Essaouira basin. Arab J Geosci 14:321. https://doi.org/10.1007/s12517-021-06540-6
El Mountassir O, Bahir M, Ouazar D, Carreira PM (2021b) Nitrate pollution in groundwater of the Ouazi Basin: case of essaouira (Southwestern Morocco). In: Abrunhosa M, Chambel A, Peppoloni S, Chaminé HI (eds) Advances in geoethics and groundwater management: theory and practice for a sustainable development. Advances in science, technology & innovation (IEREK interdisciplinary series for sustainable development). Springer, Cham. https://doi.org/10.1007/978-3-030-59320-9_49
El Mountassir O, Bahir M, Ouazar D, Carreira PM (2021c) For a better understanding of recharge and salinization mechanism of a cenomanian–turonian aquifer. In: Abrunhosa M, Chambel A, Peppoloni S, Chaminé HI (eds) Advances in geoethics and groundwater management: theory and practice for a sustainable development. Advances in science, technology & innovation (IEREK interdisciplinary series for sustainable development). Springer, Cham. https://doi.org/10.1007/978-3-030-59320-9_42
El Mountassir O, Bahir M, Ouazar D, Chehbouni A, Carreira PM (2021d) Geochemical and isotopic evidence of groundwater salinization processes in the Essaouira region, north-west coast, Morocco. SN Appl Sci 3:1–16. https://doi.org/10.1007/s42452-021-04623-3
El Mountassir O, Bahir M, Ouazar D, Chehbouni A, Carreira PM (2022a) Temporal and spatial assessment of groundwater contamination with nitrate using nitrate pollution index (NPI), groundwater pollution index (GPI), and GIS (case study: Essaouira basin, Morocco). Environ Sci Pollut Res 17:1–8. https://doi.org/10.1007/s11356-021-16922-8
El Mountassir O, Bahir M, Chehbouni A, Dhiba D, El Jiar H (2022b) Assessment of groundwater quality and the main controls on its hydrochemistry in a changing climate in Morocco (Essaouira Basin). Sustainability 14:8012. https://doi.org/10.3390/su14138012
El Ouali A (1999) Modalites d'alimentation et echanges entre aquiferes de piemont en conditions climatiques arides. Cas des systemes aquiferes du Haut Atlas/bassin Cretace d'Errachidia (Maroc) (Ph.D. Thesis). Mohammed V, Rabat, p 182
Epstein S, Mayeda TK (1953) Variations of the 18O/16O ratio in natural water. Geochim Cosmochim Acta 4:213–220
ESRI (1999) ARCVIEW GIS v.10.2.2. Environmental Systems Research, Institute Inc
Fisher RS, Mullican WF (1997) Hydrochemical evolution of sodium-sulphate and sodium-chloride groundwater beneath the Northern Chihuahuan desert, Trans-Pecos, Texas, USA. Hydrogeol J 5:4–16
Foster S, Pulido-Bosch A, Vallejos Á, Molina L, Llop A, MacDonald AM (2018) Impact of irrigated agriculture on ground water recharge salinity: a major sustainability concern in semi-arid regions. Hydrogeol J 26:2781–2791. https://doi.org/10.1007/s10040-018-1830-2

Friedman I (1953) Deuterium content of natural waters and other substances. Geochim Cosmochim Acta 4(1–2):83–103

Gibbs RJ (1970) Mechanisms controlling world water chemistry. Science 170:1088–1090. https://doi.org/10.1126/science.170.3962.1088

Gunn RH (1985) Shallow ground waters in weathered volcanic, granitic and sedimentary rocks in relation to dryland salinity in southern New South Wales. Soil Res 23:355–371. https://doi.org/10.1071/sr9850355

Hasan MM, Ahmed MS, Adnan R, Shafiquzzaman M (2020) Water quality indices to assess the spatiotemporal variations of Dhaleshwari river in central Bangladesh. Environ Sustain Ind 1(8):100068

Hem JD (1985) Study and interpretation of the chemical characteristics of natural water, vol 2254. Department of the Interior, US Geological Survey

Horton RK (1965) An index number system for rating water quality. J Water Pollut Control Fed 37(3):300–305

Jain CK, Bandyopadhyay A, Bhadra A (2010) Assessment of ground water quality for drinking purpose, District Nainital, Uttarakhand, India. Environ Monit Asses 166:663–676. https://doi.org/10.1007/s10661-009-1031-5

Jalal M, Blavoux B, Bahir M, Bellion Y, Laftouhi N, Puig JM, Mennani A, Daniel M (2001) Etude du fonctionnement du système aquifère karstique Cenomano-Turonian de l'oued Igrounzar (Bassin d'Essaouira, Maroc). J Afr Earth Sci 32:803–817

Lesch SM, Suarez DL (2009) A short note on calculating the adjusted sar index. Trans. ASABE (Am. Soc. Agric. Biol. Eng.)

Mazor E (1991) Applied chemical and isotopic groundwater hydrology. Open University Press, Buckingham

Mennani A (2001) Apport de l'hydrochimie et de l'isotopie à la connaissance du fonctionnement des aquifères de la zone côtière d'Essaouira (Maroc Occidental). PhD thesis, Cadi Ayyad University, Marrakech, Morocco

Mennani A, Blavoux B, Bahir M, Bellion Y, Jalal M, Daniel M (2001) Apports des analyses chimiques et isotopiques à la connaissance du fonctionnement des aquifères plio-quaternaire et turonien de la zone synclinale d''Essaouira (Maroc occidental). J Afr Earth Sci 32:819–835

Ministère chargé de l'Aménagement du Territoire, de l'Eau et de l'Environnement (MATEE). Etat de la Qualité des Ressources en Eau au Maroc 2000/2001 (Novembre 2003)

Meireles ACM, de Andrade EM, Chaves LCG, Frischkorn H, Crisostomo LA (2010) A new proposal of the classification of irrigation water. Rev Cienc Agron 41(3):349–357. https://doi.org/10.1590/S1806-66902010000300005

Nikpey E, Karlsen TV, Rakova N, Titze JM, Tenstad O, Wiig H (2017) High-salt diet causes osmotic gradients and hyperosmolality in skin without afecting interstitial luid and lymph. Hypertension 69:660–668

Payne (1983) Groundwater salinization. In: Guidebook on nuclear techniques in hydrology. Technical report series no 91. IAEA, Vienna, pp 351–357

Piper AM (1944) A graphic procedure in the geochemical interpretation of water analyses. Trans Am Geophys Union 25(6):914–928. https://doi.org/10.1029/TR025i006p00914

Rajmohan N, Elango L (2004) Identification and evolution of hydrogeochemical processes in the groundwater environment in an area of the Palar and Cheyyar River Basins, Southern India. Environ Geol 46:47–61. https://doi.org/10.1007/s00254-004-1012-5

Rodier J, Legube B, Merlet N, Mialocq JC, Leroy P, Houssin M, Lavison G, Bechemin C, Vincent M, Rebouillon P, Moulin L, Chomodé P, Dujardin P, Gosselin S, Seux R, Al Mardini F (2009) L'Analyse de l'eau. 9e édition, Dunod, Paris

Singh SP, Tripathi SK, Kumar V, Kumar A, Raha P (2015) Hydro- chemical investigation and groundwater quality evaluation for irrigation purpose in some blocks of Varanasi District, Uttar Pradesh, India. Int J Trop Agric 33:1653–1660

Subba Rao N, Srihari C, Deepthi Spandana B, Sravanthi M, Kamalesh T, Abraham Jayadeep V (2019) Comprehensive understanding of groundwater quality and hydrogeochemistry for the

sustainable development of suburban area of Visakhapatnam, Andhra Pradesh, India. Hum Ecol Risk Assess 636:52–80. https://doi.org/10.1080/10807039.2019.1571403
Telloli C, Rizzo A, Salvi S, Pozzobon A, Marrocchino E, Vaccaro C (2022) Characterization of groundwater recharge through tritium measurements. Adv Geosci 57:21–36. https://doi.org/10.5194/adgeo-57-21-2022
Wang Y, Jiao JJ (2012) Origin of groundwater salinity and hydrogeochemical processes in the confined quaternary aquifer of the Pearl River Delta, China. J Hydrol 438–439:112–124
WHO (2011) Guidelines for drinking-water quality. WHO Chronicle 38:104–108
Xiao J, Wang L, Deng L, Jin Z (2019) Characteristics, sources, water quality and health risk assessment of trace elements in river water and well water in the Chinese Loess Plateau. Sci Total Environ 650:2004–2012

Dr. Otman El Mountassir is a hydrogeologist specializing in water and environmental engineering. He holds a Master's degree in Water and Environmental Engineering from the Faculty of Sciences Kenitra, and a Ph.D. in Geosciences Geressources and Environment from Cadi Ayyad University under the supervision of Professor Mohammed Bahir. He has published articles in leading international journals, indexed in databases such as Scopus and Thomson. He has also acted as a reviewer for many international journals. His expertise includes Hydrogeology, Hydrology, Stable Isotopes, Hydrochemistry and Modelling. He is affiliated with the High Energy and Astrophysics Laboratory of the Semlalia Faculty of Sciences, Cadi Ayyad University, Morocco.

Chapter 12
Drought Assessment in Potwar Region, Punjab Pakistan During 1981–2019

Saira Batool, Syed Amer Mahmood, and Safdar Ali Shirazi

Abstract A severe natural disaster, drought affects the economy, the ecology, and many other areas on a global scale. For the Potwar region (PR) in Punjab, Pakistan, drought indicators (DIs), including deciles, the Standard Precipitation Index (SPI), and the Reconnaissance Drought Index (RDI), were computed using the DrinC programme. Drought conditions were predicted for the next 12, 9, and 3 months. Eight of the last 39 years have seen a severe drought, according to DIs created using the deciles technique, and such cycles repeat every 2–7 years. Similar trends were visible as deciles in both the RDI and SPI indices. However, for RDI and SPI, the extremely dry and severely dry classes only persisted for two years, whereas the years that affected the other deciles were typically and moderately dry. Deciles are less useful for assessing the severity of the drought than SPI. Based on a regression study that shows a connection between the RDI and SPI indices, the yearly RDI may be predicted if the first three months of precipitation are known. This research can help future development strategies tackle sensitive drought occurrences, their mitigation, and their socioeconomic repercussions.

Keywords Potwar region · Climate change · DrinC · SPI · Drought indices

---

S. Batool (✉)
Center for Integrated Mountain Research, University of the Punjab, Lahore, Pakistan
e-mail: saira.cimr@pu.edu.pk

S. A. Mahmood
Department of Space Science, University of the Punjab, Lahore, Pakistan
e-mail: amer.spsc@pu.edu.pk

S. A. Shirazi
Department of Geography, University of the Punjab, Lahore, Pakistan
e-mail: Shirazi.geog@pu.edu.pk

M. Behnassi et al. (eds.), *The Water, Climate, and Food Nexus*,
https://doi.org/10.1007/978-3-031-50962-9_12

1 Introduction

A common occurrence that affects many global climatic zones is drought. It is caused by inadequate water supplies for cattle, plants, and crops, which has an immediate effect on people. According to several studies (Anjum et al. 2012; Khan and Gadiwala 2013; Xie et al. 2013; Spinoni et al. 2014, 2018; Adnan et al. 2015, 2016; Haroon et al. 2016), drought first appears as a strain on water resources as a result of insufficient rainfall. Meteorological drought refers to this regular absence of precipitation.

The Potwar region in Pakistan's Punjab province's north is distinguished by its hilly topography and contains various highland ranges, including the Salt Range, with elevations ranging from 300 to 600 m. The construction of canal networks from the neighbouring Because of the topographic slopes in the area, it is impractical to cross the Rivers Jhelum in the east and the Indus in the west. Because of the geography, agriculture in this region is primarily reliant on rainfall for water retention and drainage. Researchers from a variety of fields, including agriculture, geosciences, meteorology, GIS, and environmental science, have recently given the drought in this area a great deal of attention as a huge environmental calamity. According to numerous studies (Deo et al. 2017; Bae et al. 2018; Adnan et al. 2015, 2018; Xie et al. 2013; Samo et al. 2017; Iqbal and Athar 2018; Ahmed et al. 2016), drought has significant effects on agriculture, ecology, hydrology, and the environment. This has attracted attention. Drought has a considerable impact on food output, causes water scarcity for both animal and human consumption, and affects a number of different economic sectors.

According to several studies (Jamro et al. 2019; Vasiliades et al. 2011; Krakauer et al. 2019; Hina and Saleem 2019; Jasim and Awchi 2020; Qaisrani et al. 2021, Moghbeli and Delbari 2020), there has been significant variability in the global climate over the past century. Due to this variance, the water cycle has become more intense and frequent severe events, such large-scale floods and droughts, have begun to occur (Dai 2011). According to predictions, the frequency of drought episodes is anticipated to rise as a result of the high concentrations of greenhouse gases contributing to global warming (Dai 2011). As a result, it is anticipated that this global warming would make the mismatch between water demand and availability, which is brought on by spatiotemporal changes, worse.

As a result, drought has a large negative impact on underprivileged communities throughout many different economic sectors, making it a severe natural disaster. Different academic disciplines have different interpretations for the drought. According to agricultural specialists (Adnan et al. 2016), a drought is a lack of moisture, specifically effective precipitation, that has an impact on crop productivity. Hydrologists link it to decreased water discharge while meteorologists characterise it as a prolonged rainfall shortage. These interpretations highlight the relative significance of evapotranspiration (ET), precipitation, and water runoff in the occurrence of droughts as a result of climatic variability.

The multi-dimensional and geographical effects of drought, a periodic and complicated phenomenon that impacts large regions and communities worldwide. In comparison to other natural disasters, there has been a significant increase in the intensity and frequency of droughts over the past few decades, which affect more people. Droughts can be classified according to their length, geographic scope, and intensity, with the latter being the main subject of research. It is essential to comprehend how droughts originate and evolve under various environmental and meteorological situations. Drought indices (DIs) are essential tools for determining the severity of droughts. By combining several climate factors and taking into account elements like duration, intensity, and occurrence, these indices make it possible to evaluate and analyse the severity of droughts. DIs offer important insights into both food security and overall climate variability by synthesising these intricate climate factors. Furthermore, DIs offer all important stakeholders a thorough grasp of the severity of the drought (Tsakiris et al. 2013).

The Drought Indices (DIs) results are useful tools that help a variety of users, such as researchers, academics, and bankers, make educated judgements and improve their readiness measures, which include drought monitoring and proactive management mitigation initiatives. A combination of numerous indicators appears to be more successful in the field of drought inquiry than a single index, according to literature evaluations (Qaisrani et al., 2021). Researchers looking into drought conditions have used the Standardised Precipitation Index (SPI) a lot (Mahessar et al. 2019; Zhang et al. 2015; Jamro et al. 2019; Hina and Saleem 2019; Krakauer et al. 2019; Moghbeli and Delbari 2020; Jasim and Awchi 2020; Qaisrani et al. 2021). These DIs can be calculated using a variety of methods and equations, either manually or with the aid of various models and technologies. This flexibility and adaptability help drought assessment strategies.

Using DrinC (Drought Index Calculator), the DIs for this investigation were calculated. Tigkas et al. (2014) go into detail about the calculations and sequence of steps. To compute DIs, DrinC makes use of an easy-to-understand platform and outcomes. However, a meteorological dataset may have a longer lifetime (at least 35 years) for extraordinary finds. The outputs of deciles, RDI, and SPI are used by DrinC to compute the DIs, making it simpler to assess the severity of droughts and predict the trend in future drought occurrence. DrinC is a useful instrument for drought monitoring and assessment, mapping spatial distribution, and meteorological investigations with the aim of addressing concerns about subsidy.

According to Adnan et al. (2015), Pakistan experiences droughts on average once every three to seven years. Dealing with the concerns of stakeholders, including industries, individuals, and farmers, in the wake of these frequent drought events poses significant challenges for decision-making institutions (Adnan et al. 2015, 2018; Ahmed et al. 2016; Mahessar et al. 2019; Jamro et al. 2019; Hina and Saleem 2019; Krakauer et al. 2019; Jasim and Awchi 2020; Moghbeli and Delbari 2020; Qaisrani et al. 2021).

Different mechanisms have been put in place by governments to declare droughts. However, taking into account the various agroecological zones present in Pakistan, rationalised consistent procedures are required. It's interesting to note that in Pakistan, the entire district is officially recognised as drought-stricken if a farmer plans to make a claim for crop insurance due to harvest failure brought on by water scarcity or drought (Surendran et al. 2014, 2016, 2017; Adnan et al. 2018). Therefore, Drought Indices (DIs) are essential in assisting state authorities in determining the effects of drought. As a result, DIs calculated using software like DrinC have been examined for Pakistan's Punjab province's semi-arid to dry Potwar region. Additionally, these indices can be used in other areas with comparable environmental and climatic characteristics, offering important insights into drought evaluation and management techniques.

2 Methodology

2.1 Study Area

The Potwar Plateau, which spans a sizeable chunk of Pakistan's dry agricultural area, is located between latitudes 32°1 and 35°9 N and 71°11 and 73°56 E (Fig. 1). In Pakistan's Punjab province, this plateau is made up of the four districts of Attock, Chakwal, Jhelum, and Rawalpindi. It significantly contributes to the livestock and agricultural output of the nation. Due to high production costs, unpredictable precipitation, and insufficient knowledge of current machinery and techniques, farmers in this region frequently use low inputs in agriculture (Ahmed et al. 2016). With an annual precipitation range of 950–1950 mm, the Potwar region's climate can be characterised as semi-arid to desert (Adnan and Khan 2009). The southwestern monsoon and the westerlies are often the two main weather phenomena that affect this region (Latif et al. 2017; Ahmed et al. 2016, 2019). Winter (December–March) is when the southwestern monsoon brings rain, while the monsoon (July–September) is when the westerlies have an impact. Additionally, changes in height in the area affect temperatures.

The Potwar region has a wide range of heights, from the highest peaks in the Murree formation to the southern tip of the Hazara Range, where elevations can approach 1200 m above sea level (m ASL). In contrast, the Plains of the Mighty Indus Jehlum River have the lowest altitudes, with a riverbed that drops as low as 300 m. The average heights in the Potwar region range from 300 to 600 m (Adnan et al. 2017).

According to Adnan et al. (2017), the region's mean annual maximum temperature is roughly 26.5 °C, while the minimum temperature is typically 14 °C. In the Potwar region, the two main crop-growing seasons are Rabi (October–April) and Kharif (May–September). Planting takes place for Kharif crops (such as groundnut, rice, maize, moong, soybean, and jowar) in May–June, while harvesting happens in October–November. On the other hand, Rabi are normally sown in November and harvested in March and April. The rural communities in the area depend on these cycles of agriculture.

According to Ahmed et al. (2016), (2019), Batool et al. (2021) and other researchers, the Potwar region is extremely significant in terms of agriculture,

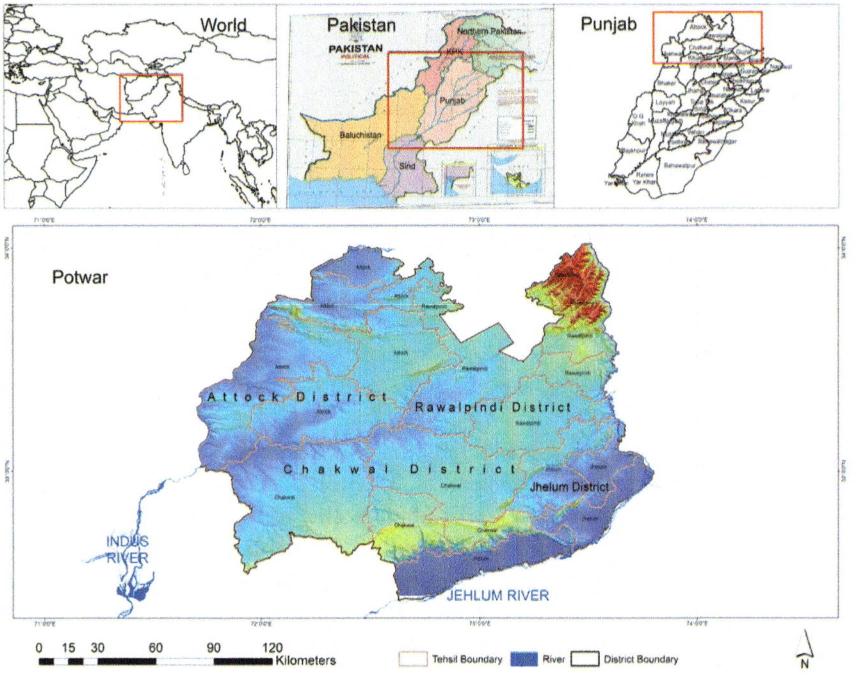

Fig. 1 Location of study area. *Source* Batool et al. (2021)

producing about 10% of the total agricultural output of the region. By lowering moisture pressure and providing the essential water supplies for Rabi crops, the easterly monsoon precipitation plays a significant role throughout the Kharif season (Adnan et al. 2018). The Gross Domestic Product (GDP) of the nation is significantly impacted by the agricultural output of the region, which is highly dependent on variations in rainfall patterns throughout the seasons (Rabi and Kharif). A significant part of the country's economy, the agriculture industry contributes over 23% of the GDP. As a result, variations in rainfall can have a detrimental impact on the GDP, underscoring the importance of reliable water supply for agricultural output and a stable economy.

The primary input datasets used during this inquiry are the lowest and maximum temperatures, daily precipitation, and meteorological ground stations in the Potwar region from TerraClimate (high resolution gridded precipitation and temperature data). TerraClimate data were used as the basis for the computation of DIs like deciles, RDI, and SPI, along with ground dataset verifications whenever possible. DrinC helps with the log-normal technique-based computation of the aforementioned DIs for periods of 12, 9, 6, and 3 months. The following is a brief description of DIs based on their severity.

Table 1 Cataloging of drought scenarios as per SPI (McKee et al. 1993)

SPI or RDI value	Drought category
2.0 or more	Extremely wet
1.5 to 1.99	Severely wet
1.0 to 1.49	Moderately wet
− 0.99 to 0.99	Near normal
− 1.0 to − 1.49	Moderately dry
− 1.5 to − 1.99	Severely dry
− 2 or less	Extremely dry

2.2 *SPI*

SPI is derived from longer-term precipitation observations, transformed, and fitted to a probability distribution (McKee et al. 1993; Tigkas et al. 2014). In order for the resulting SPI to follow a normal distribution, this conversion—which is dependent on monthly precipitation—is frequently not distributed regularly. The SPI index suggests a computed number that may deviate from the longstanding mean values by equating a randomly distributed parameters with a general distribution to multiple standard deviations. The complete procedure was thoroughly described by researchers (Tigkas et al. 2014). When the SPI index is negative it is lower than the mean rainfall, and vice versa. Higher SPI negative results signify a severe dry state. According to SPI, DIs are categorised in Table 1.

2.3 *Reconnaissance Drought Index (RDI)*

Tsakiris et al. (2013) and Tigkas et al. (2014) developed the RDI and used it to monitor and characterise the drought in order to express the water scarcity in a precise manner. RDI is equal to growing precipitation (P) and potential evapotranspiration (PET), where PET is computed and P is measured. The primary condition value (k) for RDI is calculated using time-based k-months rather than ith year, and is expressed mathematically as,

$$a_k^i = \frac{\sum_{j=1}^{k} P_{ij}}{\sum_{j=1}^{k} PET_{ij}}, \quad i = 1(1)\,N \text{ and } j = 1(1)k \tag{1}$$

The variables PETij and Pij indicate the potential evapotranspiration (PETij) and precipitation (Pij) for the jth month of the ith year, respectively, and N is the total number of years in the dataset that is currently available. Where, over various time scales and in a wide variety of spatial locations, k-values adequately track the log normal. RDI was first calculated using Eq. (2) under the assumption and use of a log-normal distribution.

$$\mathrm{RDI}_{\mathrm{st}}^{(\mathrm{i})} = \frac{\mathrm{y}^{(\mathrm{i})-\mathrm{y}}}{\sigma^{\mathrm{y}}} \tag{2}$$

where the arithmetic mean is y, the standard deviation is y, and y(i) is the ln of (ak (i)). In contrast to the regular circumstances of the investigation location, positive RDI indicate wet seasons and negative RDI indicate dry events. According to Tigkas et al. (2014), drought intensity (DI) can be divided into severe, mild, higher, and moderate classes, with corresponding borderline RDI values of (– 1.50 to – 2.0), (– 0.50 to – 1.0), (– 2.0), and (– 1.0 to – 1.50).

3 Results and Discussion

The Potwar plateau is known for its semi-humid and semi-arid climate, with its dry, scorching summers and icy winters. Between 1981 and 2019, there was an average annual rainfall of 1145 mm, with the monsoon season from July to September accounting for nearly 75% of the total. According to the deciles technique, the districts of Rawalpindi, Attock, Jhelum, and Chakwal in the Potwar plateau had eight years of severe drought within a 39-year period. This natural occurrence occurs once in every two to seven years (Fig. 2). These constant years included 1985–1986 and 1988–1992, 1998–2003, 2005, 2018–19, etc. These statistics revealed that, with the exception of a few occasions where it occurred repeatedly, such as the years 1985–1986, 1988–1992, 1998–2003, and 2018–2019, the drought is cyclical, occurring once every two–seven years repeatedly. However, there were no instances of drought from 2006 to 2017 (a 12-year span). Changes in the precipitation patterns and climatic conditions may be the main causes of this. The high number of drought incidents—seven—compared to the twenty-year period's two incidents—over the past forty years is one of the remarkable observations. The SPI demonstrates similar trends to the deciles pattern and follows it (Fig. 3).

In the context of the SPI Index, however, there were only six exceptionally dry years (1984–1989, 1993–94, and 2009–10) and one highly arid class (1999–2003, 2018–19). When viewed through the lens of deciles, the remaining drought years fell into the near-normal to moderately dry category. Tigkas et al. (2014) state that the SPI presented the degree of dryness in relation to deciles, making it simpler to quantify the impact of the drought on drinking water availability and crop yield output, among other things. As a result, SPI is a better index and pointer than deciles since it helps to determine the level of dryness more precisely.

Due to the low incidences of precipitation during this time, compared to the other drought episodes, only 3 out of 39 years correspond to an extremely dry situation. The RDI exhibits the same pattern as the SPI (Fig. 4). However, for SPI, the years 1999, 2000, and 2001 correlate to the extremely dry class and are considered to be under severe drought conditions. The DIs for the drought classes and semi-arid to humid region classes in RDI were equivalent to SPI. SPI and RDI show the degree of dryness in relation to deciles. As a result, SPI is a powerful index that analyses agricultural

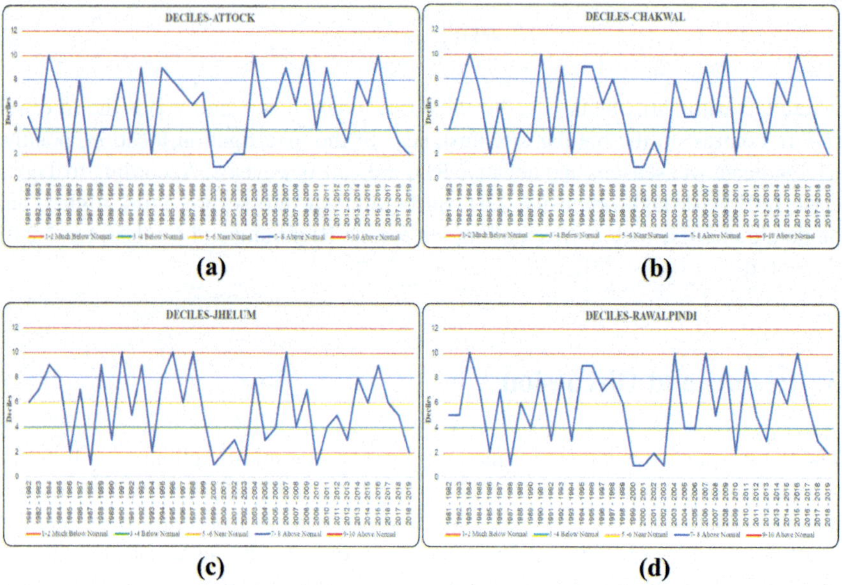

Fig. 2 Decile trends for the study area

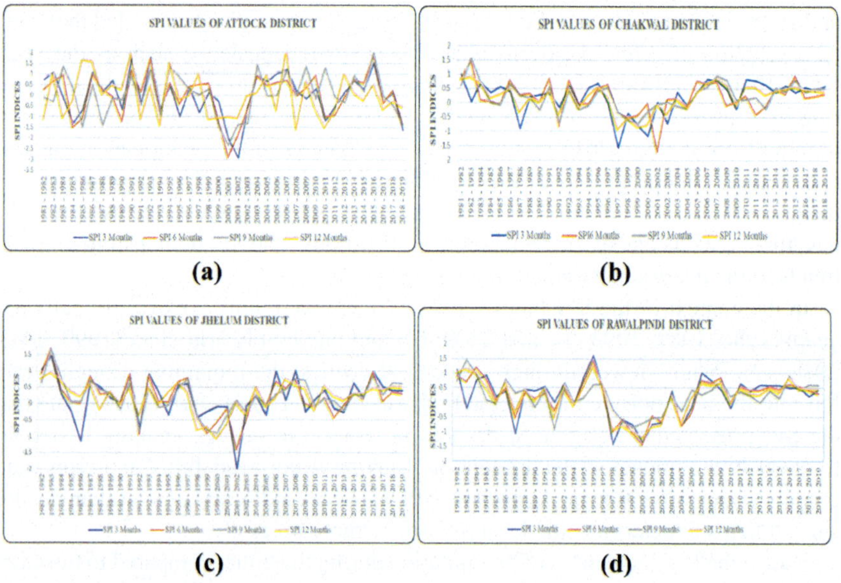

Fig. 3 SPI values for the study area

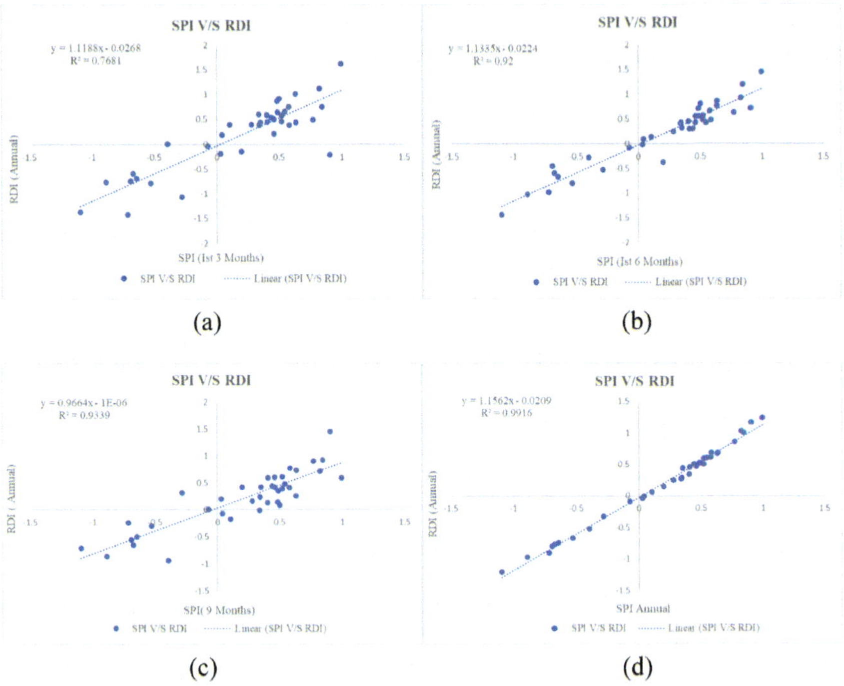

Fig. 8 Regression analysis of the period of three, six, nine, and twelve months for the Rawalpindi district

pertinent sectors that will be directly impacted by drought conditions under the current circumstances.

4 Conclusions

In order to calculate drought indices (DIs), the Drought Index Calculator (DrinC) uses a variety of parameters, including deciles, the Standardised Precipitation Index (SPI), and the Reconnaissance Drought Index (RDI). The Potwar region had severe drought on average every two to seven years throughout a 39-year period, with exceptions in some years like 1985–1986, 1988–1992, 1998–2003, 2005, and 2018–19, according to the decile approach. The RDI and SPI indexes showed similar tendencies, with highly and severely dry periods matching the deciles method. Notably, SPI was discovered to offer a better comprehension of drought severity than deciles. The results of regression analysis revealed a strong relationship between the RDI and SPI indices. It was discovered that annual RDI could be predicted using a 3-month precipitation dataset. These findings highlight the value of DIs in creating preventative plans to lessen the effects of droughts. To successfully address the socioeconomic effects of drought episodes, future development efforts, especially in susceptible places like Pakistan, must take into account the findings. The study also supports the use of satellite-based meteorological indicators for pinpointing regional drought variations, providing crucial information for agricultural drought mitigation and prediction. This concept can be applied to other parts of Pakistan, making it a useful tool for all-encompassing national drought control measures.

References

Adnan S, Khan AH (2009) Effective rainfall for irrigated agriculture plains of Pakistan. Pak J Meteorol 6(11):61–72

Adnan S, Ullah K, Gao S (2015) Characterization of drought and its assessment over Sindh, Pakistan during 1951–2010. J Meteorol Res 29:837–857

Adnan S, Ullah K, Shouting G (2016) Investigations into precipitation and drought climatologies in South Central Asia with special focus on Pakistan over the period 1951–2010. J Clim 29:6019–6035

Adnan S, Ullah K, Gao S, Khosa AH, Wang Z (2017) Shifting of agro-climatic zones, their drought vulnerability, and precipitation and temperature trends in Pakistan. Int J Climatol 37:529–543

Adnan S, Ullah K, Shuanglin L, Gao S, Khan AH, Mahmood R (2018) Comparison of various drought indices to monitor drought status in Pakistan. Clim Dyn 51:1885–1899

Ahmed K, Shahid S, Bin Harun S, Wang XJ (2016) Characterization of seasonal droughts in Balochistan Province, Pakistan. Stoch Environ Res Risk Assess 30:747–762

Ahmed K, Shahid S, Chung ES, Wang XJ, Harun SB (2019) Climate change uncertainties in seasonal drought severity-area-frequency curves: Case of arid region of *Pakistan*. J Hydrol 570:473–485

Anjum S, Saleem M, Cheema M, Bilal M, Khaliq T (2012) An assessment to vulnerability, extent, characteristics and severity of drought hazard in Pakistan. Pak J Sci 64:138–143

Bae S, Lee SH, Yoo SH, Kim T (2018) Analysis of drought intensity and trends using the modified SPEI in South Korea from 1981 to 2010. Water 10(3):327

Batool S, Shirazi SA, Mahmood SA (2021) Appraisal of soil erosion through RUSLE model and hypsometry in Chakwal Watershed (Potwar-Pakistan). Sarhad J Agric 37(2):594–606

Dai AG (2011) Characteristics and trends in various forms of the palmer drought severity index during 1900–2008. J Geophys Res 116:1–26

Deo RC, Byun HR, Adamowski JF, Begum K (2017) Application of effective drought index for quantification of meteorological drought events: a case study in Australia. Theoret Appl Climatol 128(1):359–379

Haroon MA, Zhang J, Yao F (2016) Drought monitoring and performance evaluation of MODIS-based drought severity index (DSI) over Pakistan. Nat Hazards 84:1349–1366

Hina S, Saleem F (2019) Historical analysis (1981–2017) of drought severity and magnitude over a predominantly arid region of Pakistan. Clim Res 78:189–204

Iqbal MF, Athar H (2018) Variability, trends, and teleconnections of observed precipitation over Pakistan. Theoret Appl Climatol 134(1):613–632

Jamro S, Dars GH, Ansari K, Krakauer NY (2019) Spatio-temporal variability of drought in Pakistan using standardized precipitation evapotranspiration index. Appl Sci 9:4588

Jasim AI, Awchi TA (2020) Regional meteorological drought assessment in Iraq. Arab J Geosci 13(7):1–16

Khan MA, Gadiwala MS (2013) A Study of drought over Sindh (Pakistan) using standardized precipitation index (SPI) 1951 to 2010. Pak J Meteorol 9:15–22

Krakauer NY, Lakhankar T, Dars GH (2019) Precipitation trends over the Indus basin. Climate 7(10):116

Latif M, Syed FS, Hannachi A (2017) Rainfall trends in the South Asian summer monsoon and its related large-scale dynamics with focus over Pakistan. Clim Dyn 48(11):3565–3581
Mahessar AA, Qureshi AL, Siming IA, Kori SM, Dars GH, Channa M, Laghari AN (2019) Flash flood climatology in the lower region of Southern Sindh. Eng Technol Appl Sci Res 9:4474–4479
McKee TB, Doesken NJ, Kleist J (1993) The relationship of drought frequency and duration to time scales. In: Proceedings of the 8th conference on applied climatology American Meteorological Society, Boston, 179 (184)
Mishra AM, Singh VP (2010) A review of drought concepts. J Hydrol 391:202–216
Moghbeli A, Delbari M (2020) Application of a standardized precipitation index for mapping drought severity in an arid climate region, southeastern Iran. Arabian J Geosci 13:1–16
Mohammad RK, Majid V, Amin A (2014) Drought monitoring using a soil wetness deficit index (SWDI) derived from MODIS satellite data. Agr Water Manage 132:37–45
Qaisrani ZN, Nuthammachot N, Techato K (2021) Drought monitoring based on standardized precipitation index and standardized precipitation evapotranspiration index in the arid zone of Balochistan province, Pakistan. Arab J Geosci 14(11):1–13
Samo SR, Bhatti N, Saand A, Keerio MA, Bangwar DK (2017) Temporal analysis of temperature and precipitation trends in Shaheed Benazir Abad Sindh, Pakistan. Eng Technol Appl Sci Res 7(6):2171–2176
Spinoni J, Naumann G, Carrao H, Barbosa P, Vogt J (2014) World drought frequency, duration, and severity for 1951–2010. Int J Climatol 34(8):2792–2804
Spinoni J, Vogt JV, Naumann G, Barbosa P, Dosio A (2018) Will drought events become more frequent and severe in Europe? Int J Climatol 38(4):1718–1736
Surendran U, Ramasubramoniam S, Raja P, Kumar V, Murugappan V (2016) Budgeting of major nutrients and the mitigation options for nutrient mining in semi-arid tropical agro ecosystem of Tamil Nadu, India using NUTMON model. Environ Monit Assess 188(4):1–17
Surendran U, Sushanth CM, Joseph GM, E.J. (2014) Modeling the impacts of increase in temperature on irrigation water requirements in Palakkad district—a case study in humid tropical Kerala. J Water Clim Change 5(3):471–487
Surendran U, Sushanth CM, Joseph GM, E.J. (2017) FAO-CROPWAT model-based estimation of crop water need and appraisal of water resources for sustainable water resource management: pilot study for Kollam district—humid tropical region of Kerala, India. Curr Sci 112:76–86
Tigkas D, Vangelis H, Tsakiris G (2014) DrinC: a software for drought analysis based on drought indices. Ear Sci Inf 1–13
Tsakiris G, Nalbantis I, Vangelis H, Verbeiren B, Huysmans M, Tychon B, Jacquemin I, Canters F, Vanderhaegen S, Engelen G, Poelmans L, De Becker P, Batelaan O (2013) A system-based paradigm of drought analysis for operational management. Water Resour Manag 27:5281–5297
Vangelis H, Tigkas D, Tsakiris G (2013) The effect of PET method on reconnaissance drought index (RDI) calculation. J Arid Environ 88:130–140
Vasiliades L, Loukas A, Liberis N (2011) A water balance derived drought index for Pinios river basin, Greece. Water Resour Manag 25:1087–1101
Xie H, Ringler C, Zhu T, Waqas AD (2013) A spatiotemporal variability analysis using the Standardized Precipitation Index. Water Int 38:620–631
Zhang Z, Chen X, Xu CY, Hong Y, Hardy J, Sun Z (2015) Examining the influence of river–lake interaction on the drought and water resources in the Poyang Lake basin. J Hydrol 522:510–521

Dr. Saira Batool is currently working as Assistant Professor at Centre for Integrated Mountain Research (CIMR), University of the Punjab Lahore Pakistan. She is native from the city of Layyah in Southern Punjab Pakistan, she served as lecturer (Incharge) at the department of Geography and was involved in the teaching and research in the subject of Geography at Bahahudin Zakaryiya University (BZU) Multan Pakistan. She did her PhD in Geography (Remote Sensing and GIS applications in Sustanible Development Goals (SDGs) from the Institute of Geography University of the Punjab Lahore Pakistan, with specialisation in Geospatial Applications and machine learning techniques in Climate Change, Food Security and Water Scarcity (food water nexus) and Natural Disaster Management. She has more than 40 national and international publications and conferences proceedings to her credit. She is actively engaged in various national and international Research projects in the aforementioned fields of research. She collaborated with the National Engineering Services Pakistan (NESPAK), currently engaged in "Women in Science" a research capacity building programme, Phase-I (University of Oxford, UK, John Hopkin University, USA, University of Michigan, USA, University of California, USA, University of Amsterdam, Netherland. She is involved in the project of early warning system installation in Pakistan with the collaboration of china MoST and CSU in the framework of belt and road initiative (BARI).

Dr. Syed Amer Mahmood (born December 12, 1968, Lahore), is a Pakistani-born Pioneer Space Scientist, who developed the scientific and special applications in the fields of Spatial Sciences (Remote Sensing, Machine Learning—MATLAB, Google Earth Engine, Climate Engine, R-studio, Python) GIS, Climate Change, Natural Disasters and Sustainable Development Goals (SDGs). Numerous distinctions and awards are in his credit. He did his Graduation in Space Science from Pakistan, in 1993, and joined as Lecturer in the Department of Space Science, University of the Punjab, Lahore, Pakistan. Later on, he proceeded abroad to Germany for his Ph.D., and currently serving as Professor/Chairman of the department of Space Science, University of the Punjab, being the largest and the oldest seat of learning with three Nobel Laureates to its credit. Dr. Mahmood has more than 150 national/international impact factor publications and conference proceedings to his credit. He is actively engaged in national and international research projects. He has more than 200 research supervision and 15 Ph.D.s to his credit. He has collaborated with the National space agency of Pakistan (SUPARCO), National Engineering Services Pakistan (NESPAK), NDMA, PDMA, Urban Unit Lahore, Climate Change Ministry, Forest, Fisheries and wildlife Ministry, Geological and soil surveys of Pakistan, and many Research and development (R & D) institutions both at national and international levels. Currently he is working as principal investigator in the project of early warning system installation in Pakistan with collaboration of china MoST and CSU in the framework of belt and road initiative (BARI).

Dr. Safdar Ali Shirazi graduated in 1987 from the Department of Geography, University of the Punjab (Lahore) with First Class obtained Roll of Honour. He also has a post-graduate diploma in Demography from Social Sciences Research Centre and Degree (Geography and Political Sciences) from the University of Punjab, Lahore. He was awarded PhD degree by the University of Karachi-Karachi in 2011. He completed his Post-Doctorate fellowship (funded by the Higher Education Commission, Pakistan) under the able supervision of Prof. Dr. Dagmar Haase in 2014 from the Institute of Geography, Department of Urban Ecology, Humboldt University, Berlin-Germany. In 2017, he was awarded DAAD Research Stay for Academics and Scientists for 3-month fellowship by the German Academic Exchange Service (DAAD) and he completed his second Post-Doctorate from the Institute of Geography, Humboldt University, Berlin-Germany. His teaching and research experience is more than 35 years. He is Professor of Geography since July, 2017 and Former Director of the Institute of Geography, University of the Punjab, Lahore. His research interests included Climate Change, Landscape Ecology, Ecosystem Services, Morphology of Cities, Desertification and Resilience, Population Studies, Urbanization and Sustainability.

Chapter 13
Managing Agricultural Water Productivity in a Changing Climate Scenario in Indo-Gangetic Plains

Pavneet Kaur Kingra and Surinder Singh Kukal

Abstract Climate change has significant impact on all components of the hydrological cycle. Warming scenarios and increased uncertainty in rainfall behavior may lead to increase in crop water requirements and decrease in its availability for irrigation. As a result, ground water resource is being depleted at alarming rates in many regions of the earth, especially south-east Asian region. Due to increased frequency and intensity of extreme weather events, agriculture has become highly vulnerable to climatic risks. Such scenarios are endangering food security for the burgeoning population along with over-exploitation of natural resources. Hence, most of the research is oriented towards improving/optimising crop water productivity rather than yields. Under such conditions, climate-smart agriculture seems the viable option to manage climate change impacts on water-use efficiency. On-farm water management, rainwater harvesting, groundwater development, advanced techniques of irrigation, breeding for resistance to droughts and floods as well as construction of dams for water storage are some of the practices which are immediately required to manage water scarcity. Improving soil moisture retention, changing cropping calendars, encouraging crop diversification, irrigation management such as deficit irrigation, supplemental irrigation, alternate wetting and drying in rice, etc. are very important on-farm practices for enhancing crop water productivity. In addition to this, some policy measures such as climate proofing structures, reallocation of water among different sectors, and crop insurances are also required to be implemented by governments. Improved weather forecasting can play very crucial role to minimize climatic risks in agriculture. Crop simulation and hydrological modeling are other techniques which can assist in tactical decision making for improving crop water productivity and better management of water resources. In addition to this, remote sensing and geospatial techniques can also be used successfully for improved hydrological monitoring at regional level. Hence, there is a dire need of taking quick

P. K. Kingra (✉)
Department of Climate Change and Agricultural Meteorology, Punjab Agricultural University, Ludhiana, Punjab, India
e-mail: pkkingra@pau.edu

S. S. Kukal
Punjab Water Regulation and Development Authority, Government of Punjab, Chandigarh, India

M. Behnassi et al. (eds.), *The Water, Climate, and Food Nexus*,
https://doi.org/10.1007/978-3-031-50962-9_13

actions to enhance water-use efficiency and save this precious resource for sustaining agriculture and attaining food security in future.

Keywords Climate change · Water use efficiency · Climate smart agriculture · Water management

1 Introduction

Climate change-induced fluctuations in weather patterns along with increased intensity and frequency of extreme weather events are posing major threat to crop yield sustainability (Porfirio et al. 2018). Variability in rainfall patterns have resulted in significant increase in the occurrence of droughts and floods, with adverse impacts on crop yields and water-use efficiency (Mar et al. 2018). Extreme weather events cause enormous damage to crop production. Water deficits pose a serious threat to crop productivity and food security in many parts of the world due to poor or erratic rainfall and depletion of groundwater reserves (Hussain et al. 2019). Extreme weather events such as frost, hail, heatwaves, percentile of precipitation and drought periods affected global food security, limiting rain-fed and irrigated agricultural crop production potential (Bisbis et al. 2019; Eitzinger et al. 2015; Teichmann et al. 2018; Yumurtaci 2015). The sustainability of irrigated agriculture is threatened due to adverse climate change, given future projections that every one in four people on Earth might be suffering from extreme water scarcity by the year 2025 (Nikolaou et al. 2020). Thus, improvements in crop productivity under conditions of limited water availability are vital to meet global food demand (Balyan et al. 2017).

Climate change can affect agriculture through their direct and indirect effects on the crops, soils, livestock, and pests. Although increase in atmospheric carbon dioxide has a fertilization effect on crops with C_3 photosynthetic pathway, and thus promotes their growth and productivity. But increase in temperature can reduce crop duration, increase crop respiration rates, alter photosynthesis process, affect the survival and distributions of pest populations, and thus developing new equilibrium between crops and pests, hastens nutrient mineralization in soils, decrease fertilizer-use efficiency and increase evapotranspiration (Gupta and Pathak 2016). Climate change-led increases in local and global temperatures pose a significant threat to plant growth and crop production (Priya et al. 2019). According to IPCC (2018), global temperatures are likely to further increase by 1.5 °C during 2030–2052. Under such conditions, next-generation crop plants need to be water and nutrient use efficient along with sustainable yields over a wider range of environmental conditions (Pareek et al. 2020). Thus, mitigation strategies, along with the global drivers of agricultural production, need to be adopted to combat the effects of extreme weather events (Dhankher and Foyer 2018; Schewe et al. 2019; Nutan et al. 2020).

India is a water-stressed country with an estimated availability of 1434 m^3 per person per year. Groundwater withdrawal is increasing very rapidly in India, more rapidly than in USA and China, and is about 780 billion cubic meters annually (FAO

2018). 54% of observed groundwater wells in India are reported to be overexploited, and many states show even more exploitation, such as Karnataka (80%), Maharashtra (75%), and Uttar Pradesh (73%). About 60% of the India's districts fall in water-scarce category or suffering from poor water quality (CWC 2019). It may be worth mentioning that water is likely to be a more binding constraint to Indian agriculture than even land, and therefore it is high time to change the mindset from raising agricultural productivity per unit of land to per unit of water (Sharma et al. 2018). Water is an essential part of all components of the environment and climate system. The dynamics of the water cycle are one of the key variables that determine the distribution and productivity of ecosystems (Stagl et al. 2014). The water cycle is a key process upon which other cycles of the climate system operate. It acts as an energy transfer and storage medium through the hydrological cycle. Globally, changes in water vapour content of the atmosphere, cloud cover and ice influence the radiation balance of the earth, and thus play an important role in determining the climate response to increasing greenhouse gas emissions (Bates et al. 2008). Hence, changes in climate are intricately interlinked with changes to the hydrological cycle—the most important feedback cycle in the climate system (Fig. 1).

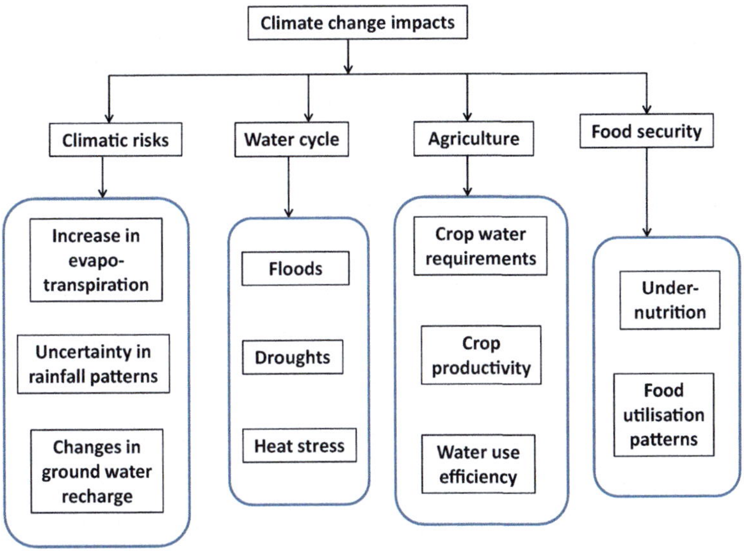

Fig. 1 Climate change impacts on water cycle, agriculture, and food security

2 Climate Change and Water Cycle

The hydrological cycle, also called water cycle, describes the continuous circulation of water between ocean, atmosphere, and land (Stagl et al. 2014). The impact of global warming on the water cycle can be extremely complex and diverse (Georgiardi 1991). The threat of climate change has increased interest in climate research, which focuses on observing, understanding, and modelling the five interconnected components of the climate system, i.e. atmosphere, oceans, land surface, cryosphere, and biosphere. Feedbacks at a variety of spatial and temporal scales shape the system's overall behaviour, and the hydrological cycle plays a critical role in this respect via two important feedback processes that involve water vapour and clouds (Davies and Simonovic 2005). Climate change can affect the amounts of soil infiltration, deeper percolation, and hence groundwater recharge. Also, rising temperature increases evaporative demand over land, which limits the amount of water to replenish groundwater (Berg et al. 2016). Over 60% of the world's food is produced on rain-fed farms that cover 80% of the world's croplands. In South Asia, where millions of smallholders depend on irrigated agriculture, climate change will drastically affect riverflow and groundwater, the backbone of irrigation and rural economy (Nellemann et al. 2009).

As air warms above the freezing point, precipitation occurs in the form of rain. However, the water holding capacity of air increases by 6–7% for each 1 °C rise in temperature. The climate and human activities are two factors that can affect the hydrological cycle in different ways. Climate change alters hydrological systems by inducing both spatio-temporal variations of regional precipitation and changes in temperature (Pereira et al. 2016; Silva et al. 2010; Woldesenbet et al. 2018). Compared with climate change, human activities are more controllable; thus, the alteration of human activities constitutes the principal measure for dealing with the potential impacts of climate change on hydrological systems (Wang et al. 2019). The hydrological cycle is expected to intensify with global warming, which likely increases the intensity of extreme precipitation events and the risk of flooding (Tabari 2020) (Table 1).

The hydrological cycle is one of the main components of the planetary system regulating human, animal, and plant life. This cycle also forms the foundation of other cycles, such as carbon cycle, nitrogen cycle, etc. Therefore, the stability of water cycle is critical for the sustainability of biological populations and ecosystem. Empirical observations allude that the stability of the hydrological cycle is being threatened by climate change (Sohoulande and Singh 2016). Changes in the hydrological cycle have become more frequent in the terrestrial ecosystems. The evidence of these changes is expressed by the abnormal frequency of water crisis and hydrologic hazards whose effects extend to economic, social, political, and cultural sectors. The alteration of the water cycle is an alarming threat for the stability and sustainability of human societies and natural ecosystems (Hanjra and Qureshi 2010).

Table 1 Climate change impact on hydrological cycle

Parameter	Effect	Reference
Rise in temperature	Increase in evaporative demand	Berg et al. (2016)
1 °C increase in temperature	Increase in water holding capacity of air by	Pereira et al. (2016)
Global warming	Increase in intensity of extreme precipitation events and flooding	Tabari (2020)
Increase in temperature, wind speed and sunshine hours	Increase in reference evapotranspiration	Goyal (2004), Wen et al. (2012), Kingra (2018)
Increase in atmospheric humidity	Decrease in reference evapotranspiration	
Increase in temperature by 1–1.5 °C	Increase in reference evapotranspiration by 4–5.4%	Hamouda et al. (2020)
Elevated CO_2 (550 ppm)	Reference evapotranspiration between − 1.1 to 2.2%	
Elevated temperature	Effect on ground water quality	Kurylyk et al. (2014), Possemiers et al. (2014)
Increased frequency of extreme weather events	Imbalances in hydrological cycle	Kaur et al. (2017)

2.1 *Increase in Evapotranspiration*

Evapotranspiration (ET) is one of the most important components of hydrologic budget, which is commonly calculated from records of climate data and geographical attributes of a meteorological station. Any change in climatic conditions will have a profound effect on crop evapotranspiration and future water resource planning (Singh 2010; Kingra 2018). The rate of ET depends on energy and water availability at the evaporating site and the vapour pressure gradient. These three conditions are highly affected by the climate change, as a result of which the rate of ET changes (Sarkar and Sarkar 2018; Kaur et al. 2017). Climate change has exerted significant impacts on eco-hydrological patterns (Portprato et al. 2002; Han et al. 2014), which in turn affect ET, causing a series of water resource problems (Li et al. 2007; Mcvicar et al. 2008). ET_o has been observed to have positive correlation with average temperature, wind speed, sunshine hours, and negative with atmospheric humidity (Goyal 2004; Wen et al. 2012), which indicates that rate of ET is likely to increase under warming scenarios, which will alter the amount of water available to plant roots, and thus the amount that is available for transpiration.

The continuous increase of atmospheric CO_2 content, mainly due to anthropogenic CO_2 emissions, is causing a rise in temperature on earth, altering the hydrological and meteorological processes and affecting crop physiology. Thus, understanding the change in evapotranspiration due to global warming is essential for better water resources planning and management and agricultural production. Simulation results indicated that, without increasing CO_2 levels (372 ppm), ET_o increased by 4–5.4%

with regard to the reference period 1981–2005, for an increase of air temperature by 1–1.5 °C. However, under the effect of elevated CO_2 levels (550 ppm) ET_o demand for all locations varied from—1.1 to 2.2% during the 2021–2050 period with regard to the reference period 1981–2005 (Hamouda et al. 2020). This shows that higher CO_2 levels moderated the increase in ET_o that accompanies an increase in air temperature. Islam et al. (2012) also showed a decrease in ET_o demand with increases in CO_2 levels, which greatly moderated the increase in ET_o due to increasing temperature. The effect of increases in CO_2 levels up to 450 ppm offset the effect of about 1 °C rise in temperature. Simulation results with projected climate change scenarios, without considering the effects of CO_2 levels, showed an 8.3, 14.7 and 21.0% increase in annual ET_o during the 2020s, 2050s, and 2080s, respectively, when simulation was carried out using an ensemble of the 112 projections. When the effect of elevated CO_2 levels was also considered in combination with projected changes in temperature, changes in annual ET_o demand varied from – 1.5% to 5.5%, – 10.4% to 6.7%, and – 19.7% to 6.6% during the 2020s, 2050s, and 2080s, respectively, depending on the different climate change scenarios considered and the relationship or equation used for estimating the effect of elevated CO_2 on stomatal resistance term in the Penman–Monteith equation.

Snyder et al. (2011) also reported that while evapotranspiration rates are known to increase with higher temperature, increasing humidity and higher CO_2 concentrations both tend to reduce transpiration and counteract the higher temperature effects on ET. As the oceans and other water bodies warm and evaporate more water into the atmosphere, global humidity is likely to increase. As CO_2 concentrations increase, leaf stomata partially close in response to maintain the CO_2 concentration inside the stomata. Thus, while climate change is likely to increase air temperature, the effect of higher humidity and CO_2 concentration could partially offset the temperature effect on ET. Thus, for efficient and sustainable water management in future, there is an urgent need to assess ET losses at regional level under changing climatic scenarios.

Agricultural crop production requires substantial amounts of water. It has been observed that 2497 L of water are required to produce 1 kg of rice (Rahaman et al. 2016). Therefore, the development of improved rice genotypes with increased water-use efficiency is essential without compromising yields (Shahane et al. 2019), and this becomes more important under climate change and global warming scenarios. Global warming may alter not only temperature, but cloudiness, windiness, and humidity also. All these factors determine the atmospheric demand for water vapour i.e. evapotranspiration. The changes in temperature are likely to have a profound effect on different components of the hydrological cycle viz., precipitation, evapotranspiration, soil moisture, etc. (Kingra and Kukal 2013). ET, being the major component of hydrological cycle, will affect the crop water requirements and future planning and management of water resources (Goyal 2004). The increase in temperature, especially during the vegetation period and the large number of years in which evapotranspiration quantitatively exceeds the precipitation, indicates the need for effective measures to regulate the water balance (Smuleac et al. 2020).

2.2 Uncertainty in Rainfall Patterns

Climate change and global warming are resulting in increased variability in global rainfall patterns. Changes in the frequency and intensity of precipitation directly affect the magnitude and timing of runoff and the intensity of floods and droughts (IPCC 2007). Due to highly erratic rainfall, there is an increased risk of drought as a result of increased prolonged dry spells, total dry days and decreased light precipitation days over India as a consequence of global warming (Mishra and Liu 2014). Erratic precipitation patterns are expected to enhance year-to-year yield fluctuations because of increased frequency and intensity of droughts and floods (Kingra and Kaur 2017). Tabari (2020) showed an intensification of extreme precipitation and flood events over all climate regions which increased as water availability increased from dry to wet regions. Similarly, there was an increase in the intensification of extreme precipitation and flood with the seasonal cycle of water availability. The connection between extreme precipitation and flood intensity changed and spatial and seasonal water availability became stronger as events became less extreme. Luhunga et al. (2018) observed variable effect of climate change on future rainfall patterns in Tanzania. Rainfall over parts of north-eastern highlands and coastal regions has been projected to increase in the range of 0.5 to 1 mm/day and 0.25 to 0.5 mm/day under RCP 8.5 and RCP 4.5 emission scenarios respectively. However, the western regions, south-western highlands and eastern side of Lake Nyasa are likely to experience decreased amount of rainfall in the range of 0.5 to 1 mm/day under both RCP 8.5 and RCP 4.5 emission scenarios.

Variation in rainfall patterns is one of the major impacts of climate change, which directly or indirectly affects the regional water sources that are rain fed/recharged (Udayashankara et al. 2016). Adefisan (2018) concluded that the higher the emission of greenhouse gases, the higher is the temperature which leads to warmer future and most likely the more rainfall, and hence likelihood of flooding, more occurrence of heat wave and other high temperature related problems. It therefore recommended that IPCC regulation to reduce emission should be strictly adhered to by all countries so that the world can have a better future to dwell in. Sipayung et al. (2018) also reported that high temperature triggered changes in rainfall patterns. Al-Ansari et al. (2014) employed two emission scenarios, used by the Intergovernmental Panel on Climate Change (A2 and B2) to study the long-term rainfall trends in northwestern Iraq and observed that all seasons consistently projected a drop in daily rainfall for all future periods with the summer season expected to have more reduction compared to other seasons. Generally, the average rainfall trend showed a continuous decrease, and they suggested the adoption of prudent water management strategies to overcome or mitigate consequences of future severe water crisis.

Climate change increases the frequency and intensity of extreme rainfall because a warmer atmosphere holds more water vapour that can rain out, sometimes over a short period. The movement of water vapour through the atmosphere, in storms, is also modified. Increases in extreme rainfall have been observed in many parts of the world. Extreme rainfall, in turn, can increase the chance of floods occurring and their magnitude in small and in urban catchments, severely impacting local populations and infrastructure. Extreme rainfall and associated flood hazards are projected to increase as global temperatures continue to rise (Blenkinsop et al. 2021). Nandargi and Barman (2018) revealed that spatio-temporal analysis of rainfall was influenced by climate change, especially 1990 onwards showing increase in percentage of dry years in Gangetic West Bengal (75–97%) although increase in rainfall is seen during 2001–2016. Even southern districts of sub-Himalayan West Bengal showed increase in dry year's percentage (69–84%), which is serious from agricultural as well as water resources planning point of view.

2.3 Changes in Ground Water Recharge

Groundwater provides critical freshwater supply, particularly in dry regions where surface water availability is limited. However, climate change impacts on groundwater storage could affect the sustainability of freshwater resources (Wu et al. 2020). Climate change will have both quantitative and qualitative effects on groundwater resources (Epting et al. 2021). Erratic rainfall patterns have direct effect on ground water recharge. Climate change induced altered rainfall distribution in the form of increased intense rainfall events is adversely affecting ground water recharge, especially in the trans-Gangetic plains of India. As a result, ground water withdrawal has exceeded recharge since last many decades, which might force the region towards desertification and might pose severe threat to food security in future. A report of Irrigation Department of Punjab on sub-soil water indicated that the average fall in water table in central Punjab was 20 cm per annum from 1980 to 1990, 25 cm from 1990 to 2000, 75 cm from 2005 to 2008, 45 cm from 2008 to 2013, and 70 cm in 2014–15. The National Aeronautics and Space Administration (NASA) has also confirmed that the northern region of India is rapidly losing its ground water. They have warned that the sub-soil water is depleting in Punjab, Haryana and U. P. states.

In the western Indo-Gangetic Plains (IGP), water is increasingly becoming scarce because agriculture is facing rising competition from urban and industrial sectors (Toung and Bhuiyan 1994). Groundwater is widely used to irrigate the plain's summer crops due to the deficit in surface-water resources. Widespread use of pressurized irrigation systems in the plains results in excessive groundwater withdrawals and, consequently, a couple of meter drops in groundwater levels each year. Such drops in groundwater levels also raise the energy costs of irrigation. Water shortages are not only a serious environmental problem but a limiting factor affecting normal crop growth and yield formation. Furthermore, the cost of pumping for irrigation, inadequate capacity for irrigation engineering and limited water sources are among the factors forcing many farmers to reduce irrigation. The two small north-west Indian states of Punjab and Haryana, located mostly in the Indus Basin, provide about half the rice and 85% of the wheat procured by the Indian government. The rice–wheat system in this region is almost entirely dependent on groundwater for irrigation (Ambast et al. 2006); however, current rates of groundwater extraction are not sustainable. The green revolution in the 1960s and 70s led to a huge increase in ground water pumping as rice and wheat production expanded and intensified. As a result, water tables declined and continue to decline at an alarming rate. The falling water table incurs greater pumping costs for farmers (Humphreys et al. 2010).

It has been observed that reduced monsoon rainfall in north India due to Indian ocean warming has led to reduced groundwater storage and increased usage of ground water for irrigation. Groundwater withdrawals in the country have increased over tenfold since 1950s from 10–20 cubic km per year in 1950 to 240–260 cubic km per year in 2009 (Prasad 2017). As a result, there is high risk of severe water stress in much of the area in Asian countries, where about half of the world's population is residing. Studies have predicted that during the next 35 years, about 1 billion more people becoming water stressed in Asia as compared to the present conditions (Charles 2016). In addition to this, elevated temperatures have complex effects on groundwater quality, including biological, chemical, and physical aspects (Kurylyk et al. 2014; Menberg et al. 2014; Possemiers et al. 2014).

3 Climatic Risks in Agriculture

Abiotic stresses are one of the major constraints to crop production and food security worldwide. The situation has aggravated due to the drastic and rapid changes in global climate. Heat and drought are undoubtedly the two most important stresses having huge impact on growth and productivity of the crops. It is very important to understand the physiological, biochemical, and ecological interventions related to these stresses for better management. A wide range of plant responses to these stresses could be generalized into morphological, physiological, and biochemical responses. Crop growth and yields are negatively affected by sub-optimal water supply and abnormal temperatures due to physical damages, physiological disruptions, and biochemical changes. Both these stresses have multi-lateral impacts and therefore, complex in mechanistic action. A better understanding of plant responses to these stresses has pragmatic implication for remedies and management (Fahad et al. 2017).

3.1 Floods

Altered rainfall distribution patterns have resulted in increased intense rainfall events. Average precipitation has also increased over many regions of earth. These changes along with warming scenarios have resulted in melting of polar ice and glaciers leading to rise in sea levels and floods over various regions (IPCC 2014). The increase in temperature, variations in precipitation, and changes in the frequency of extreme events increase the probability of flood occurrences and change the total and seasonal water supply, among other impacts (Parry et al. 2007). Several factors, including increasing impervious areas and land-cover degradations, significantly reduce infiltration. Base-flow and runoff are then affected and subsequently the risk of flood increases. Floods are damaging to the society and the natural ecosystem. The consequences of floods include crop losses, population displacement, water impairment, poor air quality, diseases expansion, wildlife mortality, and food and energy crises (McLeman and Smit 2006; Green 2004).

3.2 Droughts

Drought is a recurring natural hazard (Wilhite and Smith 2005) that can lead to widespread damage to agricultural production. Drought has significant harmful effects on socio-economic, agricultural, and environmental conditions. High and intense water-scarcity occurs in a region due to inadequate precipitation, over-exploitation of water resources or high evapotranspiration or combination of all these parameters resulting in conditions favourable for drought (Sesha et al. 2016). According to India Meteorological Department (IMD), if a meteorological subdivision receives total seasonal rainfall of less than 75% from the normal, then it is considered as affected by drought. There are many types of drought depending on the occurrence of rainfall and soil moisture stress, these are: (i) Meteorological drought, when rainfall deficiency is more than 25% than normal rainfall over an area or region; (ii) Agricultural drought, when there is stress and wilting in crop, indicating root zone soil moisture deficit which affects crop yield; and (iii) Hydrological drought, when there is depletion of ground water level resulting in drying of streams, lakes, rivers, etc. (Mishra and Singh 2010). Various methods and indices have been developed by various scientists for drought analysis, quantifying drought severity and yield impacts using many drought-causative and responsive parameters namely rainfall, soil moisture, vegetation condition, potential evapotranspiration, ground and surface water levels, etc.

Drought has become one of the major constraints to agricultural development, particularly in areas that lack water (Zhao et al. 2020). It is estimated that cultivation on the earth is only possible on 16% of the potentially arable area due to limited availability of water (Alexandratos and Bruinsma 2012). Large-scale climatic variations in the recent period have posed severe implications on rainfall patterns, and hence drought occurrence. The main reason for the occurrence of drought and flood is due to variability of distribution of rainfall with respect to space and time. The drought history suggests that India is highly vulnerable to drought due to its monsoonal climate and the inherent spatial and temporal variability of rainfall, which has further aggregated due to significant climatic changes in the recent past and are likely to be severely affected under future climatic scenario. Studies using Palmer Index suggest that there is an increased risk of drought occurrence due to prolonged dry spells and low precipitation days over India. The variability in summer monsoon rainfall in

India has been observed to be closely related to the variations in sea surface temperature over the equatorial Pacific and Indian Oceans (Gadgil et al. 2003), which might increase the drought occurrence in future.

3.3 Heat Stress

Heat waves or extreme temperature events are projected to become more intense, frequent and last longer in future than what is being currently observed (Meehl et al. 2007). The period from 1983 to 2012 was likely the warmest 30-year period of the last 1400 years in the Northern Hemisphere. The globally averaged combined land and ocean surface temperature data as calculated by a linear trend show a warming of 0.85 (0.65–1.06) °C over the period 1880–2012 (IPCC 2014). Heat stress can impair all stages of plant growth from germination to reproduction, limiting the productivity of major staple food crops (Hussain et al. 2019). All these changes are likely to have severe implications on agricultural production especially in the tropical and sub-tropical regions, including India. Hatfield and Prueger (2015) reported that warmer temperatures and more extreme temperature events will significantly impact plant productivity. The ensuing heat stress (HS) severely impacts plant growth, endangering ecosystem quality and world food security. Plant growth, physiological processes and final amount of edible products are affected by HS to an extent that reflects the physical damages, physiological commotions, and biochemical alterations incurred at various growth stages. Therefore, a better understanding of plant behaviour in response to HS has pragmatic implications for devising countermeasures, alleviation strategies, and for acknowledging the differences between HS and the companion drought stress (Hassan et al. 2020).

Along with the mean temperature increasing, the occurring frequency of extreme temperature may increase, that may abruptly affect the crop activity (Wu et al. 2006). Usually, air warming will accelerate the crop development, alter the phenological period and enhance the maintenance respiration; on the other hand, atmospheric CO_2 enrichment will increase the leaf photosynthetic rate and reduce transpiration simultaneously, adding additional carbon to the ecosystems, and hence leading to changes in the cycling of water, nutrients, and energy balance (Fuhrer 2003). Due to complex interactions between climatic and other environmental factors, the impacts of climate change on agricultural ecosystems may be interacted under diversified agronomic practices. For example, the responses of wheat yield to global warming are different between rain-fed and irrigated conditions, and between well and less fertilized conditions (Tubiello et al. 2000). High temperature is a major determinant of wheat development and growth and causes yield loss in many regions of the world (Joshi et al. 2016).

As global land surface temperature continues to rise and heatwave events increase in frequency, duration, and/or intensity, our key food and fuel cropping systems will likely face increased heat-related stress (Moore et al. 2021). Even short episodes of heat stress can reduce crop yield considerably causing low resource-use efficiency

(Siebert et al. 2014). Heat waves are predicted to increase in frequency and duration in many regions as global temperatures rise. These transient increases in temperature above normal average values will have pronounced impacts upon the photosynthetic and stomatal physiology of plants. Heat stress induced a decline in the photosynthetic capacity of the olives consistent with reduced ribulose-1,5-bisphosphate carboxylase/ oxygenase (RubisCO) activity. Damage to photosystem II was more apparent in plants subject to water deficit. In contrast to previous studies, higher temperatures induced reductions in stomatal conductance. Heat stress adversely affected the carbon efficiency of olive (Haworth et al. 2018). Heat stress is one of the major threats to wheat production in many wheat-growing areas of the world as it causes severe yield losses at the reproductive stage (Riaz et al. 2021).

Heat stress has a tremendous impact on crop growth and productivity due to its direct and indirect effects on water. Therefore, it is important to understand the morphological, physiological, biochemical, molecular, and ecological basis of heat stress in crop plants to minimize its impact on food production. Several genes alter their expression, proteins, and transcription factors to cope with heat stress. Adverse effects on macromolecules and genes change the signalling pathways for heat stress resistance. In the early phase of heat stress, plants regulate their immunity and plant nutrition management to alleviate the heat stress. However, all the self-management of plants gets fail in severe heat stress situation. Thus, a combination of conventional breeding and new technologies is required to enable plants to cope efficiently with heat stress along with maintaining/enhancing yield. A better understanding of plant responses to heat stress has practical implications for remedies and management (Verma et al. 2020).

4 Climate Change and Crops

Crop yield is reduced by heat and water stress and even more when these conditions co-occur (Luan and Vico 2021). Drought and heat are major abiotic stresses that reduce crop productivity and weaken global food security, especially under the current and growing impacts of climate change and increases in the occurrence and severity of both stress factors. Although plants have developed dynamic responses at the morphological, physiological and biochemical levels allowing them to escape and/or adapt to unfavorable environmental conditions, but, even the mildest heat and drought stress negatively affects crop yield (Table 2). Further, several independent studies have shown that increased temperature and drought can reduce crop yields by as much as 50%. Response to stress is complex and involves several factors, including signalling, transcription factors, hormones, and secondary metabolites. The reproductive phase of development, leading to the grain production is shown to be more sensitive to heat stress in several crops. Advances coming from biotechnology including progress in genomics and information technology may mitigate the detrimental effects of heat and drought through the use of agronomic management practices and the development of crop varieties with increased productivity under stress (Lamaoui et al. 2018).

Effects of higher temperature, elevated CO_2 concentration and changed precipitation are complicated (Walker and Schulze 2006). However, CO_2 fertilization can alleviate the effects of temperature and precipitation on crop yield up to some extent. Increment of atmospheric CO_2 has an obvious positive effect on photosynthetic rates, leading to enhancement of total biomass and yield of C_3 crops (de Costa et al. 2006). All these changes are having significant impact on crop and water

Table 2 Effect of abiotic stresses on crop growth and yield

Parameter	Effect	Reference
Abiotic stresses	Plant physiology, morphology and biochemical responses	Fahad et al. (2017)
Heat stress	Impairs all plant growth stages	Hussain et al. (2019)
	Physical damages, physiological commotions and biochemical alterations	Hassan et al. (2020)
	Considerable reduction in crop yield and resource use efficiency	Siebert et al. (2014)
	Adverse effect on carbon efficiency of olive, decline in photosynthetic capacity	Haworth et al. (2018)
Heat stress and water deficit	Damage to Photosysthem-II, Reduction in stomatal conductance	Haworth et al. (2018)
	Reduction in crop yield	Luan and Vico (2021)
Warmer and more extreme temperature events	Crop activity and plant productivity	Hatfield and Prueger (2015), Wu et al. (2006)
High temperature	Loss in wheat yield	Joshi et al. (2016), Riaz et al. (2021)
Heat wave	Increase in heat stress in cropping systems	Moore et al. (2021)
Increased frequency of extreme weather events	Fluctuations in crop yields	Kingra (2016a, b)
Sudden rise in temperature in March	Significant wheat yield reduction in Indo-gangetic plains	Gupta et al. (2010)
Warming trend	Reduction in winter dormancy period in northern China	Xiao et al. (2012)
Increase in temperature	Accelerated crop phenology, reduced wheat yield	Tao et al. (2014)
Peaks of high temperature	Drastic reduction in productivity of food crops	Teixeira et al. (2013)
Increase in minimum temperature by 1 °C	Wheat yield reduction by 7%	Rao et al. (2015)

productivity, especially in the tropical and sub-tropical regions. Large variations have been observed in wheat productivity during recent years. Emission of greenhouse gases from fossil fuel combustion and land use/cover changes have propelled the global climate change, which appears as a widespread rising of surface air temperatures, alteration of precipitation patterns and global hydrologic cycle, and increased frequency of severe weather events, such as drought spells and flooding. In many regions, agricultural crops are sensitive to climate change (Lobell and Field 2007).

Since land allocated for crop production cannot be increased, the ability to increase water-use efficiency is considered to be the most important tool for increasing crop production and saving water and the environment. A reasonable irrigation schedule is a key factor to help farmers increase their crop yields and save water, especially in water-deprived regions (Liu et al. 2013). A major challenge for sustainable agricultural production on irrigated croplands in Central Asia is to increase water productivity, which has been falling for decades due to high water application rate. Punjab is facing dual challenge of weather variability and over-exploitation of its ground water resources with significant impact on crop productivity in the region. A better understanding of how crops respond to increasing temperatures and limiting water availability is essential for adopting the farming practices and crop breeding to mitigate the negative effects and even taking advantage of climate change (Hoffman and Sgro 2011).

4.1 Crop Water Requirements

The potential impact of climate change on agriculture is impressible in sub-humid and semi-arid regions (Thomson et al. 2006; Tao et al. 2003). In future, evapotranspiration and water-use efficiency of crops will alter with climate change (Thomas 2008; Mo et al. 2007). Crop water requirement and the temporal and spatial changes of this important characteristic provide key information for irrigation scheduling, water resource planning, and future decision-making. The impact of weather and climate variability and change is more remarkable in the arid and semi-arid regions. The increased frequency of extreme weather events (Kaur et al. 2016) has started creating imbalances in the hydrological cycle and is resulting in large year-to-year fluctuations in crop yields (Kingra 2016a, b), and hence water productivity (Kingra and Kukal 2013) during the recent past.

4.2 Crop Productivity

Climate change induced heat and water stress has adverse impact on crop productivity, especially in the tropical and sub-tropical areas. A sudden rise in temperature during March 2010 caused significant wheat yield reductions over the Indo-Gangetic Plains (IGP) (Gupta et al. 2010). The observed warming trend in the Loess Plateau caused a delay in the onset of winter dormancy (dormancy is mainly influenced by temperature) and advanced the green-up after winter dormancy, resulting in a reduced winter dormancy period (− 3.1 days/decade). Xiao et al. (2012) observed decline in the winter dormancy period by − 2.5 days/decade in the North China Plain. They also reported that from 1981 to 2009, climate warming in the North China Plain caused the dates of green-up after winter dormancy, anthesis, and maturity of winter wheat to occur in average of 1.1, 2.7 and 1.4 days earlier per decade, respectively.

Samra et al. (2012) studied the role of temperature in regulating the wheat productivity in India and as an example, analyzed the yields of Ludhiana district, Punjab. Cold wave conditions that prevailed during *rabi* season of 2010–11 and 2011–12 coincided with flowering and seed formation stage of wheat. Over a 12-year period, 8 years were normal, two each were with heat and cold waves. Their spectral density analysis indicated that temperature during wheat growing seasons of 2010–11 and 2011–12 were significantly lower than normal. They noticed an average yield loss of 217 kg/ha (4.5%) during a heat wave year and a gain of 356 kg/ha (7.4%) during a cold wave year. Between the two continuous cold wave years, productivity gain in the Punjab state, in the relatively colder year (2011–12) was higher by 400 kg/ha. Peaks of high temperature, even when occurring for just a few hours, can reduce the production of important food crops drastically (Teixeira et al. 2013). The global average temperature change is 0.13 °C per decade, and this increasing temperature accelerates the crop phenological development, which could potentially impact the wheat production (Tao et al. 2014).

Rao et al. (2015) reported that wheat yield has become more sensitive to minimum temperature, especially during post-anthesis period. Mean wheat yields for the period 1980–2011 declined by 7% (204 kg/ha) for a 1 °C rise in minimum temperature. Exposure to continual minimum temperature exceeding 12 °C for 6 days and terminal heat stress with maximum temperature exceeding 34 °C for 7 days during post-anthesis period were observed as the other thermal constraints in achieving high productivity. The observed dates of sowing, emergence and beginning of winter dormancy were delayed by an average of 1.2, 1.3 and 1.2 days per decade, respectively. Conversely, the dates of green-up (re-growth after winter dormancy), anthesis, and maturity advanced by an average of 2.0, 3.7 and 3.1 days per decade, respectively. Additionally, the growth duration (sowing to maturity), overwintering period, and vegetative phase (sowing to anthesis) shortened by an average of 4.3, 3.1 and 5.0 days per decade, respectively. The changes in phenological stages and phases were significantly negatively correlated with a temperature increase during this time. Differently to most other phase changes, the reproductive phase (anthesis to maturity) prolonged by an average of 0.7 day per decade (Asseng et al. 2015).

4.3 Water-Use Efficiency

Zhang et al. (2015) suggested crop WUE to be an intrinsical system sensitive to climate change and agronomic measures. They opined that crops tend to reach maximum WUE (WUE_{max}) in warm-dry environment while minimum (WUE_{min}) in warm-wet environment with a difference between WUE_{max} and WUE_{min} being in the range of 29.0 to 55.5% in semi-arid area of northern China. Changes in temperature and precipitation in the past three decades jointly enhanced crop WUE by 8.1%–30.6%. Elevated fertilizer and rotation cropping would increase crop WUE by 5.6–11.0% and 19.5–92.9%, respectively. These results indicated that crop has the resilience by adjusting WUE, which is not only able to respond to subsequent periods of favorable water balance but also to tolerate the drought stress, and reasonable agronomic practices could enhance this resilience. However, this capacity would break down under the impacts of climate change and unconscionable agronomic practices (e.g. excessive N/P/K fertilizer or traditional continuous cropping).

How plants will respond to changes in temperature, precipitation, and carbon dioxide (CO_2) that affect their WUE, is highly uncertain under future climate change scenario. At the leaf level, increasing CO_2 increases WUE until the leaf is exposed to temperatures exceeding the optimum for growth (i.e., heat stress) and WUE begins to decline thereafter. Leaves subjected to water deficits (i.e. drought stress) show varying responses in WUE (Hatfield and Dold, 2019). Zwart and Bastiaanssen (2004) using a large set of experimental data, found the relationship between ET and yield for winter wheat was not straight forward ($R^2 = 0.35$), but this result integrated data from many agro-climatic zones where potential ET was quite different. Any relationship between transpiration and biomass depends on the value of potential ET; for example, achieving maximum yield in Egypt requires higher ET than maximum yield for the same crop in the UK, and a relationship between yield and ET that mixes data from both countries would not be particularly meaningful.

The results from Free Air Carbon Enrichment (FACE) experiments show that the stimulation of grain yield by CO_2 enrichment is lower than expected (Long et al. 2006). The increased air temperature and changed precipitation pattern will significantly affect crop phenological process and stomatal conductance, which lead to alteration of yield and water-use efficiency (Kattge and Knorr 2007). This discrepancy is possibly related to a fact that the crop models usually predict with non-limited supply of water and nutrition and near optimum temperature for crop growth. Usually, assessment of climate change impact is intended to seek the adaptation measures that may be the choices to mitigate the negative feedbacks to agro-ecosystems, to maintain and even increase the crop yields under future scenarios. These measures include selection of the most favourable crops, guidance for new cultivars breeding, and application of dynamic cropping (Hanson et al. 2007).

Crop water productivity (CWP) of winter wheat was calculated and analyzed in the plain of Hai Basin in northeastern China. The average CWP of winter wheat (*Triticum aestivum* L.) in the basin for 2003–2009 was 1.049 kg m^{-3}, with CWP values across the basin ranging between 0.7 and 1.4 kg m^{-3}. The spatial analysis of the relationships among CWP, yield, and evapotranspiration (ET) across the basin showed a strongly linear relationship between ET and yield ($R^2 = 0.86$). The temporal analysis showed increases in yield in the range of 100.4–211.4 kg ha^{-1} $year^{-1}$ between 1984 and 2002 at eight agro-meteorological research stations across the basin without a corresponding increase in ET, corresponding to an increase in CWP of 0.02–0.1 kg m^{-3} per year. It was concluded that the improvements in CWP have resulted from improvements in crop varieties and crop husbandry rather than reductions in water consumption (Yan and Wu 2014).

4.4 Food Security

Climate change has become one of the leading risks to food security, with droughts, floods and hurricanes expected to result in production and price volatility. Human-induced climate change is resulting in scarce and highly erratic rainfall, especially in regions where food security is very low. The poor in rural and dry areas will suffer the most and will require cheap and accessible strategies to adapt to erratic weather (Tirado and Cotter 2010). The increasing risk from unpredictable weather patterns and the resulting volatility in prices raise the probability of farmers investing less in agricultural production and are also threatening food output levels. Llyod Simon et al. (2011) also made the projection that climate change will have significant effects on future undernutrition, even when the beneficial effects of economic growth are taken into account. According to their model predictions, there will be a 62-% increase in severe stunting in South Asia and a 55-% increase in east and south sub-Saharan Africa by 2050. According to the Assessment Report of the IPCC, depending on the climate change scenario, 200 to 600 million more people globally could suffer from hunger by 2080 (Yohe et al. 2007).

Effects on food production and availability as well as the impacts of extreme climate events affect both food physical and economic accessibility. The changes in production systems induced by climate change may induce changes in dietary patterns and food utilization. Climate change will also affect the stability and resilience of food systems with consequences in terms of long-term food security. Moreover, the quest for food security, through agricultural intensification and agricultural land expansion, increases greenhouse gas emissions from deforestation and land use changes (El Bilali et al. 2020). Climate change will also have an adverse impact on the livelihoods of fishers and forest-dependent people (Ramachandran 2014). Landless agricultural labourers wholly dependent on agricultural wages are at the highest risk of losing their access to food (Schmidhuber and Tubiello 2007; Dev 2012). A recent multi-model study using IPCC's highest scenario of warming found a mean effect on yields of four crop groups (coarse grains, oil seeds, wheat and rice, accounting for about 70% of global crop harvested area) of minus 17% globally by 2050 relative to a scenario with unchanging climate (FAO 2015). To ensure the food security of future generations and to address the challenge of the 'no hunger zone' proposed by FAO, crop production must be doubled by 2050, but environmental stresses are counteracting this goal. Heat stress, in particular, is affecting agricultural crops more frequently and more severely (Janni et al. 2020).

5 Climate-Smart Agriculture—A Ray of Hope

Climate change is one of the most important global environmental challenges, with implications on food production, water supply, health and energy, etc. Addressing climate change requires a good scientific understanding as well as coordinated action at national and global scales. Each of the last three decades has been successively warmer at the earth's surface than any preceding decade since 1850. The management and planning of water resources are becoming more challenging due to the uncertainties of climate change (Ficklin et al. 2013). Strategies promoting efficient water use and conserving irrigation water are needed to attain water security to meet growing food demands (Singh et al. 2021a, b). The interactions between plant and environment require a team approach looking across the disciplines from genes to plants to crops in their particular environments to deliver improved water productivity and contribute to sustainability (Morison et al. 2008) (Fig. 2).

5.1 *On-Farm Water Management*

Water-use efficiency is strongly influenced by weather conditions affecting transpiration and assimilation by leaves, plants and crop differently (Tanner and Sinclair 1983). Van de Geijn and Goudriaan (1996) found that positive climate effects on

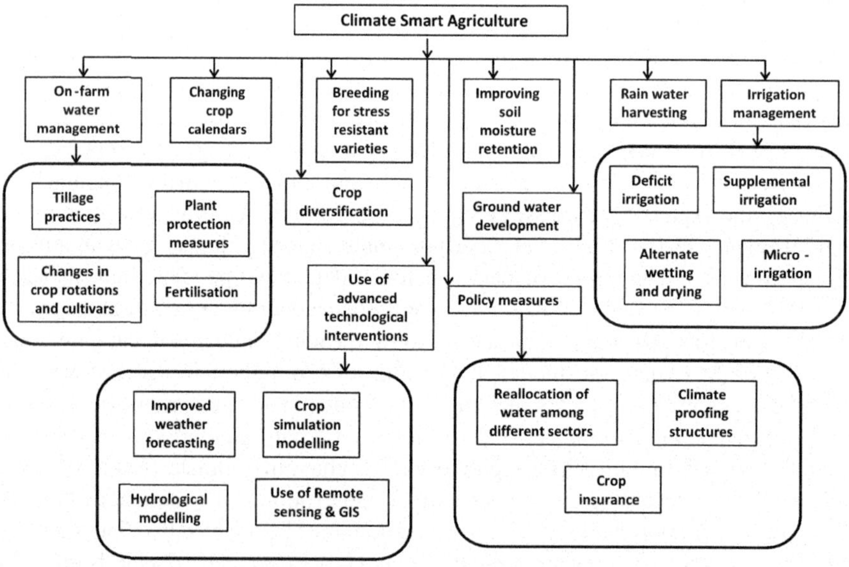

Fig. 2 Climate smart agricultural practices to manage climate change impacts on crop water productivity

crop growth can be adjusted by effective rooting depth and nutrients; meanwhile, it can improve water productivity by 20–40%. In the absence of nutrient limitations, the most critical period for yield determination in wheat generally takes place while water use (WU) is dominated by transpiration (Abbate et al. 1997).

Short-term adjustments at the farm level involve production techniques, such as changes in crop rotation and crop cultivars, changes in soil cultivation and tillage practices, a shift of sowing dates, adapted fertilization and crop protection measures (Tubiello et al. 2000). Total water use or total evapotranspiration (ET) by a crop can vary substantially due to the limited water available from soil water storage or due to limited rainfall. It can also vary as a result of variation in crop transpiration resulting from management, such as nutrient supply, sowing time, from use of different species or cultivars (Asseng et al. 2001).

Long-term adaptations, on the contrary, include major structural changes of farm production systems and need careful agro-economic planning and realization at a society level; these adaptations also involve a set of sectors and stakeholders, such as policy, research, water and land planning (Eitzinger et al. 2010). Some examples of long-term adjustments are changes in land use and landscape structure, breeding and biotechnology applications, crop substitution and changes in the farm production type (Alexandrov et al. 2002). Southworth et al. (2002) predicted the wheat responses in the Midwestern United States for 2050–2059 with atmospheric CO_2 concentration of 555 ppm, elucidating that wheat yields would increase 60–100% above current yields across the central and northern areas, but both small increases and decreases were found in the southern areas.

The importance of management to food production was also highlighted by Doos and Shaw (1999), who considered various controls on global food production and concluded that greatest changes in future production will be due to direct human factors such as improved management and the increased use of fertilizers, rather than natural and/or indirect human factors such as climate change, irrigation, salinization, waterlogging, or pests. At relatively low levels of management, the difference between the hottest and coolest growing seasons was significantly larger than at higher levels of management. This indicates that poorly managed fields are more susceptible to losses in warmer years, and similarly are able to increase their production more in cool years (Lobell et al. 2002).

5.2 *Rainwater Harvesting*

Rainwater harvesting is a relatively secure and self-sustaining way of self-preservation either at individual or community level. In the absence of adequate centralized systems of water provision, individuals can create their own systems, which they can easily manage and control through rain-water harvesting (Ertsen and Nagugi 2021). The human endeavour in the development of water sources must be within the capacity of nature to replenish and to sustain. If this is not done, costly mistakes can occur with serious consequences. The application of innovative technologies and the improvement of indigenous ones should therefore include management of the water sources to ensure sustainability and to safeguard the sources against pollution (Hatibu and Mahoo 1999). The application of an appropriate rainwater harvesting technology can make possible the utilization of rainwater as a valuable and, in many cases, necessary water resource (Jakhar et al. 2015). The water-use efficiency of protective irrigation source through small water harvesting structures in rainfed areas can be enhanced by integrating them to micro-irrigation systems, and provide live saving irrigation to the standing crop. Micro irrigation techniques not only help in water saving, but also in reducing fertilizer usage, labour expenses, and other inputs and input costs, besides sustaining soil health (GOI 2018).

Water is an essential commodity for survival and development. But the ever-increasing human population, technological, modernization, changing life patterns, and erratic monsoons are likely to lead to water crisis in this millennium. One of the solutions that can be implemented quickly is water harvesting (Pauline et al. 2020). Water harvesting is a traditional conservation technique, and there is a general agreement that conserving water will promote agricultural production, especially in arid and semi-arid regions. The high competition for water and land has led to scarcity of water resources, and it has also threatened the world's food security. Recent technological developments have led to improvements in rainwater harvesting techniques, which will help guarantee the availability of food for the growing population. However, problems always occur with the implementation of any advanced technology, and this is also the case for the water harvesting system. This system, in combination with other factors, has a great potential to improve food security, especially in developing countries (Komariah and Senge 2013).

Rainfed agriculture is a necessary way for crop production and food security. As one of the most efficient tools, rainwater harvesting can supplement rainfed agricultural and hence enhance the productivity. However, more comprehensive understanding and multidisciplinary study are needed to improve the efficiency of rainwater harvesting (Liu and Jin 2016). The in-situ and ex-situ rainwater harvesting techniques (RWHT) have shown a significant impact on improved soil moisture, runoff and ground water recharge and increased agricultural production, which, in turn, reduce risks and deliver positive impacts on other ecosystems (Tolossa et al. 2020). Tamagnone et al. (2020) revealed that RWHT may lead to a runoff retention up to 87% and to double the infiltration. Intercepting and storing runoff, RWHT increase the water content in the root zone and the right design can diminish the crop water stress. Furthermore, the results showed that adopting RWHT makes it possible to extend the growing season up to 20 days, enhancing the yield. These benefits contribute to the reduction of the climate-related water stress and the prevention of crop failure. Rain-water harvesting offers a critical and promising solution to replenish and recharge the groundwater (in areas where geologic conditions are conducive). In a typical setting, much of the rainwater is lost to surface flows. Rainwater harvesting for agriculture generally involves the creation of structures such as check dams, ponds, and percolation tanks to slow the flow of water, and to collect and hold limited quantities at a planned set of places along the flow path (Gandhi and Bhamoriya 2011).

5.3 *Groundwater Development*

Groundwater, which occurs beneath the surface of the land, filling the pore spaces of rocks called aquifers, plays a great role in the human lives as it constitutes the major portion of the world's drinking water. Groundwater is heavily relied upon for other domestic purposes, agriculture and industrial uses also. Indiscriminate and unplanned over extraction and use of groundwater has led to groundwater depletion and pollution and the groundwater reserve all over is under an intense pressure (Jose and Bose 2021). The aquifer planning at the ground level/local scale is an essential need prior to undertaking any water preservation program. The practical improvement of groundwater assets needs selection of an all-encompassing methodology, which includes examining boundaries identified with groundwater and other security dataset, like slant, geomorphology, lineament, soil, land-use, and so on. There is a need to have a well-planned extraction system thinking about future interest of water system and drinking water.

Groundwater is a hidden asset that contributes significantly to maintaining the overall quality of our environment and is a globally important source of water (Taylor et al. 2012). Hughes et al. (2021) reported that groundwater systems provide an important source of water supply as well as contributing base flow to rivers, lakes and dependent ecosystems and so the impact of climate change on these systems needs to be understood. In this direction, it is required to understand the sequencing of flooding and drought events and to the effects of soil health and land cover changes in the future analysis. Smerdon (2017) highlighted the importance of understanding groundwater recharge processes, including timing and location for the characterization of groundwater resource assessment. Studies have predicted that the average global temperature may increase by 1.4–5.8 °C and there would be substantial reduction in fresh water resources and agricultural yield by the end of the twenty-first century. Approximately 75% of the Himalayan glaciers are on retreat and will disappear by 2035. Groundwater recharge using artificial recharge structures and Soil Aquifer Treatment (SAT) systems equipped artificial set of lithologies through wastewater can help in minimizing the impact of climate change on water resources and agricultural yield (Misra 2014).

Groundwater, which represents a major fraction of the water resources functional to humans, is highly impacted by climate change because the recharge rate and the depth of the groundwater surface have been considerably affected. Applications using groundwater are increasing because it has the advantage of source recharging in a shorter period compared to that of surface water, and it is a reliable source of clean water at a low cost. However, when the quantity of groundwater exploited exceeds the water quantities obtained through rainwater or surface water, or water quality is deteriorated because of pollutant infiltration, long periods will be required for water regeneration, recovery, and restoration. Therefore, a rational water management plan must be established by considering all water quantity and quality parameters as well as by evaluating the distribution of water resources and the environmental variables of surrounding regions (Kim and Lim 2020). Efficient and effective groundwater management will require proper attention of the local authorities to the inherent interaction among various water systems. Only with enhanced cooperation, an integrated monitoring network, strengthened scientific support and active public participation can the sustainability of groundwater management be achieved (Chen et al. 2018). Groundwater management seeks to balance and mitigate the detrimental impacts of development, with plans commonly used to outline management pathways. Thus, plan efficiency is crucial, but seldom are plans systematically and quantitatively assessed for effectiveness (White et al. 2016).

5.4 *Breeding for Resistance to Droughts and Floods*

Drought resistance is a complex trait/phenomenon (Ceccarelli et al. 2007) due to too many secondary traits associated with it, inadequate knowledge about the inheritance mechanism, difficulty in predicting the magnitude of drought and lack of direct correlation of any single trait with grain yield under stress. While concerted efforts are underway in several crops to discern the mode of action, number of genes governing the trait, the promoters, quantitative trait locus (QTL) etc., plant breeders have gone ahead and improved each one of the traits (Shashidhar et al. 2013). Marker-assisted selection is expected to boost the pace of crop improvement especially for complex traits. Intensive research needs to be conducted in this direction to develop drought resistant genotypes.

Global climate change is expected to increase the occurrence and severity of drought episodes due to higher evapotranspiration and rising temperatures. Therefore, food security in the twenty-first century will increasingly depend on the release of cultivars with improved adaptation to drought conditions (Rauf et al. 2015). The recent progress in genomics offers permits to more efficiently assess and enhance diversity in germplasm collections, introgress valuable traits from new sources, and identify genes that control key traits. Marker assisted selection helps reduce the impact of environment on breeder selection. Significant advances have been made in the development of in vitro selection methods. The broader use of traits from alien species and the manipulation of heterosis and polyploidy create new perspectives for improving yield potential and adaptation to abiotic stresses. The use of the knowledge generated by these approaches should permit to clarify the functional basis of drought adaptation traits.

A major problem of climate change is the increasing duration and frequency of heavy rainfall events. This leads to soil flooding that negatively affects plant growth, eventually leading to death of plants if the flooding persists for several days. Most crop plants are very sensitive to flooding, and dramatic yield losses occur due to flooding each year. This review summarizes recent progress and approaches to enhance crop resistance to flooding. Most experiments have been done on maize, barley, and soybean. Work on other crops such as wheat and rape has only started. The most promising traits that might enhance crop flooding tolerance are anatomical adaptations such as aerenchyma formation, the formation of a barrier against radial oxygen loss, and the growth of adventitious roots. Metabolic adaptations might be able to improve waterlogging tolerance as well, but more studies are needed in this direction. Reasonable approaches for future studies are quantitative trait locus (QTL) analyses or genome-wide association (GWA) studies in combination with specific tolerance traits that can be easily assessed. The usage of flooding-tolerant relatives or ancestral cultivars of the crop of interest in these experiments might enhance the chances of finding useful tolerance traits to be used in breeding (Mustroph 2018).

5.5 Improving Soil Moisture Retention

Increasing frequencies of drought coupled with increasing populations are requiring more water for irrigated agriculture. As global populations approach 9 billion by 2050, even more water will be required to produce an estimated 60–70% more food (McKenna 2012). Production of these greater quantities of food require, at current water use efficiency rates, 50% more water (Clay 2004). Consequently, the growing demand for food and fiber combined with dwindling water supplies (in terms of both quantity and quality) available for agricultural irrigation require new soil technologies that conserve water (Kavdir et al. 2014).

Moisture is a key limitation on the productivity of soil. Three main factors which affect soil moisture content are how well soil can absorb water; how well soil can store moisture and how quickly the water is lost or used. Although these factors are strongly determined by the proportions of clay, sand and silt, good soil management also plays a critical role (Reid 2004). Irregular or insufficient rainfall can be a serious limitation to agricultural production, causing low yields and even crop failure. This is particularly true in drylands, where productivity levels are generally very low. In most cases, a great deal can be done to improve the efficiency of rainwater use. Conservation Agriculture is one way of improving soil moisture management (Benites and Castellanos 2003). Soil moisture management is, therefore, a key factor when trying to enhance agricultural production. Increasing the amount of water stored in the soil can result in improved yields (if there are also enough nutrients), reduced risk of yield losses due to drought and recharge of groundwater, securing the water level in wells and the continuity of river and stream flows.

Although these factors are strongly determined by the proportions of clay, sand and silt, good soil management also plays a critical role. Conservation Agriculture improves the physical and biological condition of the soil. A soil that is porous, absorptive, and rich in organic matter and biological activity is able to support maximum crop production for every drop of water it receives. Reducing deep percolation losses of root zone soil water is becoming a major research focus among agricultural and hydro-pedological scientists and engineers (Graham and Lin 2012). Plastic film mulching has been shown to decrease soil evaporation (Zhang et al. 2017). Similarly, protected cultivation systems (greenhouses, screenhouses) have been proved to have higher WUE values comparing with open-field cultivation (Raveh et al. 2003).

5.6 *Changing Cropping Calendars*

Early sowing has an advantage in terms of crop grain yield and WUE, particularly when practised in conjunction with a good N supply and supplement irrigation. Delayed sowing decreases intercepted solar radiation and reduces the duration of growth. Therefore, late-sown crops accumulate less dry matter. Earlier sown crops have not only increased accumulated dry matter but also reduced water evaporation from the soil surface resulting from an earlier and larger ground cover. Thus, the amount of water transpired by the crop and WUE increase with early sowing. The heat stress to which the late-sown crops are subjected can reduce the kernel number per ear (Gregory and Eastham 1995). The effects of various levels of supplemental irrigation (SI) (rainfed, 1/3 SI, 2/3 SI, full SI) and sowing time (November, December and January) on evapotranspiration (ET) and WUE of wheat were examined at International Center for Agricultural Research in the Dry areas (ICARDA), Syria. WUE was calculated for rain (WUE_r), for total water (gross: rain 1 irrigation) (WUE), and for SI water only (WUE_{SI}). ET ranged from 246 to 328 mm for rainfed crops and ET in case of irrigated crops was found to be 304 to 485 mm. The degree to which water supply limits grain yield was indicated by the ratio of pre- to post-anthesis ET. The SI treatments significantly increased WUE_g: from 0.77 to 0.83 to 0.92 kg/m^3 in November and December sowings for 1/3 SI and from 0.77 to 0.92 kg/m^3 in November sowing for 2/3 SI. The highest WUE_g and WUE_{SI} were achieved at 1/3 to 2/3 SI. Delaying sowing had a negative effect on WUE for both irrigation and rainfed conditions. In this rainfed Mediterranean environment, WUE can be substantially improved by adopting deficit SI to satisfy up to 2/3 of irrigation requirements (Oweis et al. 2000).

The effect of time of sowing on the yield of 15 wheat cultivars grown under irrigation was examined at Narrabri, New South Wales. There was evidence of a vernalization requirement for some of the winter and midseason cultivars, but, overall, photoperiod was more important environmental factor determining pre-anthesis development. Each day's delay in sowing caused a delay in 0.48 and 0.75 days in anthesis; the delays observed for spring wheat were generally greater than those reported for dryland wheat in eastern Australia. Winter cultivars generally did not show an optimum sowing or anthesis date. For spring cultivars, the optimum time of sowing was early June (range of about 3 weeks), while the optimum anthesis date was the last week of September (range of 1 week). Grain yields of spring cultivars were reduced by 6 and 16% per week's delay in sowing and anthesis, respectively (Mcdonald et al. 1983). Using a simulation model, Stapper and Harris (1989) estimated that wheat grain yield in Syria declined by 4.2% per week when sowing delayed after 1 November. Under rainfed conditions, the date of the first significant rain determines the sowing date. Early sowing of appropriate cultivars is a recognized means of increasing wheat yields in other Mediterranean-type environments, such as Western Australia (Anderson 1992).

Enhancement of sowing by 10 days in late sown and delaying of sowing by 10 days in normally sown cultivars resulted in higher yields under a modified climate, whereas reduction in yield was observed in the reversed strategies (Attri and Rathore 2003). Growth and yield of wheat at different sowing dates (10 Nov, 25 Nov and 10 Dec) with different plant densities (200, 300 and 400 plants m^{-2}) were analyzed in terms of solar radiation intercepted by the leaves during 1998–99 and 1999–2000. Leaf area index, radiation interception and biomass accumulation were measured throughout the growing seasons. The relationship between dry matter production and intercepted photosynthetically active radiation (PAR) was highly significant and linear throughout the growing season for all treatments. Results showed that the highest yields were obtained from early (November) sowings and a plant density of 300 plants m^{-2}; yield variations among treatments were caused by affecting both the amount of intercepted PAR and RUE (Wajid et al. 2004).

The observed changes in phenology for winter wheat across the Loess Plateau were likely caused by temperature increase over the recent decades, but were also partially caused by changes in sowing dates and the introduction of new cultivars with altered temperature requirements. The date of sowing is determined by farming management decisions, which can change in response to changes in climate (Estrella et al. 2007). El-Gizawy and Kh (2009) conducted a trial for 2 subsequent years to study the effect of 3 planting dates (November 1, November 15 and November 30).The results showed that the highest values of number of tillers and spikes/m^2, 1000-kernel weight, grain yield/fed and grain NPK uptake were obtained when wheat was sown on mid-November. Early or delayed planting significantly reduced forenamed traits.

A study was conducted at Adaptive Research Farm, Vehari, Pakistan during the year 2003–06 with two newly evolved wheat varieties SH-2002 and AS-2002 along with a standard Uqab-2000 subjected to different sowing times starting from November 1 to December 30 at ten days interval. The pooled data revealed that significantly higher grain yield (3826 kg/ha) was obtained from variety AS-2002 sown on November 10 followed by same variety sown on November 20 (3731 kg/ha). Each successive delay in sowing beyond November 20 progressively decreased the grain yield significantly. Yield was reduced by 27.24% in sowing of December 30 as compared to November 10. Three years results concluded that regardless of the varieties November 10 to November 20 is the optimum sowing time for wheat (Ali et al. 2010).

A warming trend over the recent decades provided additional suitable growing conditions before winter dormancy, which led farmers to postpone sowing dates accordingly. An increase in temperature of one degree resulted in an average of 0.88 days delay of sowing dates. A similar trend in delaying the sowing date of winter wheat because of climate warming has been reported (1.5 days/decade) for the North China Plain (Xiao et al. 2012). Numerous studies (Anderson and Smith 1990; Connor et al. 1992; Owiss et al. 1999; Bassu et al. 2009; Bannayan et al. 2013) have reported an increased yield with early sowing and a reduction in yield when sowing is delayed after the optimum time. These authors reported an advantage of early sowing dates when combined with cultivars that avoid frost risk at anthesis or in regions or seasons with low frost risk, aiming at high above ground biomass at flowering to maximize radiation interception. The delay in sowing date not only affects yield, but it affects the yield components and other aspects of the growth and development of wheat. The results of simulations done by Andarzian et al. (2015) at the Khuzestan province, Iran showed that the yield of early sowing dates (before 15 November) was lower than the yield of normal sowing date (e.g. 15 November) in all locations. It was because of decreasing crop growth cycle particularly the time from sowing to the anthesis stage. The high temperature in early sowing dates has resulted in accelerating crop development stages, reducing crop canopy (leaves and tillers) and decreasing biomass production which in turn reduced the yield and its components.

5.7 Crop Diversification

Water being the most critical input for agriculture, its judicious use is important to ensure sustainable agricultural development and food security. There is a need for adopting optimum cropping pattern and efficient water application (GOI 2018). Diversification of cropping and farming systems is a central agroecological principle, which may improve resource use efficiency, reduce pests and diseases, diversify income sources, and enhance the resilience of the production (Rodriguez et al. 2021). Crop diversification has a beneficial influence on ensuring water management, enhancing local food security and delivering a healthy and diverse diet (PWC 2020). The dominance of paddy-wheat crop rotation has led to over-exploitation of ground water resulting in rapid decline of water table over north-west India. Diversifying agriculture towards less water consuming crops is of great importance. Large-scale crop diversification is being recommended by experts in Punjab for the past four decades as one of the most potent solutions for attaining water and agricultural sustainability. Such a policy cannot be implemented in isolation but requires multifaceted policy action with the Government playing a key role (Bhogal and Vatta 2021). Kaur et al. (2015) reported that shift of area under maize and basmati in Punjab resulted into a significant water saving of 8%. Bautista-Capetillo et al. (2018) revealed 11.97% more water use for intensive crops (Babenero peppers and bell peppers) than extensive crops (sorghum and corns), however economic profitability for farmers was 72 times higher for intensive crops than intensive crops.

The continuous cultivation of water guzzling crops like paddy due to frequent flood irrigation has resulted into depletion of ground water in the original Green revolution States namely; Punjab, Haryana and Western Uttar Pradesh. The continuous cultivation of rice wheat cropping system has witnessed the stagnancy in crop yield, infestation of weeds, contamination of ground water, incidence of pests-diseases and deterioration of soil health. Therefore, it is essential to diversify the area from paddy to alternate crops not only to improve soil fertility and arrest depletion of ground water but also to enhance the farm income (GOI 2014). Crop diversification significantly improves efficiency and reduces income variability for farmers, so do not have to give up efficiency for income stability or vice versa. Thus, crop diversification could be an ideal Climate Smart Agricultural (CSA) strategy for promoting agricultural growth and resilience (Mzyece and Ng'ombe 2021). Diversified crop rotation (DCR) improves the efficiency of farming systems all over the world. It has the potentiality to improve soil condition and boost system productivity. Improved soil attributes such as increased soil water uptake and storage, and a greater number of beneficial soil organisms, may improve yield tolerance to drought and other hard growing conditions in a variety of crop rotations. Crop rotations with a variety of crops benefit the farmers, reduce production risk and uncertainty, and enhance soil and ecological sustainability (Shah et al. 2021).

For sustainable water use in agriculture, cropping patterns need to be recalibrated to maximise crop productivity per unit of water consumed or applied for irrigation (Sharma et al. 2018). Amarasinghe et al. (2021) used economic water productivity (EWP) and water cost curve for EWP as tools to reallocate irrigation consumptive water use (CWU) and identify economically viable cropping patterns and concluded that drought-tolerant annual crops such as fruits and/or fodder should be the preferred option in irrigated cropping patterns. Cropping patterns with orchard or fodder as permanent fixtures will provide sustainable income in low rainfall years. Orchards in combination with other crops will increase EWP and value of output in moderate to good rainfall years. Governments should create an enabling environment for conjunctive water use and allocation of CWU to achieve a gradual shift to high-value annual/perennial crops as permanent fixtures in cropping patterns.

5.8 Irrigation Management

Increased water use has led to water scarcity in many Asian countries. The gap between water demand and supply is projected to increase due to population growth and economic development as well as environmental factors, such as land degradation and climate change. It is projected that by 2080, net crop water requirements will increase globally by 25% despite the increased irrigation efficiency, attributed to changes in precipitations patterns, global warming, and extended crops' growing periods (Fischer et al. 2007). Efforts to reverse these trends should focus on irrigated agriculture, since irrigation is the largest consumer of freshwater withdrawals in almost all water-scarce regions (Van Opstal et al. 2021). Irrigation applications aimed at keeping crops under well-watered conditions could reduce canopy temperature but in most cases were unable to maintain it below the threshold temperature for potential heat damage; the benefits of irrigation in terms of reduction of canopy temperature decreased as average air temperature increased (Luan and Vico 2021).The appropriate irrigation scheduling has been considered to be the most important factor for crop growth and sustainable irrigation water management (Nikolaou et al. 2020).

Irrigation management is a key factor in attaining optimal yields, as different irrigation strategies lead to different yields even when using the same amount of water or under the same weather conditions (Kuschel-Otarola et al. 2020).

5.8.1 Deficit Irrigation

Deficit irrigation (i.e. irrigation application below evapotranspiration) save water and enhance water use efficiency (WUE) of the crops as a result of an improved ratio of carbon fixation to water consumption ratio (Farooq et al. 2019). Deficit irrigation (DI or regulated deficit irrigation RDI; i.e. the application of water at a lower rate and/ or volume than the plants evapotranspiration) has been considered as a sustainable irrigation strategy as opposed to conventional irrigation under limited water supply conditions (Álvarez et al. 2013; Farahani and Chaichi 2012). The principal attitude of DI is to increase water productivity by irrigating crops only at critical crop growth stages without causing severe yield reductions or to save water for expanding farmlands (Geerts and Raes 2009). It is a common practice for farmers to roughly double the nominal irrigated area with a given amount of water by applying DI strategy (Egea et al. 2017). Partial root zone drying (PRD), where partial half of the root system irrigated, while the remaining half is exposed to drying soil switching to the other half every 2–3 weeks (Nangia et al. 2018).

Oweis (1997) analyzed the values of water use efficiency for wheat cultivated under irrigated and non-irrigated experimental conditions in Syria and noticed that the average WUE of rainfed wheat in Syria is 0.5 kg/m^3, although with good management and favourable rainfall amounts and distribution, this average could be increased to 1 kg/m^3. In fully irrigated areas with good management, the WUE was about 0.75 kg/m^3. However, water used in supplemental irrigation could be much more efficient (WUE = 2.5 kg/m^3). This extremely high WUE was mainly attributed to the effectiveness of a small amount of water in alleviating severe moisture stress during the most sensitive stage of crop growth. Hassan et al. (2000) investigated the impact of deficit irrigation strategies on wheat yield and water savings. They reported from a 1-year study that a two-stage deficit at yield formation and ripening stage produced the highest yield, and saved 34% of irrigation water, compared to normal watering.

Singh et al. (2021a, b) reported that moving from the highest (> 80%FI) to the lowest (< 35% FI) irrigation level, the overall yield decline was 6.9–51.1% compared to FI, respectively. The WP gains ranged from 8.1 to 30.1%, with 35–50%FI recording the highest benefits. Soil texture affected the yield significantly only under the least irrigation class (< 35%FI), wherein sandy clay and loam recorded the highest (82.1%) and the lowest (26.9%) yield decline, respectively. Among the climates, temperate climate was overall the most advantageous with the least yield penalty (21.9%) and the highest WP gain (21.78%) across various DI levels. The DI application under the greenhouse caused lesser yield reduction compared to the open-field. The WP gains due to DI were also higher for greenhouse (18.4%) than open-field (13.6%). Consideration of yield penalties and the cost of saved irrigation water is crucial while devising the reduced irrigation amounts to the crops. The yield reductions under low to moderate water deficits (> 65%FI) accompanied by gains in WP may be justifiable in the light of anticipated water restriction.

5.8.2 Supplemental Irrigation

Supplemental irrigation (SI), optimally scheduled for the amount and timing of irrigation to ensure that a minimum water amount is available to the crops during the critical stages that it would permit a significant increase in the yield. Usually, SI is combined with earlier planted dates in order to prevent exposure of crops to drought stress and heat in hot areas and frost in cold areas. In the dry environments, Supplemental irrigation (SI) is a common practice and aims at improving and stabilizing yields by adding small amounts of water to rainfed crops during the times when rainfall fails to provide sufficient moisture for normal plant growth. The relationship between water regime and nitrate supply on wheat cultivation in Syria was analyzed by Oweis (1997). It was noticed that under rainfed conditions, the rate of nitrogen fertilizer needed was not high and only 50 kg/ha was sufficient. However, with higher water supply, the crop responded to nitrogen up to 100 kg/ha, after which no benefit was obtained. This rate of N greatly improved WUE. It was also important to maintain available phosphorus in the soil so that the response to N and applied irrigation was not constrained.

5.8.3 Alternate Wetting and Drying

Alternate Wetting and Drying (AWD) irrigation is a promising method in irrigated rice cultivation with dual benefits of water saving and environment saving, while maintaining rice yields at least at the same level (Yang et al. 2009). AWD irrigation is one method of managing the water so that water will not be wasted but it will aid the root growth, facilitate higher nutrient uptake and increase land and water productivity (Sarkar 2001). Alternate wetting and drying irrigation had significant effect on yield and yield contributing characters of boro rice (Nasir et al. 2014). Although the agronomic effects of AWD vary with the duration and severity of soil drying, mild soil water deficits decreased water use by 23% while yields were statistically similar to continuously flooded crops, especially if AWD was applied either during the vegetative growth phase or reproductive growth phase, but not both (Carrijo et al. 2017).

The declining soil health, increasing micronutrient deficiencies, and declining organic matter are also threatening the long-term sustainability of the conventional rice production system. In this scenario, alternate wetting and drying (AWD) irrigation system is a promising, water-saving, economically viable, and ecofriendly alternative to CF (Ishfaq et al. 2020). AWD irrigation technique can reduce the total water inputs (25–70%), CH_4 emission (11–95%), As (13–90%), and Hg (5–90%) in rice grains while maintaining similar or better paddy yield (10–20%) than the CF depending upon weather conditions, soil type, degree of dryness, crop duration and crop growth stage. The mild-AWD improves the rice grain quality by reducing the kernel chalkiness (40%) and increasing the head rice recovery (6%) and concentration of grain micronutrients (like zinc). The adoption of AWD system has contributed to significantly reduce the global warming potential (GWP) and water use (Li et al. 2011; Mazza et al. 2016).

It has been observed that moderate AWD, in which photosynthesis is not severely inhibited and plants can rehydrate overnight during the soil drying period, or plants are rewatered at a soil water potential of − 10 to − 15 kPa, or midday leaf potential is approximately − 0.60 to − 0.80 mPa, or the water table is maintained at 10 to 15 cm below the soil surface, could increase not only WUE but also grain yield (Yang et al. 2017). AWD could also improve rice quality, including reductions in grain arsenic accumulation, and reduce methane emissions from paddies. Adoption of moderate AWD with an appropriate nitrogen application rate may exert a synergistic effect on grain yield and result in higher WUE and nitrogen use efficiency. Changes in shoot phytohormone concentrations were associated with increased water and phosphorus use efficiency (WUE and PUE) of vegetative rice plants grown under AWD (Acosta-Motos et al. 2020).

5.8.4 Micro-irrigation

India's agriculture is facing acute water scarcity and a major reason for this is very low water use efficiency—only about 25 to 35% in conventional irrigation. It is of tremendous importance to improve the efficiency and in this context, the modern technology of micro irrigation (MI) which includes drip and sprinkler irrigation offers a very significant advantage. MI techniques can bring numerous benefits including not only enhanced water use efficiency, but also increase in irrigated area with the given quantity of water, enhanced crop productivity/yields, labour cost savings, electricity and energy savings through lesser pumping hours (Gandhi et al. 2021). The field water application efficiency of traditional surface irrigation methods such as, e.g., furrow, basin, or border strips is estimated to be as low as 40% (Brouwer et al. 1989) with excessive deep percolation losses and low water distribution uniformity (Walker 1983). Modernization of irrigation increases the water application efficiency (Nikolaou et al. 2020). Converting from traditional surface irrigation methods to closed pressurized pipe network systems could lead to as much as 90% of the total water savings, as is the case of trickle irrigation (Azahara Mesa-Jurado et al. 2012). The field application efficiency is about 50–70% with sprinkler system and 80–90% with surface drippers (De Pascale et al. 2013). That is because as drip irrigation system minimizes water losses due to surface runoff and deep percolation of water under difficult soil and terrain conditions (Sacks et al. 2009). Drip irrigation systems account for up to 90% application efficiency, and they have been used with success in arid and semi-arid regions for vegetable production, forage crops, and maintenance of trees (Ahmed et al. 2011; Çolak et al. 2018; El-Attar et al. 2019).

Micro-irrigation systems deliver water savings of up to 40% conventional flood irrigation methods, along with appreciable crop productivity and income enhancement (GOI 2018). Drip and sprinkler irrigation is a solution that reduces conveyance and distribution losses and allows higher water use efficiency (Ashoka et al. 2015). Over the past decades, a significant shift to pressurized irrigation was observed, with a significant component being micro-irrigation, including micro-sprays, mini-sprinklers, surface drip, and subsurface drip irrigation systems (SDI) (Borsato et al. 2020). Yield of onions almost doubled using SDI, allowing for more frequent irrigation with smaller depths of water (Enciso et al. 2015). In tomatoes, the most appropriate irrigation arrangement for optimum growth and production is considered to be SDI with plastic film mulching (Wang et al. 2018). In any case, water savings of up to 20% were recorded in olives under the SDI treatment as opposed to surface drip irrigation (Martinez and Reca 2014). The promotion of MI is extremely important in reducing the water footprint, and increase water use efficiency at the farm level (Gandhi et al. 2021).

6 Use of Advanced Technological Interventions

6.1 Improved Weather Forecasting

Weather forecasting is a complex and challenging science that depends on the efficient interplay of weather observation, data analysis by meteorologist and computers, and rapid communication system (Iseh and Woma 2013). Weather plays an important role in agricultural production. It has a profound influence on crop growth, development and yields on the incidence of pests and diseases on water needs and on fertilizer requirements. This is due to differences in nutrient mobilization as a result of water stresses, as well as the timelines and effectiveness of preventive measures and cultural operations with crops. Weather aberrations may cause physical damage to crops and soil erosion. The quality of crop produce during movement from field to storage and transport to market depends on weather. Bad weather may affect the quality of produce during transport and the viability and vigour of seeds and planting material during storage (Singh et al. 2021a, b). Access to seasonal climate forecasts can benefit farmers by allowing them to make more informed decisions about their farming practices. However, it is unclear whether farmers realize these benefits when crop choices available to farmers have different and variable costs and returns; multiple countries have programs that incentivize production of certain crops while other crops are subject to market fluctuations (Gunda et al. 2017). Advanced information in the form of seasonal climate forecasts has the potential to improve farmers' decision making, leading to increases in farm profits (Hansen and Osgood 2008).

Climate is changing and its effects on agriculture are uncertain, and to get maximize output and to improving their livelihood within the major constraint, there is need for accurate weather forecast and information. Due to this the dependency of the agriculture sector on monsoon correlates accurate weather forecasts with high demand. The key factor in all agriculture policy is the weather forecasting information which involves enhancing farm risk management. The analysis showed that a 75% accuracy of agro-meteorological information is necessary for the agro-meteorological information to be worthwhile. However, challenges are there to the uncertainty of climate forecasts and to the complexities of agricultural systems. If better predictions of climate were available three to six months ahead of time, it may be possible to modify decisions to decrease unwanted impacts and to take advantage of expected favourable conditions (Kumar et al. 2017). Developing appropriate interdisciplinary systems to connect climate, weather, and agronomic information, especially including forecasting systems, with farm management is needed if uptake of weather and climate information by farmers is to be successful. Provision of output of climate change scenario and trend information to aid long-term strategic farm management decisions needs to be considered, especially in regions where more vulnerable farming zones exist (Stone and Meinke 2006).

6.2 *Crop Simulation Modelling*

To study the impact of changing climate, a number of crop models are available in the literature but the models like decision support system of agro transfer technology (DSSAT) and cropping system simulation (CropSyst) are widely used because these models assess the effects of CO_2 and weather parameters on crop growth, biomass, water balance and nitrogen balance on daily basis (Jalota et al. 2013). Eitzinger et al. (2003) used the CERES-wheat model to assess climate change impacts on soil water balance under four climate scenarios, and the results show that the factors which affect the soil water balance also have influences on sustainable crop production and water resources in agriculture. He also utilized the CERES-wheat model to assess climate change impacts on wheat production under four climate scenarios and the results showed that the CO_2 effect maintained a great responsibility for increasing crop yield in the research area.

The sensitivity analysis indicated that beneficial effects of increasing CO_2 are cancelled out with an increase in temperature of the order of 3 °C or more. A persistent decrease in the yield of the order of 2.0–21.0% was observed in different cultivars when temperature was increased from 3 °C to 5 °C (Attri and Rathore 2003). Luo et al. (2003) discussed climate change impacts on wheat production with DSSAT 3.5 (Decision Support System for Agrotechnology Transfer) and CERES-Wheat model under all CO_2 levels in Southern Australia for 2080s, and the result shows that wheat yield will increase under all CO_2 levels and the drier sites are more suitable for wheat production but are likely to have lower wheat quality.

Anwar et al. (2007) used CropSyst version-4 to predict climate change impacts on wheat yield in south-eastern Australia and their results showed that the elevated CO_2 level can reduce the median wheat yield by about 25%. The report of the IPCC and a few other global studies indicate a probability of 10–40% loss in Indian food grain production with an increase in temperature by 2080–2100. Heat waves are likely to become more frequent with global warming (IPCC 2007).

Mo et al. (2009) at Institute of Geographical Sciences and Natural Resources Research, Chinese Academy of Sciences, Beijing, studied the response of wheat to future climate scenarios of twenty-first century projected by the GCM (HadCM3) with Intergovernmental Panel on Climate Change Special Report on Emission Scenario (IPCC SRES) A2 and B1 emissions. The results showed a rapid enhancement of crop yield in the past 56 years, accompanying with slight increment of ET and noticeable improvement of WUE. There existed spatial patterns of crop yield which stemmed mainly from soil quality and irrigation facilities. For climate change impacts, it is found that winter wheat yield will significantly increase with the maximum increment in A2 occurring in 2070 with a value of 19%, whereas the maximum in B1 being 13% in 2060. Its ET is slightly intensified, which is less than 6%, under both A2 and B1 scenarios, giving rise to the improvement of WUE by 10% and 7% under A2 and B1 scenarios, respectively.

The responses of yield and water use efficiency of wheat climate change scenarios were explored over the North China Plain by Guo et al. (2010). The impacts of increased temperature and CO_2 on wheat yields are inconsistent. As a more probable scenario, climate change under B2A is moderate relative to A2A and A1B. Under B2A in 2090s, average wheat yield will increase 9.8% without CO_2 fertilization in this region. High temperature not only affects crop yields, but also has positive effect on water use efficiencies, mainly ascribing to the evapotranspiration intensification. There is a positive effect of CO_2 enrichment on yield and water use efficiency. If atmospheric CO_2 concentration reaches nearly 600 ppm, wheat yields will increase 38% and water use efficiencies will improve 40%, in comparison to those without CO_2 fertilization.

6.3 Hydrological Modeling

Changing hydrological conditions due to climate, land use and infrastructure pose significant ongoing challenges to the hydrological research and water management communities. While, traditionally, hydrological models have assumed stationary conditions, there has been much progress since 2005 on model parameter estimation under unknown or changed conditions and on techniques for modelling in those conditions (Peel and Bloschl 2011). A properly calibrated hydrological model can provide useful information for the management and planning of water resources (Montecelos-Zamore et al. 2018). Some models allow quantifying the impacts of climate change on water resources through the simulation of hydrological processes (Ficklin et al. 2009; Hawkins et al. 2015). Hydrological models combined with climate change scenarios generated by general circulation models (GCMs) (Mango et al. 2011; Jha and Gassman 2013) or regional climate models (RCMs) (Jha et al. 2004) are commonly used. However, this combined approach has not been used yet in Cuba, at least not in peer-reviewed literature. On the other hand, climate change has been extensively studied in Cuba and the Caribbean region using GCMs and RCMs, such as RegCM4.3 (Martínez-Castro et al. 2006, 2014) and PRECIS (Centella et al. 2015).

Gayathri et al. (2015) reviewed different hydrological models and observed that variable Infiltration capacity (VIC) model performed well in moist areas and could be used in water management for agricultural purposes. Requirement of large data and physical parameters makes the use of MIKE SHE model limited to smaller catchments. Only a little direct calibration is required for SWAT model to obtain good hydrologic predictions. HBV model gives satisfactory results and TOPMODEL can be used in catchments with shallow soil and moderate topography. Many grid-based spatial hydrological models suffer from the complexity of setting up a coherent spatial structure to calibrate such a complex, highly parameterized system. There are essential aspects of model-building to be taken into account: spatial resolution, the routing equation limitations, and calibration of spatial parameters, and their influence on modeling results. Farrag et al. (2021) showed that the spatial sensitivity of the resolution is highly linked to the routing method, and it was found that routing sensitivity influenced the model performance more than the spatial discretization, and allowing for coarser discretization makes the model simpler and computationally faster. Slight performance improvement is gained by using different parameters' values for each cell. It was found that the 2 km cell size corresponds to the least model error values.

Horan et al. (2021) compared the predictive capability of three hydrological models, and a mean ensemble of these models, in a heavily influenced catchment in Peninsular India: GWAVA (Global Water aVailability Assessment) model, SWAT (Soil Water Assessment Tool) and VIC (Variable Infiltration Capacity) model. This highlighted the importance of an accurate spatial representation of precipitation for input into hydrological models, and that comprehensive reservoir functionality is paramount to obtaining good results in this region. It was demonstrated that the ensemble mean has a better predictive ability in catchments with reservoirs than the individual models, with NashSutcliffe values between 0.49 and 0.92. Therefore, the use of multiple models could be a suitable methodology to offset uncertainty in input data and poor reservoir operation functionality within individual models.

6.4 Use of Remote Sensing and GIS for Hydrological Monitoring

Adoption of modern and innovative technologies and methods is the answer to the growing need for smart use of water in agriculture. A smart irrigation system is a mix of usage of water and fertilisers based on soil types, climatic conditions and different stages of crop development, and the implementation of micro-irrigation techniques in a controlled manner through the use of sensors and controllers (PWC 2020). Information technology, remote sensing techniques and proximal data gathering and analyzing (i.e., precision agricultural systems; PA) is a key factor for efficient agricultural water management (Ferrández-Pastor et al. 2018). VRI in wheat crop based on differences of soil available water holding capacity reduces by 7% the irrigation water used as opposed to a uniform rate irrigation management application (Li et al. 2019). The future is likely to see increased use of remote sensing techniques as well as wireless communication systems and more versatile sensors to improve WUE. In many cases, water saved as a result of using efficient technologies ends up being reused to expand the area of land under irrigation, sometimes resulting in a net increase in the total water consumption at the basin scale (Koech and Langat 2018). Internet of things (IoT) helps in effective management of water in agriculture and increasing crop yield. It may also solve the issue of water use and increase the incomes of small and marginal farmers (PWC 2020).

7 Policy Measures—Need of the Hour

7.1 Climate Proofing Structures

Climate proofing is a shorthand term for identifying risks to a development project, or any other specified natural or human asset, as a consequence of climate variability and change, and ensuring that those risks are reduced to acceptable levels through long-lasting and environmentally sound, economically viable, and socially acceptable changes implemented at one or more of the following stages in the project cycle: planning, design, construction, operation, and decommissioning (ADB 2012). In the 'climate proofed' scenario (future 1) potential climate change risks are addressed prior to the emergence of the impacts. This can be based on improved knowledge, early warning systems and planning taking the potential impacts of climate change into account (UNDP 2009). The costs of climate proofing and adaptation are often seen as the main obstacle to the implementation of adaptation measures. However, it is important to bear in mind that without preventive actions to climate change, human lives and the global environment may suffer greatly. There are five main steps involved in the climate proofing process (Fig. 3).

7.2 Reallocation of Water Among Different Sectors

India, with 2.4% of the world's total geographical area and 18% of the world's population, has only 4% of the world's total freshwater resources. With about 4000 billion cubic meter (bcm) of annual rainfall, the estimated utilizable water resources is only 1123 bcm (28%), mainly due to hydrological, topographic and other physical constraints. Of the available utilizable resource, 690 bcm is from surface water sources and the remaining 433 bcm is from replenishable groundwater sources. As

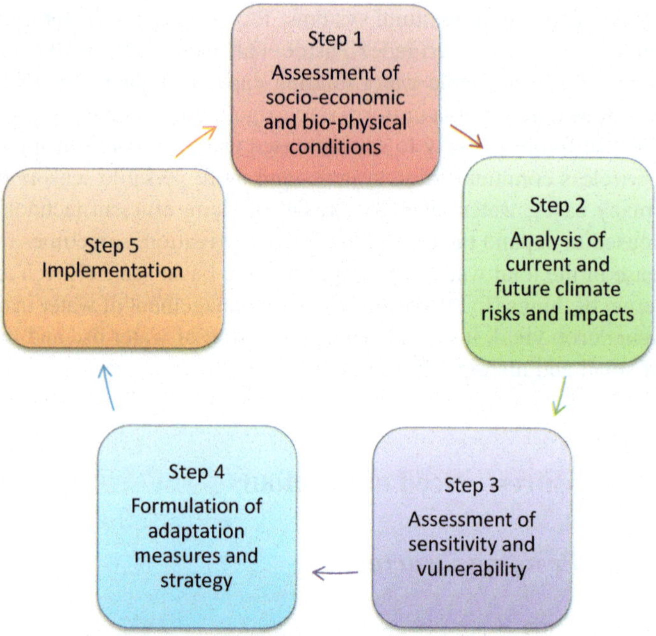

Fig. 3 Climate proofing process. *Source* UNDP (2009)

against this, the cumulative water utilization by all sectors of the economy is 702 bcm of which, agriculture sector alone consumes around 78% of the total water utilization (NABARD 2018). This is despite the fact that more than 55% of agriculture in India is rainfed and depends on the vagaries of monsoon. Addressing the challenge of water governance and water allocation between and within economic sectors requires significant efforts, especially in the context of constraints on water availability and climate change (FAO 2017).

With rapid population growth, urbanization and improvement in the living standards, the water requirement for all sectors is increasing giving a challenge for fair allocation of water. The National per capita annual water resource during 2001 was 1816 cubic meter which fell to 1544 cubic meter in 2010 (CWC 2015). As per International standard, a situation with less than 1000 cubic meter per capita is considered to be water scarcity situation. It is estimated that by 2050, the total water demand by all sub-sectors (1180 bcm) will surpass the total utilizable water resource of the country and the share of irrigation will come down to 68%. This means that improving water use efficiency is one of the key priorities of Indian Agriculture. Presently, the average efficiency in respect of surface water irrigation is 35–40% whereas the same is around 55% in the case of ground water irrigation.

The growing number of areas facing water scarcity necessitates adaptive water management strategies beyond traditional water supply and demand management methods, which are becoming increasingly difficult in many regions. Water reallocation offers a flexible water management approach to mitigate water scarcity under changing socioeconomic, climatic, and environmental conditions. In spite of the numerous benefits of reallocating water between users, examples of successful water transfers are relatively sparse and the expected benefits are rarely met in full due to several complex impediments (Marston and Cai 2016). Quantitative analysis of the real location and linkages of virtual water in the economic sector is important for the integrated water resources management in inland arid regions. Gao et al. (2020) showed that the direct blue and green water consumption of primary industry respectively accounted for 99.2% and 100% of the total water consumption in Northern Tianshan Mountains (NTM), China. Planting sector had the largest amount of Virtual Water (VW) outflow among all sectors. Animal husbandry, forestry and construction had a large pulling effect on VW outflow of planting sector, while planting sector and animal husbandry were the main sectors for VW export of blue and green water. We suggest that the government can increase the import of blue-green VW for agricultural raw materials through VW trade and develop industries such as service and electricity that have less pulling effect on the primary industry VW, so as to improve the economic added value of VW in the primary industry and reduce the loss of VW in primary industry production and trade flows in future water management. Razaee et al. (2021) reported that the economy-oriented allocation of water resources has caused many socio-environmental problems. The results showed that in the best scenario most water was allocated to the non-metal industry with a relative distance of 0.63 to the ideal solution in Iran. On the other hand, the current water allocation scenario ranked seventh, indicating that significant improvements are required to take into account the social, economic, and environmental factors for optimal reallocation of water resources among different industry users.

7.3 Crop Insurance

Agricultural production remains highly vulnerable to weather fluctuations and extreme events, such as droughts, floods and heat waves. Crop insurance is a risk management tool developed to mitigate some of this weather risk and protect farmer income in times of poor production. However, crop insurance may have unintended consequences for water resources sustainability, as the vast majority of freshwater withdrawals go to agriculture. Deryugina and Konar (2017) determined the empirical relationship between crop insurance and irrigation water withdrawals in the United States and found that 1% increase in insured crop acreage lead to a 0.223% increase in irrigation withdrawals, with most coming from groundwater aquifers. Kurdys-Kujawska et al. (2021) recognized that there is a mutual relationship between crop insurance, land productivity and the environment. Our empirical results show that the level of insurance coverage may support the increase in land productivity, indirectly affecting the environment. Farms with the highest productivity level were characterized by an average value of insurance that was double that compared to farms with the lowest productivity level. Thus, keeping in view the weather variability and socio-economic conditions of farmers, viable crop insurance schemes can be implemented to manage climatic risks in agriculture.

8 Conclusion

Climate change induced warming and variable rainfall patterns seem inevitable in future. As a result, heat stress along with floods and drought, will have adverse impacts on crop water productivity, especially in the tropical and sub-tropical regions including indo-gangetic plains. Such conditions might lead to increase in crop water requirements, thus putting severe pressure on already over-exploited ground water resources in the region along with decreased crop and water productivity. Under such conditions, climate smart agriculture needs to be adopted to sustain natural resources and crop water productivity. On-farm water management, rain water harvesting, development of groundwater and soil moisture improving strategies need to be adopted. In addition to this, diversification of cropping pattern and changes in crop calendars should be made. New crop varieties with resistance to floods, drought and heat stress should be explored. Irrigation management strategies such as deficit irrigation, alternate wetting and drying and micro-irrigation etc. should be adopted to improve water use efficiency in agriculture. Advanced technological interventions improved weather forecast, crop simulation, hydrological modeling as well as remote sensing and GIS should be used to manage climate risks in agriculture. Some policy measures viz. reallocation of water among different sectors and crop insurance etc. also need to be implemented to sustain future water resources in the region.

References

Abbate PE, Andrade FH, Culot JP, Bindraban PS (1997) Grain yield in wheat: effects of radiation during spike growth period. Field Crops Res 54:245–257

Acosta-Motos JR, Rothwell SA, Massam MJ, Albacete A, Zhang H, Dodd IC (2020) Alternate wetting and drying irrigation increases water and phosphorus use efficiency independent of substrate phosphorus status of vegetative rice plants. Plant Physiol Biochem 155:914–926. https://doi.org/10.1016/j.plaphy.2020.06.017

Adefisan EA (2018) Climate change impact on rainfall and temperature distributions over west Africa from three IPCC scenarios. J Earth Sci Clim Change 9:6. https://doi.org/10.4172/2157-7617.1000476

Ahmed EM, Abaas O, Ahmed M, Ismail MR (2011) Performance evaluation of three different types of local evaporative cooling pads in greenhouses in Sudan. Saudi J Biol Sci 18:45–51

Al-Ansari N, Abdellatil M, Ali SS, Knutsson S (2014) Long term effect of climate change on rainfall in northwest Iraq. Central Eur J Eng 4(3):250–263. https://doi.org/10.2478/s13531-013-0151-4

Alexandratos N, Bruinsma J (2012) World Agriculture towards 2030/2050. ESA Working Paper No. 12–03, 147 p

Alexandrov VA, Eitzinger J, Cajic V, Oberforster M (2002) Potential impact of climate change on selected agricultural crops in north-eastern Austria. Glob Change Bio 8:372–389

Ali MA, Ali M, Sattar M, Ali L (2010) Sowing date effect on yield of different wheat varieties. J Agric Res 48(2):157–162

Álvarez S, Bañón S, Sánchez-Blanco MJ (2013) Regulated deficit irrigation in different phenological stages of potted geranium plants: Water consumption, water relations and ornamental quality. Acta Physiol Plant 35:1257–1267

Amarasinghe UA, Sikka A, Mandave V, Panda RK, Gorantiwar S, Ambast SK (2021) Improving economic productivity to enhance resilience in canal irrigation systems: a pilot study of the Sine irrigation system in Maharashtra, India. Water Policy 1–19.https://doi.org/10.2166/wp.2021.231

Ambast SK, Tyagi NK, Raul SK (2006) Management of declining groundwater in the Trans Indo-Gangetic Plain (India): some options. Agric Water Manage 82:279–296. https://doi.org/10.1016/j.agwat.2005.06.005

Andarzian B, Hoogenboom G, Bannayan M, Shirali M, Andarzian B (2015) Determining optimum sowing date of wheat using CSM-CERES-Wheat model. J Saudi Soc Agric Sci 14:189–199

Anderson WK (1992) Increasing grain yield and water use of wheat in a rainfed Mediterranean type environment. Aust J Agric Res 43:1–17

Anderson WK, Smith WR (1990) Yield advantage of two semi dwarf compared with two tall wheats depends on sowing time. Aust J Agric Res 41:811–826

Anwar MR, O'Leary G, McNeil D (2007) Climate change impact on rainfed wheat in south-eastern Australia. Field Crops Res 104:139–147

Ashoka P, Kadasiddapa MM and Sanjey MT (2015) Enhancing water productivity through micro-irrigation technologies in Indian agriculture. Ann Plant Soil Res 17 (Special Issue):601–605

Asian Development Bank (2012) Guidelines for climate proofing investment in agriculture, rural development and food security. ISBN 978-92-9092-997-0 (Print), 978-92-9092-998-7 (PDF) Publication Stock No. TIM135541-2

Asseng S, Turner NC, Keating BA (2001) Analysis of water- and nitrogen-use efficiency of wheat in a Mediterranean climate. Pl Soil 233:127–143

Asseng S, He L, Zhao G, Wu D, Yang X, Zhuang W, Jin N, Yu Q (2015) Impacts of recent climate warming, cultivar changes and crop management on winter wheat phenology across the Loess Plateau of China. Agric Forest Meteor 200:135–143

Attri SD, Rathore LS (2003) Simulation of impact of projected climate change on wheat in India. Int J Climatol 23:693–705

Azahara Mesa-Jurado M, Martin-Ortegab J, Rutoc E, Berbel J (2012) The economic value of guaranteed water supply for irrigation under scarcity conditions. Agric Water Manage 113:10–18

Balyan S, Kumar M, Mutum RD, Raghuvanshi U, Agarwal P, Mathur S, Raghuvanshi S (2017) Identification of miRNA-mediated drought responsive multi-tiered regulatory network in drought tolerant rice, Nagina 22. Sci Rep 7:15446

Bannayan M, Eyshi Rezaei E, Hoogenboom G (2013) Determining optimum sowing dates for rainfed wheat using the precipitation uncertainty model and adjusted crop evapo-transpiration. Agric Water Manage 126:56–63

Bassu S, Asseng A, Motzo R, Giunta F (2009) Optimizing sowing date of durum wheat in a variable Mediterranean environment. Field Crops Res 111:109–118

Bates BC, Kundzewicz ZW, Wu S, Palutikof JP (2008) Climate change and water. Technical paper of the intergovernmental panel on climate change. IPCC Secretariat, Geneva

Bautista-Capetillo C, Marquez-Villagrana H, Pacheco-Guerrero A, Gonzalez-Trinidad J, Junez-Ferreira H, Zavala-Trejo M (2018) Cropping system diversification: water consumption against crop production. Sustainability 10:2164. https://doi.org/10.3390/su10072164

Benites J, Castellons A (2003) Improving soil moisture with conservation agriculture. Leisa Magazine

Berg A, Findell K, Lintner B, Giannini A, Senevirante SI, Hurk VRD, Lorenz R, Pitman A, Hageman S, Meier A, Cheruy F, Ducharne A, Malyshev S, Milly PCD (2016) Land–atmosphere feedbacks amplify aridity increase over land under global warming. Nat Clim Change 6:869–874

Bhogal S, Vatta K (2021) Can crop diversification be widely adopted to solve the water crisis in Punjab? Curr Sci 120(8):1303–1307

Bisbis MB, Gruda NS, Blanke MM (2019) Securing horticulture in a changing climate-a mini review. Horticulturae 5:56. https://doi.org/10.3390/horticulturae5030056

Blenkinsop S, Alves LM, Smith AJP (2021) Climate change increases extreme rainfall and thte chances of floods. ScienceBrief Review. In: LeQuéré C, Liss P, Forster P (eds) Critical issues in climate change science. https://doi.org/10.5281/zenodo.4779119

Borsato E, Rosa L, Marinello F, Tarolli P, D'Odorice P (2020) Weak and strong sustainability of irrigation: a framework for irrigation practices under limited water availability. Front Sustain Food Syst 4:17

Brouwer C, Prins K, Heibloem M (1989) Irrigation water management: irrigation scheduling. In Train. Manual 4; Food and Agriculture Organization, Rome, Italy

Carrijo DR, Lundy ME, Linquist BA (2017) Rice yields and water use under alternate wetting and drying irrigation: a meta-analysis. Field Crop Res 203:173–180. https://doi.org/10.1016/j.fcr.2016.12.002

Ceccarelli S, Grando S, Baum M (2007) Participatory plant breeding in water-limited environments. Expl Agric 43:411–435

Centella A, Taylor MA, Bezanilla A, Martinez D, Campbell JD, Stephenson TS, Vichot A (2015) Assessing the effect of domain size over the Caribbean region using the PRECIS regional climate model. Clim Dyn 44:1901–1918

Charles E (2016) Asia faces severe water crisis by 50—rising economic activity, population boom and climate change to blame: MIT. The Times of India 5(4):2016

Chen J, Wu H, Qian H, Li X (2018) Challenges and prospects of sustainable groundwater management in an agricultural plain along the silk road. Int J Water Resour Dev 34(3):354–368. https://doi.org/10.1080/07900627.2016.1238348

Clay J (2004) World agriculture and the environment: a commodity-by-commodity guide to impacts and practices. Island Press, Chicago

Çolak YB, Yazar A, Gönen E, Erog˘lua EC, (2018) Yield and quality response of surface and subsurface drip-irrigated eggplant and comparison of net returns. Agric Water Manage 206:165–175

Connor DJ, Theiveyanathan S, Rimmington GM (1992) Development, growth, water-use and yield of a spring and a winter wheat in response to time of sowing. Aust J Agric Res 43:493–516

CWC (2015) Annual Report 2014–15. Central Water Commission, Govt. of India

CWC (2019) Water and related statistics. http://www.indiaenvironmentportal.Org.in/files/file/water-and-related-statistics-2019.pdf

Davies EGR, Simonovic SP (2005) Climate change and the hydrological cycle. 17th Canadian hydrotechnical conference on hydrotechnical engineering: cornerstone of a sustainable environment, Edmonton, Alberta, 17–19 Aug 2005

Deryugina T, Konar M (2017) Impacts of crop insurance on water withdrawals for irrigation. Adv Water Resour 110:437–444. https://doi.org/10.1016/j.advwatres.2017.03.013

De Costa WAJM, Weerakoon WMW, Herath HMLK, Amaratunga KSP, Abeywardena RMI (2006) Physiology of yield determination of rice under elevated carbon dioxide at high temperatures in a subhumid tropical climate. Field Crops Res 96:336–347

De Pascale S, Barbieri G, Rouphael Y, Gallardo M, Orsini F, Pardossi A (2013) Irrigation management: challenges and opportunities. In: Good agricultural practices for greenhouse vegetable production in the South East European countries for greenhouse vegetable; plant production and protection paper 230; FAO, Rome, Italy, pp 79–105

Dev SM (2012) Small farmers in India: challenges and opportunities, working paper, WP-2012-014, Indira Gandhi Institute of Development Research, Mumbai. http://www.igidr.ac.in/pdf/publication/WP-2012-014.pdf

Dhankher OP, Foyer CH (2018) Climate resilient crops for improving global food security and safety. Plant Cell Environ 41:877–884
Doos BR, Shaw R (1999) Can we predict the future food production? A sensitivity analysis. Glob Environ Change 9:261–283
Egea G, Fernández JE, Alcon F (2017) Financial assessment of adopting irrigation technology for plant-based regulated deficit irrigation scheduling in super high-density olive orchards. Agric Water Manage 187:47–56
Eitzinger J, Stastna M, Zalud Z (2003) A simulation study of the effect of soil water balance and water stress on winter wheat production under different climate change scenarios. Agric Water Manage 61:195–217
Eitzinger J, Orlandini S, Stefanski R, Naylor REL (2010) Climate change and agriculture: introductory editorial. J Agric Sci Cambridge 148:499–500
Eitzinger J, Alexandrou V, Utset A, Saylan L (2015) Factors determining crop water productivity in agricultural crop production-a review. In: Marta AD, Nejedlik R, Altobelli F (eds) Water footprint applications for water resource management in agricultural. Ital J Agrometeorol 3:15–29
El-Attar HA, Merwad MA, Mostafa EAM, Saleh MMS (2019) The beneficial effect of subsurface drip irrigation system on yield, fruit quality and leaf mineral content of Valencia orange trees. Biosc Res 16:620–628
El Bilali H, Bassole IHN, Dambo L, Berjan S (2020) Climate change and food security. Agric Forestry 66(3):197–210
El-Gizawy N, Kh B (2009) Effect of planting date and fertilizer application on yield of wheat under no till system. World J Agric Sci 5(6):777–783
Enciso J, Jifon J, Anciso J, Ribera L (2015) Productivity of onions using subsurface drip irrigation versus furrow irrigation systems with an internet based irrigation scheduling program. Int J Agron 178180:1–6
Epting J, Michel A, Affolter A, Huggenberger A (2021) Climate change effects on groundwater recharge and temperatures in Swiss alluvial aquifers. J Hydrol X 11:100071. https://doi.org/10.1016/j.hydroa.2020.100071
Ertsen MW, Nagugi KN (2021) Ambivalent assets: The success of sand-storage dams for rainwater harvesting in Kitui County, Kenya. Front Water 3:676167. https://doi.org/10.3389/frwa.2021.676167
Estrella N, Sparks TH, Menzel A (2007) Trends and temperature response in the phenology of crops in Germany. Global Change Biol 13:1737–1747
FAO (2015) Climate change and food security: risks and responses. ISBN 978-92-5-108998-9
FAO (2017) Water for sustainable food and agriculture: a report produced for the G20 Presidency of Germany. ISBN 978-92-5-109977-3
FAO (2018) AQUASTAT—FAO's global information system on water and agriculture. From HYPERLINK "http://www.fao.org/aquastat/en/" http://www.fao.org/aquastat/en/
Fahad S, Bajwa AA, Nazir U, Anjum SA, Farooq A, Zohaib A, Sadia S, Nasim W, Adkins S, Saud S, Ihsan MZ, Alharby H, Wu C, Wang D, Huang J (2017) Crop production under drought and heat stress: plant responses and management options. Front Plant Sci 8:1147. https://doi.org/10.3389/fpls.2017.01147
Farahani SM, Chaichi MR (2012) Deficit (Limited) irrigation—a method for higher water profitability, irrigation systems and practices in challenging environments, Teang Shui Lee, IntechOpen. Available online: https://www.intechopen.com/books/irrigation-systems-and-practices-in-challenging-environments/deficit-limited-irrigation-a-method-for-higher-water-profitability-in-agriculture
Farooq M, Hussain M, Ul-Allah S, Siddique KHM (2019) Physiological and agronomic approaches for improving water-use efficiency in crop plants. Agric Water Manage 219:95–108
Farrag M, Perez GC, Solomatine D (2021) Spatio-temporal hydrological model structure and parameterization analysis. J Mar Sci Eng 9:467. https://doi.org/10.3390/jmse9050467

Ferrández-Pastor FJ, García-Chamizo JM, Nieto-Hidalgo M (2018) Mora-Martínez J (2018) Precision agriculture design method using a distributed computing architecture on Internet of Things context. Sensors 18:1731
Ficklin DL, Luo Y, Luedeling E, Zhang M (2009) Climate change sensitivity assessment of a highly agricultural watershed using SWAT. J Hydrol 374:16–29
Ficklin DL, Stewart IT, Maurer EP (2013) Climate change impacts on streamflow and Subbasin-scale hydrology in the upper Colorado River basin. pLoS ONE 8:8–19
Fischer G, Tubiello FN, Van Velthuizen H (2007) Wiberg DA (2007) Climate change impacts on irrigation water requirements: effects of mitigation, 1990–2080. Technol Forecast Soc Chang 74:1083–1107
Fuhrer J (2003) Agroecosystem responses to combinations of elevated CO_2, ozone and global climate change. Agr Ecosyst Environ 97:1–20
Gadgil S, Vinayachandran PN, Francis PA (2003) Droughts of the Indian summer monsoon: role of clouds over the Indian Ocean. Curr Sci 85:257–266
Gandhi VP, Bhamoriya (2011) Rainwater harvesting for irrigation in India. In: India Infrastructure Report 2011, pp 118–133
Gandhi VP, Johnson N, Singh G (2021) Improving water use efficiency in India's agriculture: the performance and impact of micro irrigation: a study of the Pradhan Mantri Krishi Sinchayee Yojana (PMKSY)—per drop more crop (PDMC). Final report. Centre for Management in Agriculture (CMA), Indian Institute of Management Ahmedabad (IIMA), Supported by Ministry of Agriculture and Farmers' Welfare, Government of India
Gao D, Long A, Yu J, Xu H, Su S, Zhao X (2020) Assessment of inter-sectoral virtual water reallocation and linkages in the northern Tianshan mountains, China. Water 12:2363. https://doi.org/10.3390/w12092363
Gayathri KD, Ganasri BP, Dwarakish GS (2015) A review on hydrological models. Aquatic Procedia 4:1001–1007. https://doi.org/10.1016/j.aqpro.2015.02.126
Geerts S, Raes D (2009) Deficit irrigation as an on-farm strategy to maximize crop water productivity in dry areas. Agric Water Manag 96:1275–1284
Georgiardi AG (1991) The change of the hydrological cycle under the influence of global warming. Hydrology for the water management of large River Basins—Proceedings of the Vienna symposium, August 1991. IAHS Publ. no. 201
GOI (2014) Crop diversification program in Haryana, Punjab & western Uttar Pradesh: sustainable agriculture with increased productivity & profitability 2013–14. Government of India, Ministry of Agriculture, Department of Agriculture & Cooperation, Crops Division, Krishi Bhavan, New Delhi, 110 011
GOI (2018) Operational guidelines of Pradhan Mantri Krishi Sinchayee Yojana (PMKSY)—micro irrigation fund. Department of Agriculture, Cooperation and Farmers' Welfare, Government of India
Goyal RK (2004) Sensitivity of evapotranspiration to global warming: a case study of arid zone of Rajasthan (India). Agric Water Manage 69:1–11
Graham C, Lin HS (2012) Subsurface flow networks at the hillslope scale: detection and modeling. In: Lin H (ed) Hydropedology: synergistic integration of soil science and hydrology. Elsevier, Waltham, MA, pp 559–594
Green C (2004) The evaluation of vulnerability to flooding. Disaster Prev Manag Int J 13(4):323–329. https://doi.org/10.1108/09653560410556546
Gregory PJ, Eastham J (1995) Growth of shoot and roots, and interception of radiation by wheat and lupin crops on a shallow, duplex soil in response to time of sowing. Aust J Agric Res 47:427–447
Gunda T, Bazuin JT, Nay J, Yeung KL (2017) Impact of seasonal forecast use on agricultural income in a system with varying crop costs and returns: an empirically grounded simulation. Environ Res Lett 12:034001. https://doi.org/10.1088/1748-9326/aa5ef7
Guo R, Lin Z, Moa X, Yang C (2010) Responses of crop yield and water use efficiency to climate change in the North China Plain. Agric Water Manage 97:1185–1194

Gupta A, Pathak H (2016) Climate change and agriculture in India. Climate change programme. Department of Science & Technology, Ministry of Science & Technology, Government of India, New Delhi

Gupta R, Gopal R, Jat HL (2010) Wheat productivity in Indo-Gangetic plains of India during 2010: terminal heat effects and mitigation strategies. Conserv Agric News Lett PACA 14:9

Hamouda GB, Tomozieu R, Pavan V, Antolini G, Snyder RL, Ventura F (2020) Impacts of climate change and rising atmospheric CO_2 on future projected reference evapotranspiration in Emilia-Romagna (Italy). Theoret Appl Climatol 46:801–820. https://doi.org/10.1007/s00704-021-03745-3

Han SJ, Tang QH, Xu D, Wang SL (2014) Irrigation-induced changes in potential evaporation: more attention is needed. Hydrol Process 28:2717–2720

Hanjra MA, Qureshi ME (2010) Global water crisis and future food security in an era of climate change. Food Policy 35(5):365–377. https://doi.org/10.1016/j.foodpol.2010.05.006

Hansen JW, Osgood D (2008) Economic value of seasonal climate forecasts for agriculture: review of ex-ante assessments and recommendations for future research. J Appl Meteorol Climatol 47:1269–1286. https://doi.org/10.1175/2007JAMC1540.1

Hanson JD, Liebig MA, Merrill SD, Tanaka DL, Krupinsky JM, Stott DE (2007) Dynamic cropping systems: increasing adaptability amid an uncertain future. Agron J 99:939–943

Hassan AA, Sarkar AA, Karim NN, Ali MH (2000) Irrigation schedule and deficit irrigation for wheat cultivation. Bangladesh J Agric 25(1/2):43–50

Hassan MU, Chattha MU, Khan I, Bhattha MB, Barbanti L, Aamer M, Iqbal MM, Nawaz M, Mahmood A, Ali A, Aslam MT (2020) Heat stress in cultivated plants: nature, impact, mechanism and mitigation strategies—a review. Plant Biosyst. https://doi.org/10.1080/11263504.2020.1727987

Hatfield JL, Dold C (2019) Water-use efficiency: advances and challenges in a changing climate. Front Plant Sci. https://doi.org/10.3389/fpls.2019.00103

Hatfield JL, Prueger JH (2015) Temperature extremes: effect on plant growth and development. Weather Clim Extremes 10:4–10

Hatibu N, Mahoo H (1999) Rainwater harvesting technologies for agricultural production: a case for Dodoma, Tanzania. In: Kaumbutho PG, Simalenga TE (eds) Conservation tillage with animal traction. A resource book of the animal traction network for Eastern and Southern Africa (ATNESA), Harare, Zimbabwe. 173p. A publication supported by French Cooperation, Namibia. http://www.atnesa.org

Hawkins GA, Vivoni ER, Robles-Morua A, Mascaro G, Rivera E, Domínguez FA (2015) Climate change projection for summer hydrologic conditions in a semiarid watershed of Central Arizona. J Arid Environ 118:9–20

Haworth M, Marino G, Brunetti C, Killi D, De Carlo A, Centritto M (2018) The impact of heat stress and water deficit on the photosynthetic and stomatal physiology of olive (*Olea europea* L.)—a case study of the 2017 heat wave. Plants 7:76.https://doi.org/10.3390/plants7040076

Hoffmann AA, Sgro CM (2011) Climate change and evolutionary adaptation. Nature 470:479–485

Horan R, Gowri R, Wable PS, Baron H, Keller VDJ, Garg KK, Majumdar PP, Houghton-Carr H, Rees G (2021) A comparative assessment of hydrological models in the upper Cauvery catchment. Water 13:151. https://doi.org/10.3390/w13020151

Hughes A, Mansour M, Ward R, Kieboom N, Seccombe D, Charlton M, Prodhomme C (2021) The impact of climate change on ground water recharge: scale assessment National-scale assessment for the British mainland. J Hydrol 598:126336. https://doi.org/10.1016/j.jhydrol.2021.126336

Humphreys E, Kukal SS, Christen EW, Hira GS, Sharma RK (2010) Halting the groundwater decline in north-west India-which crop technologies will be winners? Adv Agron 109:155–217. https://doi.org/10.1016/B978-0-12-385040-9.00005-0

Hussain HA, Men S, Hussain S, Chen Y, Ali S, Zhang S, Zhang K, Li Y, Qiwen X, Liao C, Wang L (2019) Interactive effects of drought and heat stresses on morpho-physiological attributes, yield, nutrient uptake and oxidative status in maize hybrids. Sci Rep 9:3890. https://doi.org/10.1038/s41598-019-40362-7

IPCC (2007) Summary for policymakers. In: Solomon S, Qin D, Manning M, Chen Z, Marquis M, Averyt KB, Tignor M, Miller HL (eds) Climate change 2007: the physical science basis, contribution of working group I to the fourth assessment report of the intergovernmental panel on climate change. Cambridge University Press, Cambridge, UK/New York, USA

IPCC (2014) Climate change Impacts, adaptation and vulnerability, Working group II contribution to the fifth assessment report of the Intergovernmental Panel on Climate Change. Technical report. Cambridge University Press, Cambridge, UK/New York, USA

IPCC (2018) Summary for policymakers. In: Masson-Delmotte V, Zhai PHO, Roberts D, Skea J, Shukla PR, Pirani A, Moufouma-Okia W, Péan C, Pidcock R, Connors S, Matthews JB, Chen Y, Zhou X, Gomis MI, Lonnoy E, Maycock T, Tignor M, Waterfield T (eds) Global warming of 1.5C. Intergovernmental panel on climate change, Geneva, p32. An IPCC special report on the impacts of global warming of 1.5c above pre-industrial levels and related global greenhouse gas emission pathways, in the context of strengthening the global response to the threat of climate change

Iseh AJ, Woma TY (2013) Weather forecasting models, methods and applications. Int J Eng Res Technol 2(12):1945–1956

Ishfaq M, Farooq M, Zulfiqar U, Hussain S, Akbar N, Nawaz A, Anjum SA (2020) Alternate wetting and drying: a water-saving and ecofriendly rice production system. Agric Water Manag 241:106363. https://doi.org/10.1016/j.agwat.2020.106363

Islam A, Ahuja LR, Garcia LA, Ma L, Saseendran AS (2012) Modelling the effect of elevated CO_2 and climate change on reference evapotranspiration in the semi-arid central Great Plains. Trans Am Soc Agric Biol Eng 55(6):2135–2146

Jakhar S, Chakhiya R, Dhankar M (2015) Rain water harvesting and its methods. Int J Eng Res Technol Special Issue 3(10):1–4

Jalota SK, Kaur H, Kaur S, Vashisht BB (2013) Impact of climate change scenarios on yield, water and nitrogen-balance and-use efficiency of rice-wheat cropping system. Agric Water Manage 116:29–38

Janni M, Gulli M, Maestri E, Nguyen HT, Marmiroli N (2020) Molecular and genetic bases of heat stress responses in crop plants and breeding for increased resilience and productivity. J Exp Bot 7(13):3780–3802. https://doi.org/10.1093/jxb/eraa034

Jha M, Gassman PW (2013) Changes in hydrology and streamflow as predicted by modeling experiment forced with climate models. Hydrol Process 28:2772–2781

Jha M, Pan Z, Takle SE, Gu R (2004) The impacts of climate change on stream flow in the Upper Mississippi River Basin: a regional climate model perspective. J Geophys Res Atmos 109:1984–2012

Jose VR, Bose R (2021) Groundwater management in India: problems and perspectives. Int J Water Resour 11:418

Joshi MA, Faridullah S, Kumar A (2016) Effect of heat stress on crop phenology, yield and seed quality attributes of wheat (*Triticum aestivum* L.) J Agrometeorol 18(2):206–215

Kattge J, Knorr W (2007) Temperature acclimation in a biochemical model of photosynthesis: a reanalysis of data from 36 species. Plant Cell Environ 30:1176–1190

Kaur B, Vatta K, Sidhu RS (2015) Optimising irrigation water use in Punjab agriculture: role of crop diversification and technology. Indian J Agric Econ 70(3):307–318

Kaur P, Kaur N, Singh H (2016) Near- and long-term projected changes in extreme weather events in Ludhiana district of Punjab. Agric Res J 53(3):360–366

Kaur H, Kingra PK, Singh Som Pal (2017) Climate variability impact on water requirement of *kharif* maize in different agroclimatic regions of Punjab. J Agrometeorol 19(Special Issue—AGMET 2016):285–289 (October 2017)

Kavdir Y, Zhang W, Basso B, Smucker AJM (2014) Development of a new long-term drought resilient soil water retention technology. J Soil Water Conserv 69(5):154A-160A. https://doi.org/10.2489/jswc.69.5.154A

Kim H, Lim C (2020) Strategies to assure the sustainability of groundwater resources. Sustainability 12:9183. https://doi.org/10.3390/su12219183

Kingra PK (2016) Climate variability impacts on wheat productivity in central Punjab. J Agrometeorol 18(1):97–99
Kingra PK (2016) Climate variability impact on rice productivity in central Punjab. J Agrometeorol 18(1):146–148
Kingra PK (2018) Climate variability impact on reference crop evapotranspiration computed using cropwat model. Agric Res J 55(2):265–273. https://doi.org/10.5958/2395-146X.2018.00049.2
Kingra PK, Kaur H (2017) Microclimatic modifications to manage extreme weather vulnerability and climatic risks in crop production. J Agric Phys 17(1):1–15
Kingra PK, Kukal SS (2013) Impact of climate change on evapo-transpiration and water productivity of wheat in central Punjab. J Res Punjab Agric Univ 50(3 & 4):181–183
Koech R, Langat P (2018) Improving irrigation water use efficiency: a review of advances, challenges and opportunities in the Australian context. Water 10:1771. https://doi.org/10.3390/w10121771
Komariah and Senge M, (2013) The development of water harvesting research for agriculture. Rev Agric Sci 1:31–42. https://doi.org/10.7831/ras.1.31
Kumar SN, Aggarwal PK et al (2014) Vulnerability of wheat production to climate change in India. Climate Res 59(3):173–187. https://doi.org/10.3354/cr01212
Kumar V, Bala S, Khan S (2017) Importance of weather prediction for sustainable agriculture in Bihar, India. Arch Agric Environ Sci 2(2):105–108
Kurdys-Kujawska A, Sompoiska-Rzechula A, Pawlowska-Tyszko J, Soliwoda M (2021) Crop insurance, land productivity and the environment: a way forward to a better understanding. Agriculture 11:1108. https://doi.org/10.3390/agriculture11111108
Kurylyk BL, MacQuarrie KTB, McKenzie JM (2014) Climate change impacts on groundwater and soil temperatures in cold and temperate regions: Implications, mathematical theory, and emerging simulation tools. Earth Sci Rev 138:313–334. https://doi.org/10.1016/j.earscirev.2014.06.006
Kuschel-Otarola M, Rivera D, Holzapfel E, Schutze N, Neumann P, Godoy-Faundez A (2020) Simulation of water use efficiency of crops under different irrigation strategies. Water 12:2930. https://doi.org/10.3390/w12102930
Lamaoui M, Jemo M, Datla R, Bekkaoui F (2018) Heat and drought stresses in crops and approaches for their mitigation. Front Chem 6:26. https://doi.org/10.3389/fchem.2018.00026
Li LJ, Zhang L, Wang H, Wang J, Yang JW, Jiang DJ, Li JY, Qin DY (2007) Assessing the impact of climate variability and human activities on stream flow from the Wuding River basin in China. Hydrol Process 21:3485–3491
Li X, Yuan W, Xu H, Cai Z, Yagi K (2011) Effect of timing and duration of midseason aeration on CH_4 and N_2O emissions from irrigated lowland rice paddies in China. Nutr Cycl Agroecosyst 91:293–305. https://doi.org/10.1007/s10705-011-9462-0
Li X, Zhao W, Li J, Li Y (2019) Maximizing water productivity of winter wheat by managing zones of variable rate irrigation at different deficit levels. Agric Water Manage 216:153–163
Liu Z, Jin J (2016) Review on rainfed agriculture and rainwater harvesting techniques. Adv Biol Sci Res 3:329–331
Liu X, Shao L, Sun H, Chen S, Zhang X (2013) Responses of yield and water use efficiency to irrigation amount decided by pan evaporation for winter wheat. Agric Water Manag 129:173–180. https://doi.org/10.1016/j.agwat.2013.08.002
Llyod Simon J, Sari Kovats R, Zaid Chalabi (2011) Climate change, crop yields, and undernutrition: development of a model to quantify the impact of climate scenarios on child undernutrition. Environ Health Perspect 119:1817. http://www.ncbi.nlm.nih.gov/pmc/articles/PMC3261974/
Lobell DB, Field CB (2007) Global scale climate–crop yield relationships and the impacts of recent warming. Environ Res Lett 2:014002. https://doi.org/10.1088/1748-9326/2/1/014002
Lobell DB, Ortiz-Monasterio JI, Addams CL, Asner GP (2002) Soil, climate, and management impacts on regional wheat productivity in Mexico from remote sensing. Agric for Meteorol 114:31–43
Long SP, Ainsworth EA, Leakey ADB, Nosberger J, Ort DR (2006) Food for thought: lower-than-expected crop yield stimulation with rising CO_2 concentrations. Science 312:1918–1921

Luan X, Vico G (2021) Canopy temperature and heat stress are increased by compound high air temperature and water stress and reduced by irrigation—a modeling analysis. Hydrol Earth Syst Sci 25:1411–1423. https://doi.org/10.5194/hess-25-1411-2021

Luhunga PM, Kijazi AL, Changá L, Kondowe A, Ngóngolo H, Motongori H (2018) Climate change projections for Tanzania based on high-resolution regional climate models from the coordinated regional climate downscaling experiment (CORDEX)—Africa. Front Environ Sci 6:122. https://doi.org/10.3389/fenvs.2018.00122

Luo Q, Williams M, Bellotti W (2003) Quantitative and visual assessments of climate change impacts on South Australian wheat production. Agric Syst 77:173–186

Mango LM, Melesse AM, Mclain ME, Gan D, Setegn SG (2011) Land use and climate change impacts on the hydrology of the upper Mara River Basin, Kenya: results of a modeling study to support better resource management. Hydrol Earth Syst Sci 15:2245–2258

Mar S, Nomura H, Takahashi Y, Ogata K, Yabe M (2018) Impact of erratic rainfall from climate change on pulse production efficiency in lower Myanmar. Sustainability 10:402

Marston L, Cai X (2016) An overview of water reallocation and the barriers to its implementation. Wires Water. https://doi.org/10.1002/wat2.1159

Martínez-Castro D, Da Rocha RP, Benzanilla-Morlot A, Alvarez-Escudero L, Reyes-Fernández JP, Silva-Vidal Y, Arritt RW (2006) Sensitivity studies of the RegCM3 simulation of summer precipitation, temperature and local wind field in the Caribbean region. Theor Appl Climatol 86:5–22

Martínez-Castro D, Vichot-Llano A, Bezanilla-Morlot A, Centella-Artola A, Campbell J, Giorgi F, Viloria-Holguin C (2014) The performance of RegCM4 over the Central America and Caribbean region using different cumulus parameterizations. Revista De Climatologia 16:77–97

Martinez J, Reca J (2014) Water use efficiency of surface drip irrigation versus an alternative subsurface drip irrigation method. J Irrig Drain Eng 140(04014030):77–98

Mazza G, Agnelli AE, Orasen G, Gennaro M, Valè G, Lagomarsino A (2016) Reduction of global warming potential from rice under alternate wetting and drying practice in a sandy soil of northern Italy. Italian J Agrometeorol 2:35–40. https://doi.org/10.19199/2016.2.2038-5625.035

McDonald GK, Sutton BG, Ellison FW (1983) The effect of time of sowing on the grain yield of irrigated wheat in the Namoi Valley, New South Wales. Aus J Agric Res 34:229–240

Mckenna T (2012) Feed the future deputy coordinator for development, Report before the senate foreign relations international development subcommittee, 28 Nov 2012. http://www.usaid.gov/news-information/congressional-testimony/testimony-tjada-mckenna-feed-future-deputy-coordinator

McLeman R, Smit B (2006) Vulnerability to climate change hazards and risks: crop and flood insurance. Can Geogr/le Géographe Canadien 50(2):217–226. https://doi.org/10.1111/j.0008-3658.2006.00136.x

McVicar TR, Van Niel TG, Li LT, Roderick ML, Rayner DP, Ricciardulli L, Dono-hue RJ (2008) Wind speed climatology and trends for Australia, 1975–2006: capturing the stilling phenomenon and comparison with nearsurface reanalysis output. Geophys Res Lett 35:L20403. https://doi.org/10.1029/2008GL035627

Meehl GA, Stocker TF, Collins WD, Gaye AJ, Gregory JM, Kitoh A, Knutti, Murphy JM, Noda A, Raper SCB, Watterson JG, Weaver AJ, Zhao Z (2007) Global climate projections. In: Solomon S, Qin D, Manning M, Chen Z, Marquis M, Averyt KB, Tignor M, Miller HL (eds) Cambridge University Press, Cambridge, UK and New York, NY

Menberg K, Blum P, Kurylyk BL, Bayer P (2014) Observed groundwater temperature response to recent climate change. Hydrol Earth Syst Sci 18(11):4453–4466. https://doi.org/10.5194/hess-18-4453-2014

Mishra AK, Singh VP (2010) A review of drought concepts. J Hydrol 391:202–216

Mishra A, Liu SC (2014) Changes in precipitation pattern and risk of drought over India in the context of global warming. J Geophys Res Atmos 119:7833–7841

Misra AK (2014) Climate change and challenges of water and food security. Int J Sustain Built Environ 3:153–165. https://doi.org/10.1016/j.ijsbe.2014.04.006

Mo XG, Lin ZH, Liu SX (2007) Climate change impacts on the ecohydrological processes in the Wuding River basin. Acta Ecol Sin 27(12):4999–5007

Mo X, Liu S, Lin Z, Guo R (2009) Regional crop yield, water consumption and water use efficiency and their responses to climate change in the North China Plain. Agric Ecosys Environ 134:67–78

Moore CE, Meacham-Hensold K, Lemonnier P, Slattery RA, Benjamin C, Bernacchi CJ, Lawson T, Cavanagh AP (2021) The effect of increasing temperature on crop photosynthesis: from enzymes to ecosystems. J Exp Bot 72(8):2822–2844. https://doi.org/10.1093/jxb/erab090

Montecelos-Zamora Y, Cavazos T, Kretzschmar T, Vivoni Enrique R, Corzo G, Molina-Navarro E (2018) Hydrological modeling of climate change impacts in a tropical river basin: a case study of the Cauto River, Cuba. Water 10:1135. https://doi.org/10.3390/w10091135

Morison JIL, Baker NR, Mullineaux PM, Davies WJ (2008) Improving water use in crop production. Philos Trans R Soc B 363:639–658. https://doi.org/10.1098/rstb.2007.2175

Mustroph A (2018) Improving flooding tolerance of crop plants. Agronomy 8:160. https://doi.org/10.3390/agronomy8090160

Mzyece A, Ng'ombe JN, (2021) Crop diversification improves technical efficiency and reduces income variability in northern Ghana. J Agric Food Res 5:100162. https://doi.org/10.1016/j.jafr.2021.100162

NABARD (2018) Sectoral paper on water resources. Farm Sector Policy Department, NABARD Head Office, Mumbai

Nandargi SS, Barman K (2018) Evaluation of climate change impact on rainfall variation in west Bengal. Acta Sci Agric 2(7):74–82

Nangia V, Oweis T, Kemeze FH, Schnetzer J (2018) Supplemental irrigation: a promising climate-smart practice for dry-land agriculture. Global alliance for climate smart agriculture (GACSA). Rome, Italy 1:1–8

Nasir A, Shahriar S, Rupa WS, Mehjar H, Jamal Uddin AFM (2014) Alternate wetting and drying irrigation system on growth and yield of hybrid boro rice. J Soil Nat 7(2):28–35

Nellemann C, MacDevette M, Manders T, Eickhout B, Svihus B, Prins AG, Kaltenborn BP (2009) The environmental food crisis—the environment's role in averting future food crises. A UNEP rapid response assessment. United Nations Environment Programme, GRID-Arendal, www.grida.no

Nikolaou G, Neocleous D, Christou A, Kitta E, Katsoulas N (2020) Implementing sustainable irrigation in water scarce regions under the impact of climate change. Agronomy 10:1120. https://doi.org/10.3390/agronomy10081120

Nutan KK, Rathore RS, Tripathi AK, Mishra M, Pareek A, Singla Pareek SL (2020) Integrating dynamics of yield traits in rice in response to environmental changes. J Exp Bot 71:490–506

Oweis T (1997) Supplemental irrigation. A highly efficient water use practice. ICARDA Editions, p 16

Oweis T, Zhang H, Pala M (2000) Water use efficiency of rainfed and irrigated bread wheat in a Mediterranean environment. Agron J 92:231–238

Owiss T, Pala M, Ryan J (1999) Management alternatives for improved durum wheat production under supplemental irrigation in Syria. Eur J Agron 11:255–266

Pareek A, Dhankher OP, Foyer CH (2020) Mitigating the impact of climate change on plant productivity and ecosystem sustainability. J Exp Bot 71(2):451–456. https://doi.org/10.1093/jxb/erz518

Parry ML, Canziani OF, Palutikof JP, van der Linden PJ, Hanson CE (2007) Climate change 2007: impacts, adaptation and vulnerability. In: Contribution of working group II to the fourth assessment report of the intergovernmental panel on climate change; Cambridge University Press: Cambridge, UK; New York, NY, USA

Pauline AA, Mahandrakumar K, Karthekeyan C (2020) Adoption of rainwater harvesting structures in dry land areas of Tamil Nadu, India. Int J Curr Microbiol Appl Sci 9(3):3271–3278. https://doi.org/10.20546/ijcmas.2020.903.375

Peel MC, Bloschl G (2011) Hydrological modeling in a changing world. Prog Phys Geogr 35(2):249–261. https://doi.org/10.1177/0309133311402550

Pereira DR, Martinez MA, da Silva DD, Pruski FF (2016) Hydrological simulation in a basin of typical tropical climate and soil using the SWAT Model Part II: simulation of hydrological variables and soil use scenarios. J Hydrol Reg Stud 5:149–163

Porfirio LL, Newth D, Finnigan JJ, Cai Y (2018) Economic shifts in agricultural production and trade due to climate change. Palgrave Commun 4:111

Porporato A, D'Odoroco P, Laio F, Ridolfi L, Rodriguez-Iturbe I (2002) Ecohydrology of water-controlled ecosystems. Adv Water Res 25:1335–1348

Possemiers M, Huysmans M, Batelaan O (2014) Influence of aquifer thermal energy storage on groundwater quality: a review illustrated by seven case studies from Belgium. J Hydrol Reg Stud 2:20–34

Prasad R (2017) Relationship between groundwater and monsoon rainfall over India. www.the hindu. 11.1.2017

Priya M, Sharma L, Kaur R, Bindumadhava H, Nair RM, Siddique KHM, Nayyar H (2019) GABA (γ-aminobutyric acid), as a thermo-protectant, to improve the reproductive function of heat-stressed mungbean plants. Sci Rep 9:7788

PWC (2020). Creating an ecosystem for increasing water use efficiency in agriculture

Rahaman MM, Shourov MR, Akram MN (2016) Production of rice versus groundwater depletion in Bangladesh during 1988–2014. In: Proceedings of conference on water security and climate change: challenges and opportunities in Asia. Asian Institute of Technology, Bangkok, Thailand. Springer, Berlin

Ramachandran N (2014) Persisting under nutrition in India: causes, consequences and possible solutions. Springer, New Delhi

Rao BB, Chowdary PS, Sandeep VM, Pramod VP, Rao VUM (2015) Spatial analysis of the sensitivity of wheat yields to temperature in India. Agric Forest Meteorol 200:192–202

Rauf S, Al-Khayri JM, Zaharieva M, Monneveux P, Khalil F (2015) Breeding strategies to enhance drought tolerance in crops. In: Al Khayri JM et al (eds) Advances in plant breeding strategies, volume 2: agronomic, abiotic and bitic stress traits. https://doi.org/10.13140/2.1.2343.9682

Raveh E, Cohen S, Raz T, Yakir D, Grava A, Goldschmidt EE (2003) Increased growth of young citrus trees under reduced radiation load in a semi-arid climate. J Exp Bot 54:365–373

Razaee A, Bozorg-Haddad O, Chu X (2021) Reallocation of water resources according to social, economic and environmental parameters. Nat Sci Rep 11:17514. https://doi.org/10.1038/s41 598-021-96680-2

Reid G (2004) Improving soil moisture. Agnote DPI—494 ISSN 1034-6848. www.dpi.nsw.gov.au

Riaz MW, Yang L, Yousaf MI, Sami A, Mei XD, Shah L, Rehman S, Xue L, Si H, Ma C (2021) Effects of heat stress on growth, physiology of plants, yield and grain quality of different spring wheat (*Triticum aestivum* L.) genotypes. Sustainability 13:2972. https://doi.org/10.3390/su1305 2972

Rodriguez C, Martensson LD, Zachrison M, Carlsson G (2021) Sustainability of diversified organic cropping systems—challenges identified by farmer interviews and multi-criteria assessments. Front Agron. https://doi.org/10.3389/fagro.2021.698968

Sacks WJ, Cook BI, Buenning N (2009) Effects of global irrigation on the near-surface climate. Clim Dyn 33:159–175

Samra JS, Kaur P, Mahal AK (2012) Spectral density analysis of the cold wave (2010–11 and 2011–2012) and its impact on wheat productivity in Indian Punjab. Paper presented at 3rd international agronomy congress, New Delhi 27th Nov 2012

Sarkar S (2001) Effect of water stress on growth, productivity and water expense efficiency of summer rice. Indian J Agril Sci 71(3):153–158

Sarkar S, Sarkar S (2018) A review of impact of climate change on evapotranspiration. Pharma Innov J 7(11):387–390

Schewe J, Gosling SN, Reyer C et al (2019) State-of-the-art global models underestimate impacts from climate extremes. Nat Commun 10:1005

Schmidhuber J, Tubiello FN (2007) Global food security under climate change. Proceedings of the National Academy of Sciences, 104(2007):19703. http://www.pnas.org/content/104/50/19703.full.pdf

Sesha MVR, Murthy CS, Chandrasekar K, Jeysaleen A, Diwakar PG, Dhadhwal VK (2016) Agricultural drought: assessment and monitoring. Mausam 67:131–142

Shah KK, Modi B, Pandey HP, Subedi A, Aryal G, Pandey M, Shrestha J (2021) Diversified crop rotation: an approach for sustainable agriculture production. Adv Agric Article ID 8924087.https://doi.org/10.1155/2021/8924087

Shahane AA, Shivay YS, Prasanna R, Kumar D (2019) Improving water and nutrient use efficiency in rice by changing crop establishment methods, application of microbial inoculations, and Zn fertilization. Global Challenges (Hoboken, NJ) 3:1800005

Sharma BR, Gulati A, Mohan G, Manchanda S, Ray I, Amarasinghe U (2018) Water productivity mapping of major Indian crops. NABARD and ICRIER

Shashidhar HE, Kanbar A, Toorchi M, Raveendra GM, Kundur P, Vimarsha HS, Soman R, Kumar NG, Bekele BD and Bhavani P (2013) Breeding for drought resistance using whole plant architecture—conventional and molecular approach. https://doi.org/10.5772/54983

Siebert S, Ewert F, Rezaei EE, Kage H, Grab R (2014) Impact of heat stress on crop yield—on the importance of considering canopy temperature. Environ Res Lett 9:044012. https://doi.org/10.1088/1748-9326/9/4/044012

Silva VDPR, Campos JH, Silva MT, Azevedo PV (2010) Impact of global warming on cowpea bean cultivation in northeastern Brazil. Agric Water Manage 97:1760–1768

Singh RK (2010) Sensitivity of evapotranspiration to climate change: a case study of Jaipur district in Rajasthan. The Asian Man 4:39–48

Singh AK, Singh N, Singh H, Kushwaha HS (2021a) Role and importance of weather forecasts in modern agriculture. Int J Curr Microbiol App Sci 10(1):2646–2662

Singh M, Singh P, Singh S, Saini RK, Angadi SV (2021b) A global meta-analysis of yield and water productivity responses of vegetables to deficit irrigation. Nat Sci Rep 11:22095. https://doi.org/10.1038/s41598-021-01433-w

Sipayung SB, Nurlatifah A, Siswanto B, Lilik Slamet S (2018) Analysis of climate change impact on rainfall pattern of Sambas district, west Kalimantan. Earth Environ Sci 149:012029. https://doi.org/10.1088/1755-1315/149/1/012029

Southworth J, Pfeifer RA, Habeck M, Randolph JC, Doering OC, Rao DG (2002) Sensitivity of winter wheat yields in the Midwestern United States to future changes in climate, climate variability and CO_2 fertilization. Climate Res 22(1):73–86

Smerdon BD (2017) A synopsis of climate change effects on groundwater recharge. J Hydrol 555:125–128. https://doi.org/10.1016/j.jhydrol.2017.09.047

Smuleac L, Rujescu C, Smuleac A, Imbrea F, Radulov I, Manea D, Lenciu A, Adamov T, Pascalau R (2020) Impact of climate change in the Banat Plain, western Romania, on the accessibility of water for crop production in agriculture. Agriculture 10:437. https://doi.org/10.3390/agriculture10100437

Snyder RL, Moratiel R, Song Z, Swelam A, Jomaa I, Shapland T (2011) Evapotranspiration response to climate change. In: Fernández JE, Ferreira MI (eds) Proceedings of XXVIIIth IHC—IS on water use in a changing world. Acta Hort 922, ISHS 2011, pp 91–98

Sohoulande DDC, Singh VP (2016) Impact of climate change on the hydrologic cycle and implications for society. Environ Soc Psychol 1(1):36–49. https://doi.org/10.18063/ESP.2016.01.002

Stagl J, Mayr E, Koch H, Hattermann FF, Huang S (2014) Effects of climate change on the hydrological cycle in central and Eastern Europe. In: Rannow S, Neubert M (eds) Managing protected areas in central and Eastern Europe under climate change. Adv Global Change Res 58. https://doi.org/10.1007/978-94-007-7960-0_3

Stapper M, Harris HC (1989) Assessing the productivity of wheat genotype in a Mediterranean climate, using a crop-simulation model. Field Crops Res 20:129–152

Stone RC, Meinke H (2006) Weather, climate and farmers: an overview. Meteorol Appl (Supplement): 7–20. https://doi.org/10.1017/S1350482706002519

Tabari H (2020) Climate change impact on flood and extreme precipitation increases with water availability. Nat Sci Rep 10:13768. https://doi.org/10.1038/s41598-020-70816-2

Tamagnone P, Cea L, Comino E, Rosso M (2020) Rainwater harvesting techniques to face water scarcity in African Drylands: hydrological efficiency assessment. Water 12:2646. https://doi.org/10.3390/w12092646

Tanner CB and Sinclair TR (1983) Efficient water use in crop production: Research or re-research? In: Taylor HM et al (eds) Limitations to efficient water use in crop production. ASSA, CSSA, SSSA, Madison, WI, pp 1–27

Tao FL, Yokozawa M, Hayashi Y, Erda L (2003) Future climate change, the agricultural water cycle, and agricultural production in China. Agr Ecosyst Environ 95:203–215

Tao F, Zhang Z, Xiao D et al (2014) Responses of wheat growth and yield to climate change in different climate zones of China, 1981–2009. Agric For Meteorol 189–190, 91–104.497. https://doi.org/10.1016/j.agrformet.2014.01.013

Taylor RG, Scanlon B, D¨oll P, Rodell M, van Beek R, Wada Y, Longuevergne L, Leblanc M, Famiglietti JS, Edmunds M, Konikow L, Green TR, Chen J, Taniguchi M, Bierkens MFP, MacDonald A, Fan Y, Maxwell RM, Yechieli Y, Gurdak J, Allen DM, Shamsudduha M, Hiscock K, Ye, PF, Holman I, Treidel H (2012) Ground water and climate change. Nat Clim Change 3(4):322–329.https://doi.org/10.1038/nclimate1744

Teichmann C, Bülow K, Otto J, Pfeifer S, Rechid D, Sieck K, Jacob D (2018) Avoiding extremes: benefits of staying below +1.5 °C compared to +2.0 °C and +3.0 °C global warming. Atmosphere 9:115

Teixeira EI, Fischer G, Velthuizen HV, Walter C, Ewert F (2013) Global hotspots of heat stress on agricultural crops due to climate change. Agric Meteorol 170:206–215

Thomas A (2008) Agriculture irrigation demand under present and future climate scenarios in China. Global Planet Change 60(3–4):306–326

Thomson AM, Izaurralde RC, Rosenberg NJ, He XX (2006) Climate change impacts on agriculture and soil carbon sequestration potential in the Huang-Hai Plain of China. Agr Ecosyst Environ 114:195–209

Tirado R, Cotter J (2010) Ecological farming: drought resistant agriculture. Greenpeace Research Laboratories, University of Exeter, UK. GRL-TN 02/2-10

Tolossa TT, Abebe FB, Girma AA (2020) Review: rainwater harvesting technology practices and implication of climate change characteristics in eastern Ethiopia. Cogent Food Agric 6:1724354. https://doi.org/10.1080/23311932.2020.1724354

Toung TP, Bhuiyan SI (1994) Innovations towards improving wateruse efficiency in rice. In: Paper presented at the world bank's 1994 water resource seminar, 13–15 Dec 1994, VA, USA

Tubiello FN, Donatelli M, Rosenzweig C, Stockle CO (2000) Effects of climate change and elevated CO_2 on cropping systems: model predictions at two Italian locations. Eur J Agro 13:179–189

Udayashankara TH, Sadashiva Murthy BM, Madhukar M (2016). Impact of climate change on rainfall pattern and reservoir level. J Water Resour Eng Manag 3(1):10–14. ISSN: 2349-4336

UNDP (2009) Guidelines for climate change proofing in UNDP projects and programmes in Armenia

Van de Geijn SC, Goudriaan J (1996) The effects of elevated CO_2 and temperature change on transpiration and crop water use. In: Bazzaz F, Sombroek WG (eds) Global climate change and agriculture production. Wiley, Chichester (UK), pp 1–21

Van Opstal J, Droogers P, Kaune A, Steduto P, Perry C (2021) Guidance on realizing real water savings with crop water productivity interventions. Wageningen, FAO and Future Water. https://doi.org/10.4060/cb3844en

Verma S, Kumar N, Verma A, Singh H, Siddique KHM, Singh NP (2020) Novel approaches to mitigate heat stress impacts on crop growth and development. Plant Physiol Rep 25(4):627–644. https://doi.org/10.1007/s40502-020-00550-4

Wajid A, Hussain A, Ahmad A, Rafiq M, Goheer AR, Ibrahim M (2004) Effect of sowing date and plant density on growth, light interception and yield of wheat under Semi Arid conditions. Int J Agric Biol 6:1119–1123
Walker WR (1983) Irrigation engineering—sprinkler, trickle, surface irrigation principles, design and agricultural practices: A. Benamiand A. Ofen. IESP, Haifa, in Cooperation with the International Irrigation In. Agric Water Manage 9:263–264
Walker NJ, Schulze RE (2006) An assessment of sustainable maize production under different management and climate scenarios for smallholder agro-ecosystems in KwaZulu-Natal, South Africa. Phys Chem Earth 31:995–1002
Wang J, Niu W, Li Y, Lv W (2018) Sub surface drip irrigation enhances soil nitrogen and phosphorus metabolism in tomato root zones and promotes tomato growth. Appl Soil Ecol 124:240–251
Wang X, Zhang P, Liu L, Li D, Wang Y (2019) Effects of human activities on hydrological components in the Yiluo River Basin in Middle Yellow River. Water 11:689. https://doi.org/10.3390/w11040689
Wen J, Wang X, Guo M, Xu X (2012) Impact of climate change on reference crop evapotranspiration in Chuxiong City, Yunnan Province. Procedia Earth Planet Sci 5:113–119
White EK, Peterson TJ, Costelloe J, Western AW, Carrara E (2016) Can we manage groundwater? A method to determine the quantitative testability of groundwater management plans. Water Resour Res 52:4863–4882. https://doi.org/10.1002/2015WR018474
Wilhite DA, Smith MB (2005) Drought as Hazard: understanding the natural and social context, Drought and Water Crisis. Sci Technol Manag 3–29
Woldesenbet TA, Elagib NA, Ribbe L, Heinrich J (2018) Catchment response to climate and land use changes in the Upper Blue Nile sub-basins, Ethiopia. Sci Total Environ 644:193–206
Wu DR, Yu Q, Lua CH, Hengsdijk H (2006) Quantifying production potentials of winter wheat in the North China Plain. Eur J Agron 24:226–235
Wu W-Y, Lo M-H, Wada Y, Famiglietti JS, Reager JT, Yeh Pat J-F, Ducharne A, Yang Z-L (2020) Divergent effects of climate change on future groundwater availability in key mid-latitude aquifers. Nature Commun 11:3710.https://doi.org/10.1038/s41467-020-17581-y
Xiao D, Tao F, Liu Y, Shi W, Wang M, Liu F, Zhang S, Zhu Z (2012) Observed changes in winter wheat phenology in the North China Plain for 1981–2009. Int J Biometeorol pp 1–11
Yan N, Wu B (2014) Integrated spatial-temporal analysis of crop water productivity of winter wheat in Hai Basin. Agric Water Manage 133:24–33
Yang J, Dongfen H, Hua D, Guilu T, Jianhua Z (2009) Alternate wetting and moderate soil drying increases grain yield and reduces cadmium accumulation in rice grains. J Sci Food Agric 89(10):1728–1736
Yang J, Zhou Q, Zhang J (2017) Moderate wetting and drying increases rice yield and reduces water use, grain arsenic level and methane emission. Crop J 5:151–158. https://doi.org/10.1016/j.cj.2016.06.002
Yohe et al (2007) Perspectives on climate change and sustainability", In: Parry ML et al (eds) Climate Change 2007: impacts, adaptation and vulnerability. Contribution of working group II to the fourth assessment report of the intergovernmental panel on climate change. Cambridge University Press, UK, pp 811–841. https://www.ipcc.ch/pdf/assessment-report/ar4/wg2/ar4-wg2-chapter20.pdf
Yumurtaci A (2015) Utilization of wild relatives of wheat, barley, maize and oat in developing abiotic and biotic stress tolerant new varieties. Emir J Food Agric 27(1)
Zhang J, Ren W, An P, Pan Z, Wang L, Dong Z, He D, Yang J, Pan S, Tian H (2015) Responses of crop water use efficiency to climate change and agronomic measures in the semiarid area of northern China. PLoS ONE 10(9):e0137409. https://doi.org/10.1371/journal.pone.0137409
Zhang D, Du Q, Zhang Z, Jiao X, Song X, Li J (2017) Vapour pressure deficit control in relation to water transport and water productivity in greenhouse tomato production during summer. Sci Rep 7(1):43461. https://doi.org/10.1038/srep43461

Zhao W, Liu L, Shen Q, Yang J, Han X, Tian F, Wu J (2020) Effects of water stress on photosynthesis, yield and water use efficiency in winter wheat. Water 12:2127. https://doi.org/10.3390/w12082127

Zwart SJ, Bastiaanssen WGM (2004) Review of measured crop water productivity values for irrigated wheat, rice, cotton and maize. Agric Water Manage 69:115–133. https://doi.org/10.1016/j.agwat.2004.04.007

Dr. Pavneet Kaur Kingra is serving as Professor and Head in the Department of Climate Change and Agricultural Meteorology at Punjab Agricultural University, Ludhiana, Punjab, India. She is involved in teaching, research and extension in the field of Climate Change and Agricultural Meteorology during the last 24 years. She has authored about 140 research papers and 5 text books along with several scientific reports, research bulletins and popular articles. She has conducted research in the field of climate change and its impacts on crop and water productivity as well as application of geospatial technology in agricultural meteorology. She has also contributed in the field of weather forecasting and agro-advisory services for the farmers of Punjab.

Dr. Surinder Singh Kukal has served as Professor of Soil Conservation, Additional Director of Research (Natural Resource Management) and Dean, College of Agricultureat Punjab Agricultural University for 35 years. He has done extensive research work in the field of soil and water conservation, the most notable being a novel technique of scheduling irrigation to crops using a cheap, simple and farmer-friendly, patented tensiometer. Dr. Kukal has about 400+ research publications in the field of soil and water conservation in journals of national and international repute. He has authored 2 books on environmental science and edited several books in the field of soil and water conservation. Besides, he has served as Editor-in-Chief of Indian Journal of Ecology, and presently he is Associate Editor of International Soil and Water Conservation Research. He has guided the research work of more than 30 M.Sc. and Ph.D.students as Supervisor. He has several awards to his credit, the most prestigious being Fellow, National Academy of Agricultural Sciences and Fellow, Indian Society of Soil Science; Dr. G.S. Khush Distinguished Professor award; 12th International Congress Commemoration award; Honorary Member of Watershed Management of Iran; Outstanding Teacher Award of PAU. He has delivered invited lectures as Visiting Scientist in various countries like USA, Australia, Serbia, China, Thailand, Singapore, Pakistan, etc.

Chapter 14
A Sustainable Method of Production Towards Food Security Using Aquaponics: A Case Study from Oman

Ahmed Al-Busaidi, Mushtaque Ahmed, Wenresti Gallardo, Waad Al-Aghbari, and Yahya Al-Yahyaei

Abstract In order to achieve Sustainable Development Goal (SDG) 12, Gulf countries need to look for innovative and sustainable production of food for local and expatriate populations. Opportunity exists for method of food production using limited but highly treated wastewater. A change to fish and vegetable production using treated wastewater will contribute to sustainable consumption and production (SCP) in the region. Therefore, the objective of the study conducted at Sultan Qaboos University, Oman was to evaluate the effect of tertiary treated wastewater on fish growth and later on the effect of the produced effluent coming from fish tank on grown crops. Nine tanks with dimensions of 80×40×40 cm were filled either with freshwater or a mixture of freshwater and treated wastewater (at 50:50 and 75:25 ratios). Each tank was stocked with 25 tilapias (*Oreochromis niloticus*) with an initial body weight of 49 g. Each tank was connected to another tank of same dimensions that was used to grow lettuce and bean crops on the top layer. Water was circulating between the two tanks. No fertilizer was added to all treatments and all tanks got similar amount of fish feed. It was found that tanks with treated wastewater got higher concentrations of dissolved oxygen due to algae growth and more salts content due to minerals added from treated wastewater compared to fresh water alone. Therefore, lettuce and bean growth was much better and got higher values of chlorophyll content compared to plants in control tanks. For heavy metal analysis, all waters got similar values but, in some samples, the concentrations of B, Cu, Mn and Zn were higher in treated wastewater compared to fresh water and that was reflected in lettuce roots. For the edible part, lettuce grown in treated wastewater got higher value of Fe and B compared to control. Similar concentrations were found with bean plants with higher values in treated wastewater compared to freshwater. However, low concentrations of heavy

A. Al-Busaidi · M. Ahmed (✉) · W. Al-Aghbari
Department of Soils, Water and Agricultural Engineering, College of Agricultural & Marine Sciences, Sultan Qaboos University, Muscat, Oman
e-mail: ahmedm@squ.edu.om

W. Gallardo · Y. Al-Yahyaei
Department of Marine Science and Fisheries, College of Agricultural and Marine Sciences, Sultan Qaboos University, Muscat, Oman

M. Behnassi et al. (eds.), *The Water, Climate, and Food Nexus*,
https://doi.org/10.1007/978-3-031-50962-9_14

metals were found in the edible parts of all plants in all treatments. Fish analyses showed that all tested heavy metals were within the safe limit. The positive aspect of this system is that it will help the environment by utilizing treated wastewater and reducing fertilizer applications. Moreover, farmer income will increase since both fish and crops will be produced with minimum resources.

Keywords Aquaponics · Tilapia · Lettuce · Bean · Fruit quality · Heavy metals · Treated wastewater

1 Introduction

Responsible consumption and production is the stated goal of the Sustainable Development Goal 12 (SDG 12 of the United Nations 2030 Agenda for Sustainable Development). Striving for such a stated goal will ultimately lead to sustainable consumption and production in all sectors of economy and will lead to conservation of energy and water and reduce concerns in achieving food security. In the GCC countries, water is in short supply leading to production of potable water at very high cost using the desalination technology. Because of high standard of living, water consumption is among the highest in the world. Food security is achieved through import of food as the countries are rich. One of the obvious choices for these countries would be to use recycled water (treated wastewater) multiple times for food production in controlled environment. A system using treated wastewater for growing fish and crops (commonly known as aquaponics) achieves the goal of sustainable consumption and production as it involves both water and food. In this chapter we will provide details of a research project undertaken at Sultan Qaboos University in Oman which if adopted in a large-scale will result in the realization of SDG 12.

More countries are facing serious water scarcity, which will make it difficult for these countries to feed their people (WWAP 2012). On average, global agriculture uses around 70% of the available freshwater resources. In arid climate zones such as the Middle East and North Africa, the agricultural water consumption can even be up to 90% (FAO 2005). Compared to conventional agriculture, aquaponics uses less than 10% of water, depending on the climatic conditions (Bernstein 2011). Aquaponics can reduce freshwater depletion associated with irrigation whilst encouraging sustainable farming and food production practices, which in turn reduces the freshwater consumption in countries facing water stress. System-related water losses that occur in evaporation, plant transpiration and the water content of the agricultural products can be compensated for by capturing water from air humidity (Gesellschaft 2009) or by reverse osmosis desalination plant in coastal areas (Greenlee et al. 2009).

Global demand for water is expected to grow by 50% in the year 2030. Most of this demand will be in cities and will require new approaches to wastewater collection and management (Arnell 2004). Indeed, reused wastewater may help address other challenges including food production and industrial development. Treated wastewater is a good source of different plant nutrients for agricultural lands. It can improve soil fertility and crop productivity and minimize the inputs of fresh water and synthetic fertilizers.

In Oman, "Haya Water" is a government company that is responsible for building, operating, and managing wastewater projects in Muscat Governorate. Disposing of treated wastewater in an environmentally friendly way is a major concern for Haya Water and relevant authorities in the country. Furthermore, reusing of treated wastewater produced by the current as well as the future treatment plants in sustainable, economic and environment friendly ways is very important in a country like Oman. Hence, there is a need to implement comparative and comprehensive studies on potential reuse options for the treated wastewater produced by Haya Water. These options include: urban reuse, agriculture reuse, industrial reuse, groundwater recharge, and energy generation. Moreover, there is a gap of knowledge that might affect the public's perception towards wastewater, treatment and reuse and associated risks. Therefore, there is a need to update the level of knowledge and community perceptions about treated wastewater reuse. This is followed by public awareness that can be done through meetings with farmers, home gardeners, and other community groups either by field or site visits, school visits or through the different media outlets.

In general, the use of treated wastewater in agriculture could help in improving the fertility of Oman's coarse soil, increasing the yield of different crops and reduce the usage of freshwater resources. However, the treated wastewater may contain high concentrations of salts, heavy metals, pathogens, and emerging pollutants with unknown fate and effects on the ecological system. The guidelines did not count for emerging pollutants and limited studies have been done to evaluate the presence of pathogens, heavy metals and emerging pollutants in cash crops (such as cucumber and tomato) irrigated by treated wastewater. Therefore, there is a need to evaluate the possible health risks associated with application of treated wastewater in agriculture, aquaculture and home gardens, so that a clear message with supported data can be transferred to decision makers, public and different farmers.

Oman lies in an arid region where rainfall is limited and irregular over much of the country, which makes fresh water an expensive commodity. Thus, despite an increasing demand for freshwater fish, the development of fresh-water aquaculture is slow, and it needs to be further expanded. In recent years the Oman aquaculture industry has been gaining attention from both public and private sectors. The Ministry of Agriculture and Fisheries Wealth has issued 19 licenses to investors who have met the technical criteria to set up aquaculture projects and the total investment in these projects is valued at OMR 128.8 million (1 OR equals 2.58 USD) (Zafar 2014).

Aquaculture has been identified as a sector that could further diversify the Oman's economy beyond primarily hydrocarbons. Oman Government has made the development of aquaculture sector a top priority (Prins 2014). Considering that aquaculture may cause environmental pollution if not done properly, the government is promoting the integrated system which includes aquaponics. In Oman, aquaponics will help to preserve our heritage, while integrating it with the technology to sustain natural resources by minimizing water used in aquaculture and production of fish and crop.

Aquaponics is the combination of aquaculture and hydroponics. It allows the production of fish and plants in one system with big reduction in overall water uses.

The two main components of the system are the fish tank and the grow beds with small pump moving water between the two. The water passes through the roots of the plants before draining back into the fish tank. The plants absorb the water with different nutrients coming from fish waste cleaning the water for the fish. There are number of different styles of grow bed designs. The most common designs are flood and drain, and floating raft style (Diver 2006; Goodman 2011).

In general, aquaponics system incorporates recirculating aquaculture with the soilless production of plants. The recirculating systems are designed to raise large quantities of fish in relatively small volumes of water by treating the water to remove waste products and then re-using it. As water is recycled beneficial nutrients and organic materials accumulate. These are then channeled into secondary crops in an integrated system. Plants grow rapidly with dissolved nutrients that are excreted by the fish or generated by the microbial breakdown in fish wastes (Rakocy et al. 2006). There are a number of benefits linked to aquaponics. These include: (1) reduced water requirement for intensive fish and plant production, (2) the daily supply of feed to the fish supplies a steady flow of nutrients, which are recovered, after treatment, from the fish tank effluent and used to irrigate crops, (3) shared infrastructure and operating costs, (4) reduced land requirement (Rakocy 1997).

In recent years, aquaponics has gained popularity in many countries including Saudi Arabia, Oman, Egypt and Algeria. Aquaponics in Oman is developing well and is following the ten components of agroecology being promoted by FAO (Gallardo 2019). Currently, Sultan Qaboos University is involved in a FAO project on the use of non-conventional water in support to sustainable agri-aquaculture development in desert and arid lands in the Near East and North Africa region, particularly in Algeria, Egypt and Oman. The survey in Oman shows that there are already several aquaponic farms that can be categorized into: (1) large-scale commercial farm, (2) medium-scale commercial farm, (3) small-scale (hobby/family) farm, (4) research/demonstration farm (Gallardo et al., in preparation).

Tilapia (*Oreochromis niloticus*) is the most commonly used fish in aquaponics systems due to their high availability, fast growth, stress and disease resistance and easy adaptation to indoor environment (Popma and Masser 1999). Treated wastewater (Haya) was tested as a potential replacement for tap water in the future. Lettuce and beans are good vegetables to be grown in aquaponics systems because they grow fast in response to high levels of nutrients in aquaculture water (Rakocy et al. 1997).

From other side, in water-scarce regions like Oman, the use of treated wastewater in agriculture will free up, and prevent the contamination of good quality water resources for the use in urban centers and industry. That is why the use of treated wastewater in agriculture is encouraged and promoted (Abdelrahman et al. 2011; Al-khamisi et al. 2015). The objective of the study was to evaluate the effect of treated wastewater on fish life and later on the effect of the produced effluent coming from fish tank on grown crops.

1.1 Literature Review: Aquaponics

Currently some of the global issues are: population rise, climate change, soil degradation, water scarcity and food security. Aquaponics, as a closed loop system consisting of hydroponics and aquaculture elements, could contribute positively to address some of the issues mentioned. It can provide food for the growing number of world population which is expected to reach 9.6 billion around 2050 with more than 75% living in urban areas (UN 2013). Arid regions like Oman suffering from water stress will particularly benefit from this technology being operated in a commercial environment (Tyson et al. 2011).

In a typical aquaponics unit, water from the fish tank cycles through filters to the plant grown beds and then back to the fish. Filters including biofilter remove waste materials. The biofilter provides a location for bacteria to convert ammonia, which is toxic for fish, into nitrate, a more accessible nutrient for plants. This process allows the fish, plants, and bacteria to thrive symbiotically and to work together to create a healthy growing environment for each other, provided that the system is properly balanced (Timmons and Ebeling 2010; Blidariu and Grozea 2011).

In aquaponics, the effluent is used for growing plants and there is no need for chemical fertilizers. This integration is highly sustainable and cost effective. Aquaponics is even more attractive where land and water are lacking. On the negative side, aquaponics is complicated and requires substantial start-up costs. 'Although the production of fish and vegetables is the most visible output of aquaponics units, it is essential to understand that aquaponics is the management of a complete ecosystem that includes three major groups of organisms: fish, plants and bacteria' (Timmons and Ebeling 2010; Blidariu and Grozea 2011; Yamamoto and Brock 2014).

It is clear from a literature review that aquaponics is most appropriate where land is expensive, water is scarce, and soil is poor. Examples are: deserts and arid areas, sandy islands and urban gardens are the locations most appropriate for aquaponics because it uses an absolute minimum of water. There is no need for soil, and aquaponics avoids the issues associated with soil compaction, salinization, pollution, disease and tiredness. Similarly, aquaponics can be used in urban and peri-urban environments where no or very little land is available, providing a means to grow dense crops on small balconies, patios, indoors or on rooftops (Timmons and Ebeling 2010; Blidariu and Grozea 2011).

Aquaponics combines the culture of aquatic animals and the cultivation of plants in recirculating systems, integrating aquaculture and hydroponics in a soil-less system (Rakocy 2012). Toxic ammonia produced by unutilized feed, fish faeces and excreted urea is oxidized by nitrifying bacteria (microbial breakdown) (mainly by Nitrosomonas and Nitrobacter spp.) into vital and usable nitrate for plants (Cebron and Garnier 2005). Plants absorb nitrate and other nutrients, permitting purified water recirculate back to fish tanks. Aquaponics promotes an innovative system as a solution to possible environmental impacts of aquaculture (Tyson et al. 2011), shortage of drinking water, climate change, loss of soil fertility and biodiversity.

Aquaponics system vitality and prosperity is based on fish, plant and bacterial interactions and welfare. The interrelatedness between these are highly complex and are in direct association with water quality (Yildiz et al. 2017). Aquaponics food products are chemical-free with zero use of hormones, pesticides/fungicides or antibiotics (Stathopoulou et al. 2018).

1.2 Water Quality Issues in Aquaponics Systems

It has already been stated that an aquaponics controlled environment is a closed loop system which mimics the natural ecology and is constructed in small scale, where recirculating aquaculture and hydroponic plant production systems are integrated (Pattillo 2017). The main purposes of linking both systems are to minimize water use, fish tank waste nutrient, and reduce the release of wastes to the environment, increase fish and year-round plant production in a small area thereby increasing profit potential (Pattillo 2017). Basically, the principle of this system is to utilize the excreted fish waste and convert it into nutrients, by the action of bacteria, to be absorbed by plants (Sallenave 2016). Since this system involves animals with plant production, it needs special parameters of water quality to be maintained as optimal health criteria (Sallenave 2016). Therefore, certain water quality parameters should be maintained as a guide for good health and productivity (Sallenave 2016). Turbidity, dissolved oxygen, total nitrogen concentration, pH, hardness, and water temperatures are the essential key quality parameters for aquaponics. It is very important to keep monitoring these parameters to avoid water deterioration and quality issues. Wastewater from the fish tank is rich of nutrients, organic and inorganic compounds such as ammonia, phosphorus, organic carbon and organic matter. A hydroponic system can reuse the excreted nutrients by the plants instead of discharging it via the action of nitrobacter bacteria (Khiari et al. 2020; Rogers and Klemetson 1985). These natural bacteria work as a catalyst to enhance the process of nitrification and convert ammonia and ammonium into more simple compounds for the plants as a form of nitrates and nitrites (Rogers and Klemetson 1985). Eventhough, plant uptake is not enough to remove the accumulated toxins, therefore there is a need to construct a filtration unit to be responsible for toxins removal and treat this water. The presence of large solid wastes, ammonia, phosphorus, organic carbon, organic matter, imbalance of pH are examples for of the major issues which could be solved by constructing a filtration unit.

Nitrogenous compounds are present naturally in aquaponics systems as a result of fish excretion due to metabolism. Ammonium, ammonia, nitrate and nitrite are examples of those natural nitrogenous compounds (Rogers and Klemetson 1985). Ammonia is one of the major enemies for aquaculture where it is considered to be lethal even at low concentrations (Goddek et al. 2019). Ammonia (NH_3) is an organic, colorless, corrosive, irritating malodorous with a very low odor threshold value and highly toxic gas (Rogers and Klemetson 1985). This is a naturally occurring compound especially in surface water and can be produced by the breaking down of organic matter and fish waste excretion as a by-product of their protein metabolism. In natural surface water, ammonia is present in two forms: ionized ammonia, NH^{4+}, and un-ionized ammonia, NH_3 and their presence in water is a function of pH (Su et al. 2020). Basically, un-ionized ammonia, NH_3 starts to accumulate at about pH 7.5, and its concentration increases as pH increases, consequently, ammonia's aquatic toxicity increases because of the presence of high relative proportion of unionized ammonia. In contrast, as pH decreases, the concentration of un-ionized ammonia starts to decrease until its concentration becomes 0 ppm at pH values lower than 7.5 (Tyson et al. 2011; Goddek et al. 2019). At 7.5 pH, ammonia NH^{4+} starts to accumulate and concentrates more and more due to denitrification process as pH decreases (Awad 2017). Principally, ammonia's aquatic toxicity occurs due to the presence of the un-ionized form which is NH_3 (Rogers and Klemetson 1985). This problem starts if ammonia levels exceed 0.02 mg/L causing fish death (Goddek et al. 2019). Ammonia toxicity can be eliminated either by bacterial nitrification of ammonia into nitrite and nitrate or via reconciling pH (Tyson 2007). Nitrate can be easily taken up by plants as a nutrient and is a harmless compound to fish in natural water (Rakocy 2012). The nitrification process occurs naturally however, it is a very slow process which explains the urgent need for a feasible solution (Ebeling and Timmons 2012).

In the aquaponics system, the residence time of water is high and the natural nitrification process is very slow, these will lead to accumulation of ammonia causing ammonia's fish toxicity (Rogers and Klemetson 1985). Hence, there is a need to construct a catalyst component in the system which is an Ammonia removal filter. Ammonia removal filter (submerged and non-submerged) is used to treat fish sludge generally, and convert the produced ammonia (NH_3) into nitrate (NO_3). Basically, the filtrate (which contains NO_3 and NO_2) will be used for plants to nourish them and ultimately remove NO_3 and NO_2 as well, then water will be pumped back to the fish tank (Khiari et al. 2020). This type of systems is called recirculating aquaculture system (RAS), where water will be recalculated and reused again to raise fish and grow crops too (Ebeling and Timmons 2012).

2 Materials and Methods

2.1 *System Design and Operation*

The study was done in shade house in College of Agricultural and Marine Sciences at Sultan Qaboos University, Oman. The aquaponics system consisted of 16 rectangular tanks with length of 80 cm, width of 40 cm and height of 40 cm having capacity of 100 L as shown in Figs. 1 and 2. The important elements of this aquaponics system were the fish rearing tanks, solid removal filter, a Styrofoam (to cover the tanks and as raft for the plant seedlings), PVC pipe to connect the fish tanks with the solid removal filter tanks (plants tanks). Air stones connected to air blower were installed in the culture tank to supply oxygen for the fish.

The fishes were distributed at density of 25 fish per tank which was filled either with tap water or treated wastewater in ratio of 50:50 or 75:25. Fish densities of 20

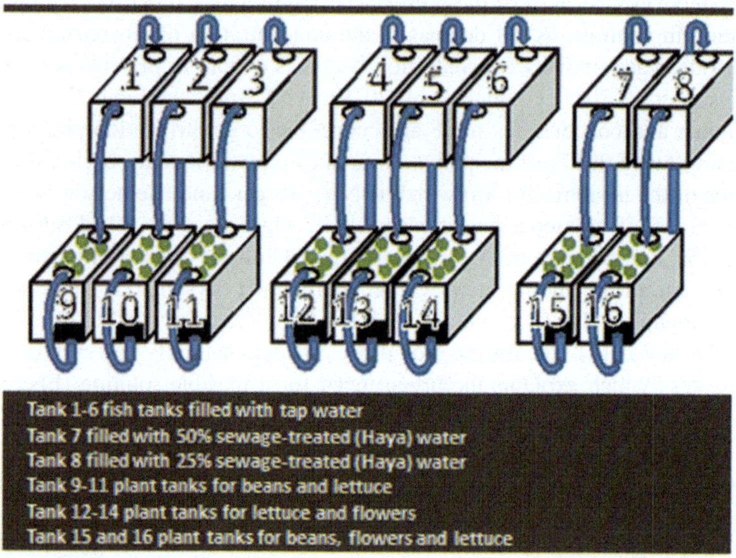

Fig. 1 Simple diagram of the system

Fig. 2 The aquaponics system

and 30 fish per tank were also used as a comparison for other study. Holes were made on the Styrofoam raft and plant (lettuce and beans) seedlings were planted on them as shown in Fig. 2.

Cotton was used to wrap the roots of the seedlings of lettuce and beans in order to aid absorption of the water from the tanks. Moreover, seedling of flowers was also used as a comparison for other study. A total of 108 seedlings were planted into the 8 tanks. Evaporation was minimized by using Styrofoam to cover each tank. A submersible pump was installed in each tank to return the filtered water back to the fish tank.

Constant aeration was provided in the fish tank. No water was added or removed during the study. The fish was fed three times a day and water was circulated between each two tanks. Weekly analyses were done for water salinity, dissolved oxygen and pH. Plant growth was monitored and data for chlorophyll using CCM-200 devise were taken. At the end of the study, plants were harvested and different parameters were measured in plants such as: ammonia, nitrate, nitrogen, microbiological analysis and other metal analysis using ICP machine.

3 Results and Discussions

3.1 Fish growth

Initial average weight of 40 randomly selected tilapias was 49.4 g and the final average weights of tilapia in both treatments (tap water and treated wastewater) are shown in Figs. 4 and 5. Figure 3 shows the average weight of fish grown during the 6 weeks of experiment. The weight of fish increased significantly among weeks but the final weight was not significantly ($P > 0.05$) different among densities. According to Roy (2002) and Rashid (2008) the relationship between weight gain and stocking density is inverse but our results show no significant difference which indicate that in aquaponics system, higher fish densities can be used. Many studies confirmed that tilapia performed well in the aquaponics system (Rakocy et al. 2004a, b).

The weight of fish increased significantly among weeks in all treatments but there was a significant difference ($P < 0.05$) between two systems (tap water and treated wastewater) in term of weight gain. We clearly observed that fish weight in treated wastewater was greater than tap water tanks (Fig. 4). This needs further investigation on the elements available in treated wastewater that may enhance fish growth.

The final weight of fish grown with beans and lettuce was the highest (100 g) in 25 fish/tank (Tanks No. 2) and the lowest (87.08 g) in tank number 1 with 20 fish/ tank (Fig. 5). Whereas, the final weight of tilapia grown with lettuce and flowers was found to be the highest (106.18 g) in 20 fish/tank (Tank No. 4) and the lowest (98.26 g) in 30 fish/tank (Tank No. 6). On other hand, the final weight of fish in treated wastewater was higher than in tap water which was 119 g in both tanks. The

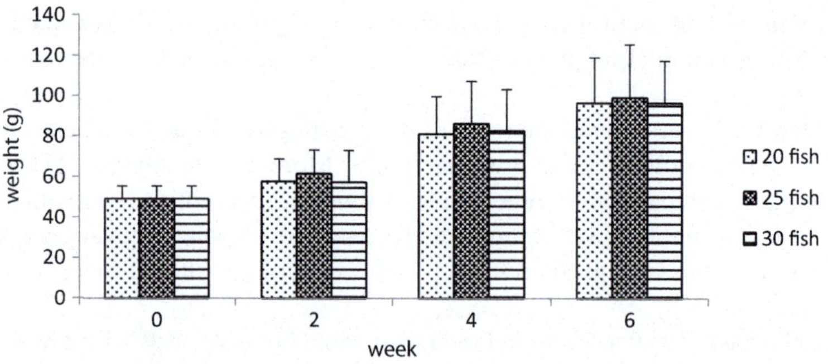

Fig. 3 Growth of fish at densities of 20, 25, and 30 fish/tank during the 6-week experiment period. Bars show the average weight and error bars are standard deviation

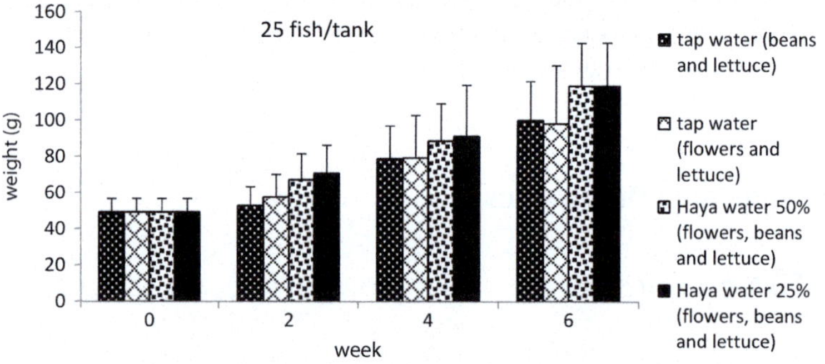

Fig. 4 Growth of fish at densities of 25 fish/tank in different culture systems: (1) tap water with beans and lettuce, (2) tap water with lettuce and flowers, (3) 50% treated wastewater with beans, lettuce and flowers, (4) 25% treated wastewater with beans, lettuce and flowers. Bars show the average weight and error bars are standard deviation

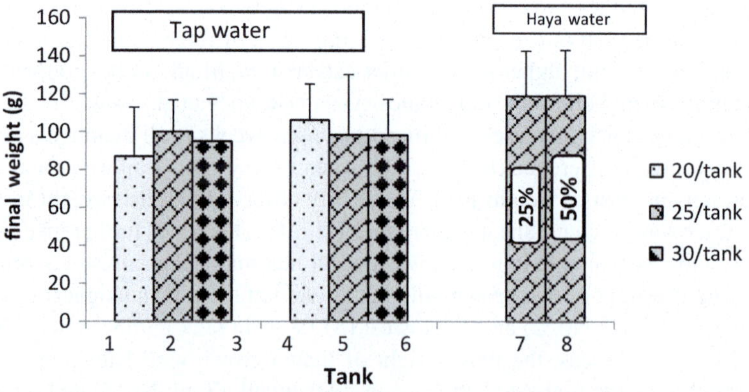

Fig. 5 Final average weight of fish in each tank after six weeks. Bars show the average weight and error bars are standard deviation

final weight of fish in tap water was not significantly different among fish density but there was significant difference between tap and treated wastewater.

The present study demonstrated that the aquaponics system used in this study was effective in fish waste treatment and water conservation in the period of 6 weeks. Growth of fish was high in all fish densities in treated wastewater growth medium which can be a potential replacement for tap water.

The percent weight gains of fish grown in combination with beans and lettuce was 100% in tank 2 (50.5 g) and lowest in tank 1 (70%). The percent weight gains of tilapia grown in combination with flowers and lettuce was highest in tank 4 (115%) and lowest in tank 6 (98.9%). On other hand, it was observed that the treated wastewater resulted in higher weight gain up to 141% than tap water (Fig. 6).

The average Specific Growth Rate (SGR, indicates daily growth rate) of fish in tap water was 1.2 with the highest in rearing tank 4 (1.4) and the lowest in tank 1 (0.9) but the treated wastewater had the highest specific growth rate compared to tap water (Fig. 7).

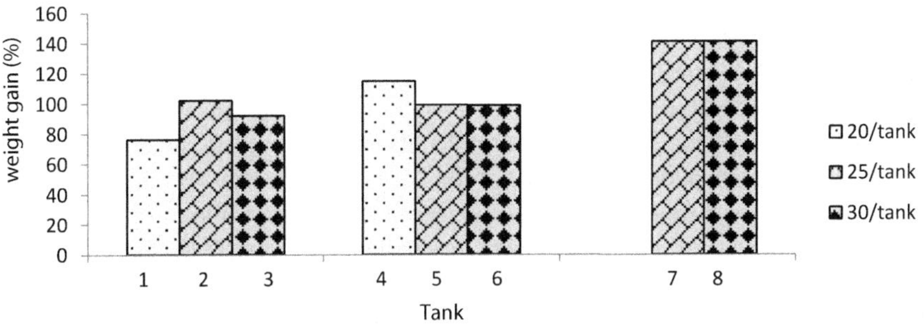

Fig. 6 Percent weight gain of fish after 6 weeks in each tank (final-initial/initial weight * 100)

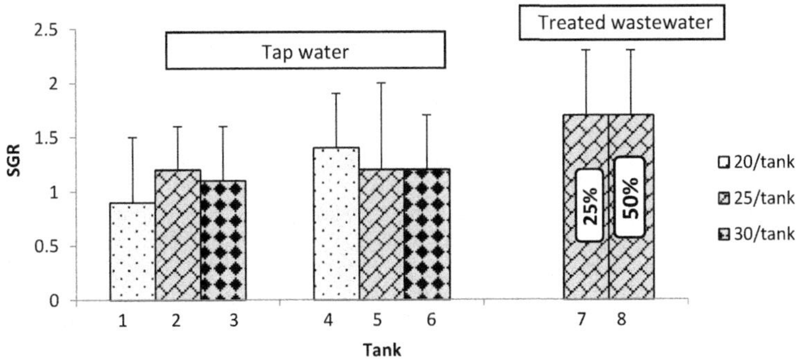

Fig. 7 Specific growth rate (SGR) of fish in 42 days at different fish density. Bars show the average SGR and error bars are standard deviation

3.2 Water Quality Parameters

Oxygen is essential for all three organisms involved in aquaponics; plants, fish and nitrifying bacteria as all need oxygen to live. Dissolved Oxygen (DO) level describes the amount of molecular oxygen within the water, and it is measured in milligrams per liter (mg/L). It is the water quality parameter that has the most immediate and drastic effect on aquaponics. Fish may die within hours when exposed to low dissolved oxygen within the fish tanks. Thus, ensuring adequate dissolved oxygen levels is crucial to aquaponics. Whenever the dissolved oxygen becomes low it is an indicative of the presence of some microbes or contaminants and such water is un-healthy for the fish. From Fig. 8, it can be seen that dissolved oxygen level varied with time and almost all treatment got similar DO concentrations. However, sometimes it was found that treatments that had treated wastewater got higher values of dissolved oxygen most likely due to algae growth that was enhancing production of dissolved oxygen through photosynthesis. However, high density of algae sometimes blocked some pipelines and reduced water circulation between tanks which affected dissolved oxygen values and that happened in tank with 50% treated wastewater.

Optimum levels of dissolved oxygen are 4–8 mg/L. Nitrification decreases if dissolved oxygen concentrations drop below 2.0 mg/L. Moreover, without sufficient dissolved oxygen concentrations, another type of bacteria grow that convert the valuable nitrates back into unusable molecular nitrogen in an anaerobic process known as denitrification (Munguia-Fragozo et al. 2015). Thus, ensuring adequate dissolved oxygen levels is crucial to aquaponics. Although monitoring dissolved oxygen levels is very important, it can be challenging because accurate dissolved oxygen measuring devices can be very expensive or difficult to find. It is often sufficient for small-scale units to rely on frequent monitoring of fish behavior and plant growth, and ensuring water and air pumps are constantly circulating and aerating the water (Timmons and Ebeling 2010; Munguia-Fragozo et al. 2015).

Fish in aquaponics system, particularly carp and tilapia, can tolerate dissolved oxygen levels as low as 2–3 mg/L, but it is much safer to have the levels higher for aquaponics, as all three organisms demand the use of the dissolved oxygen in the water (Timmons and Ebeling 2010).

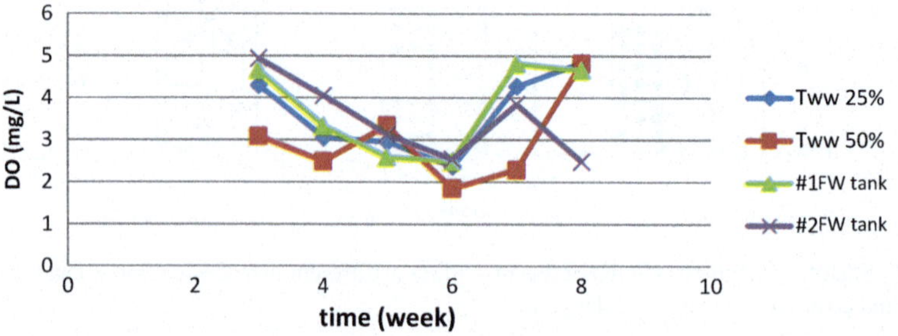

Fig. 8 Weekly changes in dissolved oxygen in all treatments

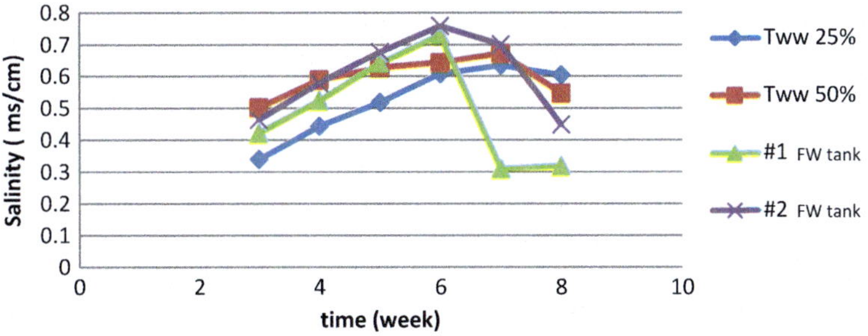

Fig. 9 Weekly changes in water salinity of all treatments

Water salinity should be constant with smal increase due to evaporation process or salts coming from fish food or fish waste. Morover, treated wastwater is rich in many nutrients so it could add some extra salts to the tanks. As it was expected and shown in Fig. 9, water salinity was increasing until plants start to utlize the salts as nutreints for their growth.

There are many biological and chemical processes that take place in an aquaponics system that affect the pH of the water, some more significantly than others, including the nitrification process, fish stocking density and phytoplankton (Timmons and Ebeling 2010; Zou et al. 2016). It can be seen from Fig. 10 that water pH ranged from 6 to 8 until week 6 then it started to decrease in week 7. The freshwater tanks had higher values of pH compared to treated wastewater tanks. However, tank of 25% treated wastewater reached the lowest values of pH 5 in week 8, which is almost acidic condition that could affect the absorbance of some plant nutrients. In all cases, no big changes in water pH was noted in this short study.

Most plants need a pH value between 6 and 6.5 in order to enhance the uptake of nutrients. The fish species Tilapia (*Oreochromis niloticus*) is known to be disease-resistant and tolerant to large fluctuations in pH value with a tolerance between pH 3.7 and 11, but achieves best growth performance between pH 7.0 and 9.0 (McAndrew et al. 2000). The ideal pH value for the system is between 6.8 and 7.0. Although root

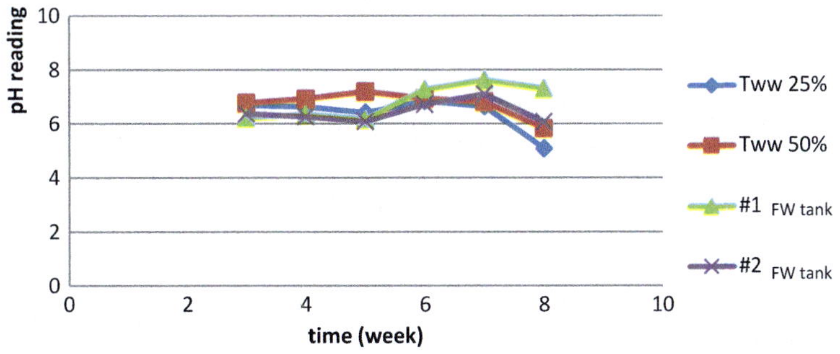

Fig. 10 Weekly changes in water pH of all treatments

uptake of nitrate raises pH as bicarbonate ions are released in exchange (Kaiser et al. 2011), the acidity producing nitrification process has a higher impact on the overall system pH, leading to a constant and slight decrease in the pH-value.

It was reported by Zou et al. (2016) that water pH has a major impact on all aspects of aquaponics, especially the plants and bacteria. For plants, the pH controls the plants' access to micro- and macronutrients. At a pH of 6.0–6.5, all of the nutrients are readily available, but outside of this range the nutrients become difficult for plants to access. In fact, a pH of 7.5 can lead to nutrient deficiencies of iron, phosphorus and manganese. This phenomenon is known as nutrient lock-out. Nitrifying bacteria experience difficulty below a pH of 6, and the bacteria's capacity to convert ammonia into nitrate reduces in acidic, low pH conditions. This can lead to reduced biofiltration, and as a result the bacteria decrease the conversion of ammonia to nitrate, and ammonia levels can begin to increase, leading to an unbalanced system stressful to the other organisms. Fish have specific tolerance ranges for pH as well, but most fish used in aquaponics have a pH tolerance range of 6.0–8.5. However, the pH affects the toxicity of ammonia to fish, with higher pH leading to higher toxicity.

In general, the ideal aquaponics water is slightly acidic, with an optimum pH range of 6–7. This range will keep the bacteria functioning at a high capacity, while allowing the plants full access to all the essential micro- and macro- nutrients. The pH values between 5.5 and 7.5 require management attention and manipulation through slow and measured means. However, a pH lower than 5 or above 8 can quickly become a critical problem for the entire ecosystem and thus immediate attention is required (Timmons and Ebeling 2010; Zou et al. 2016).

Ammonia analysis of last day of the study (Table 1) showed that treated wastewater with 25% concentration had the highest value of ammonia whereas, treated wastewater of 50% concentration got the highest values of phosphate. It was expected since treated wastewater is rich in different nutrients that can move and transform form one compound to another. However, ammonia is toxic to fish. Tilapia and carp can show symptoms of ammonia poisoning at levels as low as 1.0 mg/l. Prolonged exposure at or above this level will cause damage to the fishes' central nervous system and gills, resulting in loss of equilibrium. Other symptoms include red streaks on the body, lethargy and gasping at the surface for air. At higher levels of ammonia, effects are immediate and numerous deaths can occur rapidly. However, lower levels over a long period can still result in fish stress, increased incidence of disease and more fish loss.

Table 1 Ammonia and phosphate in aquaponics tanks

Sample	Ammonia (mg/L)	Phosphate (mg/L)
Treated wastewater 25%	1.34	3.55
Treated wastewater 50%	1.11	3.78
Fresh water #1	0.03	1.82
Fresh water #2	0.14	3.07

Table 2 Nitrate in aquaponics tanks

Sample	Nitrate (NO_3^-) (mg/L)
Treated wastewater 25%	380
Treated wastewater 50%	260
Fresh water #1	150
Fresh water #2	180

In our case the level of ammonia was above than 1.0 mg/L but no bad effects were observed in fish life. It could be due to continuous aeration and good plant growth which acted as a filter (sink) and reduced the side effects of high ammonia. From other side, ammonia could be converted to nitrate (NO_3^-) that is less toxic to aquatic life as shown in Table 2.

Nitrate is a far less toxic than the other forms of nitrogen. It is the most accessible form of nitrogen for plants, and the production of nitrate is the goal of the biofilter. Fish can tolerate nitrate levels of up to 300 mg/L, with some fish tolerating levels as high as 400 mg/l. High levels (>250 mg/L) will have a negative impact on plants, leading to excessive vegetative growth and hazardous accumulation of nitrates in leaves, which is dangerous for human health. It is recommended to keep the nitrate levels at 5–150 mg/L and to exchange/replace water when levels become higher (Timmons and Ebeling 2010; Zou et al. 2016).

When the fish biomass and biofilter size are in balance, the aquaponics unit will adequately process the ammonia into nitrate. If there are not enough plants, then the system will start to accumulate nutrients. Result will be higher concentrations of nutrients, which are not harmful to fish nor plants, but they are an indication that the system is underperforming on the plant side (Timmons and Ebeling 2010; Zou et al. 2016).

High values of heavy metals are not recommended especially in the edible part of the plant. It can be seen in Fig. 11 that all measured values had low concentrations except for B, Fe and Zn. There were higher in treated wastewater compared to fresh water. However, all concentrations for all treatments were close from each other which mean the source of those elements could be fish food or waste but not treated wastewater only. Therefore, monitoring is required to make sure that used water is not causing any health problem to fish or grown plants which later will be reflected in human health.

Usually, biological analyses are direct indicators for microbial contamination. In this study, Coliform bacteria was expected to be found because it is found in fish waste (Table 3). However, *E-coli* was not found and that was an indicator of the good quality of all used waters.

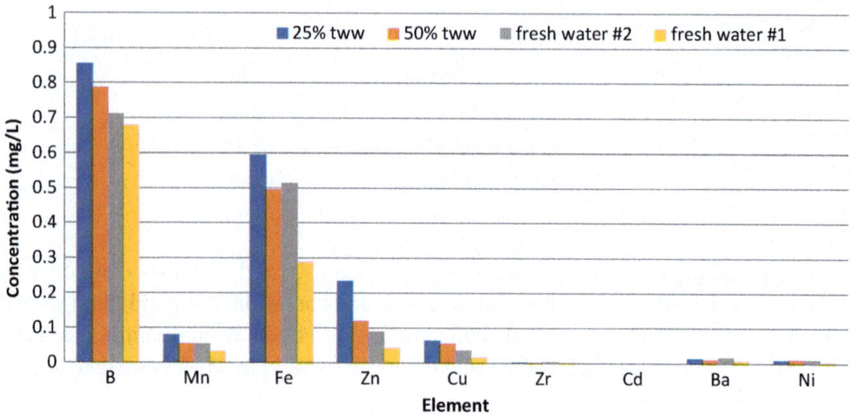

Fig. 11 Average concentrations of elements in water of all treatments

Table 3 Microbial analysis

Sample	Coliforms	MPN*/100 mL of the Coliform bacteria	*E-coli*
Treated wastewater 25%	50	≥ 200.5	0
Treated wastewater 50%	50	≥ 200.5	0
Fresh water #1	50	≥ 200.5	0
Fresh water #2	50	≥ 200.5	0

*Most probable number

3.3 *Plant Analysis*

Chlorophyll data is a good indicator for nitrogen values in plant tissues. It can be seen from Fig. 12 that measured data for all treatments had good values of nitrogen and even higher in treated wastewater treatments as shown in Table 4. This can be explained by nitrogen provided to the plants from fish waste and treated wastewater.

Since no chemical fertilizer was added to all treatments so plants were relying on nutrients coming from fish waste and those available in treated wastewater. Therefore, it can be seen from Fig. 13 that treated wastewater treatments gave better growth to lettuce plants compared to freshwater treatment.

Nitrogen is supplied to aquaponics plants mainly in the form of nitrate which is converted from the ammonia of fish waste through bacterial nitrification. Some of the other nutrients are dissolved in the water from the fish waste, but most of them remain in a solid state and are unavailable to plants. The solid fish waste is broken down by heterotrophic bacteria; this action releases the essential nutrients into the water. The best way to ensure that plants do not suffer from deficiencies is to maintain the optimum water pH (6–7) and feed the fish a balanced and complete diet, and use

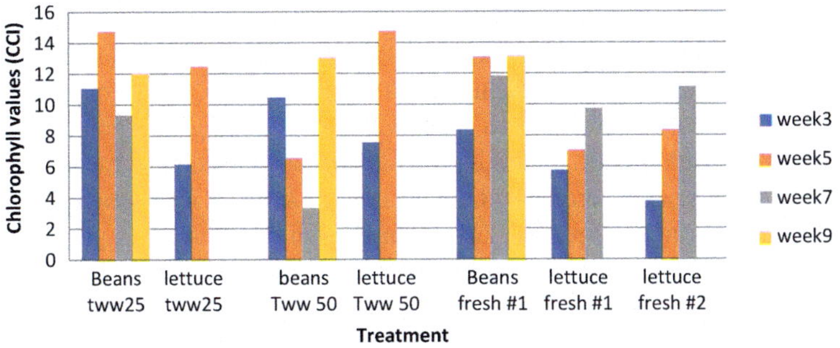

Fig. 12 Chlorophyll content in plant tissues under different treatments

Table 4 Nitrogen values in lettuce plant

Treatment	Nitrogen %
Treated wastewater 25%	3.82
Treated wastewater 50%	3.70
Fresh water	3.56

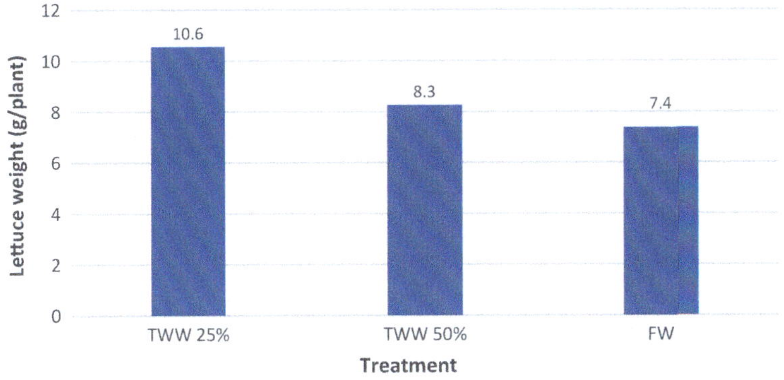

Fig. 13 Lettuce growth as affected by different treatments

the feed rate ratio (any value?) to balance the amount of fish feed to plants (Hu et al. 2015).

The most sensitive part for human consumption in each crop is the edible part. It can be seen from Fig. 14 that both crops lettuce and bean got similar values for all measured elements which is a reflection for what was found in irrigation waters. Elements of B, Fe were found in high concentrations for all treatments which could be a limited factor for crop consumption. However, other elements such as Ni, Zn and Cu can be found in higher concentrations in treated wastewater compared to fresh water but in general, the difference was small. In Selem et al. (2000) study, they observed an increase in Fe, Zn, Cu, Mn, Pb and Co with land irrigated with treated wastewater as compared to untreated soil. The accumulation of heavy metals in the edible part of some plants could adversely affect human and animal health (Abd-Elfattah et al. 2002). In present study, the increase in heavy metals with treated wastewater was small compared to fresh water and nothing was reflected in plant and fish lives. In addition to that, the grown fishes were tested and measured values related to health issues were within the acceptable limit of international standards. However, monitoring is required to avoid any health problems.

Tyson et al. (2011) reported that plants need macronutrients (C, H, O, N, P, K, Ca, S and Mg) and micronutrients (Fe, Cl, Mn, B, Zn, Cu and Mo), which are essential for their growth. In aquaponics systems, plant nutrient input from the fish tanks contains dissolved nutrient rich fish waste (gill excretion, urine and faeces), comprising of both soluble and solid organic compounds that are solubilized to ionic form in the water and assimilated by the plants. These nutrients need regular monitoring. Furthermore, bacterial population is needed for waste conversion. One of the most important microbial components is the nitrifying autotrophic bacteria consortium that is established as a biofilm on solid surfaces within the system and is principally composed of nitroso-bacteria (e.g., *Nitrosomonas* sp.) and nitro-bacteria (e.g., Nitrospira sp., Nitrobacter sp.). The ammonia within the system is converted

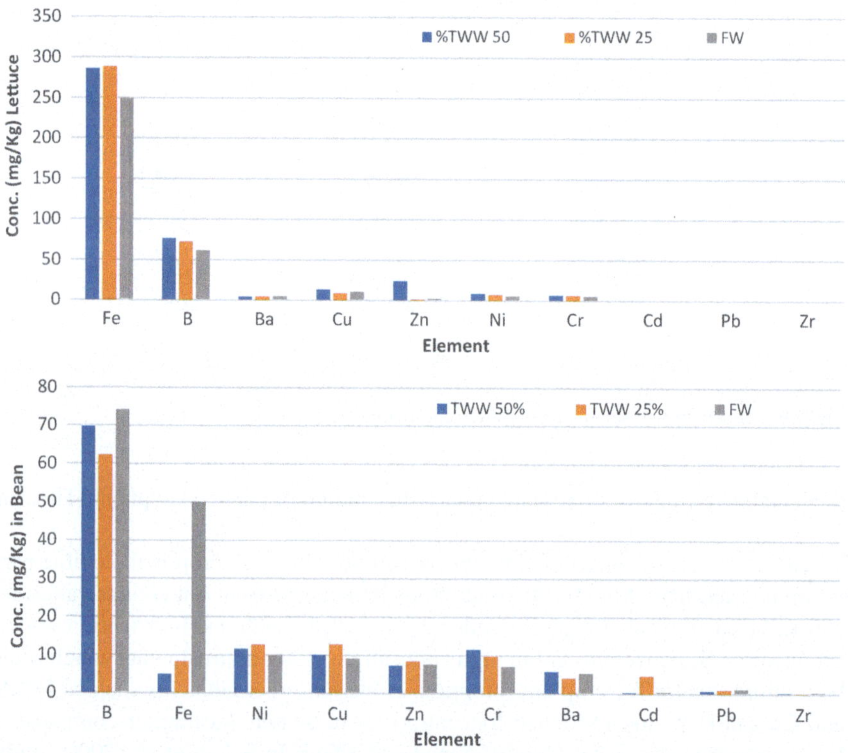

Fig. 14 Average concentration of elements in lettuce (top) and bean (bottom) shoots

into nitrite (NO_2^-) by nitroso-bacteria, before being transformed into nitrate (NO_3^-) by the nitro-bacteria (Tyson et al. 2008). The end product is nitrate, is considerably less toxic for fish and due to its bioconversion, is the main nitrogen source for plant growth in aquaponics systems (Endut et al. 2014).

4 Conclusions

The project demonstrated the possibility of producing crops in aquaponics system using both fish waste and treated wastewater. Both sources enriched the media with many nutrients needed for the plant growth. Some heavy metals accumulated in the irrigation water which was also transmitted to grown crops. The concentrations of heavy metals in fresh water and treated wastewater were close to each other indicating that diluted wastewater can be used safely in aquaponics system. Therefore, the system used in this study can provide plants with needed nutrients and produce safe crops with minimum risk. The system is simple and can be implemented in small area. It is environment friendly, help in saving fresh water and maximizing the application of treated wastewater with no or minimum need for chemical fertilizers. However, monitoring with good management is required to avoid any health issues.

References

Abd-Elfattah A, Shehata SM, Talab AS (2002) Evaluation of irrigation with either raw municipal treated waste or river water on elements up take and yield of lettuce and potato plants. Egypt J Soil Sci 42(4):705–714

Abdelrahman HA, Al-khamisi SA, Ahmed M, Ali H (2011) Effects of treated wastewater irrigation on element concentrations in soil and maize plants. In: Conference proceeding. Sultan Qaboos University, Oman

Al-Khamisi SA, Al-Wardy M, Ahmed M, Prathapar SA (2015) Impact of reclaimed water irrigation on soil salinity, hydraulic conductivity, cation exchange capacity and micronutrients. Sultan Qaboos University Research Journal of Agricultural & Marine Sciences

Arnell NW (2004) Climate change and global water resources: SRES emissions and socio-economic scenarios. Glob Environ Change 14:31–52

Awad A (2017) Evaluation of ammonia (NH3) removal through water recycling in aquaculture by aquaponic system and its effect on the growth rate of used fish and plants. http://hdl.handle.net/20.500.12358/21530

Bernstein S (2011) Aquaponic gardening: a step-by-step guide to raising vegetables and fish together. New Society Publishers, Gabriola Island, p 256

Blidariu F, Grozea A (2011) Increasing the economical efficiency and sustainability of indoor fish farming by means of aquaponics-review. Sci Pap Anim Sci Biotechnologies 44:1–8

Cebron A, Garnier J (2005) Nitrobacter and Nitrospira genera as representatives of nitrite-oxidizing bacteria: detection, quantification and growth along the lower Seine River (France). Water Res 39:4979–4992

Diver S (2006) Aquaponics—integration of hydroponics with aquaculture. Publication No. IP163. ATTRA, National Sustainable Agriculture Information Service

Ebeling JM, Timmons MB (2012) Recirculating aquaculture systems. Aquac Prod Syst 245

Endut A, Jusoh A, Ali N (2014) Nitrogen budget and effluent nitrogen components in aquaponics recirculation system. Desalin Water Treat 52:744–752

FAO (2005) AQUASTAT survey, irrigation in Africa in figures. FAO, Rome

Gallardo W (2019) Aquaponics in Oman. In: FAO 2019. Report of the special session on advancing integrated agriculture aquaculture through agroecology, Montpellier, France, 25 August 2018. FAO Fisheries and Aquaculture Report No. 1286. Rome

Gesellschaft F (2009) Research news. Fraunhofer-Gesellschaft München, Fraunhofer Press, München

Goddek S, Joyce A, Kotzen B, Burnell GM (2019) Aquaponics food production systems: combined aquaculture and hydroponic production technologies for the future. Springer

Goodman E (2011) Aquaponics: community and economic development. Massachusetts Institute of Technology, pp 97–100

Greenlee LF, Lawler DF, Freeman BD, Marrot B, Moulin P (2009) Reverse osmosis desalination: water sources, technology, and today's challenges. Water Res 43:2317–2348

Hu Z, Lee JW, Chandran K (2015) Effect of plant species on nitrogen recovery in aquaponics. Biores Technol 188:92–98

Kaiser DE, Lamb JA, Bloom PR (2011) Managing iron deficiency chlorosis in soybean. Regents of the University of Minnesota, Saint Paul, MN

Khiari Z, Alka K, Kelloway S, Mason B, Savidov N (2020) Integration of biochar filtration into aquaponics: effects on particle size distribution and turbidity removal. Agric Water Manage 229:105874. https://doi.org/10.1016/j.agwat.2019.105874

McAndrew BJ, Turner GF, Robinson RL, Beveridge MCM, Baird DJ, Ross LG, Lowe-McConnell RH, Lorenzen K, Penman DJ, Little DC et al (2000). In: Beveridge MCM, McAndrew BJ (eds) Tilapias: biology and exploitation. Springer, Heidelberg, p 508

Munguia-Fragozo P, Alatorre-Jacome O, Rico-Garcia E (2015) Perspective for aquaponic systems: "omic" technologies for microbial community analysis. Biomed Res Int 2015:480386

Pattillo DA (2017) An overview of aquaponic systems: aquaculture components

Popma T, Masser M (1999) Tilapia: life history and biology. Southern Regional Aquaculture Center (SRAC) Publication No. 283. United States Department of Agriculture, Cooperative States Research, Education, and Extension Service, Mississippi State University, Stoneville, Mississippi

Prins T (2014) Aquaculture and fisheries development in Oman. Netherlands Embassy Oman, Muscat

Rakocy JE (1997) Integrating tilapia culture with vegetable hydroponics in recirculating systems. In: Costa-Pierce BA, Rakocy JE (eds) Tilapia culture in the Americas, vol 1. World Aquaculture Society, Baton Rouge, La, pp 163–184

Rakocy JE (2012) Aquaponics: integrating fish and plant culture. Aquac Production Systems 1:343–386

Rakocy JE, Bailey DS, Shultz KA, Cole WM (1997) Evaluation of a commercial-scale aquaponic unit for the production of tilapia and lettuce. In: Fitzsimmons K (ed) Tilapia aquaculture: proceedings of the fourth international symposium on Tilapia in aquaculture at Orlando, Florida. Northeast Regional Agricultural Engineering Service, Ithaca, New York, pp 357–372

Rakocy J, Bailey D, Shultz C, Danaher J (2004a) Design and operation of the UVI aquaponic system. University of the Virgin Islands. Agricultural Experiment Station. St. Croix, U.S. Virgin Islands

Rakocy JE, Shultz RC, Bailey DS, Thomas ES (2004b) Aquaponic production of tilapia and basil: comparing a batch and staggered cropping system. Acta Hortic (ISHS) 648:63–69

Rakocy JE, Masser MP, Losordo TM (2006) Recirculating aquaculture tank production systems: aquaponics—integrating fish and plant culture. Southern Regional Aquaculture Center, Publication No. 454

Rashid MH (2008) Effect of stocking density on the growth, survival and production of mono-sex GIFT tilapia (Oreochromis niloticus L.) reared in recirculatory system in cisterns. MS thesis. Department of Aquaculture, BAU, Mymensingh, p 68

Rogers GL, Klemetson SL (1985) Ammonia removal in selected aquaculture water reuse biofilters. Aquacult Eng 4(2):135–154. https://doi.org/10.1016/0144-8609(85)90010-x

Roy R (2002) Effect of stocking density on the growth and survival of GIFT tilapia fed on formulated diet. MSc Thesis. Department of Aquaculture, BAU, Mymensingh, p 63

Sallenave R (2016) Important water quality parameters in aquaponics systems. College of Agricultural, Consumer and Environmental Sciences

Selem MM, Amir S, Abdel-Aziz SM, Kandil MF, Mansour SF (2000) Effect of irrigation with treated waste water on some chemical characteristics of soils and plants. Egypt J Soil Sci 40(1/2):49–59

Stathopoulou P, Berillis P, Levizou E, Sakellariou-Makrantonaki M, Kormas AK, Angelaki A, Kapsis P, Vlahos N, Mente E (2018) Basil and Nile tilapia production in a small scale aquaponic system. J FisheriesSciences.com 12(4):001–003

Su MH, Azwar E, Yang YF, Sonne C, Yek PN, Liew RK, Lam SS et al (2020) Simultaneous removal of toxic ammonia and lettuce cultivation in aquaponic system using microwave pyrolysis biochar. J Hazard Mater 396:122610. https://doi.org/10.1016/j.jhazmat.2020.122610

Timmons M, Ebeling J (2010) Recreation aquaculture Ithaca. Cayuga Aqua Ventures, Nutrition, NY

Tyson RV (2007) Reconciling pH for ammonia biofiltration in a cucumber/tilapia aquaponics system using a perlite medium

Tyson RV, Simonne EH, Treadwell DD, White JM, Simonne A (2008) Reconciling pH for ammonia biofiltration and cucumber yield in a recirculating aquaponic system with perlite biofilters. HortScience 43(719–724):27

Tyson RV, Treadwell DD, Simonne EH (2011) Opportunities and challenges to sustainability in aquaponic systems. Hortechnology 21:6–13

UN (2013) Human development report. http://hdr.undp.org/en/media/HDR_2013_EN_complete.pdf

WWAP (World Water Assessment Programme) (2012) The United Nations World Water Development Report 4: managing water under uncertainty and risk. United Nations Educational, Paris, p 380

Yamamoto J, Brock A (2014) A comparison of the effectiveness of aquaponic gardening to traditional gardening growth method

Yildiz H, Robaina L, Pirhonen J, Mente E, Domínguez D (2017) Fish welfare in aquaponic systems: its relation to water quality with an emphasis on feed and faeces-a review. Water 9:13–29

Zafar N (2014) MAF licenses for aquaculture projects in Oman worth over RO128mn. Muscat Daily 5 March. http://www.muscatdaily.com/Archive/Oman/MoAF-licences-for-aquaculture-projects-in-Oman-worth-over-RO128mn-2zc8#ixzz46v2amwGH. Retrieved 4 Mar 2014

Zou Y, Hu Z, Zhang J (2016) Effects of pH on nitrogen transformations in media-based aquaponics. Bioresources Technol 210:81–87

Dr. Ahmed Al-Busaidi Researcher in the Department of Soils, Water and Agricultural Engineering, College of Agricultural and Marine Sciences, Sultan Qaboos University, Oman. Obtained his Ph.D. in 2007 in Bioenvironmental Science, Tottori University, Japan. M.Sc. in 2001 in Soil and Water Management, Sultan Qaboos University, Oman and B.Sc. in 1996 in Soil and Water Sciences, Sultan Qaboos University, Oman. He acted as Principal Investigator (PI), Co-PI, and Co-Investigator on national and international funded projects. An invited Researcher in some international universities and organizations (ex: Tottori University, Japan, 2011). Invited Lecturer in some international universities and organizations (ex: University of Applied Science, Rottenburg, Germany, 2014). Invited as an instructor in some local universities and colleges (ex: Oman Tourism College, Oman, 2011). He got a Fulbright Scholar Fellowship in 2016 at the University of Florida, USA. Invited Speaker in Texas A&M AgriLife Research, Texas, USA (2016). He was involved in some international exchange programs (Japan and The Netherlands). Authoring and publishing scientific work and attending scientific meetings, workshops, symposiums and conferences. A member of different scientific committees. His research interests are soil and water science and management, salt-affected soils, saline agriculture, waste management, applications of treated wastewater, hydroponic system, controlled environment agriculture, and soil-water plant relations.

Dr. Mushtaque Ahmed obtained his Ph.D. (Civil Engineering) from Iowa State University, the USA in 1988, his M.Sc. from the University of Hawaii, the USA in 1984 (Civil Engineering), and a B.Sc. in Civil Engineering from BUET, Bangladesh in 1981. He is currently a Professor in the Department of Soils, Water and Agricultural Engineering of Sultan Qaboos University (SQU), Oman. He joined SQU in 1996 and has held different positions including Head of the Department, Assistant Dean and Director of the SQU Environmental Studies and Research Center (CESAR). Prior to joining SQU, he worked with CSIRO and New South Wales State Government in Australia. His current research interests are integrated water resources management, biosaline agriculture, managed aquifer recharge, climate change and adaptability, desalination in agriculture, wastewater treatment and reuse. He has published more than 260 scientific papers in peer-reviewed refereed journals, book chapters, conference proceedings, and manuals. He served/ serving as Editor of various publications. He is a corporate member of the Institution of Engineers, Australia. He has worked as a consultant for international organizations such as UNEP, IAEA, UNOPS, UNESCO, EU, and PRIMA. He has supervised a large number of postgraduate students and has been an external examiner for graduate students of SQU, Qatar University, Curtin University (Australia), RMIT (Australia), University of South Australia (Australia), UKM (Malaysia), UAE University, Rajshahi University (Bangladesh), and Rhodes University (South Africa). He has been a reviewer for more than 85 international refereed journals and 5 international research and funding organizations.

Dr. Wenresti Gallardo is an Associate Professor at the Department of Marine Science and Fisheries in Sultan Qaboos University (SQU), Oman. He obtained his Bachelor of Science in Fisheries and Master of Aquaculture degrees from the University of the Philippines, MSc in Fisheries and Ph.D. in Marine Science from Nagasaki University, Japan, all through scholarships. He worked as a Fishery Project Evaluator at the central office of the Philippine Bureau of Fisheries and Aquatic Resources (BFAR), and then as a Research Associate and Research Scientist at the Southeast Asian Fisheries Development Center, Aquaculture Department (SEAFDEC AQD) in the Philippines and then as a postdoc research fellow at Nagasaki University, Japan. Before moving to Oman in 2013, he worked as Assistant then Associate Professor at the Asian Institute of Technology (AIT) in Bangkok, Thailand, teaching and supervising MSc and Ph.D. students for 8 years. In 2004, he received the Outstanding Young Scientist (OYS) Award in the field of Marine Science, from the National Academy of Science and Technology (NAST) of the Philippines. He is a member of the World Aquaculture Society. He has more than 30 years of research and teaching experience. His research interests include aquaculture, aquaponics and integrated multi-trophic aquaculture (IMTA). He has published more than 30 papers in international journals and supervised and graduated 10 Ph.D. and 35 MSc students at AIT and SQU.

Waad Al-Aghbari was a student of the Department of Soils, Water & Agricultural Engineering and received a B.Sc. degree.

Yahya Al-Yahyaei graduated from Sultan Qaboos University in 2016 with a degree of Bachelor of Science in Marine Science and Fisheries. He conducted his final-year project on aquaponics under the supervision of Dr. Gallardo. He is currently working as Health, Safety and Environment (HSE) Officer in a private company.

Chapter 15
Wastewater Reuse for Agriculture, Qualitative Aspects: A Case Study of Ain Temouchent, Algeria

Chérifa Abdelbaki, Nadia Badr, Hidayat Mohammedi, Rokiatou Haidara, and Halima Belarbi

Abstract Recently, as the world facing a water shortage due to climate change, a growing number of countries are considering irrigation using reclaimed water as an appropriate solution to secure and enhance agricultural production. Among the African countries affected by water stress, Algeria is in the category of the most water-stressed countries. In this chapter, the water quality of the treated wastewater produced from Wastewater Treatment Plant (WWTP) of Ain Temouchent, a semi-arid region located in the northwestern part of Algeria, is determined including physico-chemical parameters, heavy metals and microbiological contaminants. Except of phosphate ions (PO_4^{-3}), the other physico-chemical parameters and heavy metals were below the recommended norm of WHO, FAO and Algerian standards for irrigation purposes. However, the results of the microbiological analyses indicated that the number of fecal coliforms and intestinal nematodes are above the WHO norms for treated wastewater intended for irrigation. On the other hand, according to Algerian norm, the irrigation of the fruit trees through drip irrigation technique is likely to reduce the risk of contamination while preserving the health of the consumers.

Keywords Adaptation · Agriculture · Climate change · Irrigation · Wastewater reuse

C. Abdelbaki (✉) · H. Mohammedi · R. Haidara
Pan-African University Institute for Water and Energy Sciences (PAUWES), Tlemcen, Algeria
e-mail: cherifa.abdelbaki@pauwes.dz

C. Abdelbaki · H. Belarbi
Laboratoire EOLE, Université de Tlemcen, Tlemcen, Algeria

N. Badr
Environmental Sciences Department, Faculty of Science, Alexandria University, Alexandria, Egypt

M. Behnassi et al. (eds.), *The Water, Climate, and Food Nexus*,
https://doi.org/10.1007/978-3-031-50962-9_15

1 Introduction

In regions suffering from water scarcity, wastewater is no longer considered as a waste to be disposed off but as an integral part of the water-resource potential. The use of wastewater in agricultural purposes were started in Australia, France, Germany, India, the United Kingdom, and the United States of America (USA) in the late nineteenth century (WHO 1989). It is used for irrigation in treated and untreated forms, varying by geographic and economic context (Drechsel et al. 2010).

Applications of wastewater reuse have long history and are becoming more prevalent in agriculture, industrial, household as well as urban uses (UNEP 2004). The number of countries investigating and implementing water reuse program other than the USA has increased over the past decades not only in water scarce areas—such as Mediterranean region, Middle East, Latin America—but also in densely populated areas as in Japan, Australia, Canada, and North China (EPA 2012).

In developing countries, most of the wastewater reuse is for agricultural purposes (Asano 2006). Recently, the use of treated wastewater for the irrigation of landscape, public parks, sport fields, and recreational sites has become a widespread practice in Arab Gulf (Kuwait, United Arab Emirates), Tunisia (Abu-Madi and Al-Sa'ed 2009) and Morocco (Benzine 2012; Faouzi et al. 2020).

In the Middle East and North Africa (MENA) region, Tunisia was among the first Mediterranean countries to be equipped with wastewater treatment facilities (Asano and Levine 1996). The area currently irrigated with reclaimed water is about 8000 ha, 80% of which is located around Tunis and a few other locations near Hammamet, Sousse, Monastir, Sfax, and Kairouan (Bahri et al. 2012). In Jordan, around 80% of treated wastewater is stored and conveyed to supply agriculture in the southern districts (Al-Juaidi et al. 2010). Ghneim (2011) stated that in Jordan, treating wastewater and reusing it in agriculture is increasing the amount of freshwater available for municipalities, where water with higher quality is needed.

The Ministry of Agriculture in Egypt promoted the restricted reuse of treated wastewater for cultivation of non-food crops such as timber trees and green belts in the desert to fix sand dunes (Guasch et al. 2010). In Morocco, the potential of treated wastewater is estimated to be 50 mm^3/yr from which 60% is produced in Agadir (Malki et al. 2017). In the Ben Slimane city, the wastewater plant treats about 5600 m^3/day until the tertiary level. This water is used to irrigate a golf course with an area of about 100 km^2. This allows a saving of about 3000–5000 m^3/day of water and 2 DH/m^3 (0.2 Euro/m^3) of costs in case of using fresh water (Benzine 2012).

Among the African countries affected by water stress, Algeria is in the category of the most water-stressed countries, namely below the theoretical threshold of water scarcity set by the World Bank at 1000 m^3 per inhabitant per year. Consequently, the country should dispose between 15 and 20 billion m^3 of water per year, by reserving 70% for agriculture, to achieve a satisfactory food security. This is a great challenge since the amount of water that the country dispose is about 5 billion m^3 per year. The pressure exerted on these resources will continue to grow under the combined effects of population growth and policies applied to water-consuming activities (Medkour

2003). The total volume of wastewater discharged annually is estimated at nearly 600 million m^3, whose 550 million m^3 are from the northern regions. This value would rise to almost 1150 million m^3 by 2025 (Haidara et al. 2022).

As water resources in Algeria are limited, the use of treated wastewater in agriculture has become a necessity. In this perspective, the reuse of treated wastewater in agriculture has become an important focus of the new water policy of the country. Indeed, rainwater, water from dams and groundwater will not be enough for the needs of the country, which explains the ambition to treat a billion m^3 of wastewater for the irrigation of 100,000 ha (Gharzouli 2014). The reuse of treated wastewater for irrigation should primarily concern areas with a deficit in conventional water resources (Water Resources Ministry 2012). According to the National Sanitation Office (2013), the quantities of water treated and reused in agricultural irrigation reached in 2013 a volume of 19 million m^3 for the irrigation of 12,000 ha. The research done by Bechlaghem (Boumediene 2013) concluded that the wastewater treatment plant of Chlef can be reused for irrigation, especially in arboriculture. However, an additional treatment of the wastewater with chlorine disinfection can expand this practice to other types of crops, in particular vegetable crops. With the rehabilitation of the old Wastewater Treatment Plants (WWTP) and the construction of new plants, several irrigation projects from treated wastewater are being studied or already realized. The number of WWTPs operated by the National Sanitation Office was 146 plants in 2018 (National Sanitation Office Report 2018) with a capacity of 550 million m^3/yr and tend to reach 216 WWTPs by 2025 with a capacity of 1200 million m^3/yr of treated wastewater.

Against this background, the main objective of this chapter is to evaluate the quality of the treated wastewater produced from the WWTP (domestic wastewater) of Ain Temouchent for irrigation and the possibility of its reuse for agricultural purposes.

2 Study Area

The area under investigation is the region of Ain Temouchent (Fig. 1), a semi-arid region located in the northwestern part of Algeria, a hundred kilometers from the Moroccan border. Ain Temouchent is limited by the Mediterranean Sea to the north, the city of Sidi Bel Abbes to the south, the city of Oran to the west, and the city of Tlemcen to the southeast. The region is extended over an area of 2376.89 km^2 with a population of 410,423 inhabitants. Ain Temouchent is essentially an agricultural region with an area of 180,184 ha covering more than 76% of the region's total area which spread over 8090 farms (Haidara et al. 2022). The drought occurred in Algeria over the past decades has affected the whole country, and more particularly its northwestern part (Medejerab and Henia 2011). The region of Ain Temouchent is part of the Northwestern Oran coastal river basin characterized by a semi-arid climate. The average annual rainfall varies from 249 to 389 mm/yr with an average monthly temperature varying from 12 to 26.6 °C (Baghli and Bouanani 2013). According to

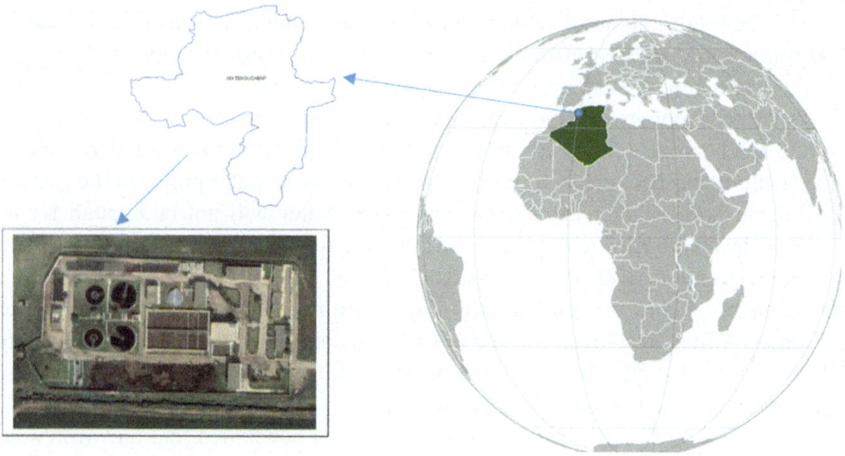

Fig. 1 Geographical location of the study area and the WWTP of Ain Temouchent

Taibi (2012), climate change has led to determinate some impacts and consequences on the river basin that is reflected by a rainfall deficit of 16–38% within the river basin.

Considering the climate vulnerability and the state of aridity of this region, and given the fact that agriculture is the main water-consuming sector, the reuse of the treated wastewater from the WWTP of Ain Temouchent for agriculture would be an adaptation measure to the water scarcity and climate change, leading to a better water resource management in the region.

3 Materials and Methods

In order to assess the quality of the treated wastewater (effluent) from the WWTP of Ain Temouchent, a series of analyses on the water quality parameters were carried out. Sampling of the treated wastewater has been done using automatic samplers where samples were collected from 10 cm below the surface of the water. The physico-chemical parameters analyzed are Temperature, pH, Total dissolved salts (TDS), Total suspended matter (TSM), Dissolved oxygen (DO), Biological oxygen demand (BOD5), Chemical oxygen demand (COD), Phosphates (PO_4^{3-}), Nitrate-nitrogen (NO_3-N), and Ammonium-nitrogen (NH_4^+-N). Temperature, TDS, and pH of surface water were measured immediately after collection using HQ 40D portable multi-parameter meters (Hach Company, USA). All analytical methods applied for other parameters had followed the standard methods for examining water and wastewater (APHA 2012). Analyses for heavy metals were carried out on the following parameters according to the method of US Environmental Protection Agency (EPA) (1994, Method 200.7). The concentrations of Hg, Cd, As, Cr, Pb, Cu, Zn, Se, F,

Al, Be, Co, Fe, Li, Mn, Mo, Ni, V were determined. Microbiological analyses of the treated wastewater concern the quantification of the following parameters: fecal coliforms and intestinal nematodes.

The water quality data were statistically analyzed and performed using Minitab 16.0 (MINITAB Inc., State College, PA, USA). To determine the relationship among the physicochemical parameters in the studied samples for selected time period, Person's correlation coefficient analysis and cluster analyses (CA) were performed.

4 Results and Discussions

Results obtained from analyses carried out on the treated wastewater are interpreted and compared to some existing recommendations and norms (Algerian, FAO and WHO). The results include the parameters usually measured in the plant: Temperature, pH, Total dissolved salts (TDS), Total suspended matter (TSM), Dissolved oxygen (DO), Biological oxygen demand (BOD_5), Chemical oxygen demand (COD), Phosphate ions (PO_4^{3-}), Nitrate-nitrogen ions (NO_3-N), Ammonium- nitrogen ions (NH_4^+-N), microbiological and heavy metals parameters. These results concern the periods 2014, 2015, 2017, and 2018.

4.1 Physico-Chemical Parameters

4.1.1 Temperature

Temperature is an important ecological factor. The analysis of this parameter is very important as it conditions many other parameters, such as electrical conductivity, dissolved oxygen, and pH, as well as the degradation and mineralization reactions of organic matter (Nasra and Zahran 2014). Also, it plays an important role in biological nitrification and denitrification. The temperature of treated water discharged from WWTP for the years 2014, 2015, 2017, and 2018 is represented in Fig. 2.

The results indicated that the temperature values varied from 7 °C in January 2017 to 28.32 °C in June 2017. Temperature variation over time is influenced by seasonal change and atmospheric temperature, as Algeria is known to have the four seasons from cold in winter and hot in summer. The temperature values were at the norms.

4.1.2 pH

The pH plays an important role in many processes of aquatic life. It reflects many biological and chemical processes that occur in the aquatic environment. The main ones are the photosynthetic activities of aquatic plants, respiration of aquatic organisms, decomposition of organic matter, precipitation, dissolution, and redox reactions

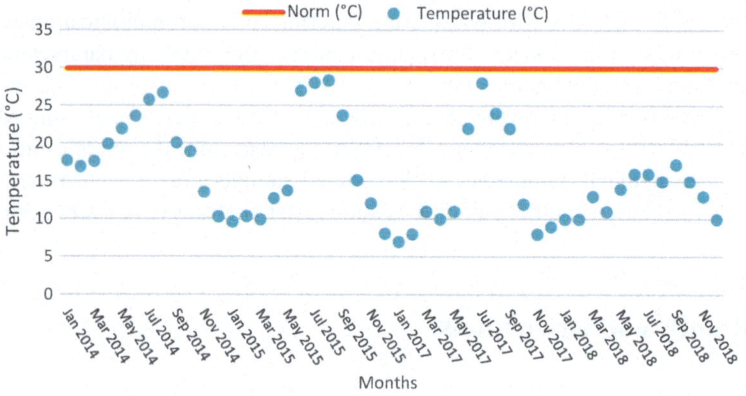

Fig. 2 **Temperature** values for wastewater after treatments (2014–2018). *Source* Developed by the authors

that occur in the aquatic environment. Living organisms are both pH-dependent and sensitive. It is not only a measure of the potential pollutant but is also initially related to the concentration of many other substances, especially weakly dissociated acids, and bases (Wootton et al. 2008).

As shown from Fig. 3, the pH values varied from 6.31 in March 2018 to 8.4 in July 2017 with an average of 7.77. This minimum value is the only value out of norms as shown in Fig. 3. It was noticed that the highest values of pH are in the last spring/summer period. This can be attributed to the elevation in temperature which generate a loss of CO_2 because of evaporation and precipitation of mono-carbonates (Wootton et al. 2008).

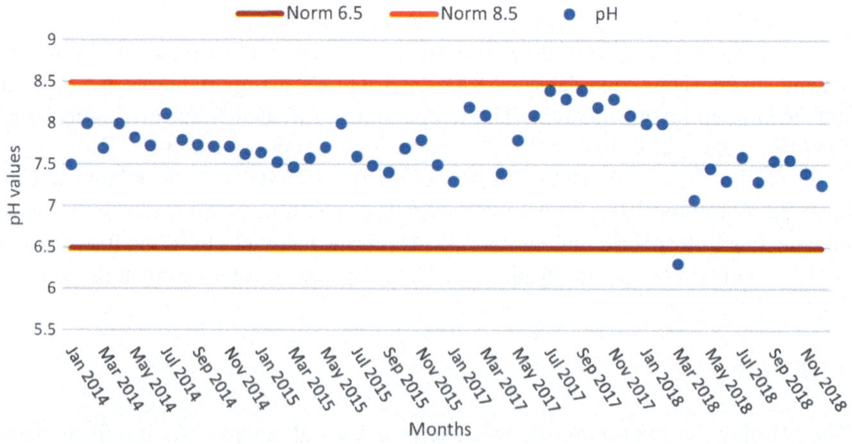

Fig. 3 pH values for wastewater after treatments (2014–2018). *Source* Developed by the authors

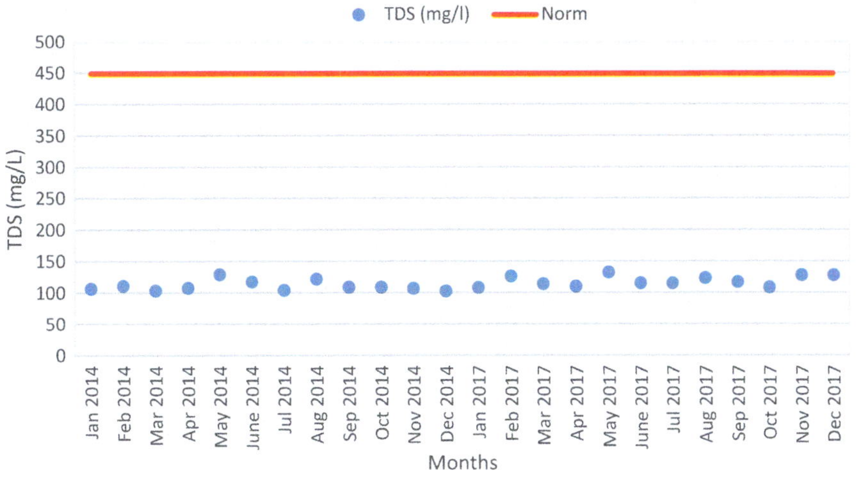

Fig. 4 ***Total*** dissolved salts (TDS) concentrations for wastewater after treatments (2014, 2017). *Source* Developed by the authors

4.1.3 Total Dissolved Salts (TDS)

The main criteria for assessing the quality of water for irrigation are its total concentration of dissolved salts (TDS). The absolute values of TDS for the treated wastewater from the WWTP of Ain Temouchent varied between a minimum of 103.68 mg/L in December 2014 and a maximum of 133.76 mg/L in May 2017 (Fig. 4).

The most influential water quality guideline on crop productivity is the water salinity represented by the amount of total dissolved salts (TDS) (Tak et al. 2010). Sources of salinity include urban and rural run-off containing salt, fertilizers and organic matter. According to Fig. 4, the seasonal variation was limited and recorded values below the standard value set by the FAO (450 mg/L). According to Camberato (2001), a salinity problem exists if the total quantity of salts in the irrigation water is high enough to cause salts to accumulate in the crop root zone to the extent that yields are affected.

4.1.4 Total Suspended Matters (TSM)

Total Suspended Matters (TSM) represent materials that are neither in a dissolved nor in a colloidal state. They are organic and/or mineral and allow a good evaluation of the degree of water pollution (Karef 2017). As shown in Fig. 5, TSM concentrations were unstable during the study period. From January 2014 to December 2015, the amount of TSM in the treated water was under the norms. However, in 2017 and 2018 increasing TSM concentrations were detected, and the levels exceed the norms with the highest values in the cold seasons. The highest concentration is recorded in

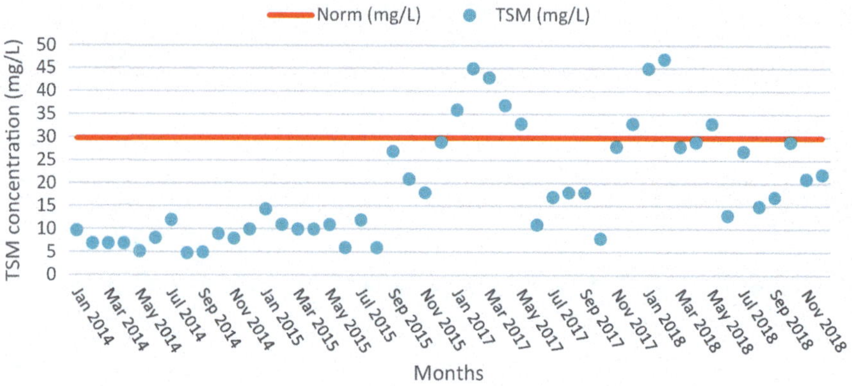

Fig. 5 TSM concentrations for wastewater after treatments (2014–2018). *Source* Developed by the authors

February 2018 with a value of 47 mg/L, while the lowest concentration of 4.8 mg/L was registered in August 2014. This can be explained by solid transport, which is very high in the wetter months and low in the warmer months, despite evaporation and algal production. For this reason, the outflow concentrations are variable that means bad decantation (Sonune and Ghate 2004).

4.1.5 Dissolved Oxygen (DO)

Dissolved oxygen is necessary in aquatic systems for the survival and growth of many aquatic organisms. It is used as an indicator of the health of surface-water bodies indicating its biological state, the predominant processes occurring in it, the destruction of organic substances and the intensity of self-purification (Lewis 2006). Oxygen content of water depends on a number of physical, chemical, biological and microbiological processes.

During this study, available data for DO for the treated wastewater from the WWTP of Ain Temouchent were for the years 2017 and 2018 as shown in Fig. 6. The maximum concentration of 9.26 mg/L was recorded in November 2017 and February 2018. However, a minimum value of 6.39 mg/L was detected in June 2017. Increased aeration derived from the blooming of active winds play an important role in increasing dissolved oxygen in the surface water especially in winter season. However, a minimum value of DO in summer may be due to higher temperature that increases the evaporation process and increasing discharges of agricultural and sewage wastes, which contain high load of organic matter that consume large amount of DO, mostly in the surface layer (Nasra and Zahran 2014).

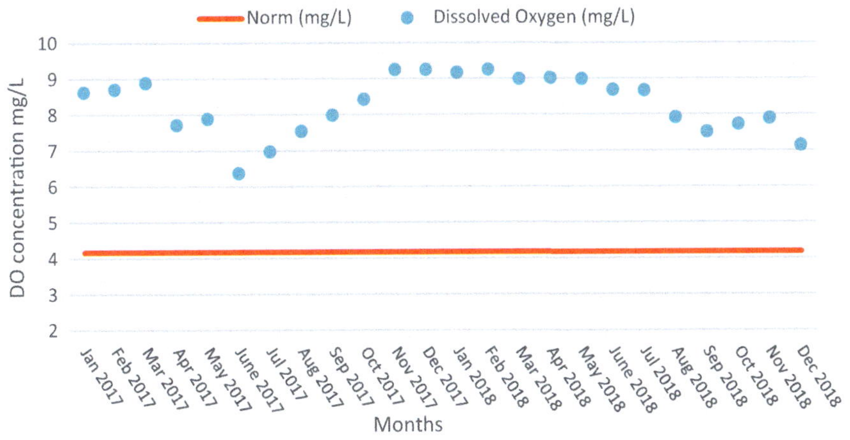

Fig. 6 DO concentrations for wastewater after treatment (2017 and 2018). *Source* Developed by the authors

4.1.6 *Biological Oxygen Demand (BOD$_5$)*

Biological Oxygen Demand is the amount of oxygen needed by aerobic microorganisms to oxidize biodegradable organic matter in water (Ponomareva et al. 2011). It is therefore a potential consumption of oxygen by biological means. This parameter is a good indicator of the biodegradable organic matter content of polluted natural water or wastewater. As shown from Fig. 7, the highest concentration of 47 mg/L was recorded in February 2018 and the lowest concentration of 2.5 mg/L was detected in September 2015.

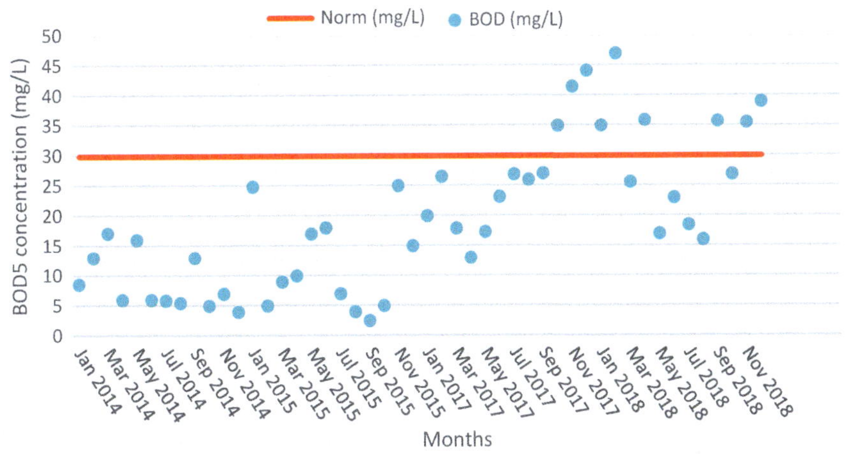

Fig. 7 BOD$_5$ concentrations for wastewater after treatments (2014–2018). *Source* Developed by the authors

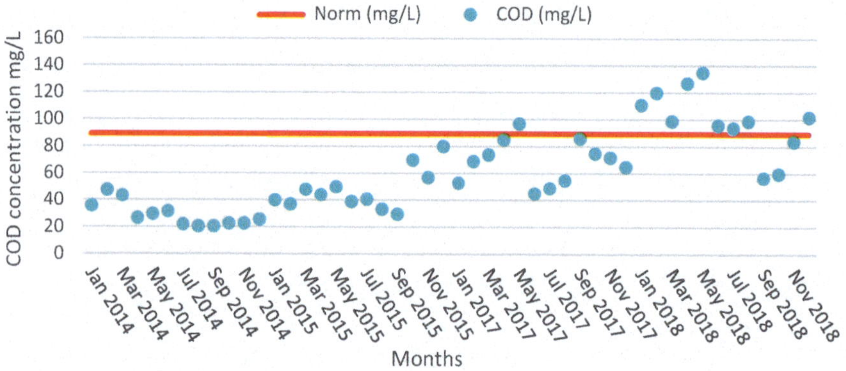

Fig. 8 COD concentrations for wastewater after treatments (2014–2018). *Source* Developed by the authors

Based on the obtained results, the amount of BOD_5 in the discharged water is higher than the norms in several months, especially in winter and autumn. These results indicated the inappropriate functioning of the biological treatment process where the metabolic process of microorganisms is not completed. This, in turn, leads to the non-stabilization of organic matters due to insufficient supply of oxygen (Von Sperling 2008).

4.1.7 Chemical Oxygen Demand (COD)

Chemical Oxygen Demand is defined as the amount of dissolved oxygen used to oxidize and stabilize wastewater when organic or inorganic matter were responsive by a strong chemical oxidant. It is considered one of the important quality control parameters of an effluent in wastewater treatment facility (Canals et al. 2002). The higher COD values, the higher the amount of pollution in the wastewater. As shown in Fig. 8, the absolute values of COD for the wastewater after the treatment varied between a minimum of 21 mg/L in September 2017 and a maximum of 135 mg/L in May 2018.

The overtaking of values over the norm is observed in the autumn of 2017 and winter of 2018. This means that the chemical compounds are not oxidized, and the process of treatment represents a problem (Schrank et al. 2004).

4.1.8 *Nitrate-Nitrogen* ($N\text{-}NO_3^-$)

Nitrogen is one of the biologically important elements in the aquatic environments where Nitrate is the final oxidation product of nitrogen compound in natural waters (Tak et al. 2010). For plants, however, nitrates are an essential source of nitrogen and in many cases appear to be the major limiting nutrient for phytoplankton growth.

that the number of fecal coliforms in the treated wastewater was equal to 1100 fecal coliform/100 mL and the number of intestinal nematodes was equal to 20 eggs/L. These values are above the WHO norms for treated wastewater intended for irrigation. This is probably due to the suspension of the chlorination disinfection treatment in the plant.

4.1.14 Statistical Analysis

Cluster analysis divides data into cluster that are meaningful. It groups data objects based only on information found in the data and describes the objects and their relationships. The goal is that the objects within a group be similar (or related to one another) and different from the objects in other groups (Tan et al. 2005). In the present study, the cluster analysis was performed on the 10 physico-chemical parameters collected monthly during the study period (Fig. 13). Nevertheless, the data of heavy metals concentrations were not included as their concentrations were very low with no effect on the quality of treated water.

The dendrograms resulted from this cluster analysis (Fig. 13) showed two major groups: The first comprises 7 parameters including Temperature, Ammonium ion (N-NH_4^+), Total suspended matters (TSM), Biological oxygen demand (BOD_5), pH, Phosphate ions (PO_4^{3}) and Chemical oxygen demand (COD); the second cluster includes 2 parameters including Nitrates-Nitrogen (N-NO_3) and Dissolved oxygen (DO). However, there was no relation between TDS with any other parameters as it shows limited variation (Fig. 13).

The highest degree of similarity is observed between Temp and NH_4^+ followed by similarity between pH and phosphate ions while the third group is observed between Nitrates -Nitrogen and dissolved oxygen with a lesser degree of similarity. This may

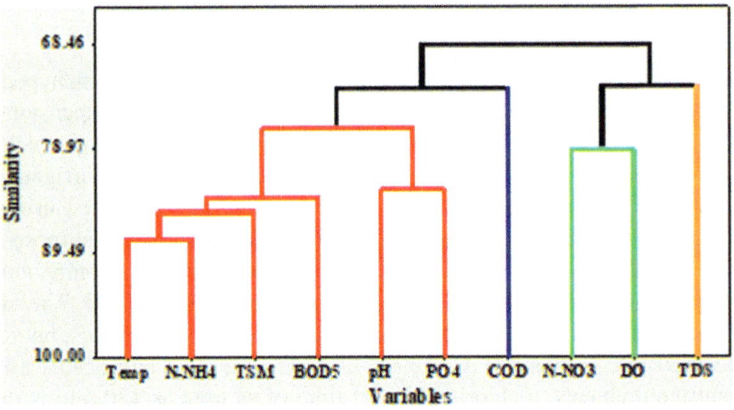

Fig. 13 Dendrogram using average linkage between groups. *Source* Developed by the authors

be attributed to slightly similar trend of distribution and all parameters are affected by the same type and amount of land-based sources (agriculture watershed area).

The calculated correlation matrix for all measured parameters during the study period (Table 2) showed significant negative correlation between dissolved oxygen and temperature ($r = -0.782$ at $p < 0.05$, $n = 36$), between temperature and N-Nitrate ($r = -0.671$ at $p < 0.05$, $n = 36$), and significant positive correlation between dissolved oxygen and N-Nitrate ($r = 0.582$ at $p < 0.05$, $n = 36$). This indicated that the increase of water temperature in hot seasons decrease the solubility of dissolved oxygen and increasing evaporation process (Saad Massoud 1978). Also, the increase in the rate of consumption of dissolved oxygen, due to the increasing activity of a decomposing bacteria, leads to a decrease in oxygen concentration and, in turn, to a decrease in nitrate concentrations and an increase in ammonia/ammonium concentrations ($r = -0.657$ at $p < 0.05$, $n = 36$). The positive significant correlation is observed between temperature and ammonia/ammonium concentrations ($r = 0.760$ at $p < 0.05$, $n = 36$). Moreover, there are positive significant corrections between BOD_5 and water temperature ($r = 0.617$ at $p < 0.05$, $n = 36$) and between BOD_5 and TSM ($r = 0.677$ at $p < 0.05$, $n = 36$). It was noted from Fig. 13 that there was a direct relation between pH and PO_4^{3-} concentration in treated water of WWTP. According to reports published by Minnesota Pollution Control Agency (2006), the precipitation process for phosphate ions vary with pH values reached the maximum at pH range of 5–7. In the area under investigation, the pH values varied between a minimum of 7.3 and a maximum of 8.4, explaining a higher concentration of phosphate ions in treated water.

5 Improvement of the Microbiological Quality

5.1 Chlorination

In Ain Temouchent WWTP, a unit of disinfection with the sodium hypochlorite is installed after the secondary treatment. However, due to the suspension of the chlorination treatment, the results of the bacteriological analyses are above the standards. To improve the bacteriological quality of wastewater to meet irrigation standards, thus minimizing health risks for crops production and consumers, an operation of the chlorine disinfection treatment is required. For disinfection purposes, chlorine is used either in the form of liquid/gaseous chlorine, or its compounds such as calcium hypochlorite, sodium hypochlorite and chlorine dioxide. The chlorine dosage depends upon the strength of the wastewater and other factors, but dosages of 5–15 mg/L are common (FAO 1992). Generally, to meet advanced wastewater treatment requirements, a chlorine contact time of as long as 120 min is required for improved irrigation uses of the treated wastewater. According to FAO (1992), the bactericidal effects of chlorine and other disinfectants are dependent upon pH, contact time, organic content, and effluent temperature.

Table 2 Correlation matrix for physico-chemical parameters in the effluent of Ain Temouchent WWTP during the study period

	Temperature	pH	TSM	N-NO_3	COD	BOD5	PO_4	TDS	DO
pH	0.539								
	0.070								
TSM	0.206	0.245							
	0.010	0.442							
N-NO_3	− 0.671	0.369	− 0.260						
	0.035	0.238	0.414						
COD	0.459	− 0.182	0.134	− 0.319					
	0.133	0.571	0.678	0.008					
BOD5	0.617	0.139	0.677	− 0.308	0.262				
	0.033	0.667	0.016	0.330	0.410				
PO_4	0.245	0.661	− 0.073	0.167	− 0.075	− 0.189			
	0.442	0.019	0.821	0.605	0.818	0.557			
TDS	− 0.132	0.323	0.130	0.453	− 0.433	− 0.495	0.298		
	0.682	0.305	0.688	0.139	0.160	0.102	0.347		
DO	− 0.782	− 0.063	− 0.505	0.582	− 0.727	− 0.307	− 0.009	0.244	
	0.003	0.845	0.094	0.047	0.007	0.332	0.977	0.445	
NH_4	0.760	0.371	0.528	− 0.285	0.243	0.433	0.027	− 0.292	− 0.657
	0.004	0.235	0.078	0.369	0.446	0.160	0.935	0.357	0.020

Source Developed by the authors

5.2 Stabilization Reservoirs

The storage of the treated wastewater in reservoirs is, in most cases, a critical link between the wastewater treatment plant and the irrigation system. The main purpose of stabilization reservoirs is to recover the treated wastewater for its efficient reuse for irrigation. Stabilization reservoirs are deep lagoons that are used for two purposes: (a) storing the treated wastewater over a long period in order to use it during a specific period of the year, under optimal and controlled conditions; and (b) improving the quality of the treated wastewater during their long residence times in the reservoirs that allow maximum removal of helminths and protozoa as well as bacterial and viral pathogens (Eme and Molle 2013).

6 Proposition of Plantation

The Algerian government has established the types of crops likely to be irrigated with treated wastewater. The interpretations of physicochemical parameters have shown that all values were below the standards except those for the phosphates. Given the microbiological results obtained, the quality of the treated wastewater from Ain Temouchent WWTP is far from acceptable level for unrestricted irrigation in general, namely for the irrigation of crops that are normally eaten raw. For these crops, according to the Algerian standards, the number of fecal coliforms bacteria must be less than 100 fecal coliforms/100 mL with an absence of intestinal nematodes eggs. Concerning the vegetables that are only eaten cooked, a number of fecal coliform bacteria less than 250 fecal coliforms/100 mL and intestinal nematodes eggs less than 0.1 egg/L is required. Therefore, the treated wastewater is not acceptable for these crops.

For fruit trees, fodder crops, shrubs, cereal crops, industrial crops, forest trees, floral and ornamental plants, a limit of 1000 fecal coliforms bacteria/100 mL and one nematode egg/liter is fixed. The number of fecal coliforms in the treated wastewater is 1100/100 mL while the number of intestinal nematodes is 20 eggs/l. These values are above the limit values mentioned above. However, according to the Algerian norms, there is no recommended standard for the treated wastewater to irrigate fruit trees and ornamental plants as the irrigation technique used is localized irrigation. In addition to that, Karef (2017) stated that the irrigation of fruit trees does not require disinfection.

Therefore, for the present study, we will propose the irrigation of a fruit tree using localized irrigation (trickle irrigation). It is a system where the water is distributed under low pressure through a piped network, in a pre-determined pattern, and applied as a small discharge to each plant or adjacent to it. There are three main categories: (a) ***drip irrigation*** where drip emitters are used to apply water slowly to the soil surface; (b) ***spray or micro sprinkler irrigation***, where water is sprayed onto the soil near individual plants or trees; and (c) ***bubbler irrigation*** where a small stream of water is applied to flood small basins or the soil adjacent to individual trees (FAO 1985).

Suitable irrigation methods can effectively mitigate negative environmental effects. Drip irrigation constitutes the environmentally friendly approach (Muchuweti et al. 2006; Abdelbaki and Medjadji 2011). It saves water and fertilizers by allowing water to drip slowly to the roots of plants, either on the soil surface or directly on the root zone, through a network of valves, pipes tubing, and emitters (Desai 2016). Although drip irrigation systems are costly, they are highly efficient in water use along with the highest levels of health protection. The clogging of drip emitters, on the other hand, may limit the use of drip irrigation systems for treated wastewater. Therefore, filtration is needed to prevent clogging of emitters (Sridhar 2016). Moreover, drip irrigation reduces water contact with crop leaves, stems, and fruit. Thus, conditions may be less favorable for disease development (Shock 2013).

7 Conclusion

Water is the major challenge facing most of the arid and semi-arid countries, including the African countries, due to population growth and climate change. In Algeria, the pressure exerted on regular water resources will continue to grow under the combined effects of such factors in addition to policies applied to water-consuming activities. Consequently, the reuse of treated water has proved to be a very promising source of irrigation water for crop cultivation. This work allowed to evaluate the physico-chemical and microbiological quality of the treated wastewater from Ain Temouchent WWTP. The findings indicated that the quality of the treated wastewater is satisfactory for the reuse in irrigation as the treatment process of the plant is effective to produce water that meets the reuse standards—WHO, FAO and Algerian standards—since it is equipped with a tertiary treatment unit. The annual volume of the treated wastewater valued through the irrigation of fruit trees will be of great benefit to Ain Temouchent. Indeed, this will contribute to a better-integrated water resources management of the region while promoting the agricultural sector.

8 Recommendations

- In a context of climate change and in order to preserve the available water resources, the water resources managers of Algeria should promote more wastewater reuse projects in agriculture for a sustainable development.
- In view of the above, the construction of sewage systems as well as treatment plants are very important to ensure that effluents from WWTP have appropriate quality from a microbiological and physico-chemical point of view. The large-scale of good practices for the recovery of wastewater in irrigation, will contribute

to the sustainability of water resources, the environment, agriculture, and human life.

- Some constraints related to economic, environmental and social issues could be overcome by following certain practices and proper public policies and regulations. Consequently, public awareness programs should be applied to overcome cultural, religious and social resistance for using treated wastewater in irrigation process.

References

Abdelbaki C, Medjadji SM (2011) Contribution à la gestion d'un réseau d'irrigation à l'aide d'un SIG—Cas du réseau d'irrigation de goutte à goutte de Safsaf, Tlemcen, Algérie, Editions universitaires européennes. ISBN: 978-613-1-58390-2

Abu-Madi M, Al-Sa'ed R (2009) Towards sustainable wastewater reuse in the Middle East and North Africa. J Sustain Dev 2:1475–1481

Al-Juaidi AE, Kaluarachchi JJ, Kim U (2010) Multi-criteria decision analysis of treated wastewater use for agriculture in water deficit regions. J Am Water Resour Assoc 46(2):395–411. https://doi.org/10.1111/j.1752-1688.2009.00409.x

APHA (2012) Standard methods for the examination of water and wastewater, 22nd edn

Asano T (2006) Water reuse: issues, technologies, and applications. McGraw-Hill, New York. https://www.accessengineeringlibrary.com/browse/water-reuse-issues-technologies-and-applications

Asano T, Levine AD (1996) Wastewater reclamation, recycling and reuse: past, present, and future. Water Sci Technol 1–14. https://doi.org/10.1016/0273-1223(96)00401-5

Baghli N, Bouanani A (2013) Gestion intégrée des ressources en eau dans le bassin Côtier Oranais: diagnostic et outils, vol 1, pp 350–353

Bahri A, Drechsel P, Brissaud F (2012) Water reuse in Africa: challenges and opportunities, (September), pp 1–16

Benzine L (2012) Réutilisation des eaux usées épurées dans le bassin de Souss Massa. Paper presented at the Journée de sensibilisation sur Les Potentialités et Perspectives de la Réutilisation des Eaux Usées Dans la Région du Sous Massa, Complexe Horticole d'Agadir

Boumediene M, Abdelbaki C, Bechlaghem A, Kellouche A (2013) Etude de la faisabilité de réutilisation des eaux usées issues de la STEP de Chlef à des fins agricoles, 3ème conférence internationale sur l'eau, 18 & 19 novembre 2013, Alger, Algérie

Camberato J (2001) Irrigation water quality: update from the 2001 Carolinas Golf Course Superintendents Association. Clemson University Turfgrass Program, p 1–13. http://www.scnla.com/Irrigation_Water_Quality.pdf

Canals A, Cuesta A, Gras L, Remedios Hernández M (2002) New ultrasound assisted chemical oxygen demand determination. Ultrason Sonochem 9(3):143–149. https://doi.org/10.1016/s1350-4177(01)00119-5

Desai A (2016) Plc based automated drip irrigation. Int J Curr Res Multidisciplinary 1(2):17–20. www.ijcrm.com

Drechsel P, Evans AEV (2010) Wastewater use in irrigated agriculture. Irrig Drainage Syst 24:1–3. https://doi.org/10.1007/s10795-010-9095-5

Eme C, Molle P (2013) Stockage des eaux usées traitées en vue de leur réutilisation

Environmental Protection Agency (EPA) (2012) Guidelines for water reuse, guidelines for water reuse, 26(September), p 642. EPA16251R-921004

FAO (1985) Land evaluation for irrigated agriculture

FAO (1992) Wastewater treatment and use in agriculture—FAO irrigation and drainage paper 47 Food and agriculture organization of the united

Faouzi E, Arioua A, Karaoui I, Ouhamchich KA, Elhamdouni D (2020) Wastewater reuse in agriculture sector: resources management and adaptation in the context of climate change: case study of the Beni Mellal-Khenifra region, Morocco. E3S Web Conf 183:02005. https://doi.org/10.1051/e3sconf/202018302005

Gharzouli R (2014) Gestion et valorisation des eaux usées en Algérie. Larhyss J 19(19):51–62. http://lab.univbiskra.dz/Larhyss/images/pdf/JOURNALN°19/4.Hannachi_et_al.pdf. Accessed 10 July 2018

Ghneim A (2011) Wastewater reuse and management in the Middle East and North Africa a case study of Jordan, assessment. http://colinmayfield.com/public/WaterHealthSept2015/course4/resources/Ghneim_Gesamt.pdf. Accessed 10 July 2018

Guasch H, Serra A, Corcoll N, Bonet B, Leira M (2010) Metal ecotoxicology in fluvial biofilms: potential influence of toxicity and algal bioassays, ASTM. 100 Barr Harbor Drive, PO Box C700, West Conshohocken, PA 19428–2959: ASTM International, 573, pp 320–343. https://doi.org/10.1520/STP39020S

Haidara R, Abdelbaki C, Badr N (2022) Feasibility of water reuse for agriculture, case study of Ain Temouchent (Algeria), sustainable energy-water-environment nexus in deserts. In: Proceeding of the first international conference on sustainable energy-water-environment nexus in desert climates. Book series: advances in science, technology and innovation (ASTI). Springer Nature (in Press). https://www.springer.com/gp/book/9783030760809)

Karef S (2017) Etude des possibilités technico-économiques de la réutilisation des boues et des eaux usées urbaines traitées. Thèse de doctorat, Ecole Nationale Polytechnique, Alger

Lewis ME (2006) Dissolved oxygen, vol 3, pp 1–48. https://doi.org/10.1094/ASBCMOA-Beer-34

Lundberg JO, Weitzberg E, Gladwin MT (2008) The nitrate–nitrite–nitric oxide pathway in physiology and therapeutics. Nat Rev Drug Discov 7(2):156–167. https://doi.org/10.1038/nrd2466

Malki M, Bouchaou L, Mansir I, Benlouali H, Nghira A, Choukr-Allah R (2017) Wastewater treatment and reuse for irrigation as alternative resource for water safeguarding in Souss-Massa region, Morocco, pp 365–371

Medejerab A, Henia L (2011) Variations Spatio-Temporelles De La Secheresse Climatique En Algerie Nord-Occidentale. Courrier Du Savoir 11:71–79

Medkour M (2003) Réutilisation des eaux usées épurées, Séminaire sur le secteur de l'eau en Algérie. Ministère des Ressources en eau

Minnesota Pollution Control Agency (2006) Phosphorus treatment and removal technologies, (June), pp 1–5. https://www.pca.state.mn.us/sites/default/files/wq-wwtp9-02.pdf

Muchuweti M, Birkett JW, Chinyanga E, Zvauya R, Scrimshaw MD, Lester JN (2006) Heavy metal content of vegetables irrigated with mixture of wastewater and sewage sludge in Zimbabwe, implications for human health. Agr Ecosyst Environ 112:41–48

National Sanitation Office (2013) http://ona-dz.org/. Accessed 11 July 2021

National Sanitation Office Report (2018) http://ona-dz.org

Nasra M, Zahran HF (2014) Using of pH as a tool to predict salinityof groundwater for irrigation purpose using artificial neural network. Egypt J Aquat Res 40:111–115. https://doi.org/10.1016/j.ejar.2014.06.005

Ponomareva ON, Arlyapov VA, Alferov VA, Reshetilov AN (2011) Microbial biosensors for detection of biological oxygen demand (a Review). Appl Biochem Microbiol 47(1):1–11. https://doi.org/10.1134/s0003683811010108

Saad Massoud AH (1978) Seasonal variations of some physicochemical conditions of Shatt al-Arab estuary, Iraq. Estuar Coast Mar Sci 6(5):503–513. https://doi.org/10.1016/0302-3524(78)90027-0

Schrank SG, Jose HJ, Moreira RFPM (2004) Schroder, Elucidation of the behavior of tannery wastewater under advanced oxidation conditions. Chemosphere 56:411–423

Shock CC (2013) Drip irrigation: an introduction. Oregon State University Extension publication. EM 8782. http://ir.library.oregonstate.edu/ xmlui/ bitstream/handle/1957/ 20206/em8782-e
Sonune A, Ghate A (2004) Developments in wastewater treatment methods. Desalination 167:55–63. https://doi.org/10.1016/3.desal.2004.06.113
Sridhar C (2016) Urban water reuse handbook, pp 525–532
Taibi (2012) Evolution de la pluviométrie au Nord Ouest Algérien.Séminaire SIGMED de Relizane, p 12
Tak HI, Bakhtiyar Y, Ahmad F, Inam A (2010) Effluent quality parameters for safe use in Agriculture. www.intechopen.com. Accessed 10 November 2021
Tan P-N, Steinbach M, Kumar V (2005) Cluster analysis: basic concepts and algorithms. Introduction Data Min (Chapter 8). https://doi.org/10.1016/0022-4405(81)90007-8
UNEP (2004) Water and wastewater reuse: an environmentally sound approach for sustainable urban water management. United Nations Environment Programme. http://wedocs.unep.org/handle/20.500.11822/8390. Accessed 10 July 2018
US EPA (1994) Method 200.7: determination of metals and trace elements in water and wastes by inductively coupled plasma-atomic emission spectrometry, Revision 4.4. Cincinnati, OH
Von Sperling M (2008) Wastewater characteristics, treatment and disposal. Choice Rev Online. https://doi.org/10.5860/CHOICE.45-2633
Wall D (2013) A2. Nitrogen in waters: forms and concerns (June), pp 1–22
Water Resources Ministry (2012) http://www.mree.gov.dz/?lang=fr. Accessed11 July 2018
WHO (1989) Health guidelines for the use of wastewater in agriculture and aquaculture. Report of a Scientific Group, Technical Report Series 778. World Health Organization, Geneva, p 74
WHO (2000) Guidelines for the microbiological quality of treated wastewater used in agriculture: recommendations for revising WHO guidelines. Bull World Health Organ 78(9):1104–1116. https://doi.org/10.1590/S0042-96862000000900006
WHO (2006) Wastewater use in agriculture in guidelines for the safe use of wastewater, excreta and greywater, p 191
Wootton TJ, Pfister CA, Forester JD (2008) Dynamic patterns and ecological impacts of declining ocean pH in a high-resolution multi-year dataset PNAS (105), pp 18848–18853. https://doi.org/10.1073/pnas.0810079105

Dr. Chérifa Abdelbaki obtained her Ph.D. degree in 2014. She is a Senior Lecturer in Water Sciences at the Department of Hydraulics, Faculty of Technology, University Abou Bakr Belkaid—Tlemcen, Algeria. She graduated from the National Higher School of Hydraulics—Blida, Algeria. She is heading the research team: Water resources and mobilization (EOLE Research Laboratory) at the University of Tlemcen, Algeria. Currently, she is working as a Water Program Coordinator at the Pan-African Institute for Water and Energy including Climate Change (PAUWES), Algeria. Her fields of research include Water resources management, Geographic information system, Water reuse and Wastewater treatment. Dr. Abdelbaki has participated in the development of several research projects in the field of water resources (mobilization, management, and optimization ...). With more than 20 years of experience in the field of Water Resources and GIS applied to the management of water resources, she has participated in several scientific events (Innovation exhibition, National and International Conferences, Doctoral Schools, etc.), supervised more than 70 topics (Engineer, Masters, and Ph.D.), and led several GIS training sessions with the ministries of transport, public works, water resources in Algeria. Consultant and expert in the field of GIS and Hydraulics, she has written more than 100 communications and more than 20 publications in these fields.

Dr. Nadia Badr obtained her Bachelor of Science in General Oceanography at the Faculty of Science, Alexandria University, Egypt, where her second major was Fresh and Marine Chemistry. She graduated in December 2002 with a Ph.D. in Chemical Engineering from the College of Engineering, Wayne State University, MI, USA, with a research project focusing on the finite element model to simulate mercury species in water column and sediment of Detroit River. She is currently working as Professor and Head of Environmental Sciences Department, Faculty of Science, Alexandria University, Egypt. Her Research and Interests Area is hazardous Waste Management. She Implemented and participated in many research projects in the field of monitoring and management of hazardous organic and inorganic wastes in water and sediments as consultant, co-principal investigator or principal investigator. She Supervised 23 M.Sc. and Ph.D. degrees in Alexandria University and different Universities and Research Institutes inside and outside Egypt. She Attended and participated by scientific papers in more than 30 conferences and workshops. She worked as a Visiting professor to PAUWES Institute in Algeria from (2016-now). She was assigned as External academic expert in EXCEED-SWINDON network project for International Network on Sustainable Water Management in Developing Countries (2018) offered from Higher Education Ministry in Germany. She published 46 papers and 4 book chapters.

Hidayat Mohammedi was born in 1994 in Tlemcen, Algeria. She holds a Bachelor in General Chemistry and a Master of Science in Chemistry of Natural Products; both from the University of Abou Bekr Belkaid-Tlemcen; and a Master of Science in Water Policy from the Pan-African University Institute of Water and Energy Sciences including Climate Change (PAUWES). She is an Alumni of Clinton Global Initiative University (cohort 2019–2020). Currently, she is teaching Physics, Chemistry and Mathematics to high school students. As a young researcher, she is passionate about science and research.

Rokiatou Haidara is a Water and Environmental Engineer from Mali. She got a Bachelor in Water Treatment and Purification from the University of Ain Temouchent in Algeria in 2015. She is graduated with a Master of Science in Water Engineering held in the Pan-African University Institute of Water and Energy Sciences (Including Climate Change) PAUWES in Tlemcen, Algeria in 2018. To acquire more knowledge in the water and environmental field, Rokiatou has completed in 2021, the Master Program oriented in Water Governance and Policy in AgroParisTech (The National Institute for Life, Industries and Environmental Sciences) in Montpellier, France. Rokiatou worked as a Project Manager at the SUEZ "Water for All" Chair in 2021 in Montpellier. She has also worked as Water and Sanitation Engineer Assistant to the National Sanitation Office of Ain Temouchent in 2018. Rokiatou was a water Quality Technician Assistant to The Drinking Water Supply and Management Company of Algeria (Algerienne des Eaux) in 2015. She has professional skills in Drinking water supply, Wastewater treatment and reuse, Water resources management, Adaptation to climate change, Water quality, and Water policy and governance.

Dr. Halima Belarbi obtained her Ph.D. degree in 2017 from University Abou Bakr Belkaid – Tlemcen, Algeria. She is an Associate Professor at the Department of Hydraulics, University Center of Maghnia, Algeria and Researcher at EOLE Research Laboratory. Her fields of research are: Hydroclimatic variability, Hydrology, Water resources management, and Climate Change.

Chapter 16
Water Quality and Its Health Impact in the Prefecture of Mohammedia, Morocco: A Review

Rachida El Morabet, Larbi Barhazi, Soufiane Bouhafa, Mohamed Behnassi, and Roohul Abad Khan

Abstract One of the major concerns of the world today is the sustainable management of water resources. The qualitative and quantitative sustainability is a necessity to meet the water demand in the face of a strong demographic growth, increased economic activities, changing consumption patterns, and global change. The Prefecture of Mohammedia is a booming region with the extension of intensive irrigated agriculture and industrial zone in addition to the accelerated urbanization and the creation of the new Zenata city within the Prefecture. This development is accompanied by an increased degradation of water resources, which affects, in turn, the socio-ecological system. In this review, we review and analyze recent studies about the qualitative and quantitative evolution of water associated with global change, with a focus on data from institutional reports, in addition to data from field surveys. The groundwater was found to have higher concentrations in indicating permeation of wastewater (presence of *E-coli*); whereas surface water sources were contaminated and crossed the permissible limits for safe drinking water quality at point of meeting wastewater discharge. Also, bacteriological concentration in groundwater validated groundwater contamination, especially in the vicinity of landfills. The results show a significant spatial variability in the quantitative (piezometry) and qualitative (physio-chemical, biological, and hydro-geomorphological quality) distribution in the Prefecture of Mohammedia. The number of water-borne diseases validated the impact of

R. El Morabet (✉) · L. Barhazi · S. Bouhafa
Department of Geography, LADES, FLSH-M, Hassan II University of Casablanca, Mohammedia, Morocco
e-mail: rachidaelmorabet@yahoo.fr

M. Behnassi (✉)
Economic, Social and Environmental Council (ESEC), Rabat, Morocco
e-mail: behnassi@gmail.com

Center for Environment, Human Security & Governance (CERES), Agadir, Morocco

Ministry of Higher Education and Innovation, Rabat, Morocco

R. A. Khan
Department of Civil Engineering, King Khalid University, Abha, Saudi Arabia
e-mail: rakhan@kku.edu.sa

M. Behnassi et al. (eds.), *The Water, Climate, and Food Nexus*,
https://doi.org/10.1007/978-3-031-50962-9_16

water quality on consumer's health. Further studies are needed to conduct an overall quality analysis of water resources based on the same parameters to serve as a reference for policy making regarding the sustainable management of water resources in the Prefecture of Mohammedia.

Keywords Water resources · Sustainability · Physio-chemical · Biological · Hydro-geomorphological

1 Introduction

Water is the foundation of life and its availability in sufficient quantity and quality to meet basic human needs is a prerequisite for sustainability and human health (World Bank Group 2017). However, in the Prefecture of Mohammedia, water resources are subject to increasing and continuous anthropogenic pressures under the effect of population growth, socio-economic development, and changes in lifestyles and consumption. These pressures affect both the availability of water resources and their quality. Indeed, the Prefecture constitutes a key economic pole since it concentrates alone nearly a half of Morocco's industries (Monographie de Mohammedia 2015). This covers various types of industry, including electronics, electricity, petroleum, pharmaceutical, food and paper industries. More precisely, the anthropogenic activities, which are the primary cause of water degradation in the Mohammedia prefecture, are as follows:

- Domestic, industrial, and other wastes (such as medical wastes) are often discharged in landfills, thus producing a significant amount of leachate, which filters and/or trickles towards rivers;
- Direct discharge of wastewater, loaded with polluting substances, in the environment without any prior treatment as a pattern of growing concern given its undesirable effects on the environment and human health;
- Effluent discharge from industry as a major cause of water pollution and degradation of water quality;
- The intensification of agriculture, which involves the overuse of water and chemical inputs—such as fertilizers and pesticides—threatens the quantity and quality of surface and underground water, therefore compromising the viability of hydrological systems.

As a consequence, demographic and industrial dynamics at the level of the Prefecture of Mohammedia are leading to a significant increase in the volumes of liquid discharges and in the flows of polluting organic matter that they generate. The geographic concentration of such discharges poses the problem of their restitution to the receiving environment (Mohammedia sea and El Maleh and Nfifikh rivers) since the self-purification capacity of the latter is declining (Kanbouchi et al. 2014).

The water quality of the Prefecture of Mohammedia has been a subject to several characterization studies in terms of determination and evaluation of the physico-chemical and bacteriological parameters of surface and groundwater as well as dam reservoirs while other studies have focused on the hydrology, hydrogeology, and water pollution and quality of the Prefecture (Laaouan et al. 2016; Idlahcen et al. 2014; Mazouni et al. 2018; Merbouh et al. 2020; Mondet et al. 2016; Serghini et al. 2013; Taleb et al. 2015).

This chapter analyzes and synthesizes the studies dealing with the issue of water quality and its health impact in the same study area. Even though there is a number of studies with respect to groundwater and surface water quality in Mohammedia, they are often restricted to a specific region or area. Also, other studies are specifically carried out around landfills for source apportionment and water pollution indexing. Hence, the objectives of this review are: first, to collect the information regarding the water quality in the study area; second, to appraise an overall water quality in terms of physio-chemical and biological characteristics; and, finally, to assess the impact of the water quality degradation on consumers' health.

In terms of methodology, this review was carried out to present an overall water quality scenario and to assess the water pollution and its health impact in the Prefecture of Mohammedia. The water quality data was obtained from published scientific literature, institutional reports, and field surveys. The collection of samples and testing have been carried out in the laboratory of the companies responsible for water quality analysis in the Prefecture of Mohammedia.

2 Water Resources in the Prefecture of Mohammedia

The Prefecture of Mohammedia is home to two rivers: Oued El Maleh, which crosses the Prefecture in the center; and Oued Nfifikh in the western border. The Prefecture is located in an area hydrologically quite poor, given the geological constitution of the region, generally composed of impermeable and very thick primary formations. This explains the lack of a significant groundwater table. The studies of Fehdi et al. (2010) concerning the hydrogeological functioning of the Mohammedia wetland complex, located downstream from Oued El Maleh, show that the area is subdivided into three sectors with different dynamics: a sector of combined marine and continental influences; a permanent water supply sector, essentially subject to underground influences; and a sector subject to strict climatic influences.

According to CESE, Morocco (2014), the hydrochemical determinism of the Mohammedia wetland complex was carried out by the spatiotemporal monitoring and analysis of a certain number of physical and chemical tracers (temperature, pH, salinity, conductivity, dissolved oxygen, biological demand for oxygen, nitrates, sulphates, and total phosphorus) of surface water during the years 2000–2002. This made it possible to identify zoning with three poles of influence: close to the mouth (outfall) and subject to combined marine and continental influences; of a temporary nature, subject mainly to climatic hazards (precipitation, floods, etc.) in addition to anthropogenic influences (wastewater); and located far from marine influences but under the influence of underground and anthropogenic water. The readings of the various tracers bear witness to organic pollution of anthropogenic origin, which clearly illustrates the extent to which the ecosystem defends itself by self-purification to fight eutrophication, especially since very high concentrations of total phosphorus have been recorded.

2.1 *Surface Water*

In hydrographical terms, the study area largely covers the sub-watersheds of temporary coastal river. The latter borrow essentially impermeable terrain at the level of the Moroccan central Meseta. They drain the Atlantic fringe of Meseta and reach the Atlantic Ocean. In this context, we noticed the following:

- The El Maleh River originates from the north of Khouribga city and the northern edges of the phosphate plateau and drains the northwestern part of this plateau. However, apart from some small tributaries such as Wadi Hessar, the waters of this Oued El Maleh are brackish outside of flooding periods. This watercourse is the source of annual inflows of around 68.8 million m^3 (Abouali et al. 2019).
- The Nfifikh River, draining a watershed of 830 km^2, is made up exclusively of impermeable formation. The flow at this eastern stream is estimated at 550 L/s.

The surface runoff accumulates in inter-dune depressions where temporary lagoons form. These lagoons, cashed in the schistous and quartzite plateaus, come in a series of strings of endorheic basins (Meseta Marocaine). The valuation of wetlands is at the center of environmental concerns in Morocco, like many countries (Dakki et al. 2015). Developing more research on the water quality in these areas is essential because it provides a useful input for the adoption of a preservation and restoration program (Serghini et al. 2013). The Mohammedia Wetlands Complex (CZHM) is a part of the important network of Moroccan coastal wetlands which, thanks to its geographical position and its biotic and abiotic characteristics, is of biological, ecological, economic, and landscape interest.

2.2 *Groundwater*

On the hydrogeological level, the Prefecture of Mohammedia does not present a generalized water table, hence the scarcity of underground water resources. They are only encountered occasionally in sand dune formations, and their durability is not guaranteed. In addition, in the low narrow terrace of the El Maleh river, an alluvial sheet is encountered linked to the flow of this River. The main groundwater table located between Casablanca and Mohammedia is called the lower Chaouia groundwater table. The north-eastern sector of the lower Chaouia constitutes a hydraulic relay of the quaternary fold underground water tables of the plain, which circulate in the Cenomanian formations. Generally, the groundwater table circulates from SE to NW, but the highly disturbed piezometer testifies to the heterogeneity of the schist-quartzite aquifer complex (Ruhard 1975). In addition, there is a hydrogeological basin limited to the west by the El Malleh River and to the east by the Nfifikh River; these two rivers prevent any lateral feeding. In the north and in the south, triastic and impermeable deposits form a natural barrier, which limits all marine influences (Ruhard 1975). A hydrogeological and hydrochemical study of groundwater around the controlled landfill of Mohammedia city (Mabrouki et al. 2019) helped determine the groundwater flow direction, which is generally from east to west. It also showed that this water table is fed by the limestone massif, as well as by the direct infiltration of rainwater.

The results of the hydrogeological and hydrochemical study of the underground water table near the controlled landfill of Béni Yakhlaf (Fig. 1) attest that the evolution of the piezometry over time shows seasonal fluctuations compared to precipitation. The relationships between Na^+, Cl^-, SO_4^{2-}, Ca^{2+} and HCO_3^-, chemical elements that characterize the geological formations of the region, show the dominance of carbonate ions (Ca^{2+}, HCO_3^-) over ions (Ca^{2+}, SO_4^{2-}) and salt ions (Na^+, Cl^-). The geological context and the spatial distribution of the chemical elements indicate that the chemical composition of the groundwater of the aquifer is strongly influenced by the dissolution of carbonates, the Triassic clay and evaporite formations, and by the hydrological parameters of the region, namely the flow direction and residence time in the plateau aquifer (Mabrouki et al. 2019).

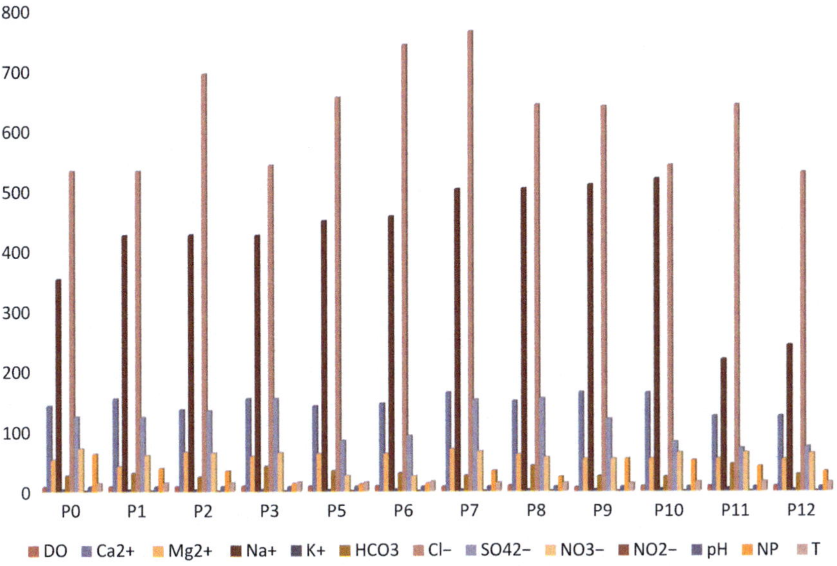

Fig. 1 Results of the physio-chemical parameters and piezometric levels of the wells studied. *Source* Mabrouki et al. (2019)

2.3 The Importance of Water Resources from a Quantitative Point of View

Morocco has succeeded in ensuring the quasi-generalization of access to drinking water (100% in urban areas and 94% in rural areas). It has successfully irrigating an area of 1.5 million hectares. At the same time, it can meet the current and increased industrial water demand (CESE 2014). The drinking water supply to the population of the Prefecture of Mohammedia is ensured through a direct tap on the pipe from the supply of the reservoir of the Sidi Mohamed Ben Abdellah dam and the Oued Bouregreg and from the pipe Fouarat-South (Lydec 2018): the branch line from the Sidi Mohamed Ben Abdellah dam adduction serves a load breeze, from which a DN 800 pipe serves the reservoirs of Mohammedia city; and another DN 500 tapping is done from the Fouarat-Sud water supply.

The transit capacity of these two connections is 1500 L/s. The water retained in the dams passes through treatment plants in order to become drinkable. This process involves several steps which are carried out by the National Office of Electricity and Water (ONEE), considered as the official drinking water producer (screening, settling, filtration, disinfection). Drinking water is then transported and stored in Lydec reservoirs, then checked again, before being distributed to households (Lydec 2018).

The useful agricultural area at the territorial level of the Prefecture of Mohammedia is 15,950 ha distributed as follows: 2550 ha as irrigated area; and 13,400 ha as *bour* area (rainfed crops). The number of farms in the Prefecture is around 3474. The Prefecture has a large land area of a likely medium quality, if properly valued, to provide the local population with an important source of income. The development of modern urban agriculture will also strengthen this income. The construction of the Hassar river dam has been launched in 2014 with the main goal to protect Mohammedia city from flooding risk in addition to the development and evolution of vegetable cultivation of the region by the irrigation of an area of 1800 km^2 (Rihane 2017). The majority of farmers in the Prefecture of Mohammedia use sprinkling water from the groundwater table and from the Maleh River as the most common irrigation technique. The use of motor-pumps is widespread in the areas, where the practice of vegetable cultivation is intense. It should also be noted that 97% of the agricultural area is devoted to *Bour* crops (NOVEC 2014).

3 Elements Impacting Water Quality in the Prefecture Mohammedia

The problem of water resources in the Prefecture of Mohammedia is not limited to the quantitative aspect, but also to the qualitative one. Indeed, the industrial and agricultural development and urbanization negatively affect the water quality in the Prefecture, therefore make it unsuitable for use (Serghini et al. 2013).

3.1 Domestic, Commercial, and Industrial Wastes

The development of human and industrial activities contributes to the increase in the production of wastes, which have harmful environmental and health impacts. Figure 2 presents the wastewater and sanitation network in the Prefecture of Mohammedia.

The Mohammedia wilderness dump, operated from 1987 to 2012 and disinfected in 2012, is located 5 km south of the city just after the bridge spanning the El Maleh River. It is an old limestone quarry of 6 ha whose soil is characterized by schists, representing cracks (Souabi et al. 2011). The landfill received all the waste from Mohammedia city and the communes of Ain Harrouda and Chellalate. The fermentation of this landfill and the contact with rainwater generated a large amount of leachate with a high flow rate, resulting in a lake about 30 m in diameter (Souabi et al. 2010). These leachates flowed towards the El Maleh River located 30 m from the landfill (Fig. 2) or infiltrate the water table, the piezometric level of which is a few meters deep (Idlahcen et al. 2014). This uncontrolled landfill, even if disinfected, constitutes a real and permanent threat to the environment.

The disinfected landfill in Mohammedia, also received industrial, medical, and slaughterhouse wastes, which increased the risk of pollution and posed environmental and health problems. Even after the closure of the landfill and site rehabilitation by residential areas, the leachate still flows and escapes to the surface and reaches the watercourse of El Maleh River, as was observed during the field survey carried out in February 9, 2020. The fermentation of this waste still generates a large quantity of blackish leachate which may contain undesirable organic and mineral elements. A part of this leachate seeps into the soil and pollutes the groundwater. Another part reaches the surface water of El Maleh River, which is located near the landfill. These leachates are too loaded with organic and mineral matter having a considerable impact on the receiving environment (Idlahcen et al. 2014).

Following the closure of the old wild dump of Mesbahiat, which was the source of several environmental problems, the controlled interprovincial landfill of Mohammedia-Benslimane was inaugurated in February 27, 2012. It is located about 8 km east of the Commune of Beni Ykhlef, at the Provincial Road RP 3313, 17 km east of the center of Mohammedia and 24 km southwest of the center of Benslimane. It is located precisely in Douar Beni M'ghit Chaâba El Hamra (Godfred 2017). Moreover, the landfill is located on the edge of the left bank of 'Chaâba El Hamra', a left bank tributary of the Nefifikh river, about 270 m to the South-East of the security perimeter of 'daya Al Hila' and 830 m north-east of the security perimeter of 'daya Halloufa'. The closest stream to the landfill site is the Nefifikh River. On an inter-annual scale, the regime is irregular. It is characterized by years with high flows and years with low flows (Godfred 2017).

This landfill—which occupies an area of 47 ha for an estimated capacity of 3.5 million tonnes over 20 years—receives only household and similar wastes from the prefecture of Mohammedia and the province of Benslimane, in addition to ordinary industrial waste from local companies. It is rather a technical landfill center built with the objective to avoid potential contamination of the groundwater table

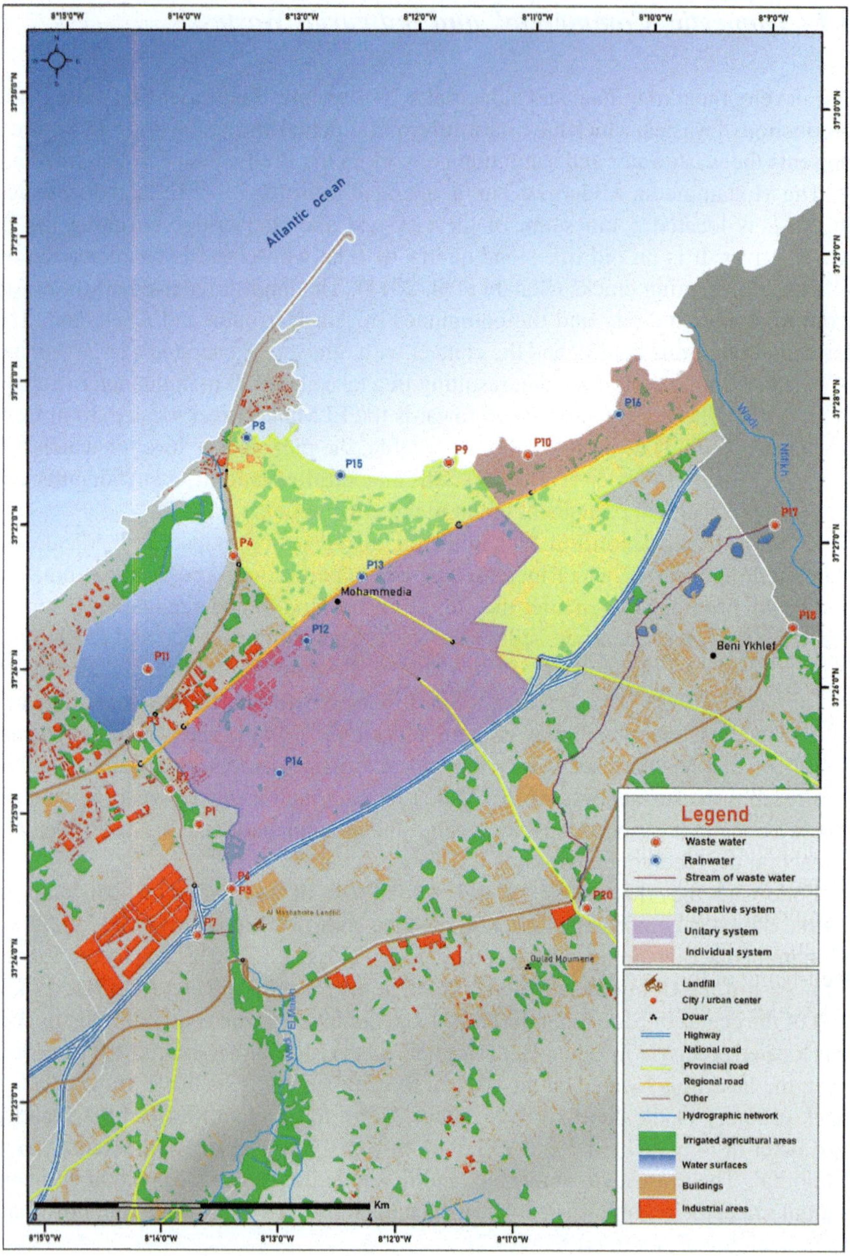

Fig. 2 Wastewater and sanitation system in the Prefecture of Mohammedia. *Source* Developed by the authors

by the leachate, which is collected and drained to a treatment station in the center (Merbouh et al. 2020).

3.2 *Discharge of Untreated Wastewater*

The population growth, urban extension, intensive agriculture, and industrial development in the study area imply excessive water consumption, which in turn increases the quantity of wastewater produced and their negative environmental impacts, especially in case this wastewater is discharged without prior treatment (Kanbouchi et al. 2014). In addition, this constitutes a key factor of pollution of the existing water sources at the level of the Prefecture of Mohammedia, mainly El Maleh and Nfifikh rivers. In this perspective, a field survey carried out during February 27 and March 03, 2020 in the northern part of the Prefecture, made it possible to locate and describe the wastewater tributaries, which constitute an environmental risk for groundwater resources and for surface water in El Maleh and Nfifikh rivers, as well as in wetlands (Fig. 2). The collected information on the discharge of wastewater into nature, either in the rivers or in the coastal area, show that there are several black spots at the level of the study area of different order of importance. This survey made it possible to update the mapping of the black spots which correspond to wastewater discharges, a priori without treatment, in the northern part of the Prefecture of Mohammedia, which is characterized by strong industrial activity and a very intense urban extension compared to the southern part.

Near the disinfected landfill of Mesbahiate (old landfill), the first point (P5 and P6: X314029,93; Y 343,286,43) is located, in which two conduits come from the Andallous residential area and discharge into El Maleh River without any treatment (Fig. 2). The wastewater evacuated by these two conduits is characterized by a significant pollutant load, at the extent to which the color of the water has changed to 'grayish water' and a nauseating odor is released affecting through the winds the well-being of the neighboring districts, such as Hassania and Essaâda.

At the level of Nfifikh River, there are two conduits discharging wastewater directly without any treatment. The first one is that of Oued Ain Tekki (P20), which is characterized by a very high flow rate, due to the number of households which are connected directly upstream. The second refers to the new allotment of Beni Yakhlef; a conduit which is characterized by a low flow. This value can be explained by the low number of households connected to this conduit. The water discharged at point P18 corresponds to the wastewater from the eastern part of the commune of Beni Yakhlaf, showing a light color with the lack of foul odors.

3.3 Pollution as a Result of Industrial Effluents

According to the field survey of March 17, 2020 (Fig. 3), several wastewater conduits originating from the ZIDE-Mohammedia industrial zone (this is the Bled Solb Logistics Industrial Park with an area of 180 ha), discharge their pollutant loads, a priori without any treatment, into the El Maleh River. The waters of the three conduits observed at the level of point P7 (X 314,019,43; Y342008,16) are characterized by a bad odor and a dark color with a significant flow due to the large number of industrial units that are connected to these three conduits and cross a long distance in the middle of the forest to finally reach El Maleh River, a priori without any prior treatment. At this point, goats, sheep, and cows were observed drinking from these waters. Three other points were identified as well (P1, P2, and P3); the last one, which originates from the oil zone of La SAMIR, is characterized by very large leaks discharging directly into the El Maleh River. The point P1, located below the railway bridge over the El Maleh River originates from the Petrom oil zone, the waters are characterized by a very dark color and a bad odor which may be due to the organic or inorganic chemical contaminants or to the biological sources and processes (e.g. aquatic microorganisms).

3.4 Inappropriate Use of Fertilizers and Pesticides Polluting Surface and Underground Water

Agriculture is an important activity in the Prefecture of Mohammedia. The useful agricultural area amounts to approximately 17,500 ha, of which 14,750 ha are in *Bour*, and 2750 ha are irrigated (Fig. 4). The rest correspond to forests (630 ha), rangelands and wasteland (10,088 ha). The present study demarcated irrigated areas of the Prefecture of Mohammedia based on the spectral bung of 'Landsat oli-8', by the calculation of two indices NDWI and NDVI. The result of the combination of these two indices made it possible to define the irrigated area with an estimate of 78%.

It can be inferred that irrigated areas are concentrated in the center of the Prefecture of Mohammedia, especially in the communes of Challalat, Béni Yakhlaf, and Sidi Moussa Al Majdoub. The number of farms in the Prefecture stands at around 3500 located in the commune of Echellalate (1150), commune of Sidi Moussa El Majdoub (670), and commune of Béni Yakhlef (450), respectively. The commune of Ain Harrouda has about 330 farms (around 10%). As mentioned earlier, intensive agriculture, as the largest user of water and the main source of pollution of land and water by nitrates, can have negative quantitative and qualitative impacts on water and land resources.

In the Prefecture of Mohammedia, the agricultural activity is characterized by:

- An 'intensive plan' which allows for the development of vegetable cultivation, especially along the El Maleh River and at the perimeters of urban centers. Such

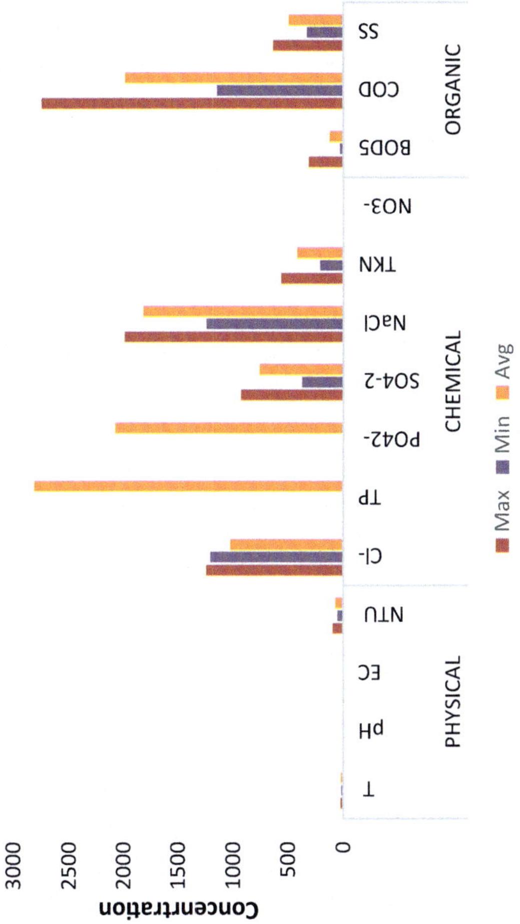

Fig. 3 Water quality of El Maleh and Nfifikh river. *Source* Data from Kanbouchi et al. (2014)

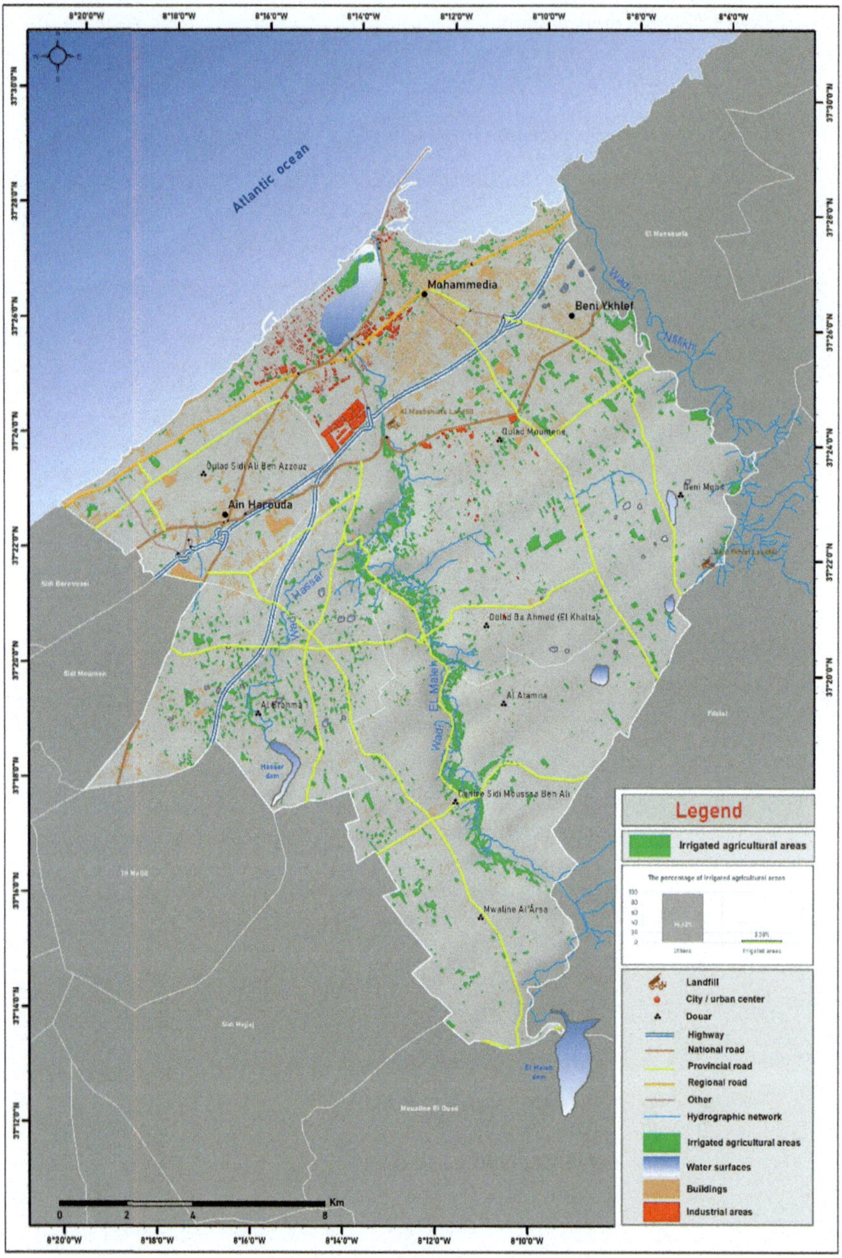

Fig. 4 Agriculture land fed through artificial irrigation, 2019. *Source* Developed by the authors

practices affect the quality of the River's surface water and the groundwater due to the misuse of fertilizers and pesticides. This situation leads as well to a decrease in water availability by reference to the amount of water used for irrigation;

- A 'semi-intensive plan' which corresponds to cereal crops (annual crops or *Bour* crops). These cultures are located outside urban centers. Due to rainfall runoff, pesticides and fertilizers used in these crops find their way into streams or seep into groundwater.

The runoff and/or infiltration of rain or irrigation water drain the soil and transport crop treatment products (fertilizers and pesticides) to rivers and groundwater points (El Assouli et al. 2007). Nitrogenous products (nitrates and nitric) and phosphates cause imbalances in the environments that receive runoff or infiltration water from agriculture. The accumulation of all these elements in the receiving environment causes the degradation of water quality (Laaouan 2018).

4 Water Quality and Pollution Source Apportionment in the Prefecture of Mohammedia

Several studies on the quality of surface and groundwater in the Prefecture of Mohammedia were carried out in order to assess and identify the contaminated areas (Fig. 5). The findings support the contamination and degradation of the quality of surface water by the direct discharge of untreated wastewater into El Malehand Nfifikh rivers and/or the lagoon (CERED 2017). The high concentration of certain chemicals in the wastewater cause leaks in the sanitation canals, which leads to infiltration of this wastewater towards the aquifers and the existing water pockets at the level of the Prefecture. The study by Taleb et al. (2015), dealing with the content of hydrogen sulphide (H2S) in the sewerage network of the lower town of Mohammedia, evaluated the production over time of H_2S in the wastewater at the level of the collection basin for all types of water in Mohammedia City as well as in the wastewater drained by sewerage networks without treatment to the sea. The finding showed that the quality of wastewater varies considerably from a temporal and spatial point of view (Fig. 6). The discharge of wastewater (industrial and domestic) into collectors without prior treatment leads to the contamination of receiving watercourses, and consequently causes significant nuisance for residents, users, and faunistic and floral resources.

Laaouan (2018) reveal through analyzes carried out at various points of Mohammedia City that the water from the studied wells is loaded with organic and mineral matter, mainly resulting from agro-food industrial activities (Fig. 5). The pollution of receiving environments can be revealed by the value of chemical oxygen demand COD and electrical conductivity, which are very high and exceed the World Health Organization (WHO) standards at the level of the studied area (receiving basin for leachate of the disinfected landfill). The study revealed that leachate puddles, from the disinfected dump of Mesbahiat in Mohammedia City which had been closed more than five years ago when this study was carried out, are heavily loaded with metal

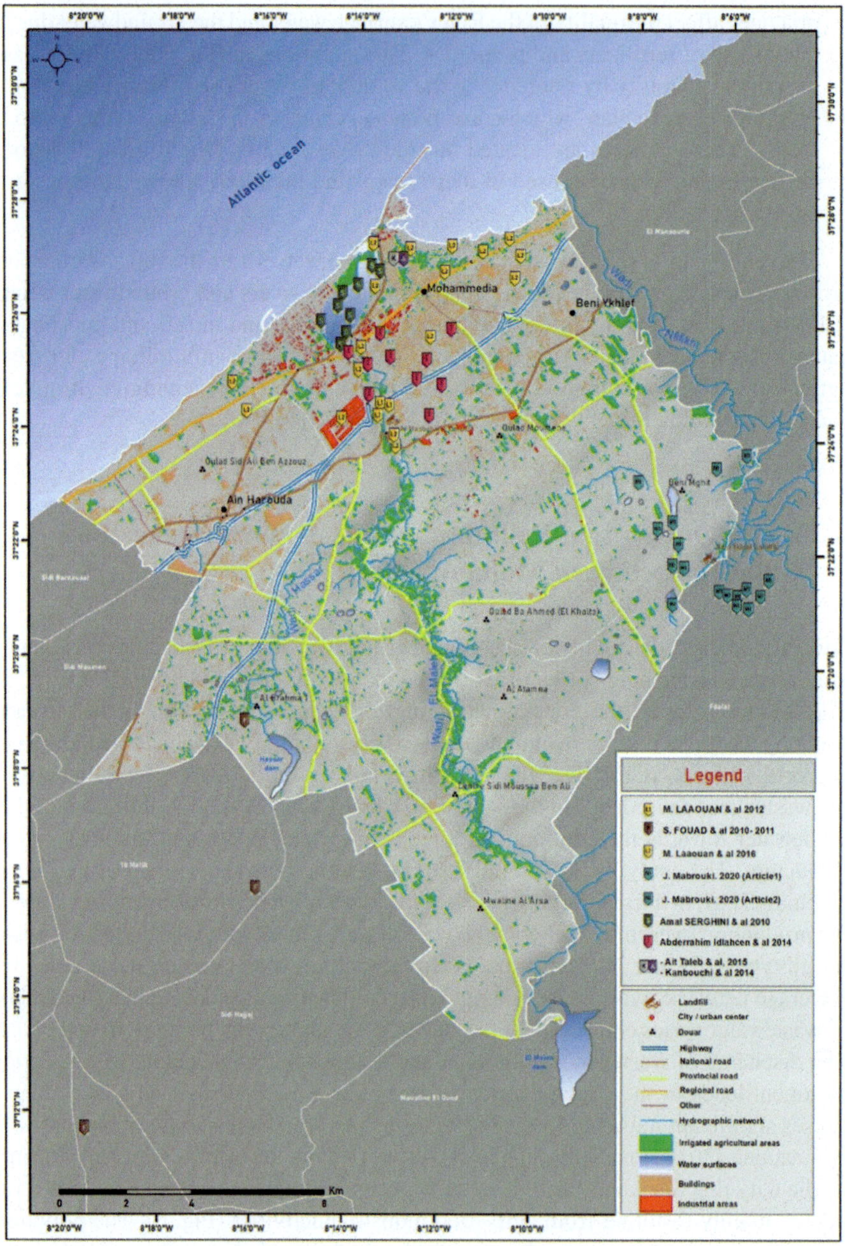

Fig. 5 Samples location considered by scientific articles reviewed. *Source* Developed by the authors

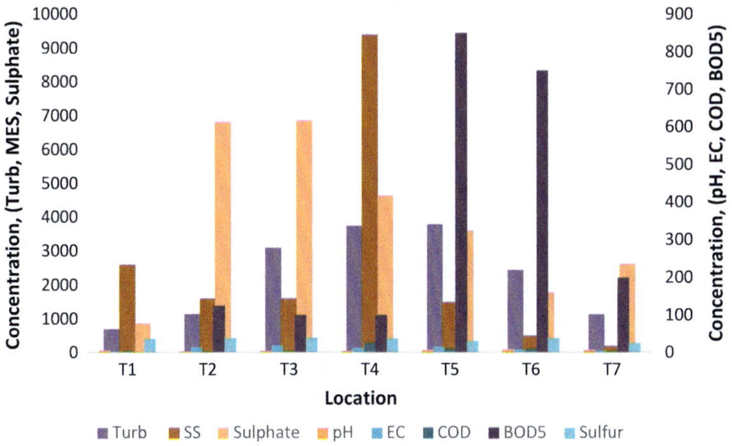

**The figure represents the average of the analysis results of the three instantaneous sampling campaigns (T1, T2, T3, T4, T5, T6 and T7) from the collection basin and during a monitored schedule. This basin collects all effluents released by different stages of the manufacturing process.*

Fig. 6 Characteristics of wastewater in the collection basin for all types of wastewaters in Mohammedia City. *Source* Data from Taleb et al. (2015)

pollutants (Fig. 7). The electrical conductivity up to 80,900 μS/cm reflects the strong mineralization of these releases. The COD is also very high (1200–7700 mgO_2/l) and constitutes a strong non-biodegradable pollution in these discharges. In addition, the concentrations of elements—such as Cd, Cr, Cu, Ni, Pb and Zn, obtained in samples L1 and L2—remain below the general discharge limit values applicable to discharges of wastewater in Morocco, and allow to conclude that the metallic pollution due to these metals and generated by the flow of these leachate decreased after the closure of the uncontrolled landfill in Mohammedia City.

To assess the impact of the Béni Yakhlef inter-municipal landfill, located in the south-east of Mohammedia City, on the quality of groundwater, the company in charge of its management 'Eco Med' carries out regular verifications. If we consider the assessments undertaken in 2019, a sampling was carried out at the landfill site and Nfifikh River in order to control the quality of surface and ground water as well as the quality of the leachate (treated and raw). The sampling points were as follows: (1) Raw leachate basin at the landfill; (2) Leachate basin treated at the landfill; (3) three boreholes close to the landfill (P1, P2, and P3); (4) Upstream and (5) downstream of Nfifikh River (Fig. 8). The physio-chemical parameters (in situ) (T °C, pH, EC, dissolved O_2), as well as other parameters (in the laboratory)—such as BOD5, COD, MES, Pt, PO4, NTK, Cl^-, NO_3^-, NH_4^+, CF, SO_4^{2-} and MO—are presented in Fig. 9.

Figure 9 shows that the quality parameters, measured in-situ, for the raw leachate have very high values, particularly for the electrical conductivity which reaches more than 57,300 μS/cm, and that the dissolved oxygen has zero values (0.8 mg/L). The parameters of the raw leachate show very high concentrations which exceed

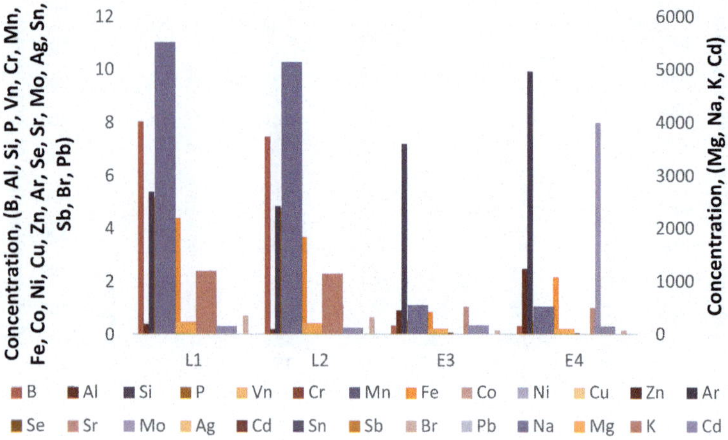

Fig. 7 **Metal** concentration in leachate puddles from the disinfected dump of Mesbahiat in Mohammedia City. *Source* Data from Laaouan (2018)

the general limit values, with a basic pH exceeding 7.5 and a BOD5/COD ratio of around 0.3 as general discharge limit values applicable to discharges of wastewater in Morocco (1.5–2).

The quality parameters, measured in-situ by ABHbc (Fig. 9), in particular for the raw and treated leachate, show values that do not comply with the standards; thus we note very high values, particularly for the electrical conductivity which reaches more than 43,000 μS/cm, and very low values for dissolved oxygen as example, which has zero values. The parameters measured in the laboratory for the raw leachate show very high concentrations, which exceed the general WHO limit values, while the concentrations of the treated leachate are generally low. Groundwater analyzes reveal variable concentrations depending on the parameter; thus we note that the nitrates reach 54 mg/L, which is a value higher than the WHO and the Moroccan Ministry of Energy, Mines, and the Environment (MEME) standards (50 mg/L) (Fig. 10).

As per Figs. 9 and 10, the impact of the Ben Yakhlef controlled landfill on the quality of groundwater resources is characterized by a high content of nitrates (NO_3^-) and chlorine (CL^-), which decreases the water quality and poses a potential danger in case of its use. These findings confirm the prevalence of a health hazard for consumers using such a water for food and sanitary purposes.

The work of Merbouh et al. (2020) determines the quality of groundwater in the vicinity of the controlled landfill installed in the region of Mohammedia-Benslimane since 2012. The measurements had been undertaken in seven water wells intended for irrigation and/or for consumption (Fig. 10). The results (Fig. 10) show that the chloride contents are very high in all these wells (158.2–845.6 mg/L) and greatly exceed the irrigation standards (105 mg/L). The COD varies between 34.6 and 67.2 mg O_2/L in five water wells; these values are high compared to the admissible value of drinking water (25 mg O_2/L). The concentrations recorded for Cu, Fe, Mn, Zn,

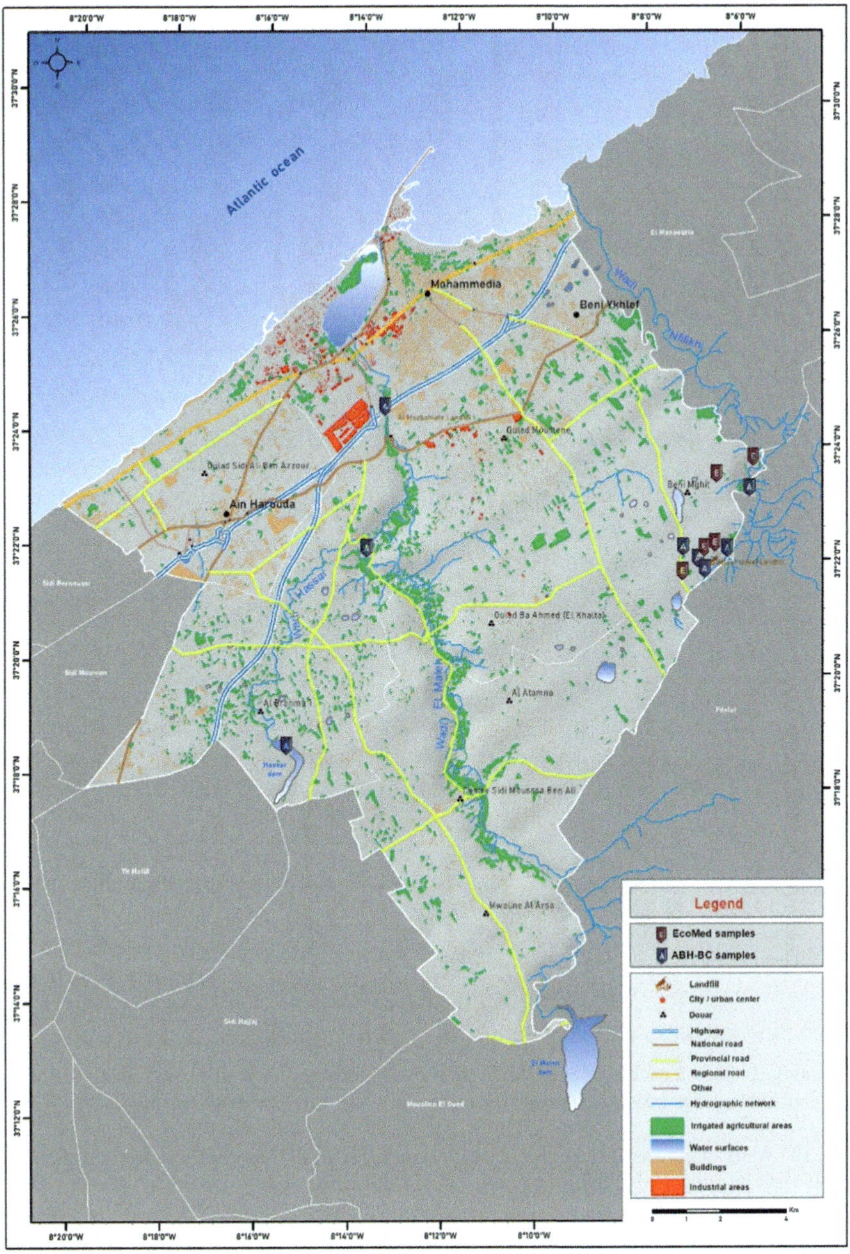

Fig. 8 Location of the sampling carried out by ECOMED and the Agency of Hydrolic Basin (ABH). *Source* Developed by the authors

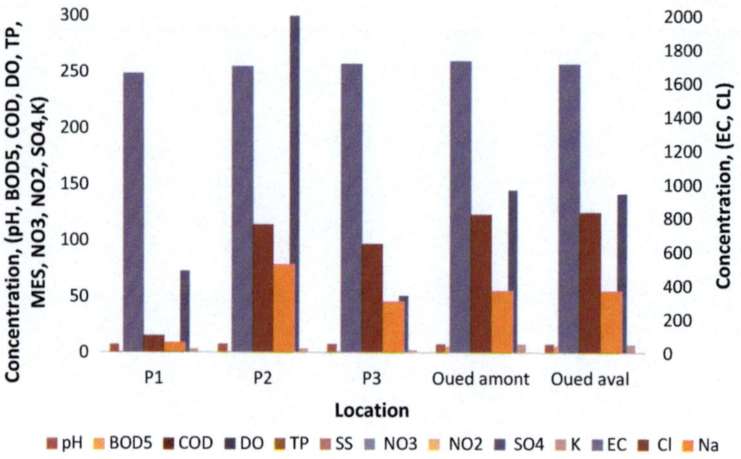

Fig. 9 The results of the analyzes carried out by Eco-Med (2019). *Source* Data from Ecomed (2019)

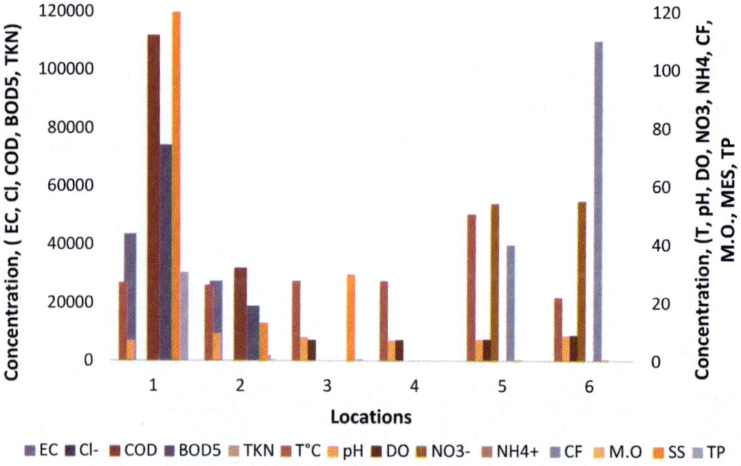

Location 1= Raw Leachate Beni yakhlef landfill, 2 = treated Leachate Beni Yakhlef landfill, 3 = Nfifikh river point of wastewater discharge, 6 = Nfifikh river downstream point)

Fig. 10 Water quality assessment for leachate originating from Beni Yakhlef landfill and landfill wastewater meeting the Nfifikh River *Source* Data from ABHbc

and Ni correspond well to irrigation and drinking water standards. However, the Pb contents vary between 0.07 and 0.14 mg/L in water wells 4 and 3, which are located downstream of the landfill; such values exceed those of the drinkability standard (0.05 mg/L).

are generally associated with socioeconomic and health disturbances. The identified sources of water pollution were the landfills and the discharge of untreated wastewater into the above-mentioned rivers.

The bacteriological contamination of water resources indicates an impact on human health. However, the water-borne disease cases, reported for the study area, validate such an impact. This not only calls for a better management of wastewater network and landfill leachate, but also necessitates immediate response mechanisms in order to improve the quality of existing water resources, therefore contributing to the improvement of public health.

The scope of the study was greatly restricted primarily due to non-uniform parameters being employed for water quality analysis. Hence, this study recommends a quality analysis of all water resources in the Prefecture of Mohammedia based on similar parameters. This will ensure true presentation of water quality, thus helping policy and decision makers and governing agencies in adopting and implementing future local planning and development frameworks, with the ultimate outcome to ensure a sustainable management of water resources capable of meeting current and future demands.

Acknowledgement The study was funded under the project scheme 'Ibn Khaldoun Research Support Program in the Area of Human and Social Sciences', Moroccan Center for Scientific and Technical Research (CNRST), Morocco, for the Project entitled "Health Security in Casablanca", Number: IK/2018/23.

References

Abouali N, Boumeaza T, Zahouily M, Yazidi Alaoui O (2019) Three-dimensional simulation of a centennial flood of Mohammedia City in the River of El Maleh—Morocco 2508, vol 4, pp 305–315

ABHbc (2018) Agence du Bassin Hydraulique du Bouregreg et de la Chaouia, Etude d'évaluation des eaux de surface et eaux souterraines de la zone d'action de l'agence du bassin hydraulique du Bouregreg et de la Chaouia, service de qualité des eaux, mission 1, p 43.

CERED (2017) Projections de la population et des menages 2014–2050. Centre d'études et de recherches démographiques (CERED), p 23

CESE (2014) La gouvernance par la gestion intégrée des ressources en eau au Maroc, Conseil Économique, Social & Environnemental

Dakki M, Menioui M, Zouhair A (2015) Stratégie nationale des zones humides 2015–2024 Stratégie—Plan d'action. Deutsche Gesellschaft für Internationale Zusammenarbeit (GIZ), 54. www.giz.de/marokko

Ecomed (2019) "GROUPE ECOMED." 2019. http://www.ecomed.ma/

El Assouli J, Kholtei S, El Amrani-Paaza N, Hilali A (2007) Impact des activités anthropiques sur la qualité des eaux souterraines de la communauté mzamza (Chaouia, Maroc). Revue Des Sciences de l'eau 20(3):309–321. https://doi.org/10.7202/016505ar

El Ouedghiri K, El Oualti A, El Ouchy M, Zerrouq F, Ouazzani Chahdi F, El Ouali Lalami A (2014) Risques sanitaires liés aux composés chimiques contenus dans l'eau de boisson dans la ville de Fès : cas des ions nitrates et nitrites. J Mater Environ Sci 5:2284–2292

Fehdi C, Abderrahmane B, Abdelkader R, Elias S (2010) Caractérisation hydrogéochimique des eaux souterraines du complexe aquifère Morsott-Laouinet (Région Nord de Tébessa, Sud-Est Algérien). Afrique Science Revue Internationale Des Sciences Et Technologie 5(2):217–231. https://doi.org/10.4314/afsci.v5i2.61732

Fouad S, Hajjami K, Cohen N, Chlaida M (2014) Qualitéphysico- chimique et contamination métallique des eaux de l'oued Hassar: Impacts Des eaux usées de la localité de Mediouna (Périurbain de Casablanca, Maroc). Afrique Sci 10(1):91–102. https://www.researchgate.net/publication/260020856

Godfred A (2017) Traitement de lixiviat par voie biologique. Mémoire de Stage de Fin d'études Traitement. Faculté Des Sciences Tétouan 70

Hébert S, Légaré S (2000) Suivi de la qualité des rivières et petits cours d'eau, Québec, Direction du suivi de l'état de l'environnement, ministère de l'Environnement, envirodoq no ENV-2001-0141, rapport n° QE-123, 24 p. et 3 annexes. http://www.environnement.gouv.qc.ca/eau/eco_aqua/rivieres/GuidecorrDernier.pdf

Idlahcen A, Souabi S, Taleb A, Zahidi K, Bouezmarni M (2014) Evaluation of pollution generated by landfill leachate public of the city of Mohammedia and its impact on the groundwater quality. Sci Study Res Chem Chem Eng Biotechnol Food Ind 15(1):35–50

Kanbouchi I, Souabi S, Chtaini A, Aboulhassan M (2014) Évaluation de la pollution des eaux usées mixtes collectées par le réseau d'assainissement de la ville de Mohammedia. Les technologies de laboratoire 8(N°34):162–170

Laaouan M, Aboulhassan MA, Bengamra S, Taleb AS, Sand Tahiri M (2016) Etude comparative de la contamination des eaux souterraines des villes de Mohammedia, Temara et Dar Bouazza par les Nitrates (Meseta Marocaine). J Mater Environ Sci 7(4):1298–1309

Laaouan M (2018) Étude de la pollution générée par la ville de Mohammedia et évaluation du coût de dégradation de l'environnement" thèse de doctorat laboratoire Traitement des Eaux/ Sciences et Technologie de l'Environnement Faculté des Sciences et Techniques Mohammedia. Université Hassan II de Casablanca. Maroc p 197

Lucas F, Servais P, Rocher V, Masnada S, Therial C, Lesage L, Mouchel JM (2012) Variabilité de la qualité microbiologique des eaux usées brutes dans une grande agglomération. LIVRE BLANC Urbanisme et Propreté TSM Numéro 4-2012-107e Année, 44–55.

Lydec Lyonnaise des Eaux de Casablanca (2018) Analyses de la qualité de l'eau. https://client.lydec.ma/site/bulletin-de-l-eau

Mabrouki J, El Yadini A, Bencheikh I, Azoulay K, Moufti, El Hajjaji S (2019) Advanced intelligent systems applied to environment towards SDI services for geological maps data, vol 3. Springer International Publishing, pp 23–33. https://doi.org/10.1007/978-3-030-11881-5

Mazouni FZ, Berri H, Maaroufi A (2018) Bulletin d'épidémiologie et de santé publique, programme national de prévention et de contrôle du diabète. Ministère De La Santé 50

Merbouh Ch, Belhsaien K, Zouahri A, Iounes N (2020) Évaluation de la qualité physico-chimique des eauxsouterraines au voisinage de la décharge contrôlée Mohammedia-Benslimane: (Étude Préliminaire). Eur Sci J ESJ 16(6). https://doi.org/10.19044/esj.2020.v16n6p455

Minister de santé (2019) Bulletin d'Epidémiologie et de Santé Publique, Volume 56, N°78.

Mondet B, Seyler T, Salem G, Gonzalez J-P (2016) l'étude des risques sanitaires liés à l'eau dans l'environnement urbain: L'exemple de la ville de Chennai, Inde Du Sud. Sociétés Environnements Santé 239–255. https://doi.org/10.4000/books.irdeditions.3616

Monographie de Mohammedia (2015) Monograpgie de la vile de Mohammedia, Faculté des Lettres et des Sciences Humaines de Mohammedia, Université Hassan II de Casablanca, en partenariat avec la Commune de Mohammedia. Dépôt légal : 2017M03963 ISBN : 978-9954-99-673-7. Editeur: Commune de Mohammedia, p 83.

NOVEC (2014) Etude d'impact sur l'environnement naturel, humain et socio-économique, Projet d'Aménagement de la ville nouvelle de Zenata, Missions 1: Description et analyse de l'état de référence, Livret des agrandissements cartographiques, Juin 2010, pp 1–36

Rihane A (2017) Importance ornithologique du barrage d'oued Hassar (région de Mohammedia). Meeting International: La valorisation des ressources en eau dans les zones arides et semi arides

26–28 Mai 2014, Faculté Des Sciences et Technique Béni Mallal-Maroc.https://www.researchgate.net/publication/319304928%0AIMPORTANCE
Ruhard J-P (1975) 2.12. Chaouia Et Plaine De Berrechid, Royaume du Maroc Ministère du Commerce, de l'industrie, des mines et de la marine marchande direction des mines, de la géologie et de l'énergie. Tome 2:185–238
Serghini A, Elabidi A, Elblidi S, Fekhaoui M (2013) Typologie Spatio-Temporelle et Fonctionnement Hydrologique Du Complexe de Zones Humides de Mohammedia (Maroc). Bulletin de l'Institut Scientifique, Rabat, Section Sciences de La Vie, 2013, N° 35, 61–65, 61–65.
Souabi S, Touzar K, Chtioui H, Khalil F, Digua K, Tahiri M (2010) Problématiques du chrome et du plomb dans les lixiviats des décharges publiques des villes de Mohameddia et de Fès. Déchets Sci Tech 58:37–44. https://doi.org/10.4267/dechets-sciences-techniques.3022
Souabi S, Touzare K, Digua K, Chtioui H, Khalil F, Tahiri M (2011) Triage et valorisation des déchets solides à la décharge publique de la ville de Mohammedia. Les technologies de laboratoire 6, N°25 :121–130
Taleb A, Kanbouchi I, Souabi S, Chtaini A (2015) Etude de la problématique de la présence de l'H_2S dans le réseau d'assainissement de la ville basse de la ville de Mohammedia. Environ Sci 6(8):2137–2147. https://www.jmaterenvironsci.com/Document/vol6/vol6_N8/260-JMES-1761-2015-Taleb-P.pdf
Tarhda Z (2018) Bulletin d'épidémiologie et de santé publique, Programme national de prévention et de contrôle du diabète. Royaume du Maroc; Ministère de la santé; N°75. ISSN 08518238 56.
Wafaa Gaour, Nade Maxime (2006) la gestion de la croissance informelle des villes du Sud: l'exemple de la ville de Mohammedia au maroc. Master I « Génie Urbain », Module Pays En Développement Responsables. l'université de Versailles Saint- Quentin-en-Yvelines. pp 1–41.
World Bank Group (2017) Le Coût de la dégradation de l'environnement au Maroc. Environ Nat Resour Glob Pract Discuss 105633

Dr. Rachida El Morabet El Morabet holds a Ph.D. in Biology from the Mohammed V University, Rabat, Morocco (2004), an Inter-University diploma of Biotechnology, Paris, France (2001) and a University Pedagogy Certificate (Teaching Practices, learning theories, models assessment practices knowledge; educational resources and distance learning) (2014). She is Associate Professor with the Department of Geography at FLSH-Mohammedia, Hassan II University of Casablanca, Morocco. In charge of the "health security in Casablanca" project, within the framework of the Ibnou Khaldoun Program. She is a member of: the Laboratory: Dynamics of Space and the Society. Research Team: The Dynamics of Natural Environments and their Impacts on Society and Territories; and the Center of Research for Environment, Human Security and Governance (CERES). Her research interests include Environmental issues in the era of global change, Socio-ecological system (vulnerability and adaptation), and Dynamics of natural systems. Rachida El Morabet received two Gold Medals with honors of International Exhibition of Invention and Innovation, Casablanca, Morocco.

Larbi Barhazi is a Ph.D. student in the LADES laboratory, FLSH-M, Hassan II University of Morocco. The subject of the thesis is: "The impact of water quality in the socioecological system and human health".

Soufiane Bouhafa is a Ph.D. student in the LADES laboratory, FLSH-M, Hassan II University of Morocco. The subject of the thesis is: "The impact of air quality in the socioecological system and human health".

Mohamed Behnassi has been nominated in July 2022 as a Senior Environmentalist Expert by the Economic Social and Environmental Council (ESEC), Rabat, Morocco. Prior to this, he served as a Full Professor (since 2004), and Head of Public Law in French Department at the College of Law, Economics, and Social Sciences of Agadir, Ibn Zohr University. He holds a Ph.D. in International Environmental Law and Governance (2003), a M.Sc. in Political Sciences (1997), and a B.A. of Administration (1995) from Hassan II University of Casablanca. He obtained a Diploma in International Environmental Law and Diplomacy from the University of Eastern Finland and UNEP, 2015. He is also an Alumnus of the International Visitors Leadership Program of the Department of State, United States of America. Dr. Behnassi is currently the Founding Director of the Center for Environment, Human Security and Governance (CERES). From 2015 to 2018, he was the Director of the Research Laboratory for Territorial Governance, Human Security and Sustainability (LAGOS). Recently, he was appointed as Expert Evaluator for the National Center for Scientific and Technical Research (CNRST/Morocco) and selected twice (2019–2024) as an Assessment Scoping Expert and a Review Editor by the Intergovernmental Science-Policy Platform on Biodiversity and Ecosystem Services (IPBES) and Member of the Mediterranean Experts on Climate and Environmental Change (MEDECC). He was also selected by The Intergovernmental Panel on Climate Change (IPCC) as Expert Reviewer of the 1st Order Draft of the Synthesis Report (SYR) of the IPCC VI Assessment Report (AR6). Accordingly, he was among the Lead Authors of the 1st Assessment Report (MAR1): Climate and Environmental Change in the Mediterranean Basin—Current Situation and Risks for the Future (MEDECC, 2021). He is a Senior Researcher in the areas of international law and politics of environment and human security where he published considerable number of scientific papers and book chapters in addition to 20 books, including recent ones on: Food Security and Climate-Climate Smart Food Systems—Building Resilience for the Global South (Springer, 2022); The Climate-Conflict-Migration Nexus from a Human Security Perspective (Springer, 2022); and Social-Ecological Systems in the Era of Risks and Insecurity - Pathways to Viability and Resilience (Springer, 2021). Dr. Behnassi serves as a reviewer for many global publishers (such as Routledge and Springer) and scientific journals with high impact factor. He has organized many international conferences covering the above research areas, managed many research and expertise projects, and is regularly requested to provide scientific expertise nationally and internationally. Other professional activities include social compliance auditing and consultancy by monitoring human rights at work and the sustainability of the global supply chain.

Roohul Abad Khan is currently working as Lecturer in Department of Civil Engineering, King Khalid University, Saudi Arabia. Prior to working in King Khalid University, he has also worked as Head of Department for Civil Engineering at Integral University in the Department of Civil Engineering (Diploma courses). Currently, he is a working member in the Committee of Research, Committee of Quality and Accreditation and Committee for Lab equipments and safety. He has a rich experience of 9 years in academic studies. Prior to working as an academician, he has serves as a Site Engineer with S.Q. Builders and Developers Pvt. Ltd. for 6 years and has been involved in completion of four multi-storeyed residential building from start to finish. His research areas are primarily focused on Water and wastewater quality and treatment, Sustainable energy, Green building materials, Air Quality monitoring, Pollution and health risk assessments. This wide range of research activities are primarily attributed to his Diploma in Architecture, Bachelor degree in Civil Engineering, and Masters in Environmental Engineering.

Chapter 17
Basic Planning Principles of Roof Precipitation Harvesting Systems

Hasan Er and Yasemin Kuslu

Abstract Water has been one of the main factors determining the fate of civilizations for centuries. Today and in the future, water scarcity is one of the biggest problems for mankind. Precipitation water harvesting can be defined as the collection and accumulation of rainwater and runoff water, the supply of water required for plant and animal production, and the supply of water required for domestic consumption. The fact that the water obtained is free of charge, protecting natural water resources can be easily integrated into the existing water network system. In addition, low operating and maintenance costs make precipitation harvesting extremely attractive for the management of water, which is a scarce resource. With the collection of precipitation water, purposes such as preventing soil erosion and floods, providing quality irrigation water, feeding groundwater, and saving network water can be achieved. Precipitation water harvesting is a water supply method that has provided drinking water to many historical cities since ancient times. Archaeological findings show that rainwater harvesting dates back to 6000 BC. Precipitation water harvesting has been practiced since humans began to live and grow crops in arid areas. While in many arid countries a large part of rainwater is lost through evaporation or turning into wastewater, worldwide awareness and importance of rainwater harvesting are increasing. This chapter describes the elements of roof precipitation water collection systems and includes the basic principles that should be considered when planning.

Keywords Climate change · Water management · Precipitation harvesting · Roof system · Planning principles

H. Er
Department of Biosystem Engineering, University of Bingöl, Bingöl, Turkey
e-mail: hasaner@bingol.edu.tr

Y. Kuslu (✉)
Department of Agricultural Structures and Irrigation, University of Atatürk, Erzurum, Turkey
e-mail: ykuslu@atauni.edu.tr

M. Behnassi et al. (eds.), *The Water, Climate, and Food Nexus*,
https://doi.org/10.1007/978-3-031-50962-9_17

1 Introduction

Water has been one of the main factors determining the fate of civilizations for centuries. If the average amount of usable water per capita in a country is less than 1000 m^3 per year, it is water-poor, if 1000–2000 m^3 is water-stressed, if 2000–8000 m^3 is water-sufficient, and if it is more than 8000 m^3, it is water-rich. Moreover, even if a country is water-rich, it is an environmental responsibility to use water resources effectively.

Although water is in a continuous cycle, it is consumed before completing its cycle due to reasons such as population growth, environmental pollution, cost, unconscious water consumption, and change in climatic conditions (Kızıloğlu et al. 2007; Hynes et al. 2020; Oskay et al. 2023). Today and in the future, water scarcity is one of the biggest problems for mankind (Oskay et al. 2022). It has become increasingly difficult for countries to find new water resources. To overcome the problems, alternative water sources have been sought and it has been seen that the precipitation water harvesting method is very important for freshwater resources (Örs et al. 2011; Haq 2017; Adler et al. 2011, 2013).

2 Harvesting Precipitation Water

Precipitation water is the main source from which other water sources are fed, and they are also used directly in the provision of drinking and utility water under some conditions. Precipitation water is collected on roofs or specially prepared surfaces and collected in tanks built on the ground, underground, or above the ground with gutters and vertical pipelines. What is meant by the collection of precipitation waters is generally rainwater. However, according to climatic conditions, the water obtained from snow and hail can also be stored and used when needed (Gwenzi and Nyamadzawo 2014; Gwenzi et al. 2015; Abdulla and Al-Shareef 2009). The characteristics of the harvested precipitation water can be listed as follows:

- Harvested precipitation is free of charge. It helps protect natural water resources.
- Precipitation collection technology is simple and easy to install; it can be simply integrated into the existing water supply system.
- The operating and maintenance costs of such systems are low.
- It is possible to collect fresh water from precipitation harvesting both in the open fields and in urban areas.
- Precipitation water can be a constant source of water for both rural and urban residents.
- Collected precipitation water can be used for agricultural purposes. This is important to get rid of the drought trend.
- Soil erosion and floods can be controlled as the surface flow is prevented depending on the way the precipitation waters are collected.

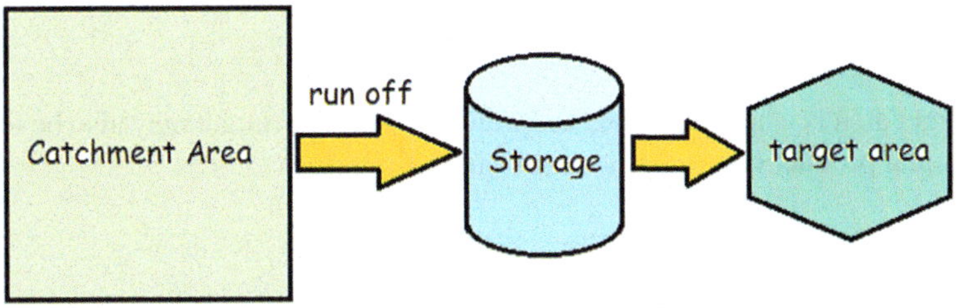

Fig. 1 Main components of precipitation water harvesting. *Source* Created by the authors

The water obtained from the precipitation can be used for irrigation of agricultural areas, the use of cooling and process water in the industry, and the use of drinking water after disinfection at different levels, in toilets, and areas such as laundry. As precipitation waters pass through the atmosphere, they are polluted with pollutants suspended in the air. If precipitation water is planned to be collected in the tanks, it should be taken into account that it will carry the pollutants and other particulate materials on the collected surfaces to the storage tanks. A treatment suitable for the intended use is necessary for sustainable water management. The main components of precipitation water harvesting are shown in Fig. 1.

2.1 *Water Collection (Basin) Area*

As a result of the topographic structure, it is the whole of the areas where the precipitation waters are stored, the waters that pass to the surface flow are collected (water collection area) and are within the impact area of these events. The catchment area can be as small as a few square meters or as large as a few square kilometers. The flow coefficient of the water to be collected varies according to the physical characteristics of the basin area.

2.2 *Water Storage*

A storage unit is where runoff water is collected and held until it is used. Storing precipitation is very important so that it can be used when needed. The most common forms of storage for precipitation storage are aboveground or underground storage. Smaller storage tanks are required in regions where precipitation distribution is relatively homogeneous throughout the year, whereas larger storage reservoirs are required in regions where precipitation distribution is not homogeneous throughout the year (Duru et al. 2013).

2.3 Target Area

The target area is where the harvested water is used. This target may also be for crop and animal production, domestic use, or industrial use.

3 Planning of Roof Precipitation Harvesting Systems

Rooftop systems can collect and store precipitation water from the roofs of houses and large buildings, greenhouses, courtyards, and similar impermeable floors such as roads. In this way, it is possible to benefit from most of the precipitation. Although these techniques are mostly used for domestic purposes, they are also suitable for agricultural use (Pelak and Porporato 2016). The water that is not suitable for use as drinking water without treatment, can be used to additional supplement resources for the garden irrigation of the houses. If the stored water is planned to be used for landscape irrigation, only a sediment filter is sufficient. However, if it is to be used as drinking water, additional purification methods are required. The water obtained after the purification process is quite safe and of high quality. Additional purification methods such as micro filtering, ultraviolet sterilization, reverse osmosis, and ozonation are used in the supply of drinking water (Ahmed et al. 2010, 2011; Evans et al. 2006; Gikas and Tsihrintzis 2012; Lim and Jiang 2013).

The components of the rooftop precipitation water collection system can be listed as the roof/catchment area, gutters and downpipes filter unit or leaf screen, storage tank distribution system, and water treatment unit.

The water collection surfaces foreseen for the collection of precipitation water should transmit the water to the tanks due to low evaporation, infiltration, and other losses. This requires that the water collection surface is smooth, its slope is high and its permeability is low (Mendez et al. 2011).

Flow coefficients for precipitation water collected from roofs are given in Table 1. These coefficients vary according to the age of the roofing material. If the material is new, higher values are taken as the flow coefficient, and if it is worn and old, lower values are taken.

To remove the precipitation water from the roof areas, it is necessary to give slopes to the surfaces. The most important way to increase the amount of water to be collected from roof surfaces is to create eaves. Eaves are the areas where rain gutters are placed where precipitation water is discharged, especially on sloped roofs. Leaf catchers and filters can be added to the gutters in case there is too much garbage and leaves on the roofs of low-rise buildings due to the wind (Meera and Ahammed 2006; Taffere et al. 2016).

The biggest investment required for roofed precipitation harvesting systems is a storage tank (Londra et al. 2015; Mitchell 2007). Galvanized steel, concrete, reinforced concrete, and durable wood can be used as warehouse construction material. But the options for plastic tanks are more common, as they are easier and more

Table 1 Flow coefficients by material type

Material	Water absorption rate (%)	Flow coefficient (cr)
Clay-based (tile etc.)	10–15	0.75–90
Cement-based	2–20	0.60–80
Metal	0	0.70–0.90
Bitumen-based	0–20	0.70–0.80
Polymers	0–1	0.8–0.90
Glass	0	0.9
Natural stone	1–3	0.20–0.60
Plant materials	15–20	0.05–0.10

Source Turkish Standards Intuition (2022)

practical to use. Water tanks will have a longer life if they are planned in a way that does not allow light and heat. The entry of sunlight into the tank causes algae to form in the water, causing deterioration of the water quality (Tsihrintzis and Baltas 2014; Haq 2017).

There are different approaches to determining the storage tank volume. These are the daily maximum precipitation approach, dry period water requirement approach, and annual total precipitation approach (Pelak and Porporato 2016; Mitchell 2007). Among these approaches, in the daily maximum precipitation approach, the storage volume is the lowest and it is suitable for regions with more wet days (over 200). In the annual total precipitation approach, the storage volume is the highest, especially in regions where the majority of the annual precipitation falls in a short period. In this approach, almost all of the annual precipitation is planned to be stored.

Approximate daily maximum precipitation approach

In this approach, the highest precipitation amount that can fall in one day is used in the calculation of the storage volume, taking into account the long-term data (Eq. 1).

$$V = AP_{max}c_r \tag{1}$$

In the equality; V = tank volume (L), A = Area (m^2), P_{max} = Daily maximum precipitation height (mm), c_r = flow coefficient.

Dry period water requirement approach

In this approach, the tank volume is equal to the volume of water required during the dry period (Eq. 2).

$$V = t\text{q} \tag{2}$$

In the equality; V = tank volume (L), t = dry season (days), q = average daily water consumption in the dry season (L/day).

Annual total precipitation approach

In this approach, the general purpose is to store all the precipitation falling during the year.

$$V = \sum_{1}^{n} PAc_r \tag{3}$$

In the equality; V = tank volume (L), A = Area (m^2), Pmax = Daily maximum precipitation height (mm), cr = flow coefficient.

A more economical approach to calculating the storage volume is to consider the annual cumulative precipitation and cumulative water consumption. In this calculation method, the entire amount of precipitation falling during the year is taken into account. However, while precipitation is being stored on the one hand, water is consumed on the other hand. The warehouse volume is determined by considering both cases (Fig. 2).

The choice of the tank location is as important as the selection of its volume. If there are legal regulations for landscaping in the location of the storage, such conditions should be taken into account primarily. In addition, the most suitable place should be determined by giving importance to the characteristics of the buildings where precipitation water is collected, the wishes of the users, aesthetic properties, and technical features (Tsihrintzis and Baltas 2014; Haq 2017; Pelak and Porporato 2016; Mitchell 2007). Each type of placement has its advantages and disadvantages. Well-planned storage ensures the distribution of water with minimum energy consumption. Technically, the following features should be sought in the selection of storage location and location:

Fig. 2 Calculation of storage volume. *Source* Created by Yasemin Kuslu

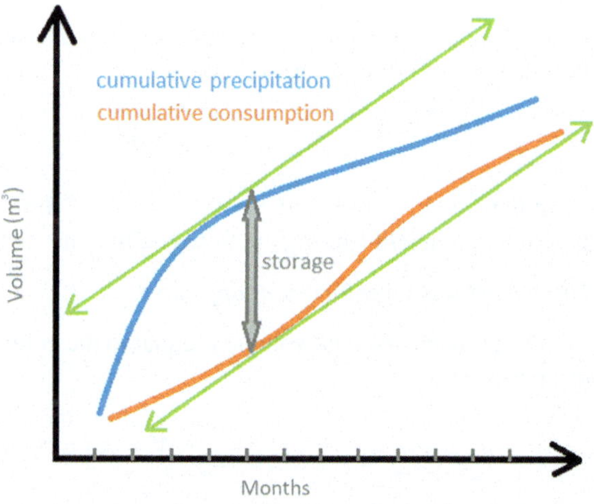

- The tank should be in a position that will not unnecessarily extend the network length and allow the use of small diameter pipes. In this case, the pressure loss must also be taken into account.
- Stores should be located as far as possible from transmission lines such as energy, communication, sewage, and drinking water.
- If possible, the tanks should be in a position to transmit the water to the target area without the need for pumping or with the least energy requirement.

The tanks can be placed inside or outside the building to collect precipitation water (Fig. 3). Generally, harvested water can be made suitable for domestic use after going through a series of processes. Depending on the nature of this process, it is possible to obtain siphon water, utility water, or drinking water. It is possible to plan the outdoor storage, as shown in Fig. 3a–c, respectively, on the ground (ground storage), underground (buried storage), and a platform above the ground (standing storage).

The ground and standing storage tanks should not spoil the aesthetic appearance of the building and be made of a light-proof material. Energy may not be needed when taking water from ground tanks for garden irrigation. Especially in the drip irrigation method, since low operating pressures are sufficient, water can be supplied to the system from the bottom of the tank. However, energy is required when sprinkling or micro-sprinkling methods are used. This requirement is for car washing for instance and other purposes. Since the standing tanks are located high from the ground, energy is usually not needed for water intake from these tanks. On the other hand, since the underground storage will not be exposed to sunlight, the material they are made of

Fig. 3 General diagram of roof precipitation harvesting systems and possible storage locations. *Source* Created by Yasemin Kuslu

doesn't have to be matt. One of the advantages of buried storage is that they are less affected by outdoor temperature changes. However, energy is required for water intake from such tanks. The water temperature in the storage located outdoors is in wide limit ranges depending on the change in the ambient temperature. Since the water temperature is more unstable, the biological oxygen requirement is affected. The most important reason for this situation is the formation of suitable ambient and temperature conditions for the proliferation of microorganisms. In case the tank location is chosen outdoors, insulation is important for the functioning and economic life of the storage system (Valdez et al. 2016; Vialle et al. 2011).

System parts

Some additional parts are used in the system according to the purpose of use of the water collected from the roof precipitation storage systems. For example, add-ons such as filters and equipment, check-valves, manometers, valves, pipes and insulation materials, and disinfection units can be used.

The quality of the water obtained by rain harvesting varies according to time and place. The quality of the collected water begins to change as it returns to the earth from the atmosphere. Precipitation waters carry dust, chemical compounds, microorganisms, and similar pollutants in the atmosphere or sweep the atmosphere to the water collection area. The quality of the harvested water varies depending on the path the water follows in its hydrological cycle (Rouvalis et al. 2009; Sazakli et al. 2007).

There is no chemically pure water in nature. The level of contamination differs between an industrial area and a rural area. It can be planned to harvest precipitation water for drinking and domestic use or purposes with high hygiene requirements. In such plans, if the annual precipitation regime of the region and the precipitation amount is suitable, the entry of the initial precipitation (the smallest value of the standard precipitation times—at least 5 minutes) to the warehouse can be prevented. Thus, the pass pollutants via contamination and sweeping effect are prevented into the water. Since this will increase the quality of the storage water, lower-cost treatment processes can be used for high-quality purposes. If the precipitation regime does not allow this, the desired quality water can be obtained with effective filtration.

Large-pore leaf screens (grids) can be used as the first stage of filtration. Undesirable elements such as leaves and plastics that can be caught in the airflow or located in the water collection area can be kept with the grids. Then, smaller-scale garbage can be removed with the help of small-mesh strainers (micro-sieves). The quality of the water to be stored can be improved by using organic, inorganic, or artificial filter materials (Fig. 4). The better the quality of the water in the tank, the more effectively it will serve the purpose of use. By using treatment systems that require high technology according to its purpose (for example, disinfection-ozonation application is required in the treatment system for drinking water), the water of the desired quality can be obtained at the entrance or exit of the warehouse. In addition, increasing the amount of oxygen in the tank by ventilating provides microbiological stabilization of the stored water.

Chapter 18
A Political Economy of Water Security: The Case of Singapore

Md Saidul Islam, Ashvini Kannan, QingLin Chen, Josephine Toh, and Lynette Loh

Abstract Water is a basic necessity, without which no individual or nation can survive. Every nation, therefore, needs to ensure its water security, which encompasses all dimensions of human health, livelihood and well-being, and food and energy production. All industrial nations first strived to ensure their water security through early substantial investments in infrastructure, institutions, and capacity to manage water and wastewater. Singapore is an important case to understand the political economy of water security. With its high dependence on neighbouring nation for water supply and with limited land to collect and store rainwater, Singapore encountered drought, flood, and pollution since its independence in 1965. Over the years, Singapore adopted integrated, effective, robust and cost-efficient approaches with strategic investments in research and technology to treat, recycle, and supply water. Today, the country is recognized internationally not only as a model city for integrated water management but also as an emerging global hydro-hub. Touching on the causes of water insecurity in the world today, this chapter delineates the Singapore story of how 'water scarcity' in the nation has been transformed into 'water opportunity'. To bring critical insights, the chapter also compares the Singapore case with two other similar nations: Saudi Arabia and Israel.

M. S. Islam (✉)
Associate Professor of Sociology, Nanyang Technological University, Singapore, Singapore
e-mail: msaidul@ntu.edu.sg

A. Kannan · Q. Chen · J. Toh · L. Loh
Environment and Sustainability Research Cluster, School of Social Sciences, Nanyang Technological University, Singapore, Singapore
e-mail: ashvini001@e.ntu.edu.sg

Q. Chen
e-mail: qchen004@e.ntu.edu.sg

J. Toh
e-mail: josephine-toh@e.ntu.edu.sg

L. Loh
e-mail: lyne0003@e.ntu.edu.sg

M. Behnassi et al. (eds.), *The Water, Climate, and Food Nexus*,
https://doi.org/10.1007/978-3-031-50962-9_18

Keywords Water security · Singapore · Climate change · Saudi Arabia · Israel · Wastewater · NEWater · Desalination

1 Introduction

In 2008, Andrew Liveris, Chief Executive of the Dow Chemical Company, declared in *The Economist* that "water is the oil of the 21st century. Like oil, water is a vital resource for the global economy". As with oil, water supplies, especially clean and easily accessible ones, are coming under enormous strain because of the growing global population and economy and climate change. One may question the need to compete over water resources, because over 70% of the Earth's surface is covered by water. In reality, however, a massive 97.5% of all water is salt water, and most of the remaining 2.5% of freshwater is inaccessible to humans (Ocean Focus 2014). Less than 1% of the world's freshwater is accessible for human use. Thus, the real issue is the limited amount of freshwater available for all.

This limited amount of freshwater is extensively used in all aspects of our lives. It is needed for businesses—such as high-tech companies, especially those that use huge quantities of water to manufacture silicon chips—and agriculture, which uses 70% of global freshwater (The Guardian 2009). It is inevitable that water will become a focus of tension and conflict among states and industries. Furthermore, water, unlike oil, has no substitute, which makes it even more precious. International efforts have been launched to address the problem of water security. However, the concept of 'water security' is not simply about dealing with dwindling water supplies. To serve as a starting point for dialogue in the UN system, the UN-Water (2013) defines this concept as "the capacity of a population to safeguard sustainable access to adequate quantities of acceptable quality water for sustaining livelihoods, human well-being, and socio-economic development, for ensuring protection against water-borne pollution and water-related disasters, and for preserving ecosystems in a climate of peace and political stability". A comprehensive definition of water security goes beyond water availability to issues of access, which involve fundamental individual rights, national sovereignty over water, equity and affordability, and the role of states and markets in water's allocation, pricing, regulation, and distribution (Gutierrez 1999).

In this chapter, we discuss the main issues of water security. First, we examine the reasons for increasing water insecurity in the world. Second, we examine how water insecurity could affect the small city of Singapore and examine the strategies that have been taken to ensure water security. Lastly, we compare the water management strategies implemented in Singapore and two other countries—Saudi Arabia and Israel—and identify room for improvement to strengthen Singapore's water security.

2 Reasons for Increasing Water Insecurity

2.1 *Effects of Climate Change*

The Intergovernmental Panel on Climate Change (IPCC) stated that the susceptibility of freshwater resources is caused by climate change, with drastic social, economic, and ecological consequences (IPCC 2008, 2012). Climate change will have major impacts on the water cycle and result in greater climatic and hydrological instability, with serious consequences for societies and their water security (Bigas 2013). The IPCC (2012) anticipated an increased frequency of droughts due to declining precipitation trends in certain regions, while others will have higher prevalence of floods and typhoons due to an increasing precipitation intensity. For example, glacial melting due to global warming causes a temporary increase in water supply to rivers, but will end once glaciers have melted completely (The Guardian 2012). In contrast, climate change in the sub-tropics will result in decreased rainfall in those already dry regions. Hence, the outcome is an increasingly extreme water cycle that will lead to more floods and droughts worldwide (The Guardian 2012).

Moreover, the rise in sea levels due to global warming will lead to the melting of polar ice caps and increased saltwater intrusion, which will in turn affect water quality (Bigas 2013). Global warming also increases the amount of water that the atmosphere can carry, therefore leading to more frequent and intense rainfall when the air cools (The Guardian 2012). Although rainfall can contribute to freshwater resources, heavy rainfall can also lead to accelerated movement of water from the atmosphere back to the sea, thereby decreasing humans' ability to store and use it. According to the World Bank (2006), fluctuation in rainfall alone could push more than 12 million people into poverty, while climate change could increase global malnutrition by up to 25% by 2080 (Fischer et al. 2002).

Changes in the hydrological cycle will threaten existing water infrastructure, making societies more vulnerable to water-related disasters and resulting in decreased water security (Bigas 2013). In 2012, the occurrence of Hurricane Sandy in the United States further reiterated that severe weather events can increase water insecurity by disrupting water infrastructure, such as hazard protection, storage and delivery capacity, pollution, and wastewater management (Bigas 2013). Marginalized and poor communities are the ones most affected by water-related disasters, but they do not have sufficient resources to cope due to factors such as poor environmental practices, rapid and unplanned urbanization in dangerous zones, and government incapacity (IPCC 2012). In other words, these communities are doubly at risk, due to a lack of resources, knowledge, and capabilities that could help resolve, improve, and manage those problems (Islam 2013).

2.2 *Irresponsible Use and Misuse of Water*

The irresponsible use and misuse of water resources is another reason for global water insecurity. The major uses of water resources are in energy production, industry, and agriculture (Hall 2012). In rapidly developing economies, industries lack modern technology for water-saving, efficiency, and pollution control. Little effort is made to ensure proper management of water resources (WWF Global 2014), as water is undervalued by humans. Therefore, concerns about climate change must be addressed in tandem with pollution issues and water management. Moreover, water resource problems are viewed as complicated, or are taken for granted, which leads to delays in intervention. Adequate management of water resources can influence every segment of society and the economy, including health, food production and security, domestic water supply and sanitation, energy, and industry, and is vital for the proper functioning of ecosystems. Therefore, the complexity of managing water resources should not be considered a valid excuse for inaction. Everyone must take responsibility to achieve water security and prevent misuse of water resources. Industries should import advanced technology to increase water efficiency, and individuals must play a role by recycling, reusing, and reducing water usage.

2.3 *Water Pollution*

Pollution is another major issue that impacts water security. Some countries with water scarcity problems resort to mining fossil groundwater, which is water stored in aquifers thousands of years ago (Gutierrez 1999). Libya for instance has spent approximately US$25 billion to pump up to 730 million cubic meters of water a year from underneath the Sahara since 1991 (Gutierrez 1999). Similarly, Saudi Arabia uses fossil groundwater to supply up to 75% of its needs, which amounts to more than 5000 million cubic meters a year (UNEP 1996). Groundwater is a non-renewable resource, and excessive use of it will lead to its depletion. This problem is linked to what Garrett Hardin called the tragedy of the commons—the idea that environmental resources are overused and eventually are depleted due to individuals acting in their self-interest to achieve immediate gains. Because the negative consequences are not felt instantly, everyone will draw water from underground and this resource will become depleted due to excessive usage.

Once groundwater is polluted, it is highly difficult to purify it due to its inaccessibility, large volume, and slow flow rates. Groundwater pollution is caused by agriculture, urbanization, and industrial activity (UNEP 1996). For example, in 1972, an irrigation system built in Yao Ba oasis in China led to economic growth due to the production of wheat and other cereals in that area (Gutierrez 1999). However, saltwater intrusion occurred as a result from a nearby alkaline lake. In the Gaza Strip in Palestine, high concentrations of nitrates and pesticides are found in the groundwater in areas with intensive agriculture (Grimble et al. 1996). In Japan, water pollution comes mainly from industrial chlorinated solvents (UNEP 1996). In developing countries, approximately 90% of wastewater is discharged without treatment, leading to outbreaks of waterborne diseases that affect the entire ecosystem (Gutierrez 1999). According to the Asian Development Bank (1998), water pollution contributes significantly to water scarcity and is the most severe environmental problem in Asia.

2.4 Food Security

Lastly, food security is threatened by climate change in every part of the world. Many countries are taking measures to ensure their future food security. This issue is especially pressing in Singapore, which lacks natural resources and thus faces a significant challenge in tackling food security. To deal with food security in Singapore, much emphasis has recently been placed on urban agriculture. For instance, a collaboration between the Agri-Food and Veterinary Authority of Singapore (AVA) and a local firm, Sky Green, popularized environmentally friendly urban farming techniques (Asia Times 2012). Sky Green has been a prominent player in the urban agriculture industry and developed vertical farming with green technologies. Their vegetable towers supply vegetables to over 230 NTUC FairPrice outlets, and this quantity will increase in future (Asia Times 2012). In addition, Singapore's first commercial rooftop farm, run by the social enterprise Comcrop, opened on the roof of the *SCAPE building in Orchard; crops such as heirloom tomatoes, Italian basil, and peppermint were harvested for sale to food and beverage outlets (Today 2014). The trend of urban agriculture will continue to grow in Singapore and other parts of the world; as a result, many states and companies will be competing for increasing amounts of limited freshwater. Thus, it is now even more important for Singapore to enhance its water security to secure its food security.

3 Political Economy of Water in Singapore

Singapore has long been on the list of water-stressed nations (Goh 2003), with a land area of 716 km^2 and a population of over 5 million people as of 2013 (Singapore Department of Statistics 2014). One of the biggest problems the Singapore government faced was how to provide clean water to its fast-growing population, which used to consume about 1.36 billion liters of water per day in early 2000s (Tortajada 2006). According to the Public utility Board (2021), water demand in Singapore is currently about 430 million gallons a day (mgd) that is enough to fill 782 Olympic-sized swimming pools, with homes consuming 45% and the non-domestic sector taking up the rest. By 2060, Singapore's total water demand could almost double, with the non-domestic sector accounting for about 70%. As Singapore experiences a tropical climate all-year round, it is never short of rainfall. However, because of its limited land area, there is a limit to the amount of land area in which rainfall can be stored (Tortajada 2006).

Singapore has relied on Malaysia for its fresh water supply through long-term agreements, even before its separation from Malaysia in 1965 (Goh 2003). When the British destroyed the causeway as a defensive strategy to prevent the Japanese from invading through Malaysia, they simultaneously cut off water supply from Johor as well (Simson 1970). Singapore suffered greatly, and this incident served as an important reminder of the importance of water security and highlighted how national security could be easily compromised (Kog 2001).

Rapid urbanization in South-East Asia has threatened environmental security, and there is an ongoing environmental crisis in the region. Over the past 20 years, rapid industrialization without proper waste and environmental management has led to the pollution of major rivers that were once sources of water for inhabitants of the region. Water security has been compromised and it is shifting from a national to a regional concern. For example, demand for water in Malaysia is increasing with population growth and urbanization, but the improper management of water resources has diminished its usable water resources. Singapore in the meanwhile has been reliant on the State of Johor in Malaysia for more than 70 years for its water supply, but this sharing of Malaysian resources is causing increasing resentment in Malaysia (Kog 2001).

Statistics show that Johor has lost around RM 3.2 million (S$1.5 million) per year by buying back treated water it has sold to Singapore (Lee 2003). The Johor Assembly also claimed that Singapore has accumulated around S$700 million by purchasing raw water from the state over a period of 40 years. Malaysia found this situation unjust and invoked a price-review clause, stating that the Chinese mainland sells to Hong Kong for almost 100 times as much. However, Singapore considered this arrangement to be fair, as it sells treated water back to Malaysia at a subsidized rate (The Economist 2003). These bilateral disputes are a cause for concern and destabilize cooperation between the two countries. Johor is set to complete construction of its own water treatment plant, and said in 2003 that it would stop buying water from Singapore when construction is completed (Lee 2003). When Johor becomes self-reliant, it can demand more money for its water supplies, which may further strain bilateral relations.

Singapore has no annual renewable water resources and is the most developed compared to neighboring ASEAN countries. With a much higher GDP and due to its affluence, Singapore leaves an ecological footprint far greater than the space it actually occupies. Thus, Singapore will need more resources and at the same time must contest over resources such as water with other countries in Southeast Asia. The need to trade for resources, such as water with Malaysia, is a source of political tension, as contestation increases and supplies diminish (Kog 2001). Because a large proportion of Singapore's fresh water supply comes from outside, ensuring water supply is crucial to maintaining its economic and military security. Hence Singapore has made efforts to find alternatives instead of being heavily dependent on Malaysia for fresh water supplies.

4 Strategies Implemented by Singapore to Ensure Water Security

Singapore has always placed a high emphasis on progress and economic development. A look at past policies demonstrates how green governance is indeed effective in ensuring sustainability. In 1992, Singapore's first environmental blueprint was created to ensure that the environment would not be compromised and that the country would be able to achieve growth for the generations to come (Ministry of Sustainability and the Environment 2020). This environmental blueprint dubbed the 'Singapore Green Plan' (SGP) was a 10-year plan towards making the country a more "clean and green" garden city.

Subsequently, a second plan, the Singapore Green Plan 2012, was developed in 2002, with a greater focus on environmental sustainability. The SGP 2012 encompasses six areas of focus, namely, Clean Air and Climate Change, Water, Waste Management, Conserving Nature, Public Health, and International Environmental Relations (Ministry of Sustainability and the Environment 2006). For each area, targets were set to be achieved by 2012. The 2008 review of the SGP 2012 revealed that Singapore's efforts were indeed effective as some of these goals had already been met or were already close to being met. To ensure that there is enough water for all, Singapore aims for at least 25% of its water demand to be constituted by non-conventional sources, such as desalination, by 2012 and this goal was already met in 2007. In addition, domestic water consumption was to be reduced to 155 L per person per day and in 2008, the domestic water consumption had already been reduced to 156 L per person per day. Another target was to increase the rate of overall waste recycling to 60% by 2012 and as of 2008, the overall waste recycling had already increased to 56%.

In order to meet Singapore's growing demand for water, technological solutions need to be optimized. Singapore has a total of four national water sources: (1) local catchment water; (2) imported water; (3) NEWater; and (4) desalinated water. Today, Singapore has a total of 17 reservoirs for stormwater collection, and a network of rivers that increase Singapore's water catchment area from half to two-thirds of Singapore's land area (Public Utilities Board 2014). Singapore aims to meet up to 80% of its water demands via NEWater and desalinated water by 2060 (Public Utility Board 2021).

4.1 *Imported Water*

Singapore has always had difficulty supplying sufficient clean water to its people. When Singapore was still a self-governing British colony in the early 1960s, it signed two long-term water agreements (in 1961 and 1962) to import fresh water from Johor. The first agreement expired in 2011 and the second will expire in 2061 (Public Utilities Board 2014). Although Singapore and Malaysia have been negotiating the possibility of extending the agreements, the outcomes have not been promising. Both countries are far from fulfilling their own water requirements. Furthermore, while Singapore has no issue with paying a higher price for the imported water from Johor to attain long-term water security well past the year 2061, its issue is with how the price will be determined by Malaysia (Tortajada 2006). As the negotiations have not yet come to a head, Singapore is definitely feeling the pressure to become as self-sufficient as possible with regard to water. Technology also plays a big part in its plans to become more water-efficient. This topic is further explored in the following parts.

4.2 Local Catchment Water

As Singapore's population density is ever-growing, the Public Utilities Board (PUB) has to strike a balance between the allocation of land area to water catchment and to urban development. Singapore's network of 48 major waterways and their catchments made it possible to harvest and store stormwater to supplement its growing demand for water. The construction of the three newest reservoirs involved not just highly skilled engineering, but also a great deal of effort in cleaning up the Marina waterway. Dams were built at the mouth of the Marina Channel, Sungei Punggol, and Sungei Serangoon to keep freshwater in their reservoirs (Public Utilities Board 2014; Chua 2014). The Marina Reservoir is now Singapore's 15th reservoir, and together with the Punggol and Serangoon Reservoirs it has increased the water catchment area from half of Singapore to two-thirds. Singapore hopes to increase the water catchment area to 90% by 2060 through harvesting water from the remaining streams and using technology to treat saline water (Public Utilities Board 2014).

4.3 Desalinated Water

In 2005, Singapore's first municipal-scale seawater desalination plant opened in Tuas (Tortajada 2006). It took a long time for the government to finally decide to use desalination as one of the national water sources, largely due to its costs (Goh 2003). Desalination is the process of treating seawater, which depends largely on electricity. The second desalination plant opened in 2013, and together, the two plants can supply enough water to meet up to 25% of current water demands of Singapore (Public Utilities Board 2014). It is hoped that technology will improve in future, and that the capacity and efficiency of the desalination plants can be ramped up to meet 25% of Singapore's water demand in 2060.

4.4 NEWater

As Singapore is under a massive amount of pressure to secure sufficient water sources for its growing needs, intensive research and development in water technologies have resulted in a success story that is now the pillar of Singapore's water sustainability. Singapore currently has four NEWater plants, the newest having started operations in 2011 in Changi. Together, they supply up to 30% of Singapore's water needs, mostly to industries that require ultraclean water (Public Utilities Board 2014).

NEWater purifies wastewater from sewage systems using advanced membrane technology that cleans the water so thoroughly that it surpasses the requirements of the World Health Organization (WHO). There are three stages that NEWater uses to process waste water so that it is safe to drink. The first stage is microfiltration, in which wastewater is passed through membranes to remove suspended solids, bacteria, and viruses. The water then goes through the second stage, reverse osmosis, in which it passes through a semi-permeable membrane that removes extremely small microbes and substances. By the end of this stage, the water is already considered 'high grade water quality'. As a safety back-up, the water then goes through the third stage of processing, in which it is disinfected with ultraviolet light, to ensure that the water is as pure as possible (Public Utilities Board 2014). After this stage, the water is so pure that it is in fact cleaner than tap water, and thus it is ideal for industrial manufacturing processes (Tortajada 2006). NEWater is also mixed into reservoirs during the dry season before undergoing conventional treatment before it reaches domestic taps (Public Utilities Board 2014).

A second Changi NEWater plant started operating in 2016 and helps generate another 50 million gallons of NEWater per day to add to Singapore's water supply (Channel News Asia 2014). With this new plant, NEWater has increased its supply to meet about 55% of Singapore's water needs.

4.5 Hybrid Management Technology

It is clear that Singapore is turning to technology to increase its self-sufficiency. While desalination utilizes the seawater that surrounds the island, NEWater reuses water already available in the water system. The two technologies complement each other to bring the total supply of water to meet 40% of Singapore's total water needs. With further technological advances and the completion of additional NEWater and desalination plants, Singapore hopes to meet 80% of its total water needs by 2061.

The country's holistic approach to water management can be distilled into three key strategies: collecting every drop of water, reuse water endlessly, and desalinating seawater. This has been captured in Fig. 1.

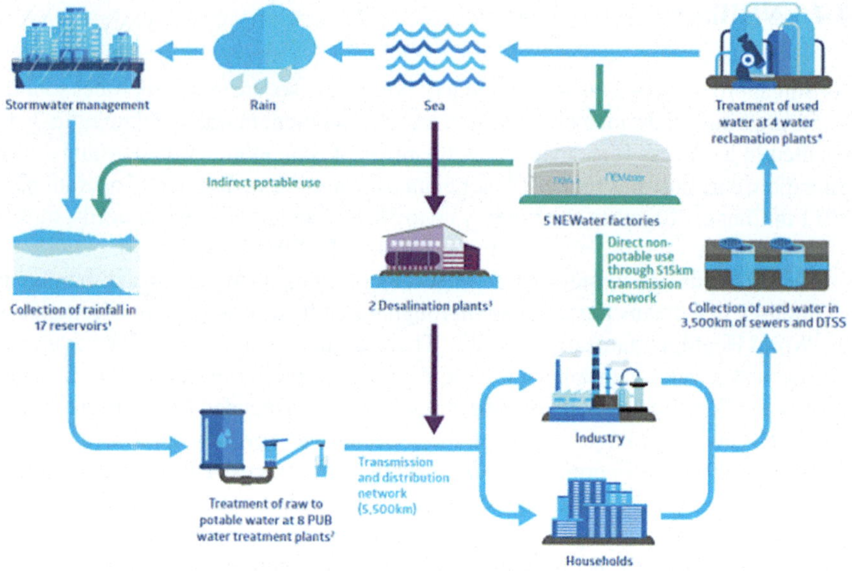

Fig. 1 Singapore's water loop. *Source* Public Utility Board (2021)

5 Comparison with Other Countries

Despite having the highest water stress ranking of 5.0, Singapore has managed to adopt exceptional water management strategies (Reig et al. 2013). However, there is definitely more to be done. Thus, we have decided to compare its position with two other countries, Saudi Arabia and Israel. These two countries, like Singapore, have a water stress ranking close to 5.0 and a relatively high GDP, and were ranked in the world's top 40 in 2013 (World Bank 2014).

5.1 Singapore Versus Saudi Arabia

Saudi Arabia has been termed the desalination nation. As it is extremely oil-rich, Saudi Arabia can afford the high cost of electricity needed for desalination. Due to the ample supply of oil, their water system is one of the cheapest for consumers. The tariff paid by consumers per cubic meter desalinated water is SAR 0.12 (about US$0.03), while the cost of production for stations that produce 20,000 cubic meters daily reaches SAR 12 (US$3.20) per cubic meter (Al-Suhaimy 2013). The low cost of water only encourages its consumption, as the public does not have to bear its cost and the price is paid by the public treasury. This situation is in contrast to Singapore, where there are public campaigns to encourage the consumers to conserve water to reduce their utility bills; PUB's slogan is '*Water for All, Conserve, Value and Enjoy*'.

The high cost of water encourages careful water consumption and prevents water wastage. The public is thus always conscious and aware of the water scarcity issue in Singapore.

Singapore is ahead of Saudi Arabia in terms of forward water management planning, because the former has realized the high cost of desalination and invested in alternative technologies like NEWater. Saudi Arabia, however, is facing a water crisis and is unable to invest in alternatives such as solar desalination plants. It is concerned about meeting existing and growing water demands, and needs to use proven and traditional desalination technologies and groundwater, instead of focusing on alternatives. Saudi Arabia is currently using 300,000 barrels of oil to operate desalination stations on the eastern and western coasts of the country. They, however, have plans to invest in solar-powered water production and aim to make Saudi Arabia an exporter of water production technologies in the future (Al-Suhaimy 2013).

5.2 Singapore Versus Israel

While Singapore is investing heavily in NEWater and alternatives, it is still far from the level of Israel. In 2009, Israel was named the most efficient recycled water user by the United Nations (Fleisher 2009). Israel recycled an astonishing 80% of household wastewater, which was four times higher than any other country (Rabinovitch 2010). Israel engages in ecological modernization through free-market environmentalism (Mol and Jänicke 2009). Businesses there are flourishing by coming up with energy efficient innovations, such an ultra-violet light purifier and a recycling system that utilizes millions of small plastic rings to breed bacteria to process organic waste (Rabinovitch 2010). Due to such innovations, water technology exports have doubled from 2005 to 2008, with 200 Israeli companies exporting $1.4 billion worth of water management technologies to over 100 countries (Fleisher 2009).

Similarly, Singapore is developing its water technology export sector, and over the next few years the government plans to attract more international companies. The Singapore government is already engaging in reformist sustainable development by integrating sustainable development into its policies. It has three funding schemes specifically for the water technologies sector: (1) the Water Efficiency Fund (WEF) of the PUB gives companies incentives to efficiently manage their water usage and encourages the stewardship of water conservation in society; (2) the Fast-Track Environmental and Water Technologies Incubator Scheme provides financial incentives to environmental and water start-ups by specialized incubators; and (3) the Innovation Voucher Scheme (IVS) by SPRING which aims to encourage local SMEs to work with public Knowledge Institutions like the Centres of Innovation (COIs) to test new technology and innovative ideas.

6 Conclusion

According to the Editor of the *Environmental Science and Technology* journal, Singapore has embraced 'best practices' for sustainability in the water industry, and has discovered that reclaiming wastewater is much cheaper than desalinating seawater (Schnoor 2009). Singapore is already in the forefront of alternative technologies compared to countries like Saudi Arabia, and now needs to implement these technologies to reach the same level of water recycling as Israel and create an advanced water technology export sector. The three goals of sustainable development are economic growth, environmental protection, and social equity. Singapore is addressing these goals and taking part in a state-led ecological modernization, coming up with better technologies to ensure water security and, in turn, water sustainability.

Every individual plays an important role in enhancing water sustainability and security. Although it may be thought of as a national issue, achieving this objective starts with everyday actions of every user. We need to understand that a simple act such as letting the tap run when we brush our teeth will not only cost us more; it will drain the taps of our world as well. For a broader sustainability that includes water security, there are three simple things that are of paramount need: reconnecting the humans with nature to realize the former is dependent on the latter; understanding that there are always limits to natural resources; and making sure that the resources claimed through technologies are distributed well across the national strata.

References

Al-Suhaimy U (2013) Saudi Arabia: the desalination nation. http://www.aawsat.net/2013/07/article55308131. Retrieved 9 Nov 2014

Asian Development Bank (1998) The Bank's policy on water, working paper. ADB Interdepartmental Water Policy Group (IWPG), Manila

Asia Times (2012) Farming in the sky in Singapore. http://www.atimes.com/atimes/Southeast_Asia/NL11Ae01.html. Retrieved 11 Nov 2014

Bigas H (2013) Water security and the global water agenda: a UN analytical brief. http://www.unwater.org/downloads/watersecurity_analyticalbrief.pdf. Retrieved 9 Nov 2014

Channel News Asia (2014) Consortium to build second NEWater plant in Changi. http://www.channelnewsasia.com/news/singapore/consortium-to-build/1367816.html. Retrieved 11 Nov 2014

Chua G (2014) How Singapore's marina barrage keeps fresh water in and the sea out. Citiscope. http://citiscope.org/story/2014/how-singapores-marina-barrage-keeps-fresh-water-and-sea-out. Retrieved 9 Nov 2014

Fischer G, Shah M, van Velthuizen H (2002) Climate change and agricultural vulnerability. Special report as a contribution to the world summit on sustainable development, Johannesburg 2002. International Institute for Applied Systems Analysis, Laxenburg, p 152

Fleisher M (2009) UN: Israel #1 world leader in water recycling. http://www.israelnationalnews.com/News/News.aspx/130565#.VGG89_SUcQQ. Retrieved 9 Nov 2014

Goh KC (2003) Water supply in Singapore: challenges and choices. Greener Manage Int 42:77

Grimble RJ et al (1996) Groundwater resource degradation: a framework for analysis, with examples from China and Palestine. In: Howsam, Carter (eds) Water policy: allocation and management in practice. E and FN Spon, London

Gutierrez E (1999) Boiling point: issues and problems in water security and sanitation. http://www.wateraid.org/~/media/Publications/integrated-water-resource-management.pdf. Retrieved 9 Nov 2014
Hall A (2012) Rio+20: water security for growth and sustainability. http://www.gwp.org/Global/ToolBox/Publications/Policy%20Briefs/12%20Rio20.%20Water%20Security%20for%20Growth%20and%20Sustainability%20%282012%29%20English.pdf. Retrieved 9 Nov 2014
IPCC (2008) Climate change and water. Technical paper of the intergovernmental panel on climate change. IPCC Secretariat, Geneva
IPCC (2012) Managing the risks of extreme events and disasters to advance climate change adaptation. In: Field CB, Barros V, Stocker TF, Qin D, Dokken DJ, Ebi KL, Mastrandrea MD, Mach KJ, Plattner G-K, Allen SK, Tignor M, Midgley PM (eds) A special report of working groups I and II of the intergovernmental panel on climate change. Cambridge University Press, Cambridge, and New York
Islam MS (2013) Development, power and the environment: neoliberal paradox in the age of vulnerability. Routledge, New York
Kog YC (2001) Natural resource management and environmental security in Southeast Asia: a case study of clean water supplies to Singapore. RSIS working paper, No. 15. Nanyang Technological University, Singapore
Lee PO (2003) The water issue between Singapore and Malaysia: no solution in sight? Institute of Southeast Asian Studies. http://www.iseas.edu.sg/documents/publication/ef12003.pdf. Retrieved 9 Nov 2014
Ministry of Sustainability and the Environment (2006) The Singapore Green Plan 2012. https://www.mse.gov.sg/docs/default-source/defaultdocument-library/grab-our-research/sgp2012_2006edition_new.pdf. Retrieved 20 Oct 2020
Ministry of Sustainability and the Environment (2020) Grab our research: Singapore Green Plan. https://www.mse.gov.sg/grab-ourresearch/singapore-green-plan-2012. Retrieved 20 Oct 2020
Mol APJ, Jänicke M (2009) The origins and theoretical foundations of ecological modernisation theory. In: Mol APJ, Sonnenfeld DA, Spaargaren G (eds) The ecological modernisation reader: environmental reform in theory and practice. Routledge, London, pp 17–27
Ocean Focus (2014) Water: the new oil of tomorrow. http://oceanfocus.org/2014/01/water-the-new-oil-of-tomorrow/. Retrieved 9 Nov 2014
Public Utilities Board (2014) The Singapore water story. http://www.pub.gov.sg/water/Pages/LocalCatchment.aspx. Retrieved 9 Nov 2014
Public Utility Board (2021) Singapore water story. https://www.pub.gov.sg/watersupply/singaporewaterstory. Retrieved 06 Jan 2021
Rabinovitch A (2010) Arid Israel recycles waste water on grand scale. http://www.reuters.com/article/2010/11/14/us-climate-israel-idUSTRE6AD1CG20101114. Retrieved 9 Nov 2014
Reig P, Maddocks A, Gassert F (2013) World's 36 most water-stressed countries. http://www.wri.org/blog/2013/12/world%E2%80%99s-36-most-water-stressed-countries. Retrieved 9 Nov 2014
Schnoor J (2009) NEWater future? Environmental Science & Technology. http://pubs.acs.org.ezlibproxy1.ntu.edu.sg/doi/pdf/https://doi.org/10.1021/es902153f. Retrieved 9 Nov 2014
Simson I (1970) Singapore—too little, too late: the failure of Malaya's Defence in 1942. Asia Pacific Press, Singapore
Singapore Department of Statistics (2014) Singapore in brief 2014. http://www.singstat.gov.sg/publications/publications_and_papers/reference/sib2014.pdf
The Economist (2003) Introducing NEWater. http://www.economist.com/node/1527091. Retrieved 9 Nov 2014
The Guardian (2009) Water shortage will damage business 'more than oil scarcity'. http://www.theguardian.com/environment/2009/feb/26/water-drought. Retrieved 9 Nov 2014
The Guardian (2012) How will climate change impact on fresh water security. http://www.theguardian.com/environment/2012/nov/30/climate-change-water. Retrieved 9 Nov 2014

Today (2014) Singapore's 1st rooftop farm aims to supply produce to F&B outlets. http://www.todayonline.com/singapore/spores-1st-rooftop-farm-aims-supply-produce-fb-outlets. Retrieved 11 Nov 2014

Tortajada C (2006) Water management in Singapore. Int J Water Resour Dev 22(2):227–240. https://doi.org/10.1080/07900620600691944

UNEP (1996) Groundwater: a threatened resource. United Nations Environment Programme, Nairobi

UN-Water (2013) What is water security? https://www.unwater.org/publications/water-security-infographic/. Retrieved 07 Jan 2021

World Bank (2006) Managing water resources to maximize sustainable growth: a country water resources assistance strategy for Ethiopia. Washington, D.C.

World Development Indicators Database, World Bank (2014) Gross domestic product 2013. http://databank.worldbank.org/data/download/GDP.pdf. Retrieved 9 Nov 2014

World Wide Fund for Nature (WWF) (2014) Environmental own goals: wasting water. http://wwf.panda.org/about_our_earth/teacher_resources/own_goals/wasting_water/. Retrieved 9 Nov 2014

Dr. Md Saidul Islam (Ph.D., York, Canada) is Associate Professor of Sociology and a Coordinator of the Environment and Sustainability Research Cluster in the School of Social Sciences and Asian School of the Environment, Nanyang Technological University, Singapore (NTU). He's a former Visiting Scholar of MIT. His research interests include international development and environmental sociology focusing on industrial aquaculture, global agro-food system, climate change, food security, and environmental sustainability. He has published 6 books and over four dozen articles and book chapters on these topics. He also taught at the College of William and Mary in the United States, and Nankai University in China.

Ashvini Kannan is an Affiliate, Environment and Sustainability Research Cluster, School of Humanities and Social Sciences, Nanyang Technological University Singapore.

QingLin Chen is an Affiliate, Environment and Sustainability Research Cluster, School of Humanities and Social Sciences, Nanyang Technological University Singapore.

Josephine Toh is an Affiliate, Environment and Sustainability Research Cluster, School of Humanities and Social Sciences, Nanyang Technological University Singapore.

Lynette Loh is an Affiliate, Environment and Sustainability Research Cluster, School of Humanities and Social Sciences, Nanyang Technological University Singapore.

Chapter 19
Water Management in Pakistan: Challenges and Way Forward

Muhammad Nawaz Khan and Adeel Mukhtar

Abstract Water management is the biggest challenge of twenty-first century being confronted by Pakistan due to inadequate water management practices, insufficient storage capacity, irrigation inefficiency, population explosion, over-exploited ground water, climate-induced water stress, and India's water hegemony as upper riparian. Continuing mismanagement of water resources is resulting in increasing water insecurity in the country. Pakistan is already a water stressed country which has not optimally managed its available water potential through adroit water conservation and storage strategies. The country would not be water secure until it implements stringent water management measures in addition to building dams. In this context, the study endeavours to answer the question as to why adequate water management practices are essential for the future water security of Pakistan as well as its socio-economic development. The study also highlights challenges faced by the water sector in the country and concludes with the doable policy recommendations.

Keywords Indus river basin · Water management · Climate change · Pakistan · India

1 Introduction

Water, a need for human existence, also has a major impact on the global economy, society, and environment. Everything that lives on the planet relies on water to function properly. In the present day, it remains a widely but unequally dispersed resource. In order to offer sustainable, cost-effective, and equitable water services based on advanced water resources, water managers must deal with many challenges. It is true that global water shortages are the main culprit, yet in certain regions there is a

M. Nawaz Khan (✉)
Islamabad Policy Research Institute (IPRI), Islamabad, Pakistan
e-mail: muhammad.nawaz@ipripak.org

A. Mukhtar
School of Public Policy, Oregon State University (OSU), Corvallis, USA
e-mail: mukhtaad@oregonstate.edu

M. Behnassi et al. (eds.), *The Water, Climate, and Food Nexus*,
https://doi.org/10.1007/978-3-031-50962-9_19

scarcity of accessible water as a consequence of ineffective water management practices. Many countries today fear that water supply will decline in the future. Effective water management may have a significant impact on water shortages when considering water-related issues such as climate change. In order to effectively manage water resources, many different sectors must work together, including government, society, and industry. This creates defined roles and responsibilities for everyone engaged in the process of water-resource creation, management, and usage.

Pakistan's severe water shortages claim the lives of hundreds of people each year. According to the WaterAid report on "The Water Gap—The State of the World's Water 2018," identified that 21 million people in a nation with a total population of 207 million lack access to clean drinking water (Kunbhar 2018). It has been shown that a lack of water is a major barrier to long-term economic growth. Water supplies are degrading, and disputes between stakeholders are becoming more common as a result of the growing disparity between water supply and demand. Climate change, population growth, a lack of political will, urbanisation, and a lack of public education all contribute to water scarcity and shortages globally, and all of these factors must be taken into consideration.

The main reason for Pakistan's water shortage is due to water mismanagement. Aside from political issues (both transboundary and provincial), other factors that contribute to flooding include poor water management practices. The social, economic, political, and environmental issues connected with water management, as well as new approaches to water management, will need new knowledge and instruments (such as technology, laws, and institutions). It is believed that Pakistan's water supply is in a state of emergency. Pakistan is at risk of food insecurity and water scarcity as a result of its fast population growth. From 5260 cubic metres in 1951 to 1000 cubic metres in 2016, the amount of surface water available to each person has decreased dramatically. Since 1951, surface water availability has declined. When it comes to water availability, Pakistan is expected to go from "water pressure" to "scarce water" by 2025, according to government estimates (Azad 2015). Water problems in Pakistan are mostly caused by insufficient water management, droughts, reduced rainfall, and pollution.

There are many water sub-sectors to consider in the area of water resources management, including hydropower, water supply, sanitation, irrigation, drainage, and environmental protection. In the opinion of many water experts, using a holistic approach to the management and development of water resources ensures that social, economic, environmental, and technical problems are taken into account when managing and developing existing water resources (Cardwell et al. 2006). Poor water management methods, limited storage capacity, and Indian water politics have made water management the most pressing problem confronting Pakistan in the twenty-first century. Continued mismanagement of water resources would jeopardise food, energy, and economic security, with grave consequences for poverty and war, according to the United Nations Environment Programme (UNEP) (Commission Proposed to Advise Govt on Water Issues 2014). Pakistan has already exhausted its supply of water cycle resources. At this moment, there are not any further water

the IWT. Pakistan claims that India is manipulating the treaty in its favour. For instance, it is estimated that the Kishanganga Dam's pond capacity is 6136 AF, whereas Islamabad thinks that just 1000 AF of storage capacity can be provided in India. Once finished, the project is expected to cut River Neelum flow by more than 20% or approximately 100 MW in the 969-MW Neelum-Jhelum hydroelectric power stations in Azad Jammu and Kashmir (AJK) (Khan 2016). Islamabad voiced worry that India had deviated from its responsibilities and has bowed to the terms and violated the IWT, thus hindering the Treaty implementation (Khan 2016).

Pakistan's difficulties in the water industry relate mainly to unequal flow and water distribution from trans-boundary rivers and technological efforts (dams, barrages, and storages) (Khan 2016). India is building many big and small hydropower projects on Indus, Jhelum and Chenab; these are causing major water shortages in Pakistan or produce flooding through mal-operation. India has 170 days of water storage, whereas Pakistan has a water storage capacity of just 30 days. India has 943 dams compared to Pakistan, whereas Pakistan has just 155 dams (Ahmadani 2018). Wilful obstruction of natural water flow is harmful to Pakistan's agricultural economy, and may lead to a serious conflict between the two nations.

India is prohibited from storing or diverting run-of-river water under the IWT. Because of India's continued construction of dams on western rivers, which has essentially shut off Pakistan's water supply, the country's fields may soon become barren, forcing it to import food and agricultural products, despite its long history of agricultural output. Water is vital for the agricultural sector of Pakistan, which accounts for 24% of GDP, 48% of jobs and 70% of exports. Pakistan is a mostly an agricultural country, with the agricultural industry accounting for 24% of GDP, 48% of its labour force, and 70% of exports (PILDAT 2011).

Pakistan alleges that India restricts the flow of river water from Kashmir by building dams and dams in the river, as well as by unilaterally diversionary water, which is a clear violation of the IWT. For example, consider the hydroelectric Kishanganga project, which is a source of contention between the two countries. Divergence began as a result of water transfer from one river to a different one. This, according to Pakistan, violates the Article (clause) 1V-3-c of the Treaty. The Neelum-Jhelum river's natural maximum water flow would be significantly decreased as a consequence of this divergence (Khan 2016).

Violations of the IWT as well as population shifts in Pakistani-controlled waters will not be allowed. They were particularly critical of the lack of transparency with which Indian hydroelectric power projects—like the Hanu Small Hydroelectric Plant, Chutak Hydroelectric Project, Nimoo Bazgo, the Baglihar Dam, the Wullar Dam, the Uri-II Hydroelectric Plant, and the Kishenganga Hydroelectric Project—were built in India (Khan 2016). In fact, India started the construction of Chutak power plant on Suru river, a tributary of Indus river, Tulbul Navigation Project and Wullar Barrage, without informing Pakistan and did not provide any data on these projects as per the requirements of Article VII(2) of IWT (Baig 2013). Rather, India stalled on the discussions over these projects with the intent to bring these projects near completion before Pakistan could choose international arbitration.

India is using water resource as a coercive tool to enhance Pakistan's water problems. Narendra Modi has already attempted to shape water dispute as a politically charged issue for his domestic consumption. "Addressing rallies in Kurukshetra and Charkhi Dadri, Modi asserted that his government will stop water flowing to Pakistan" (Singh 2019). In this context, his "nefarious strategy will be to target Pakistan's water resources" (Yousafzai 2019), and his threats should not be taken lightly.

Hypothetically, could the water conflict be escalated into wars between India and Pakistan? This requires that India should abstain from adopting unilateral measures on river running and limit the construction in line with established dialogue channels, bilateral water agreements and the current water dispute resolution system, which is necessary to preserve the biodiversity (Khan 2019a, b). Only then the possibility of water conflict getting translated into war could be avoided. If India does not act accordingly and continues to use rivers water as an Aqua Bomb, then water would become a bigger threat for the regional security. Thus, the outstanding water issues between both countries can only be resolved through effective implementation of the IWT.

3.2 *Population Explosion and Declining Per Capita Water Availability*

According to government statistics, Pakistan's population grows at a rate of about 2.4% per year (Haq 2017). It requires increasing water supplies for food production, domestic, industrial, and environmental reasons. Currently, Pakistan's population is more than 207 million (Haq 2017), which is set to "double in coming three decades" (Junaidi 2019). This means that the per capita availability of water will decrease rapidly. Countries are considered to be water-stressed according to international standards if their annual per capita supply is less than 1800 cubic metres, and the shortage gets more acute if the rate is less than 1000 cubic metres even more. In this perspective, the Asian Development Bank (ADB) has already ranked Pakistan as "one of the most water-stressed countries" (ADB 2013), with less than "1000 cubic metres (903 m^3)" available per person per year (Per Capita Water in Pakistan Shrinks to 1000 Cubic Metre 2020). Per capita water availability "at the time of independence was 5600 m^3" (Parry et al. 2017). With the present trajectory after a quintuple decrease since 1947, the per capita water supply will hit the rocky bottom by 2037 and 711 cubic metres per person are available (Azad 2015).

3.3 Water Logging and Salinity

The water table rose as a consequence of inadequate drainage infrastructure and poor water management, causing flooding and salinity in the surrounding region. According to a research done by the Hisaar Foundation, the Indus Basin collects around 24 million tonnes of salt each year (Ashir 2019). It is estimated that salinity has damaged about 27% of cultivated land and almost 43% of Pakistan's irrigated land is waterlogged (Saeed 2015).

3.4 Unregulated Private Tube-Wells, Lowering Ground Water and Brackish in Quality

The fast rise in the number of private tube-wells in Pakistan has resulted in the escalation of aquifer mining. It has been predicted by experts that mining for ground water would lead to a reduction in available water supplies in the region. The level of groundwater falls by one metre per year. Around 1.2 million tubes collect water for agricultural purposes, of which 0.8 million is in the province of Punjab (Jalil 2019). Due to a heavily over-exploited ground water, the 60% area (FAO 2011) of the Indus is having ground water of marginal quantity to brackish in quality. This serious concern has caused intrusion of brackish water into the fresh ground water zone.

3.5 Floods and Weak Financial Viability

Pakistan is witnessing floods of various magnitudes since 1950 till date. The country has suffered a cumulative financial loss of more than US $38.00 billion (Resources 2017) during the past 70 years (1947–2016). Due to 25 catastrophic floods, about 12,502 people were killed, 197,273 settlements were destroyed and damaged an area of over 616,598 km^2 (Resources 2017). Since 1950, leaving adverse impact on the irrigation system damaging most of the canals and embankments.

The whole cost of operating, maintaining, and repairing a water structure is high, and this is something that is often forgotten. The running and maintenance costs for irrigation systems are kept to a minimum for customers. The water industry's financial stability is dangerously fragile right now. Investing too little while earning too little is a major factor in our current financial predicament.

3.6 Reduced Storage Capacity and E-Flows Concerns

Our country only constructed two major dams, Tarbela and Mangla, and one important barrage, Chashma, during the independence era. When the three main reservoirs were built, there was no way to de-silt them due to a fault in the original design according to current technology. As a consequence, the reservoirs' lifespan was not anticipated to be shortened. The storage capacity of water reservoirs in Pakistan has reduced "from 15.89 MAF in 1992 (12.1 MAF in 2012) to 13.69 MAF till August 2020" (Mettis Global 2020), which is the around 14%, yet in 1976 it was 15.9 MAF. This is due to an increasing siltation and sedimentation. However, after the Mangla project in 2013, the capacity was raised up to 14.98 MAF. At the time of its construction, Mangla Reservoir had a storage capacity of 5.88 MAF; however, this capacity fell to 4.593 MAF in 2013. The dam capacity was increased by 2.90 MAF as the dam was elevated by 30 feet and additional storage was added to the dam to ensure a total storage capacity of 7.55 MAF, above Tarbela Dam's maximum live capacity of 6.58 MAF (NESPAK n.d.).

With the same pattern of siltation and sedimentation in the reservoirs, it is estimated that the loss in live storage capacity would be 8.6% or 6.45 MAF by the year 2025. If the Mangla raising project had not been completed, then the loss in the live storage capacity would have been 23.89% or 30.4 MAF till year 2020. According to the latest WAPDA data, almost 99. 10% physical work has been completed, aiming at regaining the reservoir capacity lost due to sedimentation (WAPDA n.d.).

In the process of expanding agricultural output while simultaneously meeting the needs of residential and industrial water users, environmental issues emerge and need to be addressed. Increasing public acceptance for the adoption of minimum e-flows (environmental flows) in the river system and delta ecosystems would need more water as well.

3.7 Variability and Seasonality and Lack of Optimum Use of Effective Rainfall

As the World Bank points out, seasonality is extreme in Pakistan, with Kharif season flows being five times higher than Rabi season flows (Chaudhry 2017). Due to this unpredictability and seasonality, there are substantial hydrological and physical constraints on water management from different sources (Pakistan's Water Technology Foresight n.d.). Water from a wet year cannot be stored in a dry year because there is insufficient storage space for water being transferred from the Kharif season to the Rabi season at this time. Pakistan is one of the driest countries in the world and has an annual average precipitation of less than 240 mm (mm) (Ghoraba 2015). One of the challenges for planning is to make optimum use of effective rainfall.

3.8 Aging and Outdated Irrigation Infrastructure and Innovative Knowledge Based Management

Pakistan is blessed with "one of the largest irrigation" infrastructure (Ministry of Finance 2010). However, it was designed for water requirements of the twentieth century (over 100 years ago) and not for the twenty-first century. The design of system was for 60% cropping intensity, which has crossed over 172% (Basharat 2019). In addition, considerably more time and effort must be devoted to system maintenance. Instead of using conventional techniques, we must push the limits of knowledge and innovative approaches in order to face the problems of the twenty-first century. In order to fulfil new obligations, the institutions' roles must be re-evaluated and their capacities expanded accordingly.

3.9 Low-Irrigation Efficiencies and Seawater Encroachment

As mentioned above, it is estimated that almost 45% (45.8 MAF) of water is being lost between the canal headworks to the farm gate due to poor infrastructure or poor water management (Qureshi 2017). These trends indicate that approximately 55% (56 MAF) of Pakistan's irrigation water is really being utilised by the crops, indicating an irrigation efficiency of Pakistan of about 55% (56 MAF). "Irrigation efficiency is a compound of three efficiencies including canal-head efficiency, watercourse efficiency and farm efficiency" (Sial 2013), yet unfortunately Pakistan lags behind in this field. In earlier studies, the Water Allocation and Distribution Authority (WAPDA) has demonstrated that lining tiny canals may save more than 5 irrigation MAFs and that enhancing minor water courses can save an added 3.6 MAF of irrigation (Climate n.d.).

The delta ecosystem suffers as a result of seawater encroachment, which damages up to 3.5 million acres of arable land during periods of reduced river flow, according to UNEP, because the existing storage facility is insufficient for water transfer from a wet season to a dry season.

3.10 Energy Security

The installed energy capacity is 35,972 MW (2019–20, 2020) in Pakistan. The overall capacity for the Indus river is 38,608 megawatts (MW), and the total tributary potential is 5726 megawatts (MW), of which there are a total of 44,334 megawatts (MW), and currently only 13% (576,342) of the total capacity is generated (Ishaq 2016). The reduction in the share of low-cost hydropower in the energy basket resulted in a rise in electricity costs (Mettis Global 2020). While more focus is given on producing energy from alternative sources like oil, gas, coal, etc., more attention is also needed to produce cheaper energy by investing on hydro projects.

3.11 Climate-Induced Water Stress in Pakistan

The current condition of water in Pakistan in its natural habitat is highly uncertain as a consequence of climate change (NewsDesk, October 15, 2017). Together with climate change, rapid urbanisation and increasing populations, the country's use of water is rising (Malik 2018). In response to the Indo-Pak water dispute, Pakistan's position is extreme in that, on the one hand, it jeopardises the security of the Indus Basin and, on the other hand, it is totally dependent on supplies of the Indus Basin, whereas other regional countries are able to fulfil their water demand either by rain or other sources (NewsDesk, October 15, 2017).

Another factor which affects water availability is the shifting precipitation pattern and melting of snow, which are both important to river flow (Mahbub ul Haq Centre 2013). Similarly, the unpredictability of cyclical weather patterns over many years affects the flow of rivers.[1] As for Pakistan, this pattern is extremely noteworthy as it explains a very large variation from the lowest fluvial inflows of 98.6 MAF (as obtained in 2001–02) to the maximum fluvial inflow rate of 186.8 MAF (as achieved in 2002–03), both of which are very significant. These problems of water insecurity are compounded by climate change, which may lead to a nationwide decrease in water supplies. In addition, water safety in Pakistan is jeopardised by inadequate irrigation infrastructure, inefficient water management and a lack of appropriate water storage capacities (Zaman and Abubakar 2016). As a consequence, it is another obstacle to sustainable development because of the reliance on groundwater extraction, which represents 40% of the total supply at the farm gate (Iqbal 2017).

Moreover, Pakistan has just 30 days of water storage capacity which is inadequate for the fourth largest country in the world to use water. This may be considered extremely low compared to other countries. It has been projected that Pakistan's population increased by a total of 207.774 million people between 1998 and 2018, with an annual average growth rate of 2.4% (Lehane 2015). According to recent figures, the country's population is projected to reach 245 million people in 2030 and 309 million in 2050. The growing population of the country and its accompanying demand for water have resulted in a steady drop in sustainable access to water per capita (Tariq 2020). According to 2015 UN estimations, Pakistan has less than 1000 m^3 of water per head available and was categorised as insecure in terms of water (Zaafir and Haider 2017). The population of Pakistani residents in urban areas was 38.8% in 2015, and this number is projected to reach 46.6% by 2030 and 57.5%

[1] The inter-annual variability indicator, defined by the Water Resource Institute as the standard deviation of annual total water divided by the mean of annual total water from 1950 to 2010, is calculated as follows: 0 represents the least variability, while 5 represents the most variability; the range for this metric is 0–5. This is much higher than China's index of 1.97, India's index of 1.72, Sri Lanka's index of 1.59, and Bangladesh's index of 1.59 (0.07).

by 2050. The demand for water will have risen by 15 cubic kilometres by 2050 and, when other variables are considered, a total demand of 195.2 cubic kilometres (Mukhtar 2020).

Furthermore, there is a shortage of information in the literature about future water requirements. Climate change will certainly influence water consumption, resulting in Pakistan being obliged to undertake research and development in order to satisfy the increasing demand for water. Water demand in Pakistan, for example, must be evaluated sector-by-sector in the industries, agricultural, commercial and residential use/water needs and the adaptation plans must, under UNEP, be backed by an institutional structure with legal protection. The methods for data collection, which would need greater investment and the development of competency of the relevant implementing agencies, need to be improved.

The creation and execution of provincial climate change plans and the identification and implementation of action-oriented solutions for water concerns are necessary for water governance. To accomplish this objective, the purpose of the Pakistan Climate Change Act may be extended to include the adaptation of the water sector at the local, provincial, and national levels. Provincial financing is a requirement for adaptation strategies to be implemented. Climate financing, especially for efforts to adapt to the water sector, may serve as a starting point for debate.

Enhanced water management and intelligent use of water may increase water productivity, soil productivity, irrigation efficiency and decrease the risk of Pakistan's water crisis. Water policy reforms are thus critically needed to enhance water management, water quality and allocations. In addition, improving public awareness of climate change and water shortages at the local level would help to respond quickly to the growing demand for water linked to the rapid development of urbanisation. For quite some time, government and associated organisations have been pushing the need to build a new dam. Research on the new dam sites and cooperation on the Kabul River are also available. The only thing that is needed at the moment is to implement these ideas.

4 Government's Major Steps and Polices for Water Management

4.1 Vision 2025 (2014)

In recent years, there is a realization that water, food and energy systems are becoming increasingly interdependent. Therefore, in Pakistan's 'Vision 2025', which is referred to as the 'pillars', Pillar IV is designated as Water, Energy and Food Safety, a total of seven important areas of action. A vision and roadmap have been developed for the future growth and development of the water, food and agricultural industries (NewsDesk, October 15, 2015). The key goals under this pillar are:

- **Energy**: The aim is to double output of power to 45,000 MW and to provide continuous, affordable and sustainable 'energy for everyone' with access to electricity rising from 67 to 90%, the World Bank reports (NewsDesk 2014).
- **Water**: The project will be performed in order to enhance storage capacity to 90 days and boost the efficiency of usage in agriculture by 20%.
- **Food**: Caring food security in the country for the twenty-first century.

4.2 National Water Policy of Pakistan (2018)

The first National Water Policy draft was prepared in 2004 followed by the second draft in 2014. In 2018, the National Water Policy got prepared on the following themes:

- Understanding Water Security and its Impact on National Security;
- Water-Food-Energy Nexus;
- Water Infrastructure as Engine of Growth;
- Innovative Financing of Water Infrastructure;
- System Efficiency for Urban Utilities;
- Agriculture Water Productivity;
- Ground Water Sustainability;
- Water Supply and Sanitation [Urban and Rural];
- Water Technology for the 21st Century;
- Climate Change Impacts on Water Availability (Droughts, Floods and Glaciers);
- Research and Education;
- Water and Health;
- Industrial Pollution (Resources 2018).

4.3 Ten Years National Flood Protection Plan-IV (2015)

The main objective of this plan is to establish a comprehensive flood protection strategy, which will involve the development of an investment schedule based on the lessons gained from past floods (NewsDesk n.d.). The strategy involves greater resources for reservoir operations, procedures, inspections and training, schedules for embankments' maintenance and reinforcement and bank-breaching plans, expansion and modernisation of data collection techniques, including satellite monitoring, flood forecasting as well as implementation of flood warning system.

4.4 National Climate Change Policy 2012

The objective is to ensure that climate change is incorporated into economically and socially sensitive areas of the economy and to lead Pakistan towards a climate-resilient future (Naeem 2013).

4.5 National Drinking Water Policy 2009

The policy's goals consist of:

- providing equitable, efficient and long-term access to clean drinking water at affordable cost for the entire population (Ministry of Environemnt 2009); and
- ensuring a reduction in the frequency of mortality and illness linked to waterborne infections (Ministry of Environment 2009).

4.6 Future Storage Projects

Governments across the globe have been compelled to take conservation measures due to water shortages and drought. Pakistan urgently needs dam construction to help alleviate water shortages. To meet future water and energy needs, the present government is working on projects that will expand storage capacity. In Table 2, you can find information on the various storage options that have been considered.

5 Recommendations

The following recommendations may be considered to ensure effective water management for water security in Pakistan:

- **Transnational Water Dispute**: PM Modi plan to target Pakistan's water resources should not be taken lightly. Modi's statement of diverting the Eastern Rivers waters to Haryana and Rajasthan (clear violation of IWT) should be analysed in the backdrop of his illegal action to abrogate the articles 370 and 35a. This possible diversion of Eastern River waters would further encourage Modi for completely diverting the Western Rivers waters towards India. Pakistan needs to prepare his homework without wasting further time and simultaneously start lobbing in World Bank (guarantor of IWT), in order to justify the Indian possible violation of IWT by diverting the Eastern Rivers waters.
- **Reduction in the Water Wastage and Effective Water Storage Measures**: Agricultural uses only one-half of the Indus River System's total water withdrawals. As an example, the country must restore its canal system to minimise water loss

Table 2 Future water storage projects in Pakistan

Name of project	Storage capacity (MAF)		Installed capacity (MW)	Status
	Live	Gross		
Diamer Basha Dam (WAPDA, Diamer Basha Dam)	6.40	8.10	4500	Construction work initiated
Kalabagh Dam (Climate)	6.10	7.90	3600	Project on termination
Badin Zai Dam (WAPDA, Badin Zai Dam Project)	0.15	0.20	–	"Feasibility study, detailed engineering design, tender documents, and PC-I amounting to Rs. 305.271 million submitted"
Bhimber Dam Project (WAPDA, Bhimber Dam Project)	0.28	0.04	2	"PC-II for feasibility study amounting to Rs. 121.128 million submitted"
Chiniot Dam (WAPDA, Chiniot Dam Project)	0.85	0.90	80	"PC-II for detailed engineering design, preparation of tender documents & PC-I amounting to Rs. 533.303 million submitted"
Daraban Zam Dam (WAPDA, Daraban Zam Dam)	0.04	0.06	0.75	"Tenders will be re-invited after detailed engineering design of the project and subsequently, the revised PC-I will be prepared"
Akhori Dam (WAPDA, Akhori Dam Project)	7.0	7.6	600	"CDWP has approved the PC-II for feasibility study of Akhori Dam project"
Munda Dam (WAPDA, Mohmand Dam Project)	0.67	1.23	740	Process of land acquisition is under way
Naulong Dam (WAPDA, Naulong Dam)	0.19	0.24	4.4	"Collecting data/detailed working of command area development works of the project"
Bara Dam (WAPDA, Bara Dam Project)	0.06	0.08	06	Feasibility study completed
Tank Zam Dam (WAPDA, Tank Zam Dam Project)	0.16	0.34	25.4	Feasibility study completed
Shyok Dam (WAPDA, Shyok Dam Multipurpose Project)	5.50	8.50	640	Draft feasibility study is to be submitted
Total	27.4	35.19	10,198.55	–

as a consequence of leakages and unlawful canal breaches in order to maintain this water supply. If a rapid water delivery system is put in place, leakage may be reduced by at least a third. However, efficient canal systems cannot be put in place immediately.

- **Groundwater Conservation**: By carefully controlling the abstraction and irrigation of the water supply, groundwater conservation methods may be accomplished. When it comes to water demand management, the maintenance and efficiency of the systems currently in place require a special consideration. Increasing supply is the primary focus of current supply-chain strategy. Efficiency implies reducing waste, whereas productivity refers to the amount of effort put in while utilising less resources.
- **Water Resource Conservation Plan**: reducing the amount of water we use may help in reducing the negative impacts of climate change on water availability. If unexpected precipitation—precipitation that is above average for a given year, or a trend toward decreasing rainfall—creates water shortages, then water storage capacity needs to be expanded. On the other hand, relying only on storage as a long-term solution is not an option. National conservation strategies and irrigation systems ought to be built in the same manner, taking irrigation water release and storage into account as well as water restoration during the dry season, among other things. This is suggested. Aside from that, small and medium-sized dams have a major potential to increase the capacity of present storage sites while also extending the usable life of such sites. Expanding the water reservoir increases the amount of water available for agriculture and other uses. The loss of carbon sinks in the oceans, the devastation of ecosystems, and the eviction of low-income people are all issues that may arise from water conservation, especially if it is done through large dams.
- **Adoption of New Technologies**: It is imperative that new technologies such as wastewater treatment and water reuse be used in Pakistan in order to solve the country's water shortage issue. Various parts of the country will benefit from better water efficiency as a result of this initiative. In terms of seawater recycling, utilising GIS technology to build a useful water inventory, as well as monitoring and managing the present actual water supply, a similar method would be beneficial. This may include investments in programmes that assist the development of vision and demand management as well as agricultural production, among other things. Increased water sector innovation would be encouraged to the maximum degree possible.
- **Implementing an Integrated Water Management (IWRM) Approach**: Water management is evolving from a sectorial to an integrated approach. In Pakistan, as previously mentioned, an integrated water strategy is required to address the issue of water scarcity for the long term. There is a need to integrate all water stakeholders (private–public) in order to achieve this, and this may entail anything from efficient water usage to water conservation to a national agreement on how to manage water for future generations. Water resources are integral to a wide range of local government duties, from water supply and sanitation to land use planning and economic development. The way water resources are managed affects all of

these functions, or they have an impact on downstream water use in some way. Because of this, local governments are crucial to IWRM. Local governments in Pakistan continue to rely on technical know-how and financial resources to execute IWRM effectively. This area needs to be taken seriously by government for effective water management in the country.

- **Modernization of Agriculture Sector**: On a worldwide scale, agricultural technology is poised for change. Technology has changed the way agriculture and farming operations work dramatically in the last several decades. This is especially true for sensor-enabled operations and farms. Small farmers in Pakistan are disproportionately affected by climate change because of a lack of information, unpredictable weather changes, and inadequate water resource management. In Pakistan's attempts to enhance agriculture and water management, the adoption of modern technology is, therefore, much more important than previously believed.

6 Conclusion

In rural Pakistan, more than 70% of people depend on natural resources for their livelihood, including some of the most disadvantaged individuals in the country. Effective water management is required if water, a precious resource, is being wasted on a regular basis. It is imperative that the problem of water scarcity be thoroughly studied before a feasible solution be developed and implemented. The increasing number of towns built along riverbanks has led to an increase in freshwater abstraction, reducing the quality of the river's water. This is why a reliable long-term water source requires attention to water balance, water quality, and statistics. The government also needs to undertake successful land reforms, since large land is more productive than tiny land, and land ownership must be extended to address this issue, among other things.

The agriculture sector's water management efficiency will undoubtedly increase as a consequence of effective and efficient methods. The goal of water policy is to guarantee long-term sustainability of water resources by promoting effective usage. Increasing the supply of water by improving canal water efficiency (reducing leakages and illegal canal breaches) by 33–90%, similar to industrialised countries, has a considerable potential to increase supply levels. Reducing water leakage should be encouraged, along with effective water metering.

Pakistan's ability to utilise water from the Indus Basin is controlled by the Indus Water Convention, which is governed by international law. The arrangement has been in place for decades despite environmental and economic demands from the other side of the border. As a result of climate change having an impact on surface water use and water resources, it is essential for policy makers to take additional measures to address this problem. In this regard, the ideal situation is that Pakistan, in cooperation with India, has to develop regional water management plans.

The whole water distribution system in Pakistan has to be redesigned to better manage and protect the country's water resources. There has not been enough government investment in water resource management, either. Global water problems can

Chapter 20
Climate Change, Water Variability, and Cooperation Along Transboundary River Basins in Perspective of Indus Water Treaty

Faraz Ul Haq, Ijaz Ahmad, and Noor Muhammad Khan

Abstract Rivers do not follow political borders while flowing through their natural courses. Transboundary rivers can be the cause of water conflict between nations. Treaties and agreements have been made to settle these conflicts. In the Indian subcontinent, western rivers are becoming a matter of concern for Pakistan in terms of water quantity. Various dams are constructed by India on western rivers and many more are under construction/planned. In this research work, a comparison has been made between the viewpoints of both countries regarding transboundary interventions. India's point of view is that variations in river flows and water shortages in Pakistan are because of her own negligence. India is doing nothing to violate the Indus water treaty and is utilizing the water allocated to her on western rivers. According to Pakistan, India is stealing water supplies from Pakistan and constructing its water-controlling structures only as a political exercise to improve political power. Maintenance of transboundary aquifers and groundwater management, transboundary cooperation in watershed management and integrated water resources management can lead to sustainable development between both countries. The outcomes of this

Faraz Ul Haq · I. Ahmad (✉) · N. M. Khan
Centre of Excellence in Water Resources Engineering, University of Engineering and Technology, Lahore, Pakistan
e-mail: dr.ijaz@uet.edu.pk

Faraz Ul Haq
e-mail: fuhaq@memphis.edu

N. M. Khan
e-mail: noorkhan@uet.edu.pk

N. M. Khan
Civil Engineering Department, University of Engineering and Technology, Lahore, Pakistan

Faraz Ul Haq
Center for Applied Earth Science and Engineering Research, The University of Memphis, Memphis, United States

Department of Civil Engineering, The University of Memphis, Memphis, United States

M. Behnassi et al. (eds.), *The Water, Climate, and Food Nexus*,
https://doi.org/10.1007/978-3-031-50962-9_20

work will provide information related to transboundary conflicts, treaties, and future concerns to the researchers and policymakers.

Keywords Transboundary water · Indus water treaty · Climate change · Transboundary management · Food security

1 Introduction

Rivers do not follow political borders while flowing through their natural courses. Many transboundary river basins in the world are shared by different countries. Due to an increase in the world's population and water scarcity, conflicts among countries begin to rise over limited available freshwater resources (Chaturvedi 2013). The population of the world in 1995 was 5.7 billion (Crisp et al. 2017) and is predicted to reach 9.7 billion by the year of 2050 (UN 2019). Limited freshwater resources will be shared to meet food and fiber requirements of an ever-increasing population, which will ultimately lead to the decline of per capita water availability to chronic water stress conditions. Besides the rapid growth in the world's population and consequent decline in water distribution, various additional factors are also responsible for the global water shortage. For instance, due to variations in precipitation, groundwater recharging rates, and the availability of water resources will lead to over-abstraction of groundwater resources (Yotova 2009). Moreover, climate change and glacier melting also cause floods and droughts, which are responsible for distress in water supplies (Parry 2007). Therefore, considering the above factors, it can be concluded that water conflicts between countries with shared river basins may rise in the future.

The disputes over water resources can be traced back several years (Chatterji et al. 2008) which may cause encounters between nations over the distribution of water resources (John 2007). Approximately, more than 310 river basins share water resources with more than one state/country (Dombrowsky 2007). In the catchment area of these river basins, almost 40% of the world's population resides (Miller et al. 2010) which indicates that a minimum of hundred nations are close to disputes over water resources (Yamamoto et al. 2013). During the last sixty years, around 40 water disputes have been reported and most of them were in the Middle East. These conflicts among nations due to limited water resources to meet their ever-increasing water demands may compromise the world's peace (Chellaney 2013).

In the previous century, several water disputes were reported around the globe including the river Nile, river Sanaga, Helmand River, Jordan river, Indus River basin, etc. Water-sharing disputes between the United States and Mexico in the Tijuana River basin were over water quality. Bosnia, Herzegovina, and Croatia have water conflicts over hydropower generation on the Neretva River. In Europe, flooding and flood management structures are the reason of water disputes (Wolf 2007). Similarly, in South Asia, the Ganges and Brahmaputra are disputed between India and Bangladesh, India, and Nepal. River Kabul is also contested between Pakistan and

Afghanistan. Moreover, India has serious concerns about the interventions of China on the headwaters of the Indus, Brahmaputra and Ganges in Tibet (Holslag 2011).

Various water treaties have been signed internationally including the Mexico-US water treaty, the Columbia River treaty signed between the US and Canada for the development of dams on the upper Columbia River basin (Cohen et al. 2000), and the treaty of Zonhoven signed between the Netherlands and Belgium. Moreover, treaties were also signed between South-Asian countries regarding transboundary water disputes, including Gandak treaty (1959), Kosi treaty (1954), and Mahakali treaty (1997) between India and Nepal, the Gangas treaty between India and Bangladesh, and the Indus water treaty between India and Pakistan. The Gandak and Kosi treaties have been revised whereas, in the Gangas treaty, a solution has not been found yet regarding the flow of the river during dry seasons (Ahmad and Iqbal 2016; Muhammad Imran Mehsud 2019). Indus, Ganga and Nile water treaty focuses on the water division while Danube and Mekong treatise focuses on sustainable development and protection of river basins. Effective conflict resolution mechanism is defined in the Indus, Mekong and Danube basins treaties which is missing in other internationally signed treaties and agreements.

Among all these treaties, the Indus Water Treaty can be seen as one of the most successful international treaties; it has survived frequent tensions and has provided a framework for irrigation and hydropower development for more than half a century. This treaty may be referred to as successful from the implementation point of view (Wheeler 2009). To settle the water disputes between India and Pakistan, an agreement was signed in 1960 between both countries known as Indus Water Treaty (IWT) (Ali 2008). It is a water distribution agreement signed between India and Pakistan, which was orchestrated by the World Bank. Under this treaty, eastern rivers including rivers Ravi, Sutlej and Beas were allocated to India (Khaki 2018) while rivers Indus, Jhelum, and Chenab were given to Pakistan with a share of 3.6 MAF for specified domestic, non-consumptive and agricultural use permitted to India.

However, western rivers have become a serious concern for Pakistan in terms of water quantity. Various dams are constructed by India on western rivers including Salal Dam, Dul Hasti Dam, Baghlihar Dam while many more are planned and under construction. India's point of view is that variations in river flows and water shortages in Pakistan are due to its own negligence and climate change. India is doing nothing to violate the Indus water treaty and is utilizing the water allocated to on western rivers. According to Pakistan, India is stealing water supplies from Pakistan and constructing its water-controlling structures only as a political exercise to improve political power. The outcomes of this work will provide information related to transboundary conflicts, issues, treaties, and future concerns to researchers and policymakers in the perspective of climate change.

2 The Indus Waters Treaty: Key Features

The Indus River basin is one of the world's largest transboundary river basins with a total drainage area of about 1.08×10^6 km^2, shared by Pakistan (56%), India (26.6%), China (10.7%), and Afghanistan (6.7%) (Wolf 2007), which makes it a geopolitically intricate region. The Indus River originates from Mansarovar lake in the Tibetan plateau and flows through Jammu, Kashmir, and Pakistan before draining into the Arabian Sea.

Pakistan is highly dependent on Indus, Jhelum, and Chenab Rivers for its water supplies for various purposes such as irrigation, hydropower generation, industrial and domestic uses, etc. Therefore, these rivers are considered the lifeline of Pakistan. Since these rivers flow to Pakistan through India and do not originate from Pakistan, Pakistan is under the constant threat of famine and drought. Western rivers, i.e., the Chenab and Jhelum, originate from India itself while river Indus originates from China and flows to Pakistan through Indian-controlled Kashmir (India Today 2019) (Fig. 1).

In 1947, a well-established irrigation system was disrupted due to the partition of the sub-continent, which divides the Indus system. The disturbance of flows to Pakistan in April 1948 exposed Pakistan's susceptibility to Indian control over the headwaters of the Indus system. The Indus Water Treaty was signed in 1960 between India and Pakistan with mitigation of The World Bank.

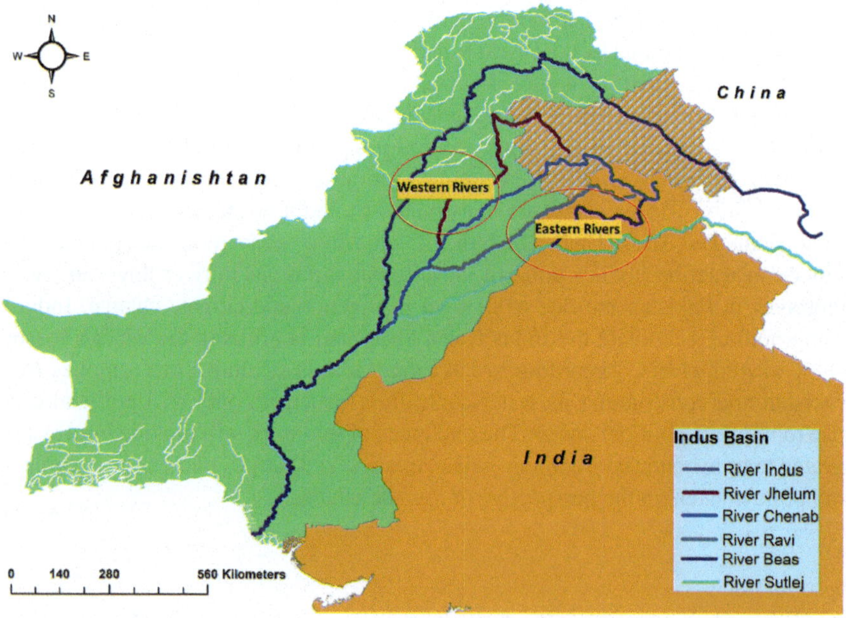

Fig. 1 Allocation of river water under Indus Water Treaty (IWT)

Table 1 Allocation of western rivers' waters to India (Million Acre Foot, MAF)

River system	General storage	Power storage	Flood storage	Total (MAF)
Indus	0.25	0.15	Nil	0.40
Jhelum (excluding Jhelum main)	0.50	0.25	0.75	1.50
Jhelum main	Nil	Nil	As in paragraph 9, annexure E	
Chenab (excluding Chenab main)	0.50	0.60	Nil	1.10
Chenab main	Nil	0.60	Nil	0.60
Total	1.25	1.60	0.75	3.6

Source United Nations (1962)

IWT comprises twelve articles and eight (A-H) annexures. According to Article 1, all the waters of the eastern rivers, i.e., Sutlej, Beas, and Ravi will be offered to India. According to Article 3 (I), Pakistan shall accept unobstructed use of all waters of the western rivers. While in article 3 (II), it is stated that India shall be responsible to let the flow of all the waters of the western rivers, except for controlled uses provided in annexure C&D. In Article 4 and annexures C, D and E of IWT, conditions for the use of waters of the western river to India are listed, i.e., for drinking purpose, agriculture, hydro-power development, etc. (United Nations 1960). The distribution of western rivers' waters to India is given in Table 1.

3 Western Rivers

3.1 The Indian Perspective

The option of completely stopping the flow of water for India seems simple but it has many complications. It can be considered a human rights violation case because water and sanitation had been recognized as a basic human right by the United Nations General Assembly (UNGA) on July 28, 2010 (United Nations 2010). Considering the insufficient storage capacity of India on western rivers, completely stopping the flow seems to be unrealistic. A convention had been passed by the UNGA regarding the use of international watercourses on May 21, 1997, and India is a signatory of this convention (Chaurvedi 2013). China has voted against this convention and India expressed its concerns over the unilateral action of China to distract the flow of the river Indus. China was hindering the Xiabugu river to construct of a dam as a part of the Lalho hydropower project (Government Forms High-Level Task Force on Indus Water Treaty 2016). Moreover, the construction of a dam on the Indus River near Demchok, Ladakh will practically stop the downstream flow, which can cause a water shortage at Tarbela in Pakistan (Sering 2010).

Another option is that a *status quo ante* is maintained. According to this option, India will be allowed to continue the flow of water to Pakistan. Various hydroelectric projects are in progress. These projects can help India generate 18,652 MW of power generation from western rivers. Out of this capacity, projects having 2324 MW power generation are near to be commissioned, whereas 659 MW projects are in various phases of construction. If the above option is followed, then the construction of these projects will have to stop and new projects cannot be taken up, which is the right of India per IWT. India has also shown its concerns over the wastage of water in Pakistan. The first concern is the inadequate storage capacity in Pakistan. According to Indus River System Authority (IRSA), the storage capacity in Pakistan is only thirty days while India's storage capacity is 190 days (Baloch 2018). The storage capacity of Ethiopia is almost the same as that of Pakistan; however, it has less water availability compared to Pakistan (Ahmad et al. 2014; Ahmad 2012; Anwar and Bhatti 2018). Groundwater is depleting at a higher rate in Pakistan, but the government is doing nothing to resolve the issue (Klugelman 2015). Consequently, due to seepage and evaporation losses, only 30% of surface water reaches crops and resulting in loss of surface water (ACE and Halcrow 2001).

Moreover, variations in streamflow may be due to climate change impacts. In the Indus River basin, most of the rivers are glacier-fed. According to the International Center for Integrated Mountain Development (ICIMOD), glaciers in the western Himalayas are declining (Akhtar 2017). However, in the crucial part of the upper Indus basin in the Karakoram range, these are advancing. This process is known as the 'Karakoram Anamaly'. The dependence of feed into the western rivers of the Indus River Basin is given in Table 2, clearly shows that water availability in Pakistan from the western rivers primarily depends on snow and health of glaciers, and there is no role of India in stopping the flow of water towards Pakistan.

According to this, the portion of water from glaciers may not be falling currently but may do so in the future, and there is a chance that the reduction of water in western rivers does happen. It may be noted that Kalahoi, which is the biggest glacier in Kashmir and is the core source of water in the Jhelum River, is melting faster over the last 10 years compared to other glaciers of the Himalayan region (Mushtaq 2009). Moreover, the Siachen glacier has also shrunk by 1.7 km in 17 years (Kalpakian 2015). It can be predicted that glaciers will carry on retreating in the next 50 years. Faster melting of these glaciers will result in increased flows but in the next 100 years, the current flow rate would drop by 30–40%. In western rivers, the flow has already come down. In 1960, the flow was 119 MAF while in 2011, it was reduced to 102

Table 2 Feed dependence of western rivers of Indus River Basin (Janjua S. et al., 2021)

Location	Snow (%)	Rain (%)	Glacier (%)
Indus above Tarbela	30–35	5–10	60–80
Jhelum above Mangla	65	35	–
River Kabul above Nowshera	20–30	20–30	30–35

MAF. So, in the future overall water resources, and particularly in Sindh, will be adversely affected.

Water quality is also a matter of concern in Pakistan. It has been estimated that only 28% area of Sindh province and 79% area of Punjab province is fit for groundwater-based irrigation (Waseem et al. 2014).

Deforestation is another factor in water scarcity. According to FAO, just 2.2% of Pakistan is woodland. Pakistan has lost approximately 33.2% of its forest cover from 1990 to 2010 (FAO 2010). The Upper Indus Basin has only 0.4% of the total forest cover. As a result of rainfall, due to less forest cover, the topsoil washed away. This soil is the cause of silting in rivers and deprivation of water bodies.

Another factor contributing in water scarcity is population growth, which is considered the highest in the region (Pakistan Bureau of Statistics 2017). Pakistan's intensity rate of water per unit GPA is highest in the world. It is estimated that Pakistan will reach absolute water scarcity by 2025 (Baloch 2018). Various projects, like Kalabagh dam, Bhasha Diamar, etc. are being restricted by natives. Likewise, the raising of Mangla dam is feeling resistance from the people of Pakistan. Uneven distribution of water and interprovincial problems in Pakistan are also playing a role in water issue of Pakistan.

As per the third option, India does not violate the IWT but goes ahead with plans to fully utilize the waters entitled to it by the treaty. IWT has entitled India full rights to store 3.6 MAF of water from western rivers. The practice of that right may not create differences with the World Bank, which is the guarantor to the IWT. Pakistan is doing nothing to address its water crises. Tarbela and Mangla reservoirs have already touched their dead levels due to extreme silting. Another reason for low flows in these reservoirs is the unusually low temperature in the Skardu Gilgit-Baltistan region of Pakistan as reported by IRSA. Moreover, India believes that the Neelum-Jhelum and the Diamer-Bhasha dam projects are in territories under Indian claims (Muhammad Imran Mehsud 2019).

From the above discussion, it can be concluded that India's point of view is that the river flow variations and water shortages in Pakistan is because of her own negligence. Various factors including climate change, population growth, water usage, groundwater depletion and canal losses are also causing water issues. India is doing nothing to violate the IWT and is utilizing the water allocated to it on western rivers. In fact, whenever Pakistan has gone to arbitration, neutral experts or the International Court of Justice (ICJ), every time India's right to use waters of the western rivers for the purposes stipulated in the IWT has been upheld. Swain (2017) suggested to expand the Indus Water Treaty (IWT) to the whole basin and exclude Afghanistan and China by dividing the rivers. According to Swain and Parvaiz (2021) and Parvaiz (2021), this treaty needs to be reviewed again to incorporate environmental climate change challenges, which are missing. Indeed, environmental flows are not addressed in the treaty, which need to be incorporated.

3.2 Pakistan's Perspective

Pakistan shares its water resource with India and Afghanistan. After the independence of Indian subcontinent, India stopped the flow of River Sutlej across borders into west Punjab and threatens the agriculture of Pakistan (Frenken 2012). According to Pakistan's point of view, India is stealing water supplies from Pakistan and constructing its water controlling structures only as a political exercise to improve political power by practicing hydro-domination. According to Pakistan, discharge below Salal dam fluctuates and is decreasing with the passage of time. Annual discharge has been decreased from 40 BCM to 22 BCM as obtained from the analysis of inflow data of 35 years. There is an average decrease of 0.5 BCM per year (Ahmad and Iqbal 2016). According to one estimate by the Punjab Irrigation Department, a loss of almost 37 billion Pakistani Rupees was experienced by the economy of Pakistan due to blockade of water from the Chenab River by India (Mustafa 2008).

The first objection raised by Pakistan under the IWT was on the construction of Salal hydropower project in July 1970. This dam was constructed on the River Chenab. PIC was unable to solve the issue; however, the issue was resolved on a governmental level through bilateral talks in 1978 when India agreed to make necessary changes in design (Ahmar 2001). Wullar Barrage, also known as the Tulbul navigation project, was the second disputed project of India. This barrage was designed on River Jhelum and, as per IWT, India had to inform Pakistan prior to construction on western rivers. Its construction was stopped by India in 1987 but now construction has resumed (India Rejects Pak Objections, Work on Tulbul Navigation Project in Full Swing 2016). Pakistan has objected that its construction will stop the water flow in River Jhelum during kharif season and can exceed the storage capacity. Water conflict over this barrage is still unresolved after several talks (Qureshi 2018).

An issue on the construction of Baglihar dam was raised by Pakistan in 1992. This issue was resolved by a neutral expert in 2007 and asked India to make certain changes in design. The Kishanganga dam project is another disputed project between both countries. According to Pakistan, the design parameters of this hydro-power production plant breached the IWT (Khan 2014; Mustafa 2007). This issue was resolved by court of arbitration and this project was completed in 2018.

Ratle dam, which is to be constructed on River Chenab, 55 km upstream of Baglihar dam in Kishtwar district is now a matter of concern for Pakistan. Pakistan believes that its construction will reduce the flow of River Chenab by 40% at Head Marala (Kiani 2018). Pakistan wants the formation of a court of arbitration while India has no objection if a neutral expert is appointed. The World Bank paused the process for the constitution of a Court of Arbitration (COA). However, according to the report, India has again started its construction in 2019 (Mustafa 2019). Another issue raised by Pakistan is the construction of the Pakal Dul dam built on the River Chenab. This dam is under argument for the last several years at the Indus Waters Commission level (Ahmad and Iqbal 2016). Pakistan has also elevated a protest to the formation of Lower Kalnai hydro-power projects by India.

Bursar dam will be the first storage project of India on western rivers with a huge storage capacity of 2 MAF, which can regulate the downstream flows. Its construction can violate the treaty since it is much greater than the permissible limits on River Chenab (Ahmad 2011). Pakistan has asked India to provide the details of Bursar dam. Its project report was completed by the National Hydroelectric Power Corporation (NHPC) (NHPC Limited: Projects: Under Survey and Investigation Stage: Bursar, 2018), India and environmental clearance is given by the Indian Ministry of Environment and Forests (MoEFF) (Yen Yin 2018). Gyspa Dam is another dam with huge storage capacity of around 0.8 MAF and is to be constructed on River Chenab at Himachal Pradesh (Ahmad and Iqbal 2016). This dam is facing opposition from local public (Chauhan 2015). Its construction is set to begin in 2020 and is opposed by Pakistan (Digwani 2020).

India has planned several mega projects other than the existing projects on the River Chenab. The list of constructed and planned dams by India is given in Table 3 and is shown in Fig. 2.

3.3 Consequences of Climate Change

The biggest concern about Indus Water Treaty is the climate change. The treaty is outdated since the changing environment due to climate change has not been initially considered. During that time, the climate change was not a prevalent issue; however, owing to current climate challenges there is a need to re-negotiate the treaty. The melting of glaciers from the western Himalayas contributes 40% of the annual water flow within the river basin. It is predicted that in the next 50 years, the flow of the rivers will increase further because of glacier melting. Due to this, the glacier reserves will be depleted followed by 30–40% decrease in water flow of the Indus River, which is the main river of Indus River Basin (Husnain and Zehra 2010). There is also a prediction that the average rainfall will also be affected in the upcoming years due to climate change. This will lead to droughts or larger floods (de Stefano et al. 2012). For example, in Pakistan the number of rainy days is decreasing but extreme precipitation events are increasing. The availability of fresh water will be reduced due to climate change and ultimately conflict will occur.

The threshold of irreversibility is being crossed by climate change. It means that even if the political and economic issues are resolved, issues due to changing climate will continue to be a burden on society. Ultimately, the effectiveness of treaty will be affected in case climate change is not mainstreamed and the ability of basin states will be affected to meet their commitments in the treaty. Food availability, human and economic livelihood, and the environment integrity will also be affected due to adverse impacts of climate change. As discussed earlier, the Indus Water Treaty is an example of successful mediation; however, the unforeseen threat and impact induced by a changing climate is challenging the authority of the treaty. The agreements which worked well in the past may not be helpful in the future as climate change is impacting

Table 3 Completed/under construction/planned projects of India on Western Rivers

Name and location	Status	Type/capacity (MAF)
Salal dam: Reasi district Jammu and Kashmir, 72 km upstream of Marala on Chenab River	Completed in 1987	Runoff river project, 0.228
Wullar barrage (Tulbul navigation project): Jhelum River, Bandipura district Jammu and Kashmir	Construction resumed in 2016	Non-power generation/0.3
Baghlihar dam: Ramban district Jammu and Kashmir, on Chenab River about 147 km upstream of Marala Headworks	Phase-1 2008, Phase-II 2015	Runoff river project, 0.321
Kishanganga dam (Jehlum River): Bandipore in Jammu and Kashmir	Completed in 2018	Runoff river project, 0.014
Chutuk project: Suru River, tributary of Indus, Kargil district	Completed in 2013	Runoff river project
Dul hasti dam: Kishtwar district, Jammu Kashmir, Chenab river	Completed in 2007	Runoff river project, 0.008
Ratle dam: Drabshalla in Kishtwar district, Chenab River, 55 km upstream of Baglihaar dam on Chenab River	Construction resumed in 2019	Runoff river project, 0.0081
Pakal dul dam: On Marusadar River, right bank tributary of Chenab River	2023 expected completion date	Power generation project, 0.087
Bursar dam: First storage dam, Village Pakal, Marusudar river tributary of Chenab river	Project report completed	Water storage project/2
Jyspa dam: LahaulSpiti valley of Himachal Pradesh	Construction began in 2020	Water storage, 0.8
Kiru dam: Chenab River	Expected completion date-2023	Runoff river project/0.033
Kwar dam: Chenab River	Expected completion date-2023	Runoff river project/0.03
Sawalkot dam: Chenab	Construction started	Runoff river project, 0.429

the condition of the environment and waterways. Therefore, the treaty needs to be upgraded to support these changing conditions of the environment.

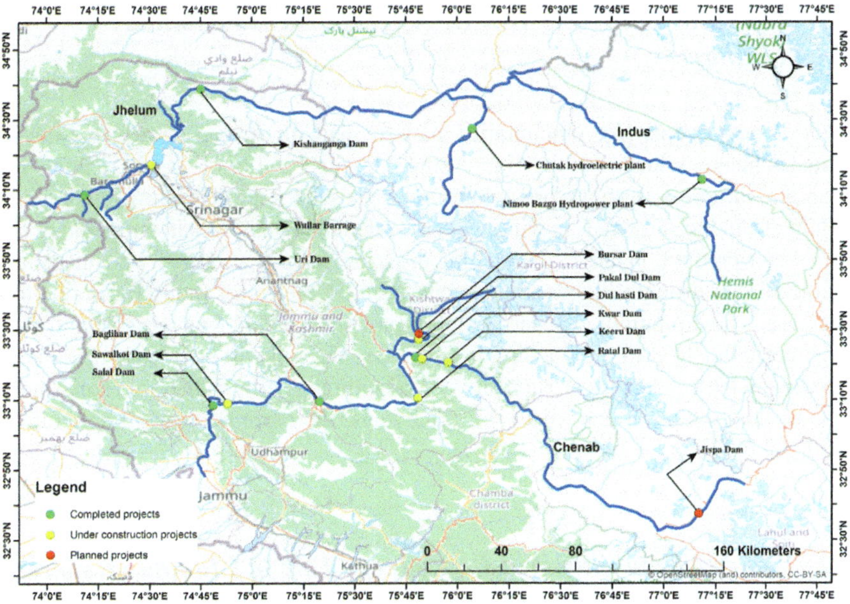

Fig. 2 Constructed/under construction/planned dams on Western Rivers

4 Why Should Pakistan and India Need to be Concerned about the Changing Climate and Environment?

The relationship between India and Pakistan is associated with the dispute over Kashmir. This is so prevalent that the conflict over water resources may not be their highest concern. However, the conflict about the even distribution of water resources is unique in such a sense that water resources are affected due to environmental conditions. The utilization of water for two larger countries like India and Pakistan is much greater than the water utilization for two smaller countries. This utilized water cannot be renewed except if it is replenished by the hydrologic cycle. However, climate change can alter the hydrologic cycle. Water conflicts will increase as the climate change persists. Both countries are in a perilous position about the surrounding water availability. India and Pakistan are facing the shortage of water and it is expected that this shortage will increase in the future. Therefore, it will be beneficial for both countries to recognize their cooperative potential and combine their expertise for mutual benefits.

4.1 Hydrological Cycle Will Be Affected Due to Climate Change

Hydrological events like precipitation patterns and runoff can be affected due to climate change, which results in increasing the vulnerability of certain regions. For example, variations in winter precipitation can be used for the determination of runoff volume from the snowmelt in winter. On the other hand, changes in temperature can vary water production due to the melting of glaciers (Yu et al. 2013). Scientists have predicted that drought like conditions can occur in the lower Indus Basin due to a decrease in precipitations. The farms and communities residing on the riverbanks can be affected due to extreme flooding in the basin. The soil will be transformed due to flooding, which can cause erosion and degradation in nearby vulnerable areas more. Due to this, pollution can affect the waterways, therefore decreasing agricultural production.

4.2 Water Availability Will Be Reduced Due to Climate Change

Due to high altitudes and inaccessibly of rugged Himalayan mountains, there is a limitation to analyze the rate of glacial melt that is contributing to Indus River Basin (Yu et al. 2013). However, hydrological modelling and glacier mass-balance calculations showed that there is a decrease in upstream water supply. Most of the glacial melting into the tributaries of the Indus River Basin occurs in the ablation zone. This is approximately equal to 18% of the total flow of 110 MAF from the headwaters coming to Pakistan from the mountains of the Indus River. According to a report, the flows in the Indus River Basin will be reduced to 8% by the year 2050 due to the shrinkage of glaciers. This will be harmful due to a decrease in water flow; the downstream portion of the Indus Basin will be mostly affected where this flow act as primary source of water (Rajbhandari et al. 2015). Moreover, another source is groundwater whose availability will be reduced to 750 m^3 per capita by 2050.

4.3 Agricultural Production Will be Affected Due to Climate Change

According to the report of Intergovernmental Panel on Climate Change (IPCC) (2007), due to increased precipitation and variation in its trends owing to climate change, the sources of freshwater system are more vulnerable. This will result in increased risk of floods and droughts, which ultimately affect food security and water infrastructure. As per Pakistan's point of view, this situation is alarming because of its dependence on the Indus River System, and there will be no alternative if Indus

through the installation of the telemetry system. It may include flood warning and weather forecasting telemetry system. Data sharing should be transparent on Indian projects on western rivers. Assessment should be done on the downstream flows due to construction projects on western rivers. Sharing the environmental impact assessment (EIA) of Indian projects on western rivers will also help build trust.

References

ACE, Halcrow (2001) Exploitation and regulations of fresh groundwater

Ahmad A (2011) Indus waters treaty a dispassionate analysis. Policy Perspect 8(2):73–83

Ahmad S (2012) Water insecurity: a threat for Pakistan and India

Ahmad S, Iqbal J (2016) Transboundary impact assessment of Indian dams: a case study of Chenab River Basin in perspective of Indus Water Treaty. Water Policy 18(3):545–564. https://doi.org/10.2166/wp.2015.137

Ahmad S et al (2014) Drought mitigation in Pakistan: current status and options for future strategies (No. 3; Drought Series 3). International Water Management Institute (IWMI)

Ahmar M (2001) The challenge of confidence building in South Asia. Har Anand Publications

Akhtar S (2017) Trans boundary water cooperation in the Indus Basin: challenges and opportunities. Transboundary water cooperation in the context of sustainable development goals (SDGs) in South Asia and beyond

Ali SH (2008) Water politics in South Asia: technocratic cooperation and lasting security in the Indus Basin and beyond. J Int Aff 61(2):167–182

Anwar AA, Bhatti MT (2018) Pakistan's water apportionment Accord of 1991: 25 years and beyond. J Water Resour Plan Manage 144(1):05017015. https://doi.org/10.1061/(ASCE)WR.1943-5452.0000831

Baloch SM (2018) Water crisis: why is Pakistan running dry. DW—made for minds

Certain Activities Carried Out by Nicaragua in the Border Area (Costa Rica v. Nicaragua) (2013)

Chatterji M et al (2008) Conflict and peace in South Asia (contributions to conflict management, peace economics and development, 5th edn). https://doi.org/10.1016/S1572-8323(08)05027-3

Chaturvedi AK (2013) Water: a source for future conflicts. Vij Books India Pvt Ltd.

Chauhan K (2015) Jispa dam project faces opposition: The Tribune India. The Tribune

Chaurvedi A (2013) Water a source of future conflicts. Vij Books India Pvt Ltd.

Chellaney B (2013) Water, peace, and war: confronting the global water crisis. Rowman & Littlefield

Cohen SJ, Miller KA, Hamlet AF, Avis W (2000) Climate change and resource management in the Columbia River Basin. Water Int 25(2):253–272. https://doi.org/10.1080/02508060008686827

Crisp J et al (2017) Population and human relations: year in review 1995. In: ENCYC. BRITANNICA, 2017th edn.

de Stefano L, Duncan J, Dinar S, Stahl K, Strzepek KM, Wolf AT (2012) Climate change and the institutional resilience of international river basins. J Peace Res 49(1). https://doi.org/10.1177/0022343311427416

Digwani M (2020) Work on Jispa Dam project starts, has been opposed by Pakistan. LOKUS

Dombrowsky I (2007) Conflict, cooperation and institutions in international water management. Advances in ecological economics series

Finlay L et al (1927) Jurisdiction of European Commission of Danube between Galatz and Braila, Advisory opinion

Food and Agriculture Organization of the United Nations (2010) GlobalfForest resources assessment

Frenken K (2012) Irrigation in Southern and Eastern Asia in figures. Food and Agriculture of the United Nations

Gabcikovo-Nagymaros Project (Hung. v. Slovk.), Judgment (1997)

Government Forms High Level Task Force on Indus Water Treaty (2016) The Economic Times

Holslag J (2011) Assessing the Sino-Indian water dispute. J Int Aff 64(2):19–35
Husnain, Zehra M (2010) The Indus water treaty in light of climate change. http://www.caee.utexas.edu/prof/mckinney/ce397/Topics/Indus/Indus_2010.pdf
India Rejects Pak Objections (2016, 7 Oct) Work on Tulbul navigation project in full swing. The Northlines
India Today (2019). Indus water treaty: everything you need to know. India Today
Indus Waters Kishenganga Arbitration (Pak. v. India) (2010)
IPCC (2007) AR4 climate change 2007: The physical science basis
Janjua S, Hassan I, Muhammad S, Ahmed S, Ahmed A (2021) Water management in Pakistan's Indus Basin: challenges and opportunities. Water Policy 23:1329–1343. https://doi.org/10.2166/wp.2021.068
John VJ (2007) Water struggle, 4th edn. Indian Society for Promoting Christian Know
Kalpakian J (2015) Climate change and national security: a country level analysis. Contemp Secur Policy 36(2):400–401. https://doi.org/10.1080/13523260.2015.1061770
Khaki SB (2018) Indo-Pak and hydro-politics
Khan AH (2014) Water sharing dispute in Pakistan: standpoint of provinces. Berkeley J Soc Sci 4:1–17
Kiani K (2018) Pakistan approaches World Bank after India builds Kishanganga on Neelum. Dawn
Klugelman M (2015) Water scarcity 'is Pakistan's worst night-mare. DW—made for minds
Mikaberidze A (2013) Atrocities, massacres, and war crimes, 1st edn. ABC-CLIO, LLC
Miller TG et al (2010) Environmental science, 13 edn. Yollanda Cossio
Muhammad Imran Mehsud TAK (2019) Water war thesis: perspective from South Asia. J Polit Stud 105–118
Mushtaq S (2009) Kashmir's main glacier melting at alarming speed. REUTERS
Mustafa D (2007) Social Construction of hydropolitics: the geographical scales of water and security in the Indus Basin. Geogr Rev 97(4):484–501. https://doi.org/10.1111/j.1931-0846.2007.tb00408.x
Mustafa K (2008) Blockade of Chenab River inflicts Rs37bn loss to agriculture. The News
Mustafa K (2019) WB pause gives India free rein to complete Ratle project. The News
NHPC Limited (2018) Projects: under survey and investigation stage. Bursar
Pakistan Bureau of Statistics (2017) 6th population and housing census
Parry M (2007) Change, climate change 2007, impacts, adaptation and vulnerability. Cambridge University Press
Parvaiz A (2021) Renegotiate Indus water treaty with Pakistan to factor in climate change: parliamentary panel. Mongabay
Qureshi W (2018) Dispute resolution mechanisms: an analysis of the Indus waters treaty. Pepperdine Dispute Resolut Law J 18(1)
Rajbhandari R, Shrestha AB, Kulkarni A, Patwardhan SK, Bajracharya SR (2015) Projected changes in climate over the Indus river basin using a high resolution regional climate model (PRECIS). Clim Dyn 44(1–2). https://doi.org/10.1007/s00382-014-2183-8
Sering SH (2010) China builds Dam on Indus near Ladakh. J Defence Stud 4(2):136–139
Swain A (2017) Expand the Indus waters treaty to make peace. The Third Pole
Swain A, Parvaiz A (2021) India, Pakistan cross-border water treaty needs climate change revision. Nature India
The Rhine Chlorides Arbitration concerning the Auditing of Accounts (The Netherlands/France) (2004)
United Nations (1960) Indus waters treaty, India and Pakistan. https://www.internationalwaterlaw.org/documents/regionaldocs/IndusWatersTreaty1960.pdf
United Nations (1962) Indus water treaty. United Nations-Treaty Series
United Nations (2010) The human right to water and sanitation: resolution/adopted by the General Assembly
United Nations (2019) Department of Economic and Social Affairs

United Nations Human Rights Council (UNHRC) (2015) Commissions of inquiry and fact-finding missions on international human rights and humanitarian law

Waseem A, Arshad J, Iqbal F, Sajjad A, Mehmood Z, Murtaza G (2014) Pollution status of Pakistan: a retrospective review on heavy metal contamination of water, soil, and vegetables. Biomed Res Int 2014:1–29. https://doi.org/10.1155/2014/813206

Wheeler W (2009) The water's edge. Good Mag 58–64

Wolf AT (2007) Shared waters: conflict and cooperation. Annu Rev Environ Resour 32(1):241–269. https://doi.org/10.1146/annurev.energy.32.041006.101434

Yamamoto K et al (2013) Southeast Asian water environment, 5th edn. International Water Association (IWA)

Yen Yin SH (2018) Stop the Dam(ned) project: outrage over the Bursar Hydroelectric Project. Glacier Hub

Yotova A (2009) Climate change, human system and policy, 2nd edn. Encyclopedia of Life Support System

Yu W, Yang Y-C, Savitsky A, Alford D, Brown C, Wescoat J, Debowicz D, Robinson S (2013) Indus Basin of Pakistan: impacts of climate risks on water and agriculture, vol 2. World Bank Group

Faraz ul Haq is Engineer working as a Lecturer at the Centre of Excellence in Water Resources Engineering, University of Engineering and Technology, Lahore, Pakistan since 2020. He received his Masters degree from UET, Taxila. He had worked in various national and international organizations including PIET and QCET. He co-authored over 30 peer-reviewed publications in international journals and proceedings. His research mainly focuses on remote sensing, climate change and its impacts on water resources, surface and groundwater hydrology and modeling.

Dr. Ijaz Ahmad is an Assistant Professor at the Centre of Excellence in Water Resources Engineering, University of Engineering and Technology Lahore, Pakistan since 2016. He received his Ph.D. degree from Hohai University, Nanjing, China. He had worked in various national and international organizations including Pakistan Engineering Services, Shanghai, Design and Research Institute, and Water Management Research Centre. He co-authored over 100 peer-reviewed publications in international journals and proceedings. He has supervised more than 10 postgraduate students. His research mainly focuses on extreme climatic events analysis, modeling and prediction, climate change and its impacts on water resources, surface and groundwater hydrology and modeling, reservoir operation optimization and forecasting, and water allocation optimization under uncertainty.

Dr. Noor Muhammad Khan had his B.Sc. Civil Engineeirng (with honors) from UET Lahore, in 1992. He Joined UET as Lecturer in 1992. Later, he joined WAPDA in 1993, and worked in Planning Division. In 1994, proceeded for higher education to Asian Institute of Technology Thailand, and did Masters with specialization in River Engineering in 1996. He joined back Wapda in Planning Division, and was involved in review of irrigation and drainage projects at the national level. He joined UET in 2001 as Assistant Professor and proceeded for Ph.D. to AIT Thailand in 2004. He Completed his Ph.D. in 2008. He is Currently working as Professor in Civil Engineering Department, and involved in teaching, research, and advisory services.

Postface

Climate change is expected to have a significant and widespread impact on water resources and food systems. Inconsistent rainfall, rising temperatures, and changes in hydrological cycles will have direct and indirect impacts on water availability, food production, and resource management.

One of the main consequences of climate change will be variations in rainfall patterns. Mid-latitude areas are expected to experience fluctuating rainfall, leading to droughts, water scarcity, and drier soils. These conditions can significantly undermine crop growth and agriculture productivity. High-latitude areas expect to see an increase in rainfall patterns, which can pose both challenges and opportunities for agriculture depending on local conditions. Moreover, rising temperatures and inconsistent precipitation in many regions will reduce the viability of rainfed farming. Consequently, there will be a greater need for water, leading to an increased demand for irrigation. Widespread irrigation practices have already depleted surface and groundwater and devastated aquatic ecosystems. Furthermore, the growing demand for irrigation will often outpace the declining water resources available for such a use.

The consequences of climate change on agriculture include the proliferation of invasive species and changing agroecosystems. In mid and low latitudes areas, increasing temperatures will likely reduce land productivity and undermine crop growth, resulting in declining yields. While Europe and North America may experience a temporary increase in crop productivity, the other impacts of climate change will negate these short-term benefits.

Hydrological cycle changes and climate variability will also affect aquaculture and inland fisheries. Changes in water salinity, temperature, and nutrient levels and rising sea levels can interrupt aquatic ecosystems and breeding cycles, reducing the productivity of fisheries. For this reason, FAO is especially concerned as fisheries provide a crucial livelihood source for some of the poorest populations.

At the same time, global food systems are also contributing significantly to climate change and water insecurity. Global agriculture produces 19–29% of total GHG

M. Behnassi et al. (eds.), *The Water, Climate, and Food Nexus*,
https://doi.org/10.1007/978-3-031-50962-9

emissions and uses roughly 70% of the global freshwater resources. Land degradation and food waste and loss (1/3rd of food produced globally) further exacerbate the threat to our planet.

Climate change, food insecurity, and water scarcity are interconnected challenges with global consequences. We are at a crucial juncture where a failure to properly explore the intricacies of this nexus can undermine the economic, environmental, and societal systems of the world. "The Water, Climate, and Food Nexus: Linkages, Challenges, and Emerging Solutions" book provides a comprehensive analysis of the nexus interlinkages, tradeoffs, and synergies. It emphasizes the importance of understanding and employing the nexus analysis to increase efficiency and encourage synergistic approaches to water, land, and resource management.

The book also explores sustainable practices and policies that have the potential to balance the competing demands of food production, resource use, and conservation. Changes in water management are at the heart of many of these approaches. The book explores the potential for wastewater treatment in India and Saudi Arabia, and the use of aquaponics in water-scarce Oman. The analysis emphasizes the broad applicability of climate-smart agriculture in reducing water wastage, improving food security and nutrition, and practicing environmental conservation. The book gives special attention to adaptation and mitigation strategies in the global South as developing countries are expected to carry 70–80% of the costs of climate change, including resource scarcity and human insecurity.

These case studies highlight a multi-sectoral approach when determining best practices, otherwise attempts to address one sector can inadvertently worsen the conditions in another sector. For instance, implementing climate adaptation strategies without considering food production and water conservation can further marginalize vulnerable communities. Addressing the complex nexus requires collaborative efforts between governments, civil society, and the private sector. "The Water, Climate, and Food Nexus" book provides key insights and can serve as a valuable tool for scientists, researchers, academics, planners, and policymakers seeking to successfully integrate resource management techniques in the context of a changing climate in the perspective of fostering human security (especially food and water security). Increased consideration of the solutions and recommendations presented in this book can help promote a more equitable and resilient global society.

Aisha Baig
Mohamed Behnassi
Mirza B. Baig

The manufacturer's authorised representative in the EU is Springer Nature Customer Service Centre GmbH, Europaplatz 3, 69115 Heidelberg, Germany. If you have any concerns regarding our products, please contact ProductSafety@springernature.com

Printed and bound by CPI Group (UK) Ltd, Croydon, CR0 4YY
01/10/2026
02225902-0019